The chemistry of cyclo-octatetraene and its derivatives

The chemistry of cyclo-octatetraene and its derivatives

G. I. FRAY
School of Chemistry, University of Bristol

R. G. SAXTON
Head of Chemistry Department, Repton School

CAMBRIDGE UNIVERSITY PRESS
CAMBRIDGE
LONDON · NEW YORK · MELBOURNE

CAMBRIDGE UNIVERSITY PRESS
Cambridge, New York, Melbourne, Madrid, Cape Town, Singapore, São Paulo, Delhi

Cambridge University Press
The Edinburgh Building, Cambridge CB2 8RU, UK

Published in the United States of America by Cambridge University Press, New York

www.cambridge.org
Information on this title: www.cambridge.org/9780521105651

First published 1978
This digitally printed version 2009

A catalogue record for this publication is available from the British Library

Library of Congress Cataloguing in Publication data
Fray, Gordon Ian, 1928–
The chemistry of cyclo-octatetraene and its derivatives
Bibliography: p. 429   Includes index
1. Cyclo-octatetraene   I. Saxton, Roy Gerald, 1945–joint author   II. Title
QD305.H9F7   547'.413   76-57096

ISBN 978-0-521-21580-0 hardback
ISBN 978-0-521-10565-1 paperback

Contents

Foreword by W. Baker   *page* vii

Preface   ix

1 Cyclo-octatetraene

  1 Formation   1
  2 Purification   4
  3 Physical properties   6
  4 Structure   8
  5 Thermal oligomerisation   11
  6 Thermolysis, photolysis and radiolysis   17
  7 Oxidation, and reactions with electrophiles   19
  8 Reduction, and reactions with nucleophiles   28
  9 Reactions with free radicals   32
 10 Polymerisation   34
 11 Carbonylation   34
 12 Cycloadditions   35
 13 Reactions with metals and their derivatives   48

2 Substituted cyclo-octatetraenes

  1 Monosubstituted derivatives   79
  2 Disubstituted derivatives   109
  3 Annulated and bridged derivatives   123
  4 Trisubstituted derivatives   133
  5 Tetrasubstituted derivatives   134
  6 Pentasubstituted derivatives   151
  7 Hexa- and heptasubstituted derivatives   152
  8 Octasubstituted derivatives   152

3 Further reactions of compounds derived from cyclo-octatetraenes

  1 1,2-Didehydrocyclo-octatetraene   164
  2 Bicyclo[4.2.0]octa-2,4,7-trienes   165

vi   *Contents*

3 Bicyclo[3.3.0]octa-2,6-dienes, bicyclo[3.3.0]octa-1,3,6-trienes and
   bicyclo[3.3.0]octa-1,3,7-trienes   165
4 Bicyclo[3.3.0]octa-1,4,6-triene   168
5 Semibullvalenes   169
6 Metal derivatives   172
7 Homotropylium species   214
8 Cyclo-octa-1,3,5-triene, bicyclo[4.2.0]octa-2,4-diene and cyclo-octa-
   1,3,6-triene   214
9 Substituted cyclo-octa-1,3,5-trienes, cyclo-octa-1,3,6-trienes and
   bicyclo[4.2.0]octa-2,4-dienes   233
10 Cyclo-octa-2,4,6-trienones   254
11 Bicyclo[6.1.0]nona-2,4,6-trienes   262
12 Bicyclo[4.2.1]nona-2,4,7-trienes   293
13 Bicyclo[6.2.0]deca-2,4,6-trienes   300
14 Bicyclo[6.3.0]undeca-2,4,6-trienes   306
15 Bicyclo[6.4.0]dodeca-2,4,6-trienes   307
16 9-Azabicyclo[6.1.0]nona-2,4,6-trienes   308
17 9-Oxabicyclo[6.1.0]nona-2,4,6-trienes   310
18 9-Phosphabicyclo[6.1.0]nona-2,4,6-trienes   317
19 9-Thiabicyclo[6.1.0]nona-2,4,6-trienes   318
20 9-Azabicyclo[4.2.1]nona-2,4,7-trienes   318
21 9-Silabicyclo[4.2.1]nona-2,4,7-trienes   320
22 9-Phosphabicyclo[4.2.1]nona-2,4,7-trienes   322
23 9-Thiabicyclo[4.2.1]nona-2,4,7-trienes   324
24 9-Aza-10-oxobicyclo[4.2.2]deca-2,4,7-trienes   326
25 Tricyclo[4.2.2.0^{2,5}]deca-3,7-dienes   327
26 Tricyclo[4.2.2.0^{2,5}]deca-3,7,9-trienes   342
27 Tricyclo[4.2.2.0^{2,5}]deca-7,9-dienes   351
28 9,10-Diazatricyclo[4.2.2.0^{2,5}]deca-3,7-dienes   353
29 Oligomers of cyclo-octatetraenes   358

Appendix   377
References   429
Author index   463
Subject index   479

Foreword

The story of cyclo-octatetraene began in 1911–13 with Willstätter's multistep preparation from pseudo-pelletierine, surely one of the most hopeful synthetical projects ever undertaken in organic chemistry. Although for many years doubt was cast on the validity of this work it was proved in 1947 to have been correct. Yet in view of its inaccessibility, for thirty years after its discovery the likelihood of ever being able either to solve the fascinating theoretical speculation as to whether cyclo-octatetraene was an aromatic compound akin to benzene, or to investigate its properties in any detail, seemed wholly remote.

The writer recalls the almost incredible news reaching this country in 1945 that Reppe had prepared cyclo-octatetraene in kilogram quantities by the polymerisation of acetylene, and his wonder on seeing this beautiful yellow liquid in bulk. Since 1945, and indeed a few years before in Germany, the investigation of this remarkable compound has been pursued in chemical laboratories throughout the world, and has revealed an astonishing variety of chemical behaviour, little of which could have been foretold by application of chemical theory or by analogy with the behaviour of other substances, and in which, in contrast with benzenoid compounds, the cyclo-octatetraene nucleus seldom retains its original structure.

It has long been evident that a comprehensive account of the chemistry of cyclo-octatetraene was needed, not only to summarise the great volume of work that has been done with it but, on the basis of the information disclosed, to lay the foundations of a real knowledge of the chemistry of this hydrocarbon. This timely book by Dr G. I. Fray in collaboration with Dr R. G. Saxton fulfils both these objectives, and generations of chemists who will continue to investigate this versatile hydrocarbon will be grateful to be able to 'look it up in Fray and Saxton'.

W. BAKER

For
Joyce and Patricia

Preface

Z,Z,Z,Z-Cyclo-octatetraene (COT) (I) has played an outstanding role in many aspects of theoretical and synthetical chemistry. As [8]annulene, the next higher vinylogue of benzene, it is of fundamental importance for the understanding of cyclic alternating π-systems. As a medium-ring polyene, it undergoes a wide variety of reactions which are often accompanied by skeletal transformations, and it is the progenitor of a large number of interesting species including, for example, cyclobutadiene (II), semibullvalene (III), the homotropylium cation (IV), the dianion (V), 1H-azonine (VI), basketene (VII), bullvalene (VIII), [16]annulene (IX) and triamantane (X).

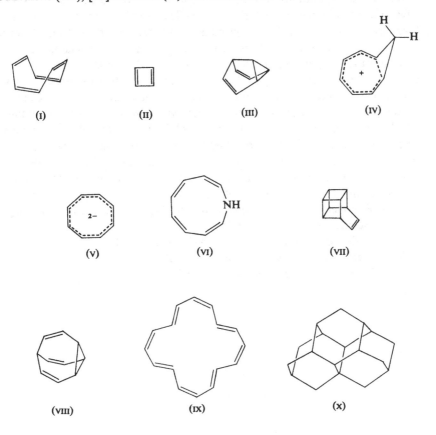

(xi)   (xii)   (xiii)

COT also forms an exceptional variety of complexes with transition metals. In addition to the interest, from the viewpoint of bonding theory, of these multifarious derivatives (some of which exhibit intriguing fluxional behaviour), modifications of the attached C_8H_8 ligand may occur, leading to systems such as pentalene (xi). Moreover, reactions of metal-coordinated COT can provide routes to systems which are not readily available by the conventional transformations of organic chemistry; examples include the dihydrotriquinacene-derivative (xii) and the COT dimer (xiii).

We feel that our interest in such a molecule needs no apology, and the rapid development of its chemistry during the last decade provides some justification for a new summary of existing knowledge. In chapter 1, the formation, physical properties and chemical reactions of COT are outlined, while chapter 2 deals with substituted (and annulated) derivatives of the parent compound; in order to keep the length of the book within bounds, benzo-derivatives and analogues containing hetero-atoms, e.g. azocines, have been omitted. Chapter 3 attempts to cover, as far as possible, the known chemistry of those compounds which are immediately derivable from COTs. By means of an appendix, prepared after the main typescript had been submitted, we have been able to include additional material from the later literature (up to the end of 1976).

We are indebted to previous reviewers of this field, notably the following:

L. E. Craig, *Chem. Rev.*, 1951, **49**, 103

R. A. Raphael, in *Non-Benzenoid Aromatic Compounds* (ed. D. Ginsburg), Interscience, 1959, chapter VIII

G. Schröder, *Cyclooctatetraen*, Verlag Chemie, Weinheim, 1965

H. P. Figeys, *Topics in Carbocyclic Chem.*, 1969, **1**, 269

L. A. Paquette, *Tetrahedron*, 1975, **31**, 2855.

July 1977   G.I.F.
                                                                           R.G.S

1

Cyclo-octatetraene

1. Formation

Cyclo-octatetraene was first described in 1911 by Willstätter and Waser,[1] who obtained it from pseudo-pelletierine (1), an alkaloid from the bark of the pomegranate tree, by a lengthy degradation proceeding *via* cyclo-octa-1,3,5-triene (see also ref. 2). The resulting sample of COT probably contained *ca.* 30 % of styrene,[3] and its authenticity did not go unchallenged (for a review of the evidence, see ref. 4). However, the original reaction sequence was repeated some thirty-six years later by Cope and Overberger[3, 5] (scheme 1), and Willstätter's claim was completely vindicated.

Scheme 1

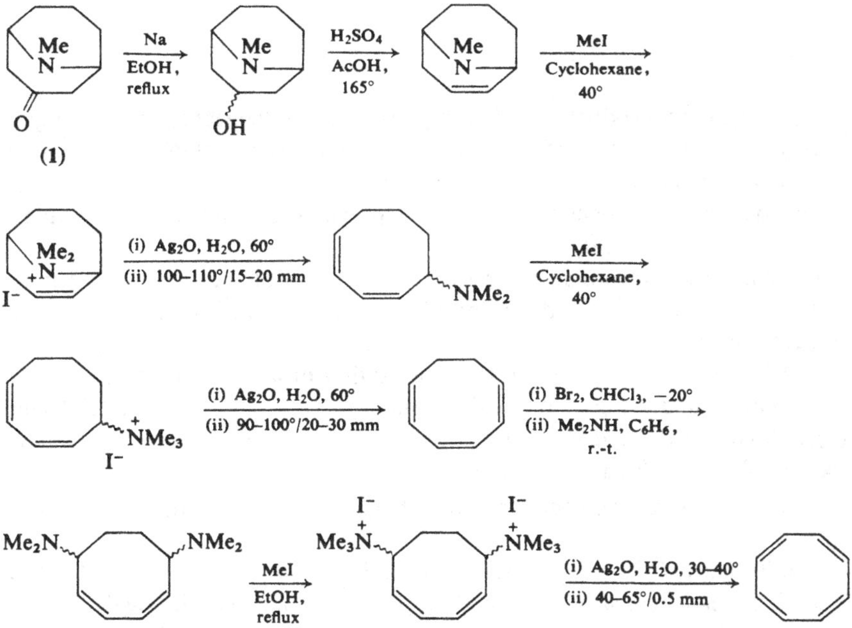

This type of approach to COT may be simplified by starting from the readily available cyclo-octa-1,5-diene, but the introduction of the additional double bonds by means of the Hofmann elimination procedure,[6,7] or by dehydro-bromination,[8-10] results in poor yields of a product which is contaminated with styrene, benzocyclobutene etc.

In 1940, the important discovery that COT could be produced by the cyclo-tetramerisation of acetylene was made in the laboratories of Badische Anilin- & Soda-Fabrik AG, by Reppe *et al.*[11] The reaction is best carried out (up to 70% conversion) in dry tetrahydrofuran or dioxan at 85–90° and a pressure of 15–25 atmospheres, in the presence of nickel(II) compounds such as the cyanide or chelate 'ato' complexes from acetylacetone, acetoacetic esters, salicylaldehyde, *N*-alkylsalicylaldimines etc.[11-15] (a much less efficient conversion results from the use of $(CH_2{=}CH.CN)_2Ni$[16]). The process has been reviewed,[17-19] and Schrauzer[15,18] has proposed a mechanism involving an intermediate octahedral nickel complex (2) (but see appendix). (For a discussion of bonding in this type of complex, see refs. 20, 21. An earlier hypothesis proposed the intermediacy of a cyclobutadiene complex (3)[22].)

Among the by-products of the Reppe process are benzene, styrene, naph-thalene, azulene, *Z*-1-phenylbuta-1,3-diene, vinylcyclo-octatetraene, and a $C_{12}H_{12}$ fraction of unknown constitution.[23-26]

Octadeuteriocyclo-octatetraene may be prepared by similar cyclotetramer-isation of dideuterioacetylene.[27]

COT has been identified as one of the products of the thermal polymer-isation of acetylene,[28,29] and very low yields of COT result from u.v. irradiation of acetylene.[30,31]

COT is also formed by various isomerisation processes occurring in other C_8H_8 compounds. Reactions which generate bicyclo[4.2.0]octa-2,4,7-triene (4), unless carried out at low temperatures,[32] lead to COT *via* valence isomerism (see p. 10); examples are given in scheme 2.

Other skeletal rearrangements leading to COT are outlined in schemes 3 and 4.

Additionally, COT is a minor photo-product of basketene (6)[55] and of the dimethyl acetylenedicarboxylate adduct (7),[56] and it has been detected amongst the pyrolysis products of the β-lactone (8)[57] and of α-cellulose.[58]

Finally, COT has been identified as one of the constituents responsible for the odour of tomatoes,[59] and may therefore be a natural product!

Scheme 2

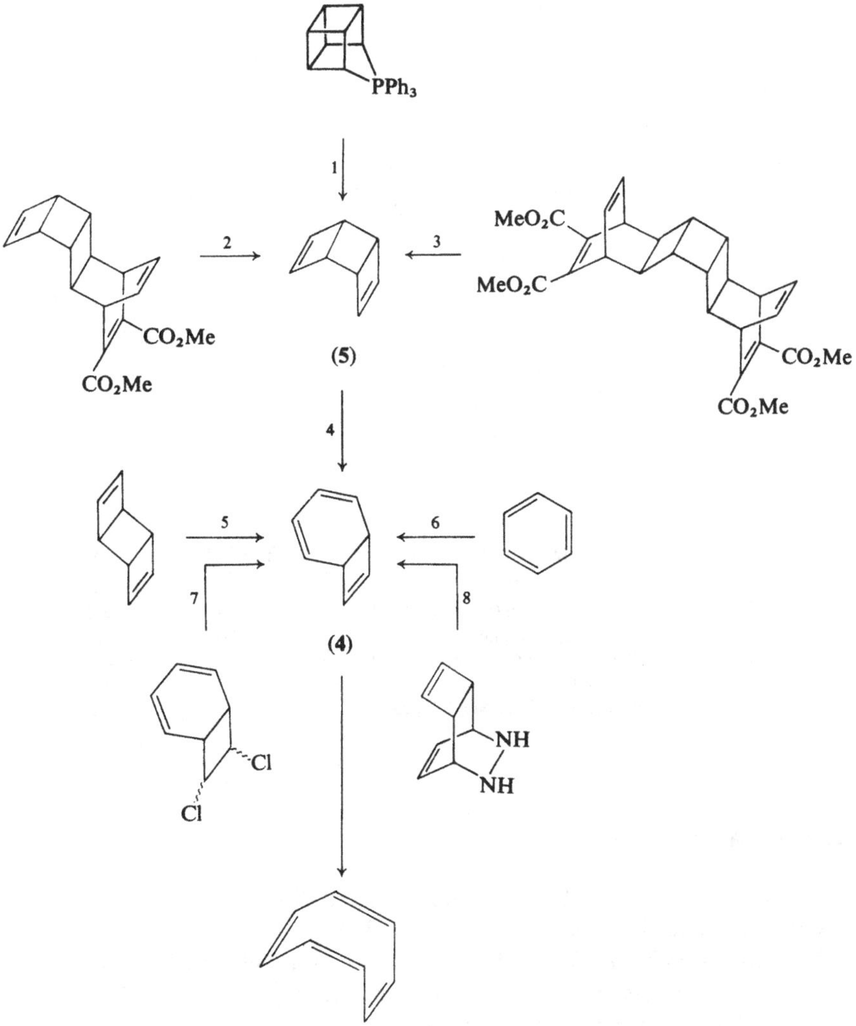

Reaction	Reagents and conditions	Yield (%)	Ref.
1	120°	85[a]	33
2	140–160°	—	34
3	*ca.* 150°	—	34
4	(>100°), 90–121°	—	(35), 36
5	{ *o*-Dichlorobenzene, 140°	100	35
	{ or e.g. AgBF$_4$, Me$_2$CO, reflux	100	37
6	HC≡CH, *hv*	(Very low)	38
7	NaI, NaHSO$_3$, Me$_2$CO, r.-t. → 50°	35	39
8	MnO$_2$, n-hexane, r.-t.	75–80	40

[a] Mixture of (5) and COT (4:1).

Reaction	Reagents and conditions	Yield (%)	Ref.
1	AgNO₃, MeOH(aq.), 80° or 427°/30 mm (flow system)	100 56	41 (42), 43
2	*hv*, methylcyclohexane	20	44
3	*hv* (Pyrex), e.g. isopentane, −60°	Up to 29	45
4	400–500° or e.g. AgClO₄, Me₂CO, r.-t.	— —	46 46

2. Purification

The purification of COT by fractional distillation and low-temperature crystallisation has been described in detail.[60]

For separation by liquid–solid column chromatography, see ref. 44.

For the use of gas–liquid chromatography, see e.g. refs. 9, 40, 44, 49; for other gas-chromatographic data see refs. 61, 62.

COT may also be purified *via* its silver nitrate complex (COT)₂(AgNO₃)₃ (see p. 51), from which it is readily regenerated by treatment with aqueous sodium chloride.[63]

COT forms an inclusion complex with thiourea,[64] but no use of this property appears to have been made.

COT is somewhat sensitive to air and light, and is best stored in the dark below room-temperature, in the presence of a free-radical inhibitor such as hydroquinone. Even so, samples of COT which have been kept for some time inevitably contain dimeric and polymeric material (see pp. 12–13, 34).

Scheme 4

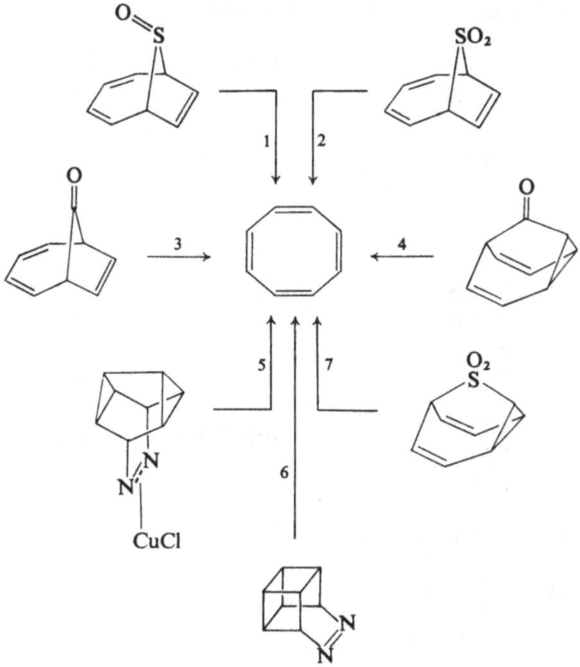

Reaction	Reagents and conditions	Yield (%)	Ref.
1	*hv*	—	47
2	{ 345° (g.l.c.)	100	48
	{ or *hv* (Corex), Me₂CO, Et₂Oᵃ	—	—
3	*hv* (Pyrex), Et₂O or THF	80–82	(50), 51, 52
4	*hv*	—	50
5	25°	100	41
6	*hv*	(Low)	53
7	240° (flow system)	100	54

ᵃ Conditions used for 1,4-dideuteriocyclo-octatetraene (40%).[49]

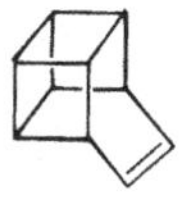

(6) 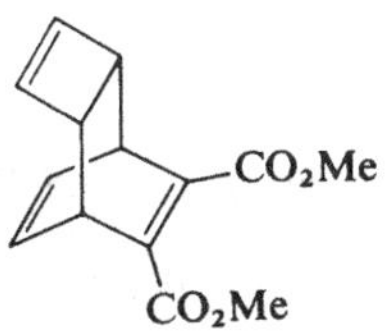   **(7)** 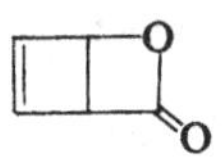   **(8)**

3. Physical properties

COT is a yellow liquid, with a strong distinctive odour.

Boiling-point (°C)	142–143	Pressure (mm Hg)	760	Ref. 11
	48		31	65
	42–42.5		17	11

Melting-point (°C)	−4.7			27, 66
	−4.5 to −3.5			63

Triple-point (K)	268.48			66

Density (g cm^{-3})	0.9382	Temperature (°C)	0	11
	(0.9206), 0.9209		20	(11), 60
	0.9196		25	60
	0.9117		30	60

Viscosity (cP)	1.42	Temperature (°C)	20	60
	1.30		25	60
	1.18		30	60

Heat capacity (12–330 K): see ref. 66

Vapour pressure (0–75 °C): see ref. 66

Heat of fusion (cal mol^{-1}) 2694.6	66

Heat of combustion (25 °C) (kcal mol^{-1})	−1084.9	67
	−1086.5	68

Heat of hydrogenation (25 °C) (kcal mol^{-1}) −97.96	69

Refractive index (Na$_D$)[a]	1.5379	Temperature (°C)	20	60
	(1.5348), 1.5350		25	(63), 60, 69
	1.5323		30	60

[a] For measurements at other wavelengths, see ref. 60.

Dielectric constant (20 °C) 2.74	11

First ionisation potential (adiabatic)[a] (eV)	7.99	70
	8.0	71
	8.04	72
	8.06	73

[a] For higher ionisation potentials, see refs. 71, 72.

Electron affinity (kcal mol^{-1}) 13.3	74

Half-wave reduction potentials: see e.g. refs. 65, 75–82

Magnetic susceptibility (cm^3 mol^{-1}) −0.0000539	83, 84

Magnetic rotation (μrad) 1009	85

The following have been calculated from thermochemical data:

Heat of vaporisation (25 °C) (cal mol⁻¹)		10 300	Ref. 66

Entropy (cal deg⁻¹ mol⁻¹)	(Liquid, 25 °C)	52.65	66
	(Gas, 25 °C, 1 atm)	78.10	66

Heat of formation (kcal mol⁻¹)	(Liquid, 25 °C)	60.82	68
	(Gas, 25 °C)	71.0	86, 87
		71.1	88, 89
		71.12	90
		71.3	91
		71.9	69

Heat of isomerisation to styrene (kcal mol⁻¹)	(Liquid, 25°C)	−36.10	68
	(Gas, 25 °C)	−36.3	69

Empirical (thermochemical) resonance energy (stabilisation energy) (kcal mol⁻¹)	2.4	69
	3.0	92
	3.3	93
	3.6	69
	4.8	67

For other calculations, see table 5, p. 11.

U.v. spectrum. COT exhibits a broad weak absorption band with a maximum near 280 nm (table 1), which tails into the visible region;[94,95] there is strong 'end-absorption', with a shoulder at *ca.* 205 nm[95,96] (ε 20 000[97]). The absorption curve is reproduced in refs. 60, 94, 98. (Measurements have also been made in the gas phase[94,95].)

Table 1

Solvent	$\lambda_{max.}$ (nm)	ε	Ref.
MeOH	280	350	94
EtOH	280	435	94
n-Heptane	280	235	94
CHCl₃	282	200	99
Cyclohexane	283	255	94
Iso-octane	283	250	98
CCl₄	288	320	94

I.r. spectrum. The C=C stretching vibration in COT gives rise to an absorption band with $\nu_{max.}$ (liquid film) 1634,[99] 1635[27] cm⁻¹. The spectrum is reproduced in refs. 27, 60; for complete lists of the principal absorption maxima, see refs. 27, 99. (Measurements have also been made in the gas phase[27,99].)

Raman spectrum. See refs. 27, 100, 101.

N.m.r. spectra. The ^{1}H n.m.r. spectrum of COT shows a singlet resonance at -341 Hz relative to tetramethylsilane, i.e. at τ 4.32 (60 MHz; neat liquid).[102] For the p.m.r. of solid COT, see ref. 103.

The ^{1}H n.m.r. spectrum is temperature-dependent with respect to the ^{13}C satellites of the proton signal, owing to the conformational mobility of the molecule (see p. 9). At very low temperatures these satellites are essentially double doublets; for the coupling constants, see table 2.

Table 2

$J(H^1–H^2)$ (Hz)	$J(H^1–H^8)$ (Hz)	Ref.
11.4	*ca.* 2.5	104
11.48	3.87	105
11.8	—	106

From the ^{13}C n.m.r. spectrum, $J(^{13}C–H) = 155$ Hz.[107]

The presence of a weak paramagnetic ring-current in COT has been inferred from measurements of solvent shifts.[108]

Mass spectrum. The major ions in the mass spectrum of COT, together with their relative abundances and appearance potentials, are listed in ref. 73.

For the doubly-charged ion mass spectrum of COT, see ref. 109.

For the collisional activation spectrum of field-ionised COT, see ref. 110.

Photo-electron spectrum. See refs. 71, 72, 111, 112.

Trapped electron impact spectrum. See ref. 113.

For electron attachment studies see refs. 114, 115.

Magnetic circular dichroism. The MCD curve of COT is reproduced in ref. 98.

4. Structure

X-ray[116] and electron diffraction[117–119] studies of COT show that the eight-membered ring exists in a non-planar 'tub' form (D_{2d} symmetry), with the bond lengths and angles listed in tables 3 and 4.

Although the non-planarity of COT in its ground state prevents *substantial* overlap of the adjacent π-orbitals, certain through-bond and through-space interactions may exist.[72,85,87]

The observed u.v. absorption (see p. 7) has been explained on the basis of interactions between the excited π-systems.[121,122]

Table 3

Bond	Length (Å)	Ref.
C=C	1.330	116
	1.334	118
	1.340	119
C—C	1.456	116
	1.462	118
	1.476	119
C—H	1.090	118
	1.100	119

Table 4

		Ref.
Angle $C\overset{C}{\underset{C}{\diagup}}$	126.1°	119
	126.5°	118
	126.8°	116
Angle $C\overset{C}{\underset{H}{\diagup}}$	117.6°	119
	118°	118
Torsion angle around $C—C$	56°	116
	58°	120

Conformational inversion and valence tautomerism. At ordinary temperatures, the tub-shaped COT molecule undergoes rapid ring-inversions (see e.g. ref. 123).

This process may be studied in substituted COTs by variable-temperature n.m.r. spectroscopy (see p. 94); for calculations of the energy barrier, see refs. 124–127.

In addition to this conformational change, two kinds of valence tautomerism occur. The first involves bond-shift in the eight-membered ring, the single and double bonds exchanging their positions.

For measurements of the rate constant k_2, see refs. 104, 106, 128, 129 ($E_a = 10.9$,[129] 14.5[104] kcal mol^{-1}; $\Delta G^{\ne} = 13.3$,[104] 13.7[106] kcal mol^{-1}).

It is generally assumed that the ring-inversion and bond-shift processes occur *via* the planar transition states (*A*) and (*B*) respectively, (*A*) containing alternate double and single bonds (alternating bond lengths, D_{4h} symmetry), but (*B*) having D_{8h} symmetry (equal bond lengths). Theoretical studies[125,127,130-133] predict that structure (*A*) should be more stable than the

(*A*)   (*B*)

delocalised form (*B*) (which is 'anti-aromatic'). It follows that inversion should proceed more rapidly than bond-shift, and indeed this is found for substituted COTs in which it is possible to discriminate between the two processes (see p. 94).

The second type of valence tautomerism displayed by COT results in a low concentration (*ca.* 0.01 % at 100°) of bicyclo[4.2.0]octa-2,4,7-triene (4)[134-136] (a disrotatory 6π electrocyclisation is thermally allowed by the Woodward–Hoffmann rules).

$$k_3 \atop k_{-3}$$

(4)

For values of k_3, see refs. 128, 136 ($E_a = 27.2$ kcal mol^{-1};[128] $\Delta H^{\ne} = 27.4$,[135] 28.1[136] kcal mol^{-1}); for k_{-3}, see ref. 128 ($E_a = 18.7$ kcal mol^{-1} [32]).

The theoretical importance of COT as [8]annulene has led to numerous calculations of various aspects of its structure and properties (table 5) (for general discussions, see e.g. refs. 137–139). Calculation of the properties of the lowest $\pi \rightarrow \pi^*$ triplet state of COT leads to the conclusion that it may be regarded as 'aromatic'.[154]

Table 5

	Ref.
Molecular geometry	124, 125, 127, 140–142
Symmetry force constants	143
Mean amplitudes of vibration	143
Strain	95, 130, 144–147
Thermodynamic functions	90
Bond energies	142
Orbital energies	72
Total energy	125, 145, 148
Bond length–bond overlap correlation	149
π-Binding energy	150
Resonance	92, 95, 130, 150, 151
Index of aromaticity	152
U.v. spectrum	92, 95
π-Electron transition energies	96, 153
Heat of formation	88, 111, 124, 144, 145
Ionisation potential	111, 124

5. Thermal oligomerisation

The thermal dimerisation of COT was first observed by Reppe *et al.*,[11] and Jones[155,156] later showed that four different products of formula $C_{16}H_{16}$ could be isolated. At *ca.* 100° (or at 20° under high pressure[157]) the two main dimeric products are compounds of m.p. 52–53°[156,158] and 76–77°[156,159] respectively (for the best preparative procedure, see ref. 160); at higher temperatures (130–170°) two other dimers, melting at 41.5°[11,155] and 38.5°[155] respectively, are produced. The structures of these products are now known to be (9), (10a ⇌ 10b), (11), and (12) severally.

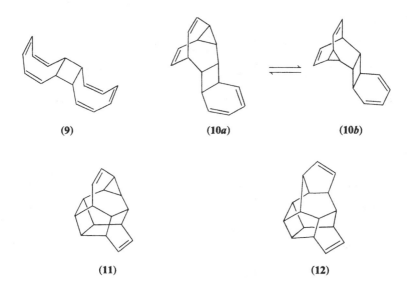

(9)   (10a)   (10b)

(11)   (12)

Scheme 5

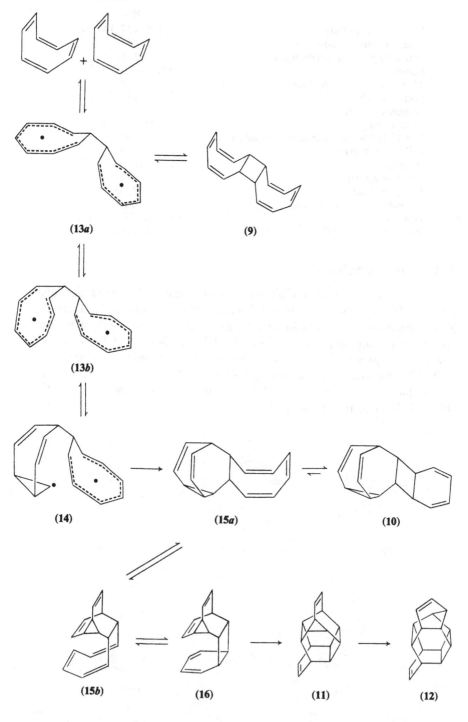

(13a)   (9)

(13b)

(14)   (15a)   (10)

(15b)   (16)   (11)   (12)

Structure (12) for the dimer of m.p. 38.5° was determined by an X-ray crystallographic examination of its silver nitrate complex;[161, 162]* the co-dimer, m.p. 41.5°, may be assigned structure (11) from spectral evidence.[165] The temperature-dependent p.m.r. spectrum of the dimer of m.p. 76–77° reveals the presence of a fluxional 3,4-homotropilidene system, in which rapidly reversible valence isomerism occurs *via* the Cope rearrangement;[159, 166] its structure is therefore represented as (10a ⇌ 10b) (for the ^{13}C n.m.r. spectrum, see ref. 167). The remaining thermal dimer, m.p. 52–53°, possesses structure (9),[34, 168] oxidative cleavage to *cis,cis,cis*-cyclobutanetetracarboxylic acid indicating the stereochemistry of the ring-fusions.[34]

Schröder and Oth[169] have proposed the following mechanism (scheme 5) for the formation of these remarkable products. Dimer (9), which is formed slowly even at room-temperature,[156] is presumed to result from a non-concerted [2 + 2] cycloaddition of two molecules of COT, the reaction intermediate being represented as the diradical (13a). If dimer (9) is heated at *ca.* 100°, COT is regenerated,[159] but formation of dimer (10) also occurs (and may be made quantitative by using high pressure[157]). It is proposed that inversion of one of the eight-membered rings in (13a), leading to the conformer (13b), is followed by ring-closure of the rearranged diradical (14) to give the valence tautomer (15a) of dimer (10). At higher temperatures the cage-like dimer (12) accumulates, and has been shown to result from dimer (11).[158, 165] Inversion of the cyclohexa-1,3-diene system in dimer (10), which is possible *via* the conformationally mobile valence tautomer (15a ⇌ 15b), would lead to a structure (16) capable of an intramolecular Diels–Alder reaction with the formation of dimer (11). Conversion of this product into dimer (12) may then proceed *via* a vinylcyclopropane → cyclopentene rearrangement, as suggested by Moore.[165] This mechanistic scheme suffers from the defect that it provides no explanation for the *stereospecific* formation of dimer (9). The conversion of dimer (11) into dimer (12), rather than into the alternative rearrangement product (17), also requires explanation. However, inspection of molecular models indicates that the shaded six-membered ring in structure (12) is in the form of a half-chair,

(12)

(17)

---

* Although it was not realised at the time, this assignment was not completely unambiguous, because of the possibility of silver-catalysed rearrangement (see e.g. ref. 163). N.m.r. spectral studies show, however, that there is no skeletal rearrangement when dimer (12) is converted into its silver nitrate complex.[164]

whereas in structure **(17)** the corresponding ring is forced to adopt a distorted boat-form;[164] this difference may account for the exclusive formation of dimer **(12)** in the vinylcyclopropane rearrangement, if the transition states leading to the alternative structures **(12)** and **(17)** are closely related to the respective products.

Alternative suggestions concerning the formation of the COT dimers have been made by Iwamura and Morio.[170] On the basis of a theoretical study it was concluded that the conversion of COT into semibullvalene **(18)** is symmetry-allowed as a thermal process ($[_\pi 4_a + _\pi 2_a]$).

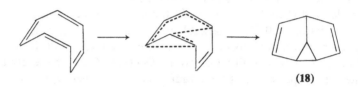

(18)

It was then claimed that dimers **(10)**–**(12)** could result from thermally-allowed cycloadditions of semibullvalene and bicyclo[4.2.0]octa-2,4,7-triene, *via* the transition states **(C)**–**(E)** (scheme 6) which were shown to contain Hückel arrays of molecular orbitals (see ref. 171). However, closer examination of the structure which would be obtained by reaction of the components in mode **(E)** shows that it would represent, not dimer **(12)**, but an impossibly strained system **(19)**. As an alternative possibility, it was proposed that dimers **(9)** and **(10)** could arise from an intermediate dimer **(20)** (*cf.* **(193)**, p. 68) *via* the transition states **(F)** and **(G)** (scheme 7) respectively.[170] Again a correction is necessary; transition state **(F)** would not lead to dimer **(9)**, but to the *anti*-isomer **(21)** · (see p. 320). Another hypothetical intermediate **(22)**, derived from **(20)**, was also visualised as a possible precursor of dimers **(11)** and **(12)**[170] (scheme 8).

For the co-dimerisation of COT and monosubstituted COTs, see pp. 101–2.

Residues from the distillation of dimerised COT contain a tetramer, $C_{32}H_{32}$,[159] which also results when dimer **(10)** is heated[172] (see p. 368). The tetramer molecule possesses two fluxional homotropilidene systems, and is formulated as the Diels–Alder adduct **(23)** derived from two molecules of dimer **(10)**[172] (only one of the four interconverting forms is shown). The illustrated stereochemistry has not been proved, but represents that which would result from [4 + 2] cycloaddition on the unhindered face of each component and in the *endo*-manner.

Scheme 6

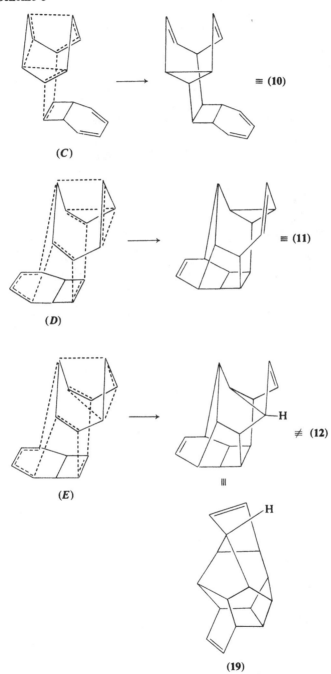

(C)

≡ (10)

(D)

≡ (11)

(E)

≢ (12)

‖

(19)

Scheme 7

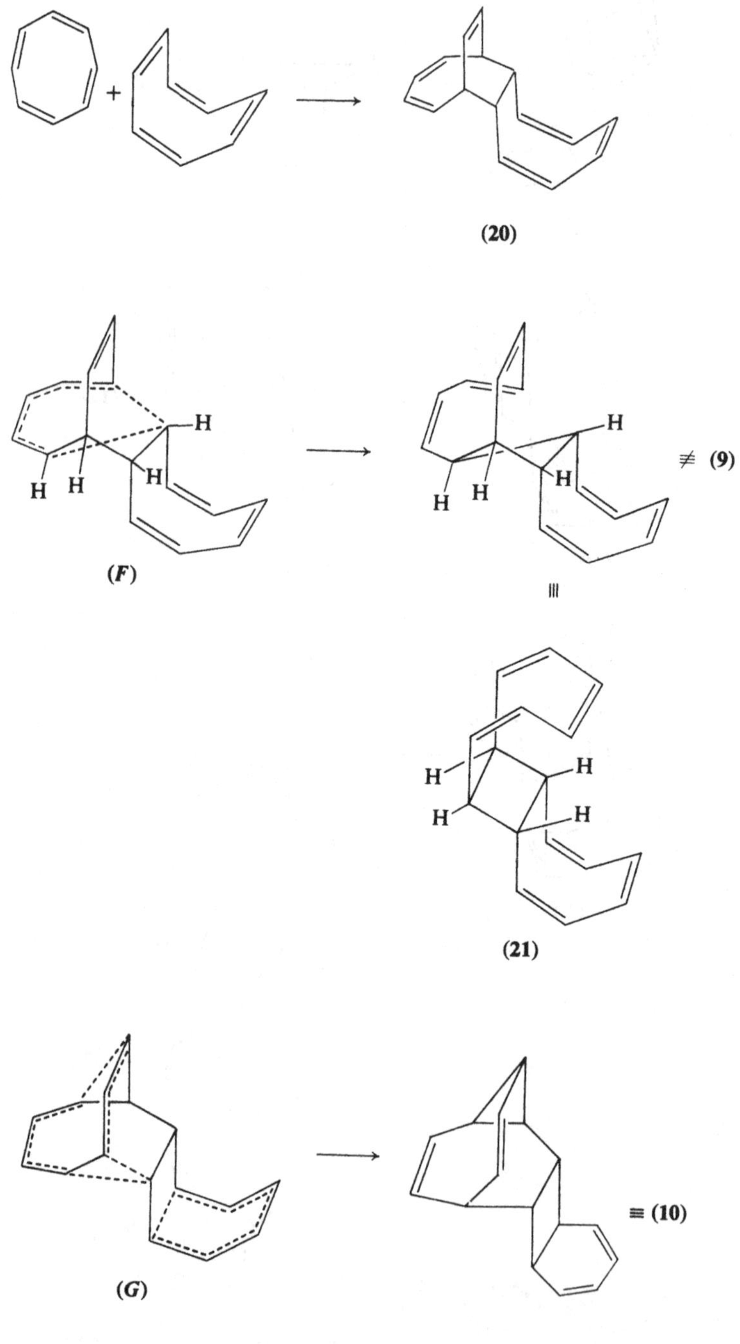

(20)

(F)   ≢ (9)

|||

(21)

(G)   ≡ (10)

Scheme 8

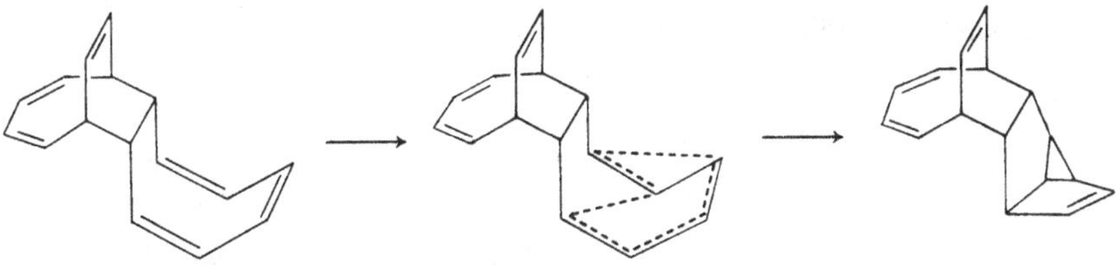

(20)   (22)

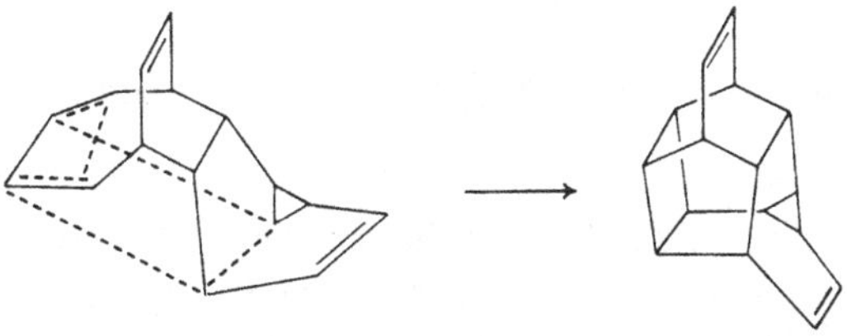

≡ (11)

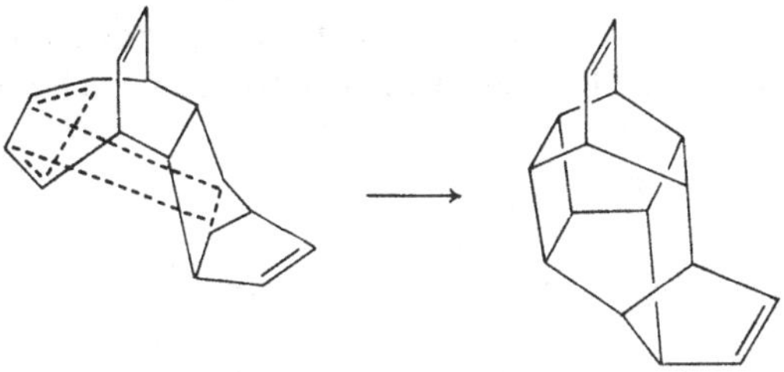

≡ (12)

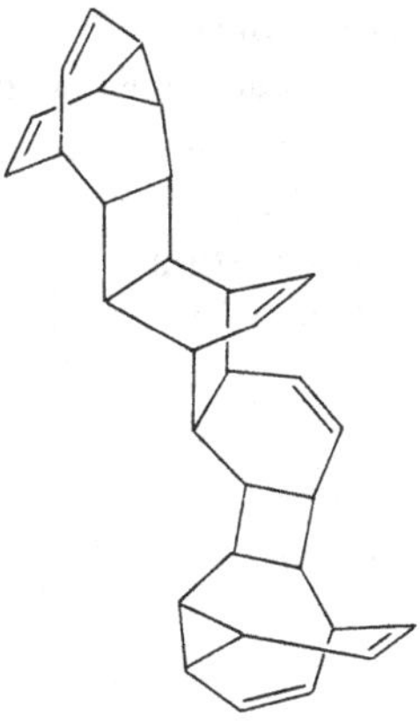

(23)

6. Thermolysis, photolysis and radiolysis

Thermolysis. At 520°/30 mm (flow system; residence time *ca.* 1–3 s), COT undergoes a degenerate structural rearrangement, detectable by the observation of deuterium scrambling in 1,4-dideuteriocyclo-octatetraene[49] (see also p. 118).

Under slightly more vigorous conditions, *ca.* 510–550° (flow system; residence time *ca.* 7.5 s), the dihydropentalene (**24**) is formed (up to *ca.* 40%), possibly *via* the diradical (**25**).[173]

510–550°
(Flow system)

(**25**)   (**24**)

At higher temperatures, in the gas or liquid phase, styrene, benzene, acetylene, etc. are produced.[174]

In the presence of a palladium catalyst, at 400° the main product (50%) is ethylbenzene.[175]

Photolysis. Direct photolysis of COT in the gas phase,[176–179] in solution,[180] or in a frozen matrix,[181] affords styrene, benzene and acetylene. Some evidence has been adduced[180, 181] for the intermediate formation of bicyclo[4.2.0]octa-2,4,7-triene (**4**), which in principle could result from a photochemically-allowed disrotatory closure of the cyclobutene ring in a 4π electrocyclic process.

hv

(**4**)

(The above evidence has been questioned,[182] but u.v. irradiation of the bicyclic triene (**4**) at −65° in the presence of acetone is known to give benzene[183].)

In isopentane containing acetone, at −60°, u.v. irradiation of COT leads to semibullvalene (**18**) (*ca.* 42% yield based on unrecovered COT); it has been postulated that initial photo-isomerisation to *Z,Z,Z,E*-cyclo-octatetraene (**26**) could be followed by a photochemical intramolecular $[_\pi2_s + _\pi2_s + _\pi2_a]$ cyclo-addition.[183, 184]

hv
Me₂CO, isopentane,
−60°

E

(**26**)   (**18**)

COT effects efficient 'quenching' of triplet benzophenone,[185] and is widely used as a specific triplet-state quencher in dye-laser studies (see e.g. refs. 185–187). (For fluorescence quenching, see ref. 188.)

Radiolysis. COT is remarkably stable to γ-radiation,[189] and has a powerful retarding effect on the radiation-induced polymerisation of vinyl monomers.[190] (For one-electron oxidation and reduction, see refs. 191, 192; for radiolysis studies in the presence of COT, see e.g. refs. 193–195.)

7. Oxidation, and reactions with electrophiles

Oxygen. COT polymerises in the presence of air,[11] but the nature of the initiation reaction has not been elucidated. COT apparently does not undergo dye-sensitised photo-oxygenation (singlet oxygen).[196]

Oxidation of COT with air in the presence of a mixed metal oxide catalyst at 370° gives benzoic acid (70%).[11]

Ozone. Ozonolysis of COT in chloroform gives, after work-up with acetic acid–aqueous hydrogen peroxide, a mixture containing oxalic acid, glycollic acid, ethane-1,1,2,2-tetracarboxylic acid and succinic acid.[197] When ozonolysis is carried out in methanol, however, the product is glyoxal.[198]

Peracids. COT may be converted into the mono-epoxide (27)[199] (52–55%) by the action of peracetic,[200] perbenzoic,[11,201] or monoperphthalic[202] acid.

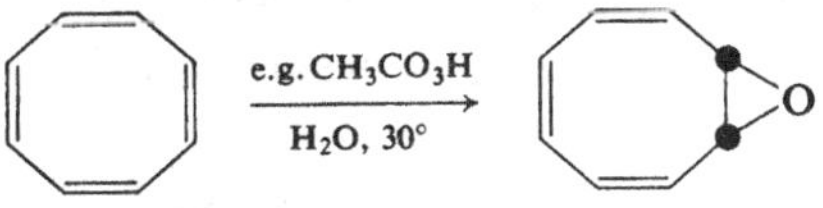

(27)

With performic acid, the products are phenylacetaldehyde (20%) and phenylacetic acid (49%).[39]

Potassium permanganate, chromic acid. Aqueous potassium permanganate oxidises COT to benzaldehyde and benzoic acid, whereas chromic acid gives in addition to these products, terephthalic acid.[11,203]

Scheme 9

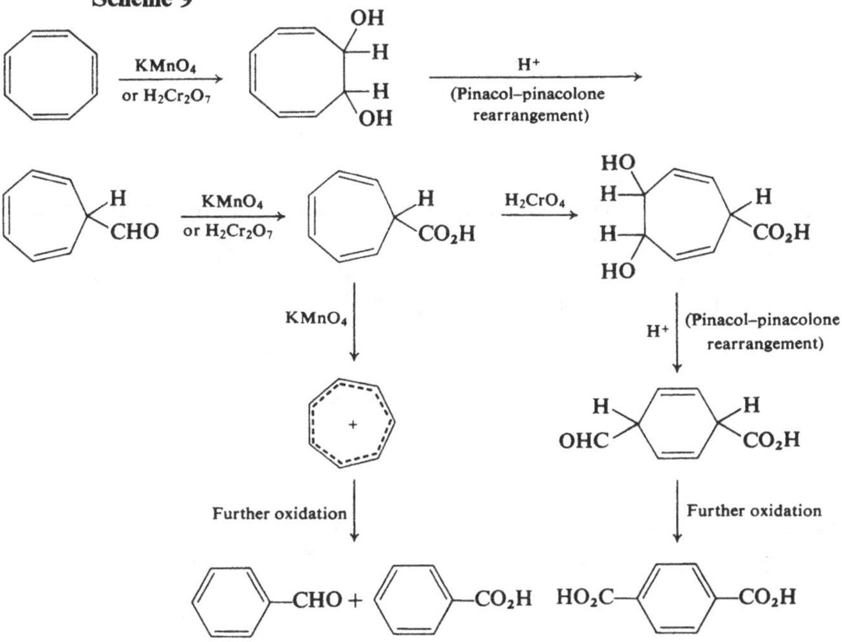

In an oxidation using permanganate in aqueous sulphuric acid containing acetone, Ganellin and Pettit[204] were able to obtain evidence for the intermediate formation of tropylium sulphate, and suggested that acid permanganate and chromic acid oxidations might proceed according to scheme 9.

Sodium hypochlorite. Alkaline hypochlorite solution oxidises COT to benzaldehyde, benzoic acid and terephthaldialdehyde.[11]

Selenium dioxide. Oxidation of COT with selenium dioxide gives (in poor yield) terephthalic acid, together with *o*-phthalaldehyde, phenylglyoxal and cyclo-octatrienone.[205]

Mercury(II) and lead(IV) salts. A suspension of mercury(II) sulphate in water oxidises COT to phenylacetaldehyde (up to 70%).[11,206]

Mercury(II) acetate in acetic acid yields the bicyclic diacetoxy-compound (28) (up to 84%),[206,207] but in methanol the product is the acetal (29) (88%).[206]

Lower yields of the products (28) and (29) are obtained with lead(IV) acetate in acetic acid (or benzene) and methanol respectively;[208] however, with lead(IV) acetate in acetic acid containing boron trifluoride, the main product is the diacetate (30) (up to 34%).[208,209]

(30)

In these reactions, the initial step is presumably an electrophilic attack on COT by $(AcO)Hg^+$ or $(AcO)_3Pb^+$; for a possible mechanism for the ring-contraction processes, see ref. 210.

Cobalt(II) *salts*. The oxidation of COT with cobalt(II) ions in a rapid-mixing flow system gives rise to a radical cation (**31**), the e.s.r. spectrum of which supports a non-planar structure.[211]

$$\text{COT} \xrightarrow[\text{CF}_3\text{CO}_2\text{H}]{\text{Co}^{2+}} \text{COT}^{\ddagger} \quad (\textbf{31})$$

(For theoretical calculations on $COT^{\ddagger}$, see ref. 124.)

For the production of $COT^{\ddagger}$ by γ-irradiation of COT in a frozen matrix, see ref. 191.

Electrolytic oxidation. The electrochemical oxidation of COT at a mercury anode gives phenylacetaldehyde in good yield.[212]

In acetic acid containing sodium or potassium acetate, with a carbon anode, a mixture of the diacetoxy-compounds (**28**), (**32**) and (**30**) is produced; the use of a platinum anode leads, additionally, to the acetoxy-methyl-derivatives (**33**) and (**34**) in low yields.[209]

(**28**)   (**32**)

(**30**)

(**33**)   (**34**)

Protonation. Dissolution of COT in concentrated sulphuric acid generates the homotropylium cation,[213,214] the p.m.r. spectrum of which provides compelling evidence for the non-classical structure (**35**);[214-217] this formulation is also supported by the u.v. spectrum[218] and by volume diamagnetic susceptibility data.[219] Comparison of this species with the protonated complexes of COT with certain transition metal carbonyls is instructive.[214] The same cation is obtained in a 'super-acid' medium.[220]

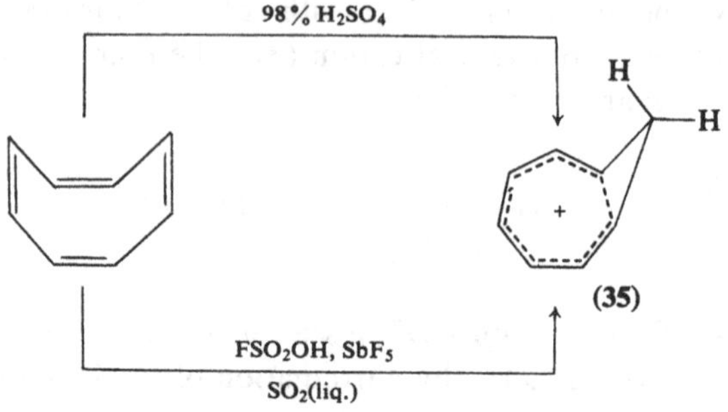

(35)

Summaries of the structural evidence for (35) have been given by Winstein[221,222] (for theoretical discussions, see refs. 223, 224).

At $-10°$ in D_2SO_4, initially *ca.* 80%* of the incoming deuterium appears with *endo*-stereochemistry, but eventually becomes distributed uniformly between the *endo*- and *exo*-positions through ring-inversion[218] (a circumambulatory rearrangement may be excluded;[226] for a theoretical study, see refs. 223, 227).

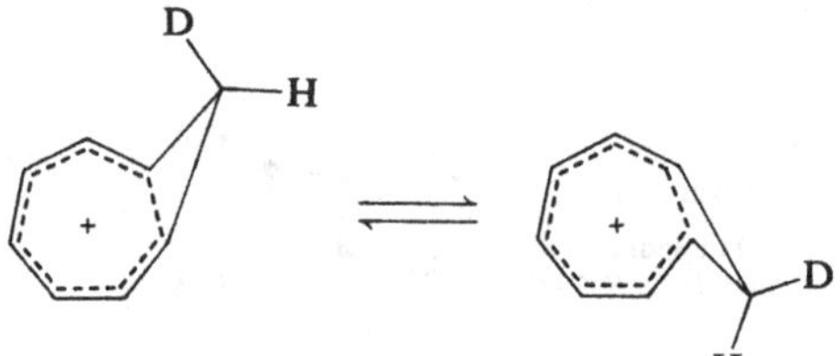

$(\Delta G^* = 22.3 \text{ kcal mol}^{-1})$ [218]

If the ring-inversion proceeds *via* the planar classical cyclo-octatrienyl cation (36), then the above value of ΔG^+ represents the difference in free energy between the classical and non-classical cations.[218]

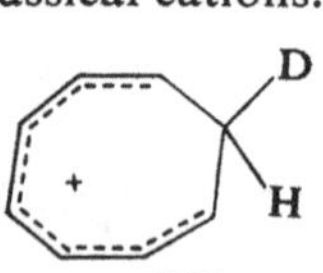

(36)

A crystalline homotropylium salt (37a) may be formed from COT by the action of hydrogen chloride and antimony(v) chloride.[213]

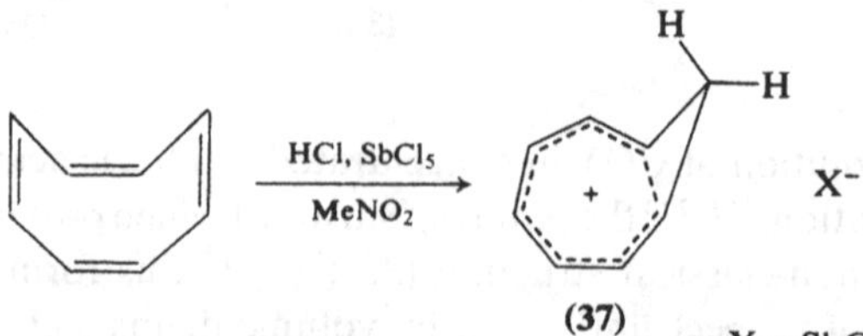

(37)

a: X = $SbCl_6$

b: X = $SbBrCl_5$

Using hydrogen bromide, the analogous salt (37b) is produced.[213]

* In FSO_2OD at $-70°$, 75%.[225]

Halogenation. Chlorine and bromine add to COT with the formation of di-, tetra- and hexahalides;[11] no octahalide has been reported.

Reaction of COT with one equivalent of chlorine gives a mixture of the stereoisomeric bicyclic dichlorides (38) and (39) (*ca.* 4:1, total yield up to 82%);[11,39,228] rearrangement of the *cis*-isomer (38) to the *trans*-compound (39) is rapid above 70°, and is catalysed by acids.[228]

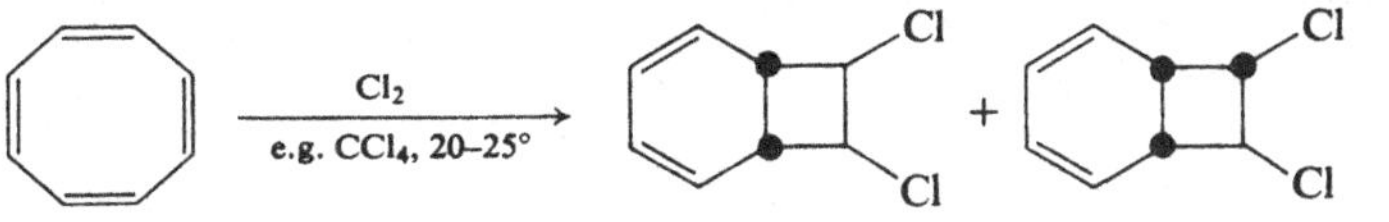

(Chlorination of COT may also be accomplished with sulphuryl chloride (see p. 34).)

Careful examination of the chlorination reaction has been made by Huisgen *et al.*,[225,229,230] the multi-stage reaction being shown in scheme 10. The initial reaction apparently proceeds *via* the *endo*-8-chlorohomotropylium ion (40), which undergoes exclusive *endo*-attack by chloride ion to form the monocyclic *cis*-dichloride (41*a* ⇌ 41*b*). Slow epimerisation of the *cis*- to the *trans*-

Scheme 10

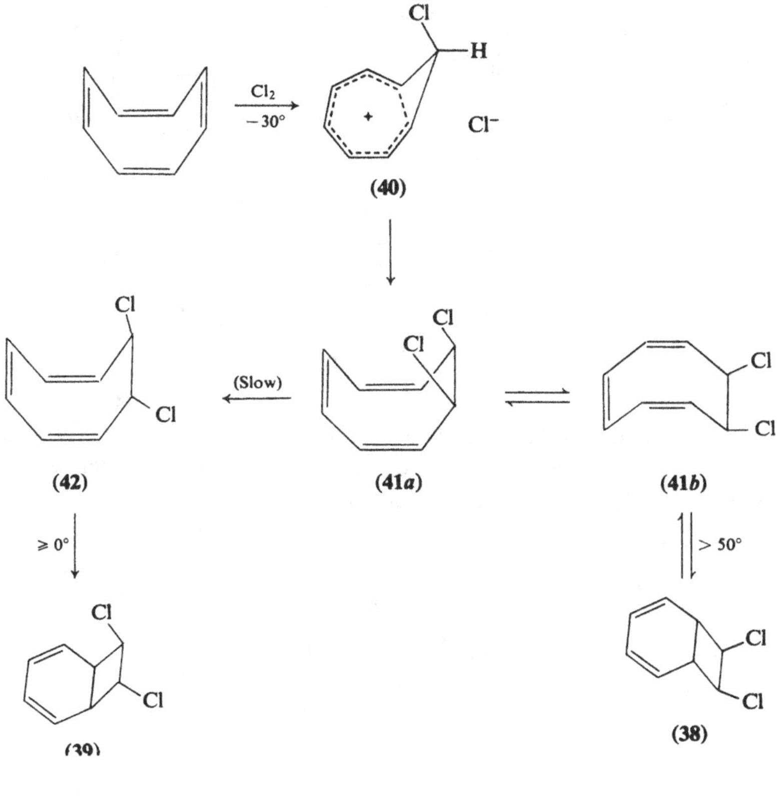

dichloride (**42**) occurs, and at temperatures $\geqslant 0°$ irreversible valence isomerism leads to the bicyclic *trans*-dichloride (**39**). Above 50° the valence isomerisation (**41***b* $\rightleftharpoons$ **38**) is mobile, the equilibrium mixture containing 80% of the bicyclic *cis*-dichloride (**38**) at 70°.

   In connection with these studies, it was found that COT reacts with two equivalents of antimony(v) chloride at low temperature to give the *endo*-8-chlorohomotropylium salt (**43**) (94%).[230]

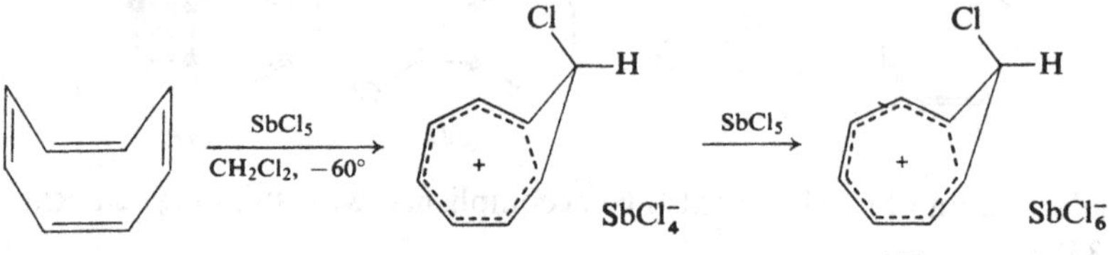

(**43**)

   For tetra- and hexachlorides of COT, see ref. 11.

   In the analogous bromination of COT, no bicyclic *cis*-dibromide is formed, the exclusive product above 0° being the *trans*-dibromide (**44**)[231] (scheme 11).

Scheme 11

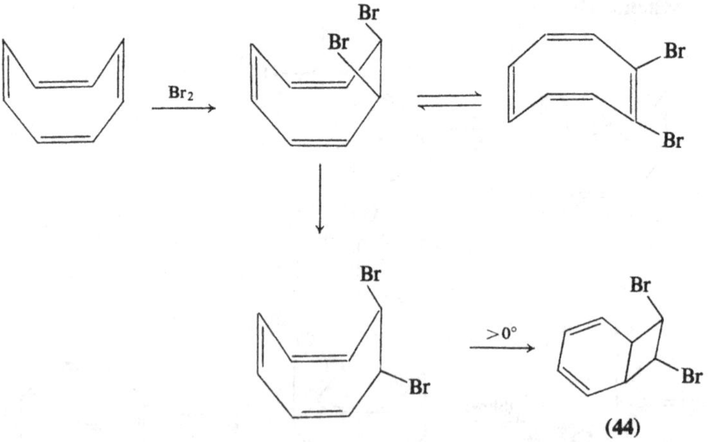

(**44**)

   Further reaction with bromine gives tetra- and hexabromides,[11] the structures of which have been partially solved, and are illustrated below (each compound is likely to be a mixture of stereoisomers).[232]

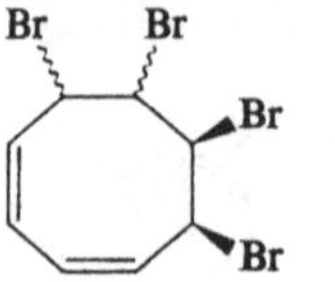

m.p. 144–145°

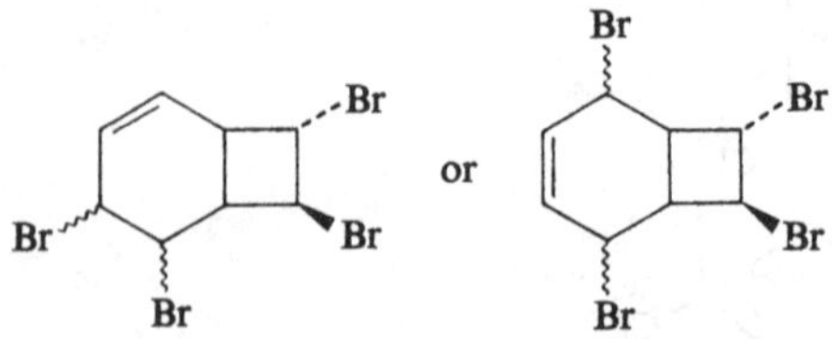

or

m.p. 87–88°

m.p. 150–151°   m.p. 97–98°

Hydrogen bromide. Reaction of COT with hydrogen bromide in acetic acid at 0° gives 1-bromoethylbenzene (**45**) (70%).[11] However, in benzene at −10° the product is 7-bromocyclo-octa-1,3,5-triene (**46**) (53%), isolated in admixture with its valence tautomer (**47**) (30%); addition of a further molecule of hydrogen bromide yields the dibromocyclo-octa-1,3-diene (**48**) (52%).[233] The

results obtained by Overberger *et al.*,[234] who employed two equivalents of hydrogen bromide (in benzene, or in the absence of solvent), and then treated the product with dimethylamine, may now be clarified (scheme 12). The principal product (68–69%) must have been derived from the dibromide (**48**), and may be assigned structure (**49**); the minor product (5–6%) was shown to have structure (**50**), arising from the benzylic bromide (**45**).

Scheme 12

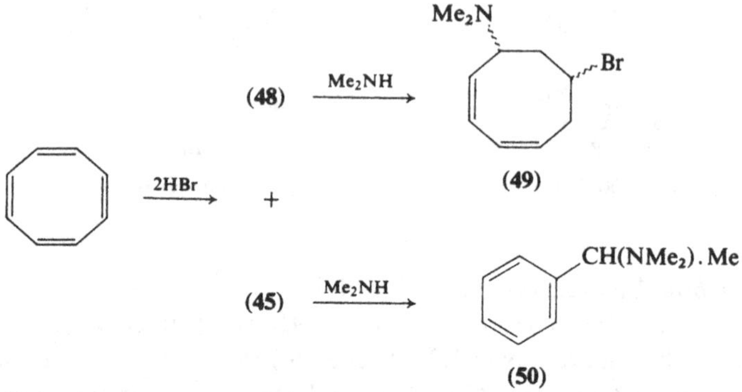

(48) $\xrightarrow{\text{Me}_2\text{NH}}$ (49)

$\xrightarrow{\text{2HBr}}$ +

(45) $\xrightarrow{\text{Me}_2\text{NH}}$ (50)

Acylation. Reaction of COT with acetyl bromide or propionyl chloride in the presence of aluminium chloride results in the formation of *o*-alkylcinnamaldehydes (51) in very low yields (2–3 %).[235]

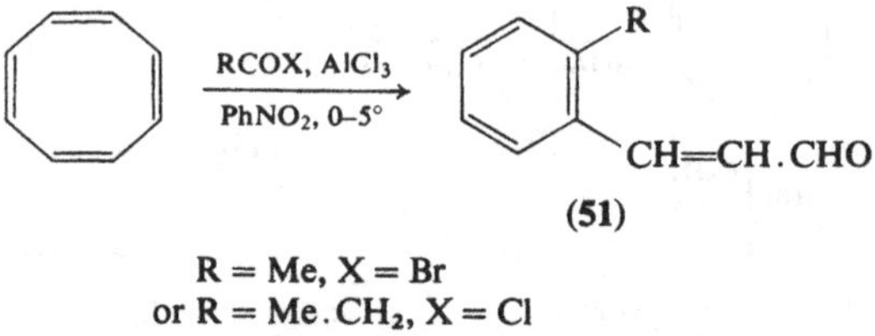

(51)

R = Me, X = Br
or R = Me.CH_2, X = Cl

Extensive polymerisation of COT is inevitable under these conditions. (For substitution of the iron tricarbonyl complex of COT, see pp. 82–3, 202.)

Iodine azide. Iodine azide, prepared *in situ* from iodine monochloride and an excess of sodium azide in acetonitrile, reacts with COT to form the *cis*-1,2-diazide (52) (80–90 %); this is thought to arise from an initial *trans*-addition followed by S_N2 displacement of iodide ion.[236] The product (isolated as the

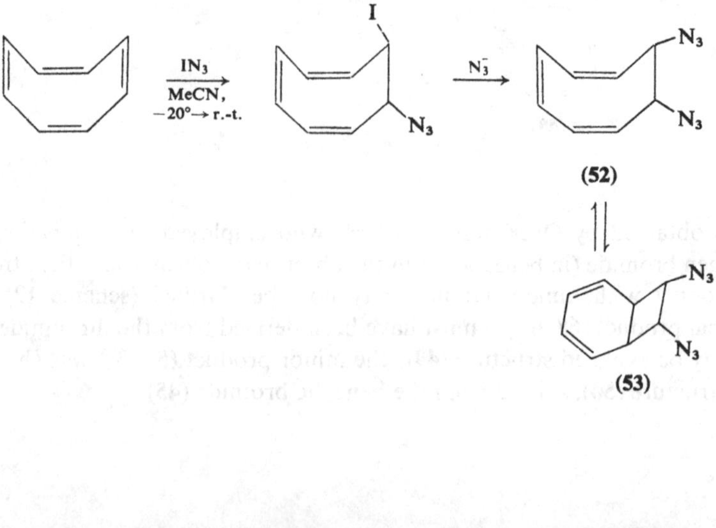

(52)

(53)

bicyclic valence isomer (**53**)) is an explosive oil, and was characterised as the adduct (**54**) resulting from 1,3-dipolar addition of the azide groups to dimethyl acetylenedicarboxylate.

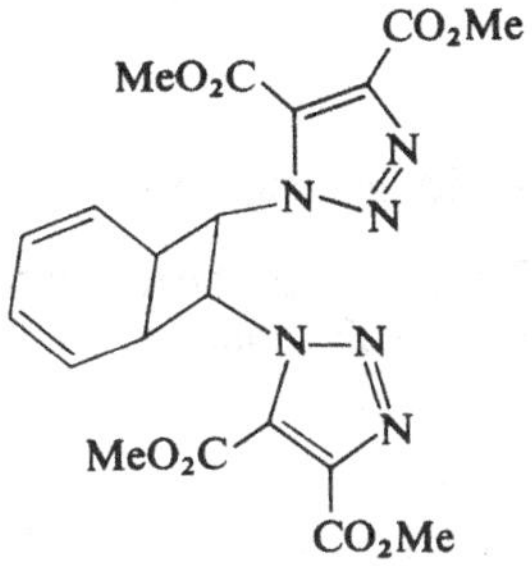

(**54**)

Sulphur dichloride. The addition of sulphur dichloride to COT at low temperature affords the sulphur-bridged product (**55**) (*ca.* 30 %).[237, 238] Reaction of this product with a second molecule of sulphur dichloride yields the dithia-adamantane (**56**) (20 %).[237, 238]

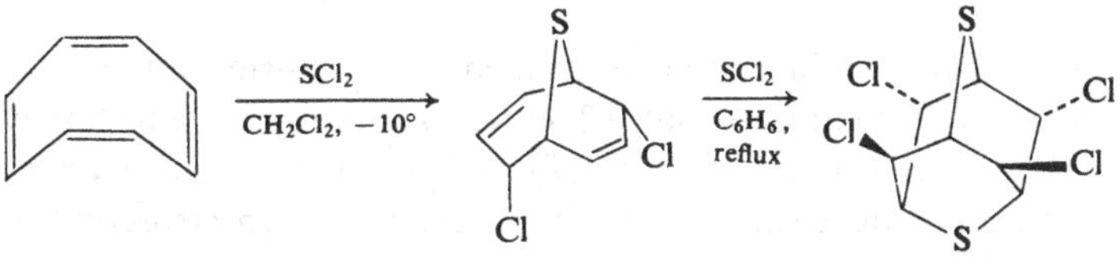

(**55**)   (**56**)

Alkanesulphenyl chlorides. COT reacts with methanesulphenyl chloride in chloroform at $-10°$ to give, in very high yield, a product formulated as (**57**).[239]

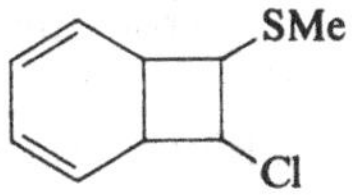

(**57**) 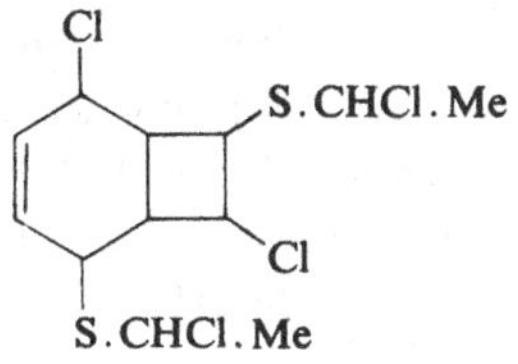   (**58**)

With two equivalents of 1-chloroethanesulphenyl chloride in tetrahydrofuran at room-temperature, there was obtained a crude product for which structure (**58**) was suggested.[240] In the absence of structural evidence, these formulations are doubtful (*cf.* ref. 241).

Boron trichloride. Boron trichloride reacts with COT at low temperature to yield the styryldichloroborane (**59**) (34 %).[242]

(59)

8. Reduction, and reactions with nucleophiles

Reduction with metals. COT reacts with the alkali metals in e.g. ethers[11,243,244] or liquid ammonia[63,99] to form the planar 'aromatic' dianion (60) (see pp. 48–9).

M = Li, Na, K etc.

(60)

In the presence of less than two equivalents of alkali metal, an equilibrium is established with the radical anion (61).[243-246]

$$COT + COT^{2-} \rightleftharpoons 2COT^{\cdot-} \quad (61)$$

(The position of this equilibrium is controlled by ion-pairing, and is therefore dependent on the solvent (see p. 49).) The radical anion (61), which may also be generated by electrolytic reduction of COT (see p. 29), likewise appears to be planar, on the evidence of e.s.r.[243,246,247] and u.v.[248] spectroscopy, and of polarography[65,249] (see, however, ref. 75).

For the e.s.r. spectrum of the monodeuterio-system, see ref. 250.

For the formation of $COT^{\cdot-}$ by γ-irradiation of COT in a frozen matrix, see ref. 191.

The electron-transfer processes occurring in the reduction of COT to its dianion (60) have been examined in some detail (see e.g. refs. 65, 246, 251–255), and the species (60) and (61) have been the subject of numerous theoretical studies (see e.g. refs. 70, 122, 124, 125, 248, 256–258).

Protonation of the dianion (60), using water or alcohol, leads to mixtures of the cyclo-octatrienes (62) and (63); the systems which have been employed to generate COT^{2-} for this purpose include lithium–ether,[11,63] triphenylmethyl-sodium–ether,[259] sodium–liquid ammonia,[11,63] sodium–liquid ammonia–ether,[260] sodium–*N*-methylaniline–ether[7] and sodium–*N*-ethylaniline–ether.[261]

(62)   (63)

For the deuteriation of COT^{2-} with deuterium oxide, see p. 173.

For the reduction of octadeuteriocyclo-octatetraene, see ref. 262.

The relative proportions of the trienes (62) and (63) obtained after work-up are variable, since the 1,3,6-triene (63) may be converted into the 1,3,5-triene (62) in the presence of base.[63, 261] If the products are heated during work-up, further complications may arise from the thermal conversion of (63) into (62), which equilibrates with bicyclo[4.2.0]octa-2,4-diene (64) (see pp. 214–15).

(63)   (62)   (64)

Sodium in boiling alcohols reduces COT to a mixture of the cyclo-octadienes (65)–(67);[261, 263] the relative proportions of the products depend on the reaction temperature and the duration of heating, possibly owing to base-catalysed isomerisations.[263]

(65)   (66)   (67)

(The dienes (65)–(67) can be separated by selective extraction with aqueous silver nitrate[263].)

Zinc in aqueous–alcoholic sulphuric acid,[263] or in aqueous–alcoholic sodium hydroxide,[264] gives a mixture of the cyclo-octatrienes (62) and (63) in which the 1,3,6-triene predominates (up to 84 % yield[265]). In aqueous–alcoholic potassium hydroxide, however, the main product is apparently the 1,3,5-triene.[264]

Electrolytic reduction. The reduction of COT in aqueous ethanol, using a mercury cathode, affords a mixture of the cyclo-octatrienes (62) and (63), the 1,3,6-isomer predominating.[261] A controlled-potential electrolysis may be achieved in aqueous lithium chloride.[266] In dry *N,N*-dimethylformamide the almost exclusive product (after the addition of water) is the 1,3,5-triene (62).[65]

For the electrochemical behaviour of COT in aqueous sulphuric acid, see ref. 267.

The radical anion (61) may be generated electrochemically, and in e.g. dimethylformamide this species is stable to disproportionation (in the absence of alkali metal ions).[65, 268]

Mechanistic studies on the electrochemical reduction of COT are described in refs. 65, 75, 77, 79–82, 249, 269–273.

Catalytic hydrogenation. Hydrogenation of COT in the presence of platinum,[2, 3] palladium-on-carbon in glacial acetic acid,[11] Raney nickel in methanol,[274] or with a chromium–nickel catalyst in methanol under pressure,[11] yields cyclo-octane. It is possible to limit the uptake of hydrogen to three equivalents, however, since the double bond of Z-cyclo-octene is unusually resistant to hydrogenation.[275,276] Thus with palladium-on-calcium carbonate in methanol,[11,277] Raney nickel in ethanol,[278] or Raney nickel alloy in aqueous sodium hydroxide,[264] Z-cyclo-octene can be isolated in good yields.

Hydrogenation of COT in the presence of $[(CPD)Cr(CO)_3]_2$* affords a mixture of the cyclo-octadienes (65)–(67) in the ratio 46:29:25.[279]

For the hydrogenation of COT in the presence of $[Co(CN)_5]^{3-}$, see ref. 280.

Catalytic deuteriation of COT in the presence of $(Ph_3P)_3RhCl$ leads to the dideuteriocyclo-octa-1,3,5-triene (68).[281]

(68)

Sodium hydrazide. Reduction of COT with sodium hydrazide in the presence of hydrazine gives cyclo-octa-1,3,5-triene (62) (85%), apparently uncontaminated with the 1,3,6-triene (63).[282]

$$\xrightarrow[\substack{H_2N.NH_2,\ Et_2O,\\34°}]{NaNHNH_2}$$

(62)

Alkali metal alkoxides. Although COT is stable towards alkali metal alkoxides at room-temperature, reaction begins to occur above *ca.* 110°.[283]

With potassium t-butoxide in diglyme, COT affords a mixture of products containing benzocyclobutene (69) (up to 44%), styrene and cyclo-octa-1,3,5-triene.[8,9]

$$\xrightarrow[\substack{(MeOCH_2CH_2)_2O,\\120-160°}]{KOBu^t}} \quad + \quad + $$

(69)

Reaction of COT with potassium triethylmethoxide in triethylmethanol at 140° gives cyclo-octa-1,3,5-triene (up to 30%), together with polymeric material.[283,284] Schröder[283] has proposed that deprotonation of the COT dimer now known to possess structure (9) is followed by electron-transfer to COT, protonation of the dianion (60) then leading to the cyclo-octatriene (62) (scheme 13).

* CPD = cyclopentadienyl.

Scheme 13

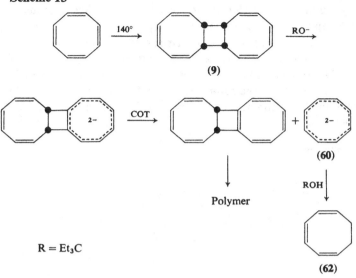

(9)

R = Et₃C

(60)

Polymer

(62)

If a mixture of COT and potassium tripropylmethoxide in Pr_3^nCOD, together with cyclo-octa-1,3,5-triene, is kept at 65° for a prolonged period, the recovered COT is found to be deuteriated; base-catalysed deuterium-exchange in cyclo-octa-1,3,5-triene, followed by proton-abstraction, is thought to result in deuteriated COT^{2-}, which by electron-transfer yields deuteriated COT^{284} (scheme 14). (This deuteriation of COT is also catalysed by dimer (9)[284].)

Scheme 14

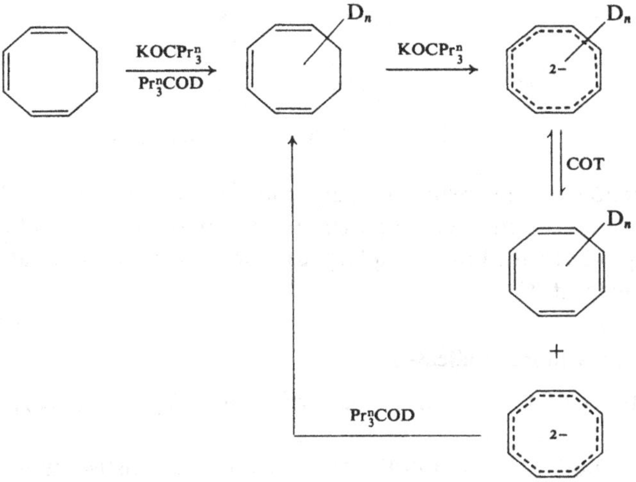

KOCPr₃ⁿ
Pr₃ⁿCOD

KOCPr₃ⁿ

COT

Pr₃ⁿCOD

Organo-compounds of lithium, sodium and magnesium. Alkyl- or aryl-lithiums, or phenylsodium, react with COT in ether to afford, after aqueous work-up, low yields (up to *ca.* 30%) of alkyl- or arylcyclo-octatetraenes (70) (see p. 80), together with cyclo-octatrienes and their substituted derivatives.[285,286] With n-butyl- or s-butyl-lithium in pentane, only butylcyclo-octatrienes are obtained (18% and 24% respectively).[286] These reactions may proceed *via* the substituted dianion (71), which by electron-transfer to COT could generate substituted COTs(70), and by protonation could form substituted cyclo-octatrienes (72) and (73) (scheme 15).

Scheme 15

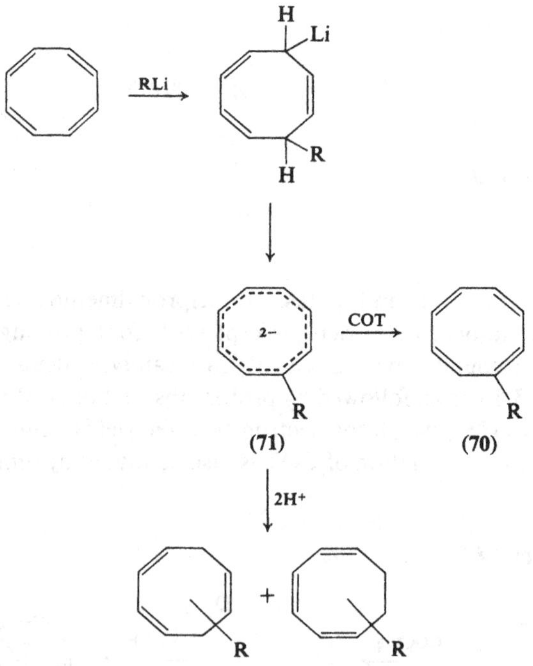

R = Et, Bu^n, Bu^s, Ph, *p*-Me$_2$N.C$_6$H$_4$

The reaction with phenylmagnesium bromide in ether gives a 45% yield of biphenyl (together with cyclo-octatrienes); in a control experiment only 6.6% of biphenyl was formed by a coupling reaction during the preparation of the Grignard reagent.[285]

9. Reactions with free radicals

The reactivity of COT towards methyl[287] and ethyl[288] radicals has been measured.

With α-cyanoisopropyl radicals, generated from azobis(isobutyronitrile) at

CMe$_2$CN

$$\text{COT} \xrightarrow[80°]{\underset{|}{\text{Me}_2\text{C}}-\text{N}=\text{N}-\underset{|}{\text{CMe}_2},\ \overset{\text{CN}}{\ }\ \overset{\text{CN}}{\ }}$$

CMe$_2$CN

(74)

80°, COT forms the 5,8-disubstituted cyclo-octa-1,3,6-triene **(74)** (5.5%).[289]
Reaction with dinitrogen tetroxide at low temperature produces the dinitro-compound **(75)** (33%).[290]

NO$_2$

$$\text{COT} \xrightarrow[\text{Et}_2\text{O},\ -70°]{\text{N}_2\text{O}_4}$$

NO$_2$

(75)

The analogous product **(76)** (65–70%) is obtained by the action of difluoro-amino radicals, generated from tetrafluorohydrazine.[291]

NF$_2$

$$\text{COT} \xrightarrow[\substack{\text{Me}_2\text{C}-\text{N}=\text{N}-\text{CMe}_2 \\ |\qquad\qquad| \\ \text{CN}\qquad\text{CN} \\ \text{PhCl},\ 75\text{-}80°}]{\text{F}_2\text{N}-\text{NF}_2}$$

NF$_2$

(76)

The stereochemistry of the products **(74)–(76)** is not certain, but the substituent groups are likely to adopt a *trans*-relationship (see refs. 289–291).

SiHEt$_2$

$$\text{COT} \xrightarrow[\substack{\text{Pt-Al}_2\text{O}_3, \\ \text{toluene},\ 250°}]{\text{SiH}_2\text{Et}_2}$$

(77)

SiEt$_3$

$$\text{COT} \xrightarrow[\substack{\text{Pt-Al}_2\text{O}_3, \\ \text{toluene},\ 170°}]{\text{SiHEt}_3}$$

(78)

The reaction of diethyl- or triethylsilane with COT in the presence of a platinum catalyst results in similar 1,4-addition to give the products (77) (26 %) and (78) (21 %) respectively (together with other, ill-defined compounds).[292]

Trichlorosilane, however, affords the 1,2-addition product (79a) (33 %) characterised as the triethyl-derivative (79b) obtained by reaction with ethylmagnesium bromide.[292]

SiHCl₃
Pt–Al₂O₃,
200°

(79)

a: R = Cl
b: R = Et

The chlorination of COT with sulphuryl chloride, which is assumed to proceed *via* a free-radical mechanism, gives a mixture of the bicyclic dichlorides (38) and (39) (*ca.* 4:1, total yield 68.5 %).[11,228]

SO₂Cl₂
CCl₄ or CH₂Cl₂,
20–30°

(38)   (39)

Further reaction with sulphuryl chloride yields tetrachloro-derivatives.[11]

10. Polymerisation

Unwanted polymers are frequently encountered in reactions of COT, which forms polymeric material on heating,[11] u.v. irradiation,[179] or mere storage[1,2,11] (even under nitrogen[60]). A wide variety of reagents, including air,[11] aluminium chloride,[235] tetranitromethane[293] and sodium,[294] can induce polymer formation.

COT inhibits the bulk polymerisation of vinyl monomers (initiated by dibenzoyl peroxide); a long induction period is followed by the formation of polymer in which COT is incorporated.[288]

11. Carbonylation

Reppe *et al.*[11] reported that COT reacts with carbon monoxide in the presence of Ni(CO)₄ and water (at 270° and 200 atm) to give a mixture of carboxylic acids, but the products were not characterised.

Hydroformylation of COT, using carbon monoxide and five equivalents of hydrogen in the presence of a cobalt catalyst, affords cyclo-octylmethanol (80).[11]

(80)

Carbon monoxide in the presence of ethanol and $(Ph_3P)_2PdCl_2$ gives the bicyclic ester (81) (33%).[295]

(81)

12. Cycloadditions

The following cycloaddition reactions of COT are classified according to the scheme devised by Huisgen.[296]

$2 + 1 \rightarrow 3$

Carbenes, nitrenes etc. The reaction of carbenes $:CR^1R^2$ with COT affords

(82)

a: $R^1 = R^2 = H$	e: $R^1 = H, R^2 = Cl$
b: $R^1 = R^2 = Cl$	f: $R^1 = Cl, R^2 = H$
c: $R^1 = R^2 = Br$	g: $R^1 = H, R^2 = Me$
d: $R^1 = R^2 = CN$	h: $R^1 = CO_2Et, R^2 = H$
	i: $R^1 = H, R^2 = CO_2Et$
	j: $R^1 = CO_2Me, R^2 = H$
	k: $R^1 = H, R^2 = CO_2Me$

bicyclo[6.1.0]nona-2,4,6-trienes (82) (2:1, 3:1 and 4:1 adducts may result from further addition of the carbene (see below)).

The parent hydrocarbon (82a) is best produced by using di-iodomethane in the presence of a zinc-copper couple;[297, 298] an alternative procedure employs diazomethane in the presence of copper(I) chloride (yield 25–30%).[299]

Dichloro- or dibromocarbene, generated from chloroform or bromoform and potassium t-butoxide, gives the adduct (82b) or (82c) respectively, allegedly in good yield.[297] The use of chloroform and aqueous sodium hydroxide in the presence of a phase-transfer catalyst leads to a mixture of the 1:1 (82b) (22–24%),[300, 301] 2:1 (83a) (16–25%),[300, 301] 3:1 (84) (10%)[301] and 4:1 (85) (3%)[301] adducts (see also ref. 302). The 2:1 dibromocarbene-adduct (83b) has

(82b) +

(83)

a: X = Cl
b: X = Br

(84)   (85)

been prepared by a similar procedure.[303]

Thermally-generated dicyanocarbene (from dicyanodiazomethane) in COT forms the adduct (82d) (70%),[304] together with a smaller yield of the isomer (86);[305] these 1,2- and 1,4-additions apparently result from singlet and triplet dicyanocarbene respectively.[305]

(82d) +

(86)

COT reacts with dichloromethane in the presence of methyl-lithium to give a *ca.* 3:1 mixture of the *syn-* and *anti*-compounds (82e) and (82f) (30–56%).[306, 307] It was reported, however, that these products were formed only when the methyl-lithium solution had been prepared from methyl bromide; if the reagent was derived from methyl iodide, the reaction product (obtained in low yield) was the *syn*-9-methyl-derivative (82g).[307]

The copper-catalysed decomposition of ethyl diazoacetate in COT gives mainly the *anti*-9-ethoxycarbonyl-derivative (82h) (40%);[308-311] the ratio of *anti*- to *syn*-product, (82h):(82i), is 19:1.[312] The corresponding methyl esters, (82j) and (82k), may be prepared similarly.[313, 314]

(87)   (88)

Table 6

	R¹	R²	R³	R⁴	Yield (%) of (88)	Ref.
a	H	H	H	H	40–50	315
b	H	Ph	Ph	Ph	23	316

Photolysis of the diazocyclopentadienes (87) in COT leads to the spiro-compounds (88) in moderate yields[315,316] (table 6). Similar products are formed from the diazo-compounds (89)–(91)[316] (table 7).

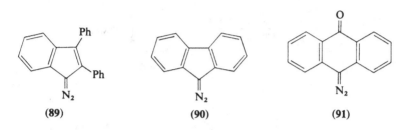

(89)   (90)   (91)

Table 7

Carbene precursor	Yield (%) of adduct	Ref.
(89)	32	316
(90)	42	316
(91)	27	316

Cyanonitrene, generated thermally from cyanogen azide at *ca.* 80°, reacts with COT to form a mixture of the labile 1,2-addition product (92a) and the more stable 1,4-adduct (93).[317,318] (In a side reaction, cyanogen azide itself reacts with COT to give compound (94) (see p. 42).) The product composition varies with the solvent, and with concentration, and it is believed that the 1,2-adduct results from the singlet nitrene while the 1,4-adduct is produced by a two-step process from triplet nitrene[317] (*cf.* the addition of :C(CN)₂, p. 36).

(92)   (93)   (94)

a: R = CN
b: R = CO₂Et

c: R =

Ethoxycarbonylnitrene (from *N*-(*p*-nitrobenzenesulphonyloxy)urethan and triethylamine) and COT furnish the 1,2-adduct (**92b**) (35–40%).[319]

Lead(IV) acetate oxidation of *N*-aminophthalimide in an excess of COT produces the analogous product (**92c**) (42%).[320]

COT reacts with the disilane $MeOSiMe_2.SiMe_2OMe$ at 550° to give the products (**95**) (30%) and (**96**) (15%); the reaction probably occurs *via* the :$SiMe_2$-adduct (**97**)[321] (*cf.* the thermal rearrangement of bicyclo[6.1.0]nonatrienes (**82**), p. 264).

(97)

(95)   (96)

With hexachlorodisilane, at the same temperature, a mixture of the products (**98**) (8%), (**99**) (13%), (**100**) (17%) and (**101**) (7%) is obtained.[322]

(98)   (99)   (100)   (101)

Sulphur. Sulphur atoms, generated by photolysis of carbonyl sulphide, allegedly add to COT to form the thi-iran (**102**)[323] (but see appendix).

(102)

2 + 2 → 4

Dimerisation. The thermal [2 + 2] dimerisation of COT has already been described (see p. 11). 'Mixed' dimers are formed in very low yields by the reaction of COT with monosubstituted COTs (see pp. 101–2).

Polyhaloethylenes. 1,1-Dichloro-2,2-difluoroethylene[324] and 1-chloro-1,2,2-trifluoroethylene[325] add to COT at *ca.* 115° to give low yields of the bicyclo-[6.2.0]deca-2,4,6-trienes (103) (table 8).

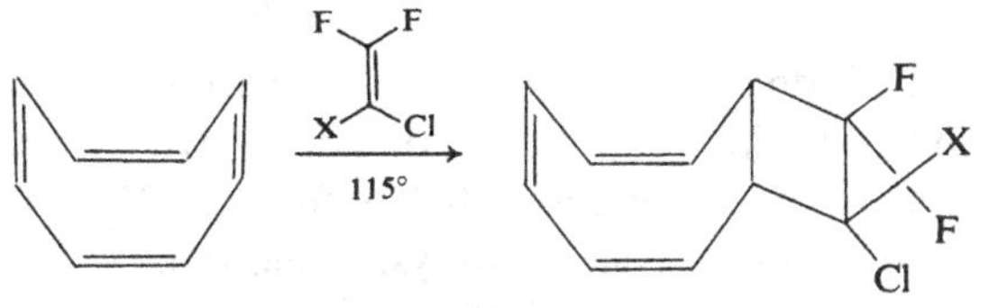

(103)

Table 8

	X	Yield (%)	Ref.
a	F	8	325
b	Cl	7	324

p-Benzoquinone. The photo-addition of *p*-benzoquinone to COT gives a product formulated as the oxetan (104) (up to 30%),[326,327] although this structure has been disputed (see p. 46). A by-product (*ca.* 3% yield) of the photo-reaction is the cage-compound (105), presumably arising from the thermal Diels–Alder adduct (106) (see pp. 42–3).

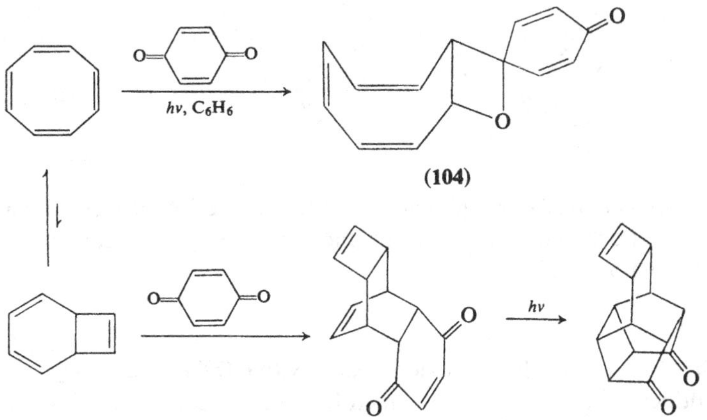

(104)

(106)   (105)

4 + 1 → 5

Sulphur monoxide. Sulphur monoxide, generated by the thermolysis of ethylene sulphoxide, reacts with COT to form the 1,4-adduct (107) (single stereoisomer; 15–30%).[47,328]

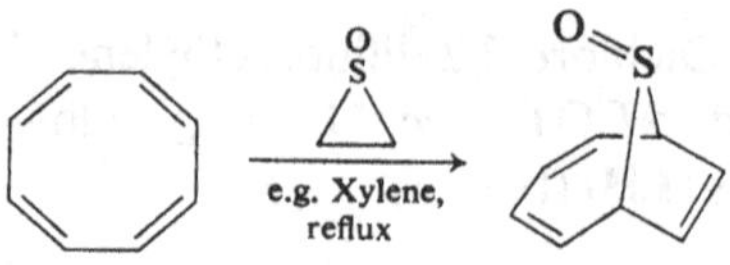

(107)

Sulphur dioxide. COT does not undergo reaction with sulphur dioxide under normal conditions, but in the presence of antimony(v) fluoride in liquid sulphur dioxide at $-70°$, the adduct (108) is formed (95%) (after aqueous work-up).[48] Careful control of the hydrolysis conditions enables up to 12% of 9-thiabarbaralane 9,9-dioxide (109) to be isolated.[54] The suggested mechanistic routes to these products are shown in scheme 16.

Scheme 16

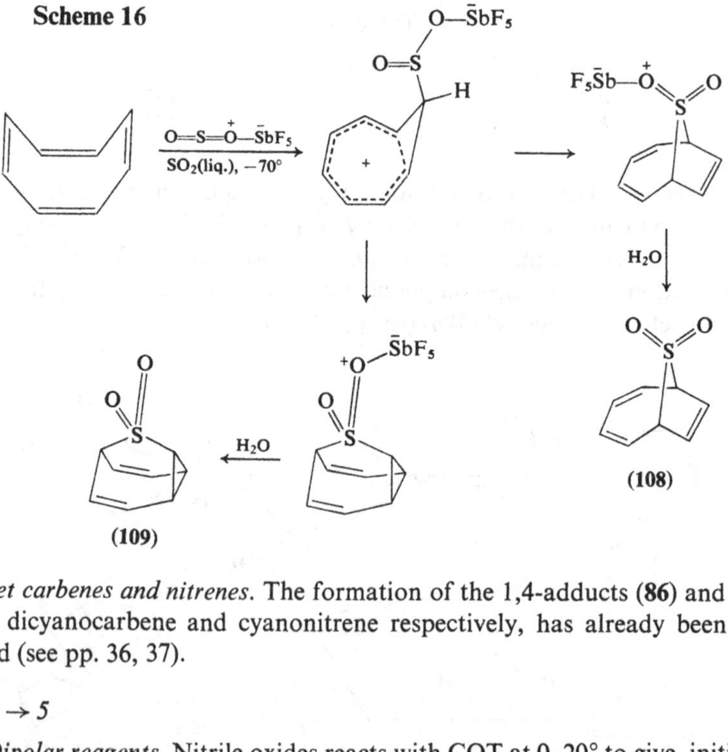

(109)   (108)

Triplet carbenes and nitrenes. The formation of the 1,4-adducts (86) and (93), from dicyanocarbene and cyanonitrene respectively, has already been described (see pp. 36, 37).

$2+3 \rightarrow 5$

1,3-*Dipolar reagents.* Nitrile oxides reacts with COT at 0–20° to give, initially, 1:1 adducts of structure (110), which by valence isomerisation yield the tricyclic products (111)[329–333] (table 9). Adduct (111c) reacts further with

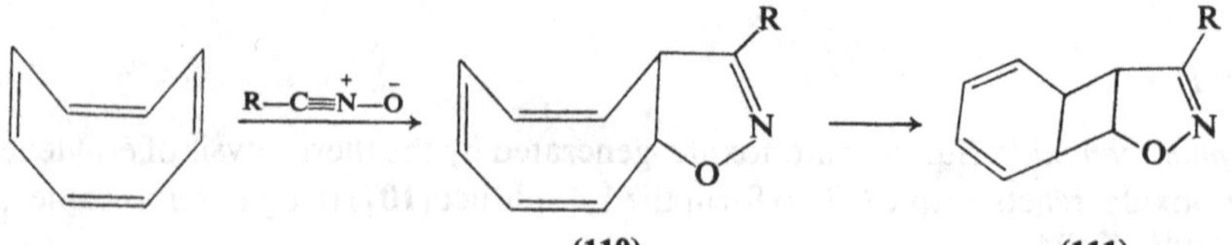

(110)   (111)

Table 9

	R	Yield (%) of (111)	Ref.
a	H	8	329, 333
b	Me	38	331
c	Ph	83, (50)	330, (331), 332
d	*p*-Br.C$_6$H$_4$	70	331
e	*m*-NO$_2$.C$_6$H$_4$	75	331

benzonitrile oxide; structure (112) has been suggested for the major product.[331]

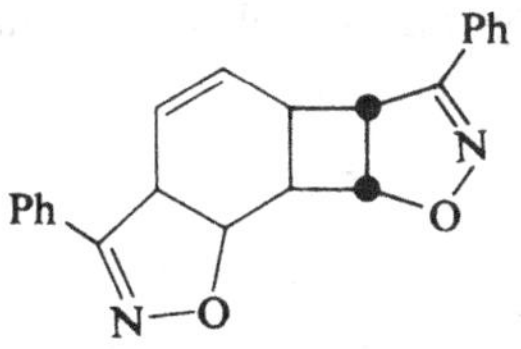

(112) 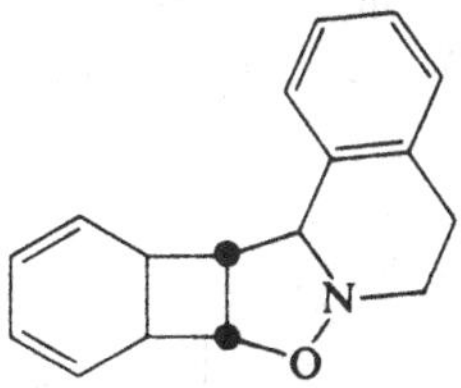   (113)

(For other reactions of the system (111), see refs. 331, 332, 334.)

Reaction of COT with the nitrone 3,4-dihydroisoquinoline *N*-oxide, at 65°, gives the adduct (113) (50%).[335]

Nitrile imines react in a similar fashion, affording 1:1 adducts (114); however, 2:1 adducts (115) result unless a large excess of COT is present[336] (table 10).

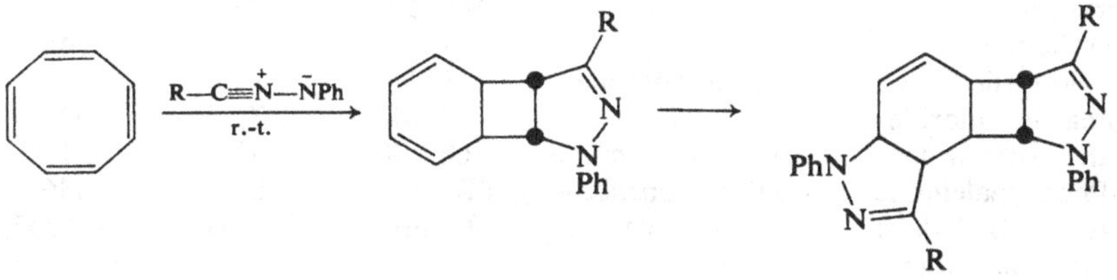

(114)   (115)

Table 10

	Yield (%)	
R	(114)	(115)
Ph	62	65
p-Cl.C$_6$H$_4$	68	50[a]

[a] Two isomeric 2:1 adducts isolated.

COT reacts with tetracyanoethylene oxide at 80° to form the adduct (116) (28%).[337]

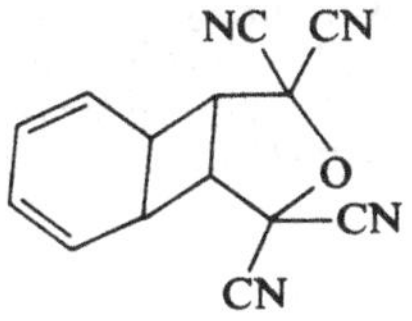

(116) 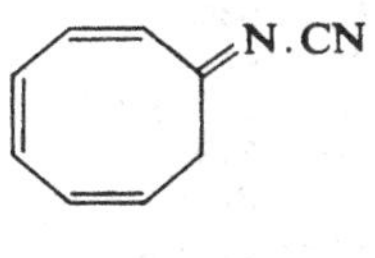   (94)

Cyanogen azide reacts with COT at room-temperature, with slow evolution of nitrogen, to give compound (94) (73%);[317] the initial reaction is almost certainly a 1,3-dipolar addition (*cf.* ref. 338).

4 + 2 → 6

Dienophiles. Although the 'tub' conformation of COT does not provide the near-planar diene system necessary for a concerted [4 + 2] (Diels–Alder) cycloaddition, such a system is available in the valence tautomer bicyclo[4.2.0]-octa-2,4,7-triene (4);* reaction of COT with e.g. maleic anhydride thus leads

(4)

(117)

Table 11

Dienophile	Solvent	Temp. (°C)	Yield (%)	Ref.
Acrylic acid	—	180–200	29	11
'Acrylic ester'	—	—	—	345
Acrylonitrile	Chlorobenzene	180	13	39
Fumaroyl chloride	—	130–140	73	340
Maleic anhydride	*o*-Dichlorobenzene	160–170	82	11
N-Phenylmaleimide	*o*-Dichlorobenzene	Reflux	87	346
Tetracyanoethylene[a]	Benzene	Reflux	92, (63)	134, (347)
Dicyanomaleimide	Dioxan	100	—	134, 135
p-Benzoquinone[b]	*o*-Dichlorobenzene	140	—	11
2,3-Dichloro-5,6-dicyano-*p*-benzoquinone	Benzene	Reflux	—	350
1,4-Naphthoquinone	Xylene	Reflux	16[c]	11
Benzoylacetylene	—	140–150	30[d]	351
Dimethyl acetylene-dicarboxylate	{—	150–155	51	352
	{—	100	82	353
Diethyl acetylene-dicarboxylate	—	160–170	—	11
Dicyanoacetylene	THF	56	17	354

[a] For the reaction of 1,4-dideuteriocyclo-octatetraene with tetracyanoethylene, see ref. 49.
[b] At 140–150° in benzene under 30 atm of nitrogen a 2:1 adduct is formed;[11] since the *endo*-stereochemistry of the 1:1 adduct (106) has been proved by photo-caging[348] (see p. 330), the stereochemistry of the 2:1 adduct may be represented as (118), by analogy with related compounds (*cf.* ref. 349).
[c] The isolated product was the dehydro-derivative (119).
[d] The isolated product was benzophenone (see pp. 343–4).

* For reactions of dienophiles with isolated bicyclo[4.2.0]octa-2,4,7-triene, see refs. 32, 40.

(106)   **(118)**   **(119)**

to the adduct (117).[11,135] The stereochemistry of this product, the evidence[135,339-343] for which has culminated in an X-ray structure determination,[344] is that which would be expected to result from *endo*-addition of the dienophile to the less hindered face of the diene system.

Similar adducts, possessing the basic tricyclo[4.2.2.0^{2,5}]decane skeleton of compound (117), are produced from the dienophiles listed in table 11.

U.v. irradiation of maleic anhydride in an excess of COT at 8° gives a 12% yield of the Diels–Alder adduct (117).[355,356] In a similar dark (thermal) reaction the yield of the product is 3%, and the difference has been ascribed to the production of bicyclo[4.2.0]octa-2,4,7-triene (4) by photo-isomerisation of COT.[356]

The reaction of benzyne (generated from benzenediazonium-2-carboxylate at *ca.* 40°) with COT is complex. The products (combined yield 25%) include the 1:1 adducts (120), (121) and (122), phenanthrene and 9,10-dihydro-9-phenylphenanthrene (123) (major product).[357] The proposed routes to these products are outlined in scheme 17. The reaction appears to be very sensitive to metal catalysis, and in the presence of e.g. silver fluoroborate the adduct

Table 12 (see scheme 17)

Catalyst	Ratio (123):(122)	(120)	(121)	(122)	Relative yield (%) Phenanthrene	Biphenylene
—	26:1	28	31	12	14	14
AgBF₄	1:11	6	8	80	1	1

(122) predominates[358] (table 12). It has been suggested[359,360] that a benzyne-silver complex is involved in the silver-catalysed reaction, and the steps leading to the adduct (122) have been visualised as shown in scheme 18.[360]

When COT is heated with poor dienophiles, such as vinylene carbonate or citraconic anhydride, its rate of dimerisation exceeds that of the addition to the valence tautomer (4), and the isolated products are derived from COT dimers (see pp. 365, 369).

Scheme 17

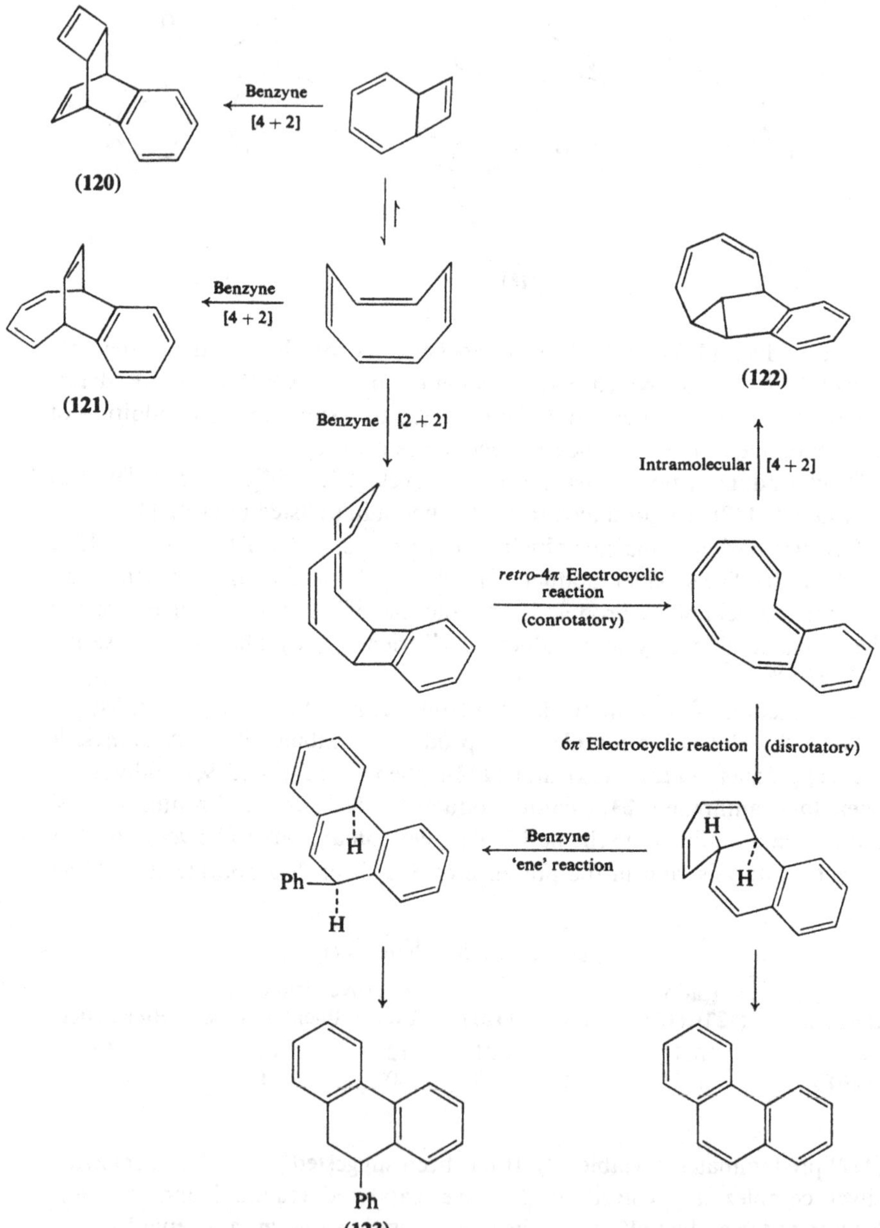

(120)

(121)

(122)

Benzyne
[4 + 2]

Benzyne
[4 + 2]

Benzyne   [2 + 2]

Intramolecular   [4 + 2]

retro-4π Electrocyclic
reaction
(conrotatory)

6π Electrocyclic reaction   (disrotatory)

Benzyne
'ene' reaction

Ph

H

Ph

H

(123)

Ph

COT reacts slowly at 100° with bis(trifluoromethyl)ketene to give the adduct (124).[361]

With 4-phenyl-1,2,4-triazoline-3,5-dione at 100°, a 1:1 mixture of the iso-

Scheme 18

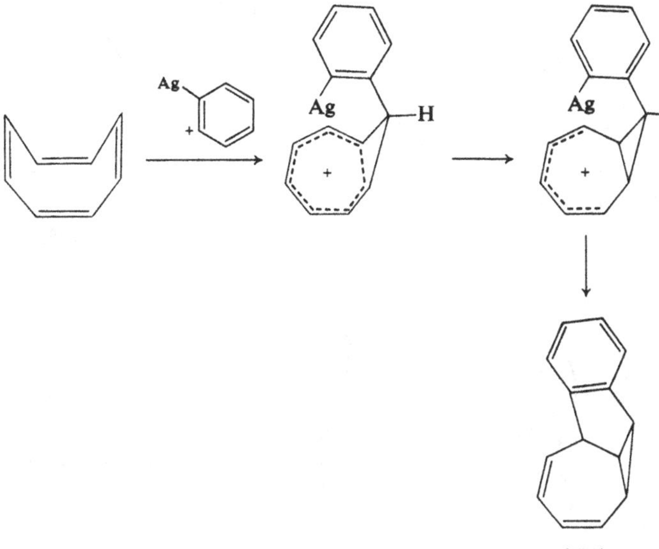

(122)

meric adducts (125a)[362] and (126a)[363] is produced (total yield up to 63%[364]). At room-temperature in benzene or acetone, however, the adduct (126a) is formed almost exclusively[363] (37%[364]). Huisgen *et al.*[364] have postulated

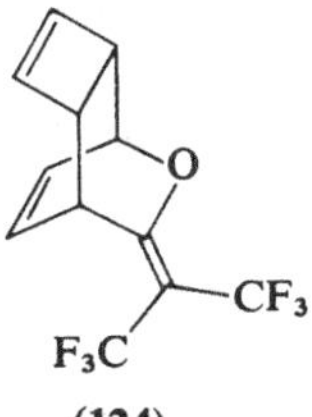

(124)

(scheme 19) that the formation of the 1,4-addition product (126a) proceeds *via* the zwitterion (127a); at room-temperature the production of adduct (125a), which requires the intermediacy of bicyclo[4.2.0]octa-2,4,7-triene (4), is suppressed.

Exclusive 1,4-addition to COT occurs with chlorosulphonyl isocyanate, which affords the adduct (128) (up to 73%).[365-367] Initial electrophilic attack on COT is again proposed.[366]

A similar reaction takes place with the *N*-sulphonylurethan (129), generated from the oxathiadiazine (130) at 80°; the product (formed in low yield) is the 1,4-adduct (131).[368]

Scheme 19

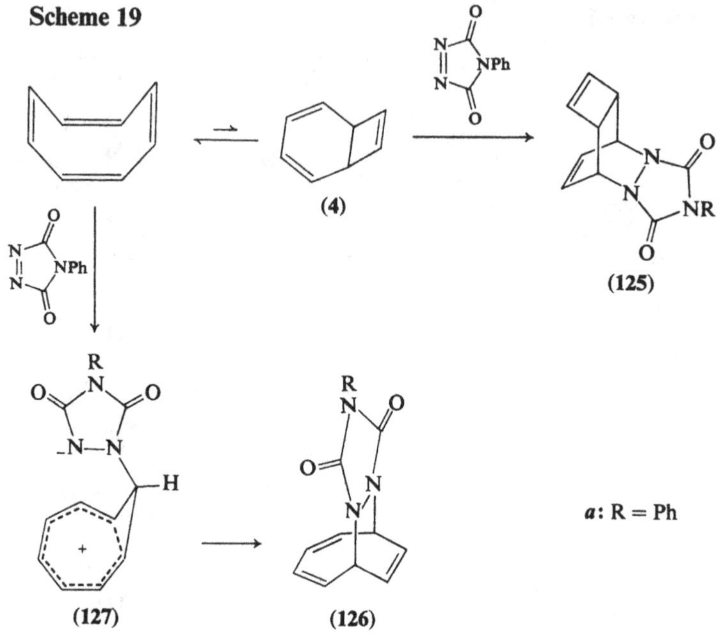

NPh
(4)
NR
(125)
R
H
(127)
(126)
a: R = Ph

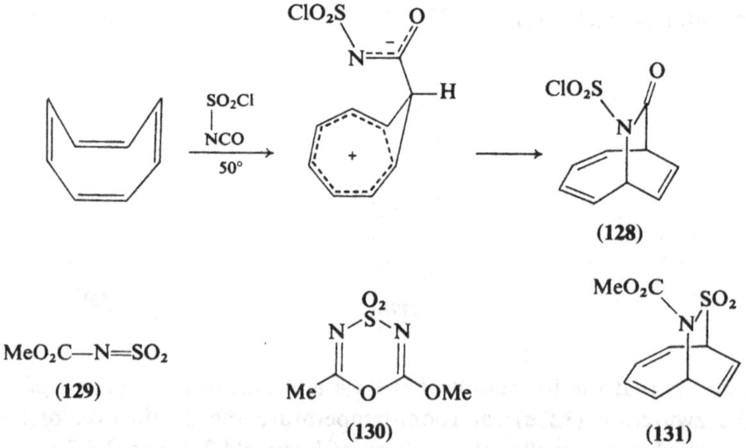

ClO₂S
SO₂Cl
NCO
50°
H
(128)
MeO₂C—N=SO₂
(129)
Me
OMe
(130)
MeO₂C
SO₂
(131)

Photo-addition. The photo-addition of *p*-benzoquinone to COT, using an argon ion-laser, yields the adduct **(132)** (77%), and the authors[369,370] dispute structure (**104**) for the product obtained by conventional irradiation (see p. 39); it is not certain, however, that the two photo-adducts are identical. In the presence of air, the adduct **(132)** (31%) is accompanied by the peroxide **(133)** (49%); in glacial acetic acid the products are **(134)** (58%) and **(135)** (26%).[369,370]

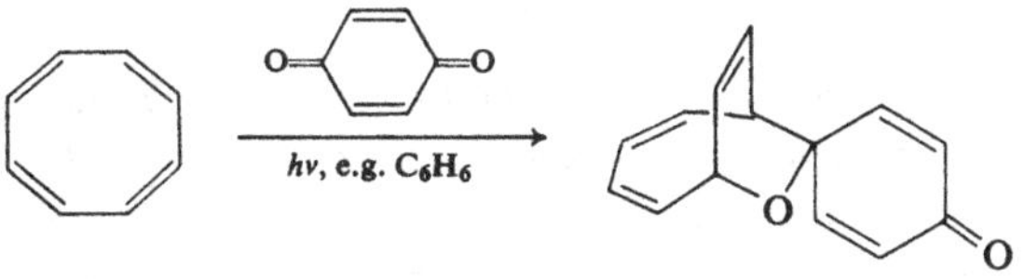

(132)

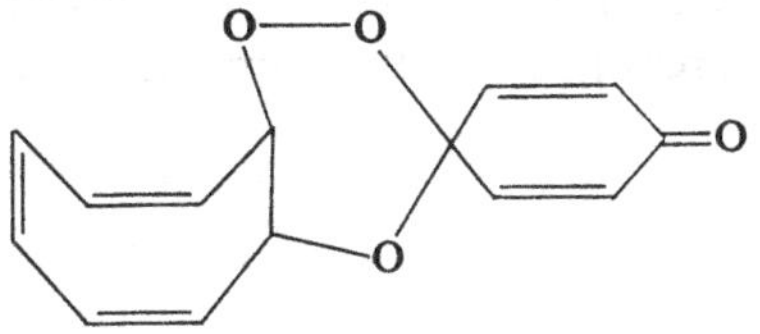

(133)

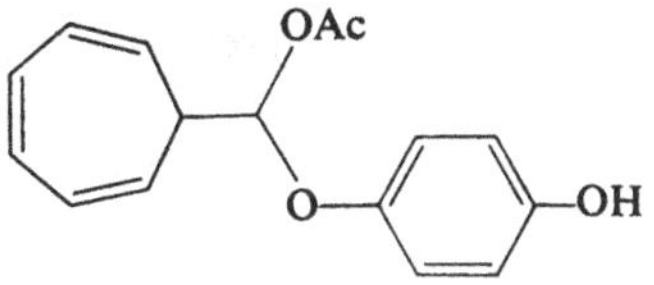

(134)

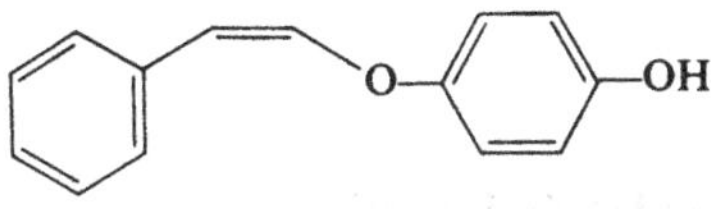

(135)

2 + 4 → 6

Electron-deficient dienes. COT functions as a dienophile towards electron-deficient diene systems, probably *via* its valence tautomer (4).

With 1,3-diphenylbenzo[c]furan, two isomeric products (presumably *endo-* and *exo*-adducts) of structure (136) are formed.[371]

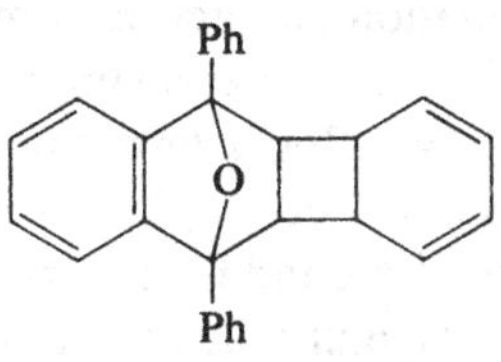

(136)

Tetrachlorocyclopentadienone ketals at 100–150° give 2:1 adducts of structure (137) (table 13); no 1:1 adducts could be isolated.[372]

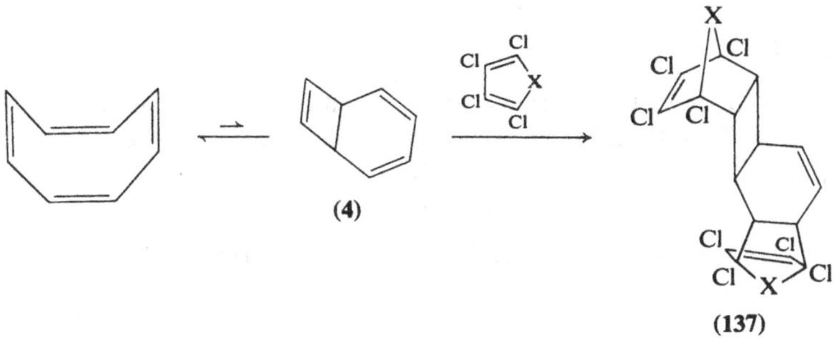

(137)

Table 13

	X	Conditions	Yield (%)
a	C(OMe)$_2$	Xylene, reflux	27
b	(cyclic carbonate structure)	*ca.* 100°	65

Reactions of COT with octachlorofulvalene (**138**),[373] and with the tetrazine (**139**),[374] have been reported, but the structures of the products are not known.

(**138**)

(**139**)

13. Reactions with metals and their derivatives

COT can bond to metals in an unparalleled variety of ways. It may act, not only as an electron-donor, but also as an acceptor, and its high affinity for electrons is reflected in its reactions with electropositive metals to form ionic compounds containing the dianion COT^{2-}. In some transition metal complexes, the eight-membered ring flattens completely to become an *octahapto*-system which approximates to the planar dianion. In other cases, however, COT retains its 'tub' conformation and acts as a mono-olefin, or a chelating 1,5-diene (1,2,5,6-*tetrahapto*)- system. Yet again, the COT ligand may adopt intermediate conformations in which the ring is partly flattened, and behave as a 1,3-diene (1,2,3,4-*tetrahapto*-) or a 1,3,5-triene (1,2,3,4,5,6-*hexahapto*-) system. In addition, examples of σ-bonding and π-allyl bonding are known in COT–metal compounds. In its multinuclear complexes containing metal–metal bonds, more complicated bonding systems may occur, and the C$_8$H$_8$ ligand is sometimes transformed from the monocyclic tetraene into a different species.

Group IA

The alkali metals. COT reacts with lithium,[11,63,244] sodium,[11,63] potassium[99,244,375] and rubidium[376] in ethers, liquid ammonia or hexamethyl-phosphoramide to form salts of type (**140**).

$$\text{COT} \xrightarrow[\text{e.g. Et}_2\text{O, THF, or NH}_3\text{(liq.)}]{2M} 2M^+ \text{ COT}^{2-} \quad (\mathbf{140})$$

M = Li, Na, K, Rb

(For preparative procedures, see e.g. refs. 377, 378 (dilithium salt in ether); 379 (dipotassium salt in 1,2-dimethoxyethane); 380, 381 (dipotassium salt in tetrahydrofuran); 382 (dipotassium salt in liquid ammonia).)

Solutions of these alkali metal derivatives are also produced by electron-transfer to COT from the species (141)[259] and (142).[383]

$$Na^+ (Ph_3C)^- \qquad\qquad 2K^+ (Ph_2C—CPh_2)^{2-}$$
$$\text{(141)} \qquad\qquad\qquad \text{(142)}$$

The alkali metal cyclo-octatetraenides may be obtained as crystalline solvates (see e.g. refs. 244, 375, 384–386); in solution (in various ethers) they appear to exist as contact ion-pairs.[376] (For the solubilities of the sodium and potassium salts in tetrahydrofuran and hexamethylphosphoramide, see ref. 387.)

The spectra of the cyclo-octatetraenides are consistent with the presence of a planar monocyclic dianion (D_{8h} symmetry),[99, 244, 245, 375] and X-ray crystallographic studies confirm the planarity of COT²⁻ in the solvated derivatives (143).[385,386]

a: M = K
b: M = Rb

(143)

The COT dianion obeys the Hückel ($4n + 2$) rule; moreover, it sustains a diamagnetic ring-current.[244] The addition of two electrons to the non-planar COT molecule is thus attended by ring-flattening, the electrons entering the non-bonding π-orbital of a planar 'aromatic' Hückel-system, the resonance energy of which must be more than sufficient to compensate for the compressional strain involved in flattening the ring. (For theoretical studies on COT²⁻, see e.g. refs. 122, 124, 125, 248, 257.)

Group IB

Copper, silver, gold. The formation of crystalline addition compounds of COT by the action of aqueous silver nitrate, or ammoniacal copper(I) chloride, was first mentioned by Reppe *et al.*[11]

COT reacts readily with finely-ground copper(I) chloride and bromide (best in the presence of e.g. benzonitrile), or with solutions of these halides in acrylonitrile. Under these conditions copper(I) bromide affords the 1:1 complex (144a), while copper(I) chloride forms the 1:2 compound (145a).[388]

CuBr → (COT)CuX   (144)   *a*: X = Br
                              b: X = Cl

COT

CuCl → (COT)(CuX)₂   (145)   *a*: X = Cl
                              b: X = CF₃CO₂

A 1:2 complex (145*b*) results (65%) from the reaction of COT with copper(I) trifluoroacetate.[389]

$$COT \xrightarrow[C_6H_6, \text{ r.-t.}]{Cu(OCOCF_3)} (145b)$$

Reduction of copper(II) halides with sulphur dioxide in the presence of COT gives complexes of the type (144)[390] (table 14). However, slight alterations

$$COT \xrightarrow[EtOH, 0°]{CuX_2, SO_2} (144)$$

Table 14

	X	Yield (%)
a	Br	76
b	Cl	55

in the reaction conditions may lead to the formation of type (145) products;[390] complex (145*a*) (79%) results when triphenyl phosphite is employed as the reducing agent.[391]

$$COT \xrightarrow[MeOH, \text{ r.-t.}]{CuCl_2 . 2H_2O, (PhO)_3P} (145a)$$

The structure of (COT)CuCl (144*b*) contains chains of –Cu–Cl– units, with only one double bond of COT interacting closely with each copper atom: the carbon atoms of the co-ordinated double bond, the associated copper atom and the attached chlorine atoms are approximately co-planar.[392]

(144*b*)

In contrast, the i.r. spectrum of (COT)(CuCl)₂ (145*a*) indicates that all four of the COT double bonds are equivalent;[391] the suggested structure for (COT)[Cu(OCOCF₃)]₂ (145*b*) incorporates bridging or chelating trifluoroacetate groups.[389]

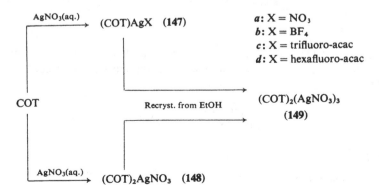

(145*b*)

Reaction of COT with the benzene complex of copper(I) trifluoromethane-sulphonate affords the cationic species (146) (86%).[393, 394] The structural

$$COT \xrightarrow[\text{EtCOMe}]{\text{(Benzene) [Cu(OSO}_2\text{CF}_3\text{)]}_2} [(COT)Cu]^+ \ ^-OSO_2CF_3$$

(146)

evidence indicates that in (146) the COT forms a 1,2,5,6-*tetrahapto*-system, but that rapid intermolecular ligand-exchange occurs in solution at normal temperatures.[393–395]

(146)

$^-OSO_2CF_3$

COT reacts readily with silver nitrate to give 1:1 (147*a*) or 2:1 (148) complexes; both of these products disproportionate when heated, forming a 2:3

COT

AgNO₃(aq.) ⟶ (COT)AgX (147)

a: X = NO₃
b: X = BF₄
c: X = trifluoro-acac
d: X = hexafluoro-acac

Recryst. from EtOH ⟶ (COT)₂(AgNO₃)₃
(149)

AgNO₃(aq.) ⟶ (COT)₂AgNO₃ (148)

complex (**149**).[63] With silver tetrafluoroborate, only the 1:1 complex (**147b**) is produced, the reaction being insensitive to the COT:salt ratio.[396]

$$\text{COT} \xrightarrow[\text{e.g. Toluene, } 0°]{\text{AgBF}_4} \quad (\textbf{147}b)$$

If COT is treated with aqueous silver nitrate followed by the sodio-derivative of trifluoro- or hexafluoroacetylacetone, the complex (**147c**) (52 %) or (**147d**) (34 %) is precipitated.[397]

X-ray studies show that $(\text{COT})\text{AgNO}_3$ (**147a**) contains a tub-shaped COT ring, bound to silver *via* two non-adjacent double bonds.[398, 399] In the crystal,

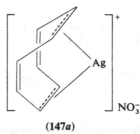

(**147a**)

these units form infinite chains by longer-range interactions. (For spectra of (**147a**), see ref. 99; for ^{13}C n.m.r. shifts of COT in the presence of silver nitrate, see ref. 400.) The p.m.r. spectra of the complexes (**147c**) and (**147d**) show a singlet resonance for the COT protons, presumably due to rapid exchange of COT in solution.[397]

Treatment of COT with gold(I) or gold(III) chloride affords the complexes (**150**) (84 %) or (**151**) (75 %).[401]

$$\text{COT} \begin{cases} \xrightarrow[\text{CH}_2\text{Cl}_2,\ -45°]{\text{AuCl}} & (\text{COT})\text{AuCl} \quad (\textbf{150}) \\[2em] \xrightarrow[\text{SO}_2(\text{liq.}),\, -78° \rightarrow -40°]{\text{AuCl}_3} & (\text{COT})\text{Au}_2\text{Cl}_4 \quad (\textbf{151}) \end{cases}$$

Group IIA

Magnesium. COT reacts with magnesium in tetrahydrofuran, in the presence of a catalytic amount of magnesium bromide; the solvated cyclo-octatetraenide (**152**) separates.[402]

$$\text{COT} \xrightarrow[\text{THF, } 20°]{\text{Mg, MgBr}_2} \quad (\text{COT})\text{Mg} \cdot 2\tfrac{1}{2}\text{THF} \quad (\textbf{152})$$

Group IIIA

The lanthanides. COT reacts with dipyridinium hexachlorocerate(IV), with the elimination of hydrogen chloride, to give the compound (153) (40%)[403] (*cf.* titanium, zirconium and hafnium (below) and molybdenum (p. 59)).

$$\text{COT} \xrightarrow[\text{C}_6\text{H}_6,\text{ reflux}]{(\text{PyH}^+)_2[\text{CeCl}_6]^{2-}} (\text{C}_8\text{H}_7)_2\text{CeCl}_2 \quad (\textbf{153})$$

The addition of COT to a solution of europium or ytterbium in liquid ammonia yields the cyclo-octatetraenide (154).[404]

$$\text{COT} \xrightarrow[\text{NH}_3(\text{liq.})]{2\text{M}} (\text{COT})\text{M} \quad (\textbf{154})$$

$$a: \text{M} = \text{Eu}$$
$$b: \text{M} = \text{Yb}$$

The presence of M^{2+} ions in these products is indicated by the e.p.r. spectrum of (154a) and the diamagnetism of (154b).[404]

The actinides. At *ca.* 150°, COT reacts with finely-divided thorium, uranium or plutonium (prepared from the hydrides) to afford the bis(cyclo-octatetraene) compound (155) (up to 57%); the reaction is catalysed by mercury vapour.[405]

$$\text{COT} \xrightarrow[\text{Hg, 150°}]{\text{M}} (\text{COT})_2\text{M} \quad (\textbf{155})$$

$$\text{M} = \text{Th, U, Pu}$$

(For the structure of complexes of this type, see p. 186.)

Group IVA

Titanium, zirconium, hafnium. The tetrachlorides of titanium, zirconium and hafnium react thermally with COT to form supposedly σ-bonded bis(cyclo-octatetraenyl) species (156) (92–95%), hydrogen chloride being eliminated[406] (*cf.* cerium (above) and molybdenum (p. 59)).

$$\text{COT} \xrightarrow[\text{C}_6\text{H}_6,\text{ reflux}]{\text{MCl}_4} (\text{C}_8\text{H}_7)_2\text{MCl}_2 \quad (\textbf{156})$$

$$\text{M} = \text{Ti, Zr, Hf}$$

The electrochemical reduction of titanium(IV) chloride in tetrahydrofuran containing COT, using an aluminium anode, leads to the solvated complex (157a) (68%); with pyridine as the solvent the same compound may be isolated as the pyridinate (157b) (77%), which when heated under reduced pressure loses its co-ordinated solvent to form the dimeric species (158).[407]

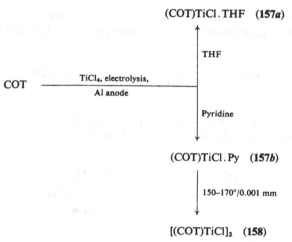

(COT)TiCl.THF   **(157a)**

↑

THF

COT $\xrightarrow{\text{TiCl}_4,\ \text{electrolysis},\ \text{Al anode}}$

Pyridine

↓

(COT)TiCl.Py   **(157b)**

↓ 150–170°/0.001 mm

[(COT)TiCl]₂   **(158)**

Preliminary electrochemical reduction of titanium(IV) chloride, followed by further electrolysis in the presence of COT, produces bis(cyclo-octatetraene)-titanium **(159)** (45%).[407]

COT $\xrightarrow[\text{Al anode, THF}]{\text{TiCl}_3,\ \text{electrolysis}}$ (COT)₂Ti   **(159)**

In this product **(159)** the titanium atom is attached symmetrically to one (planar) eight-membered ring, but is bonded to the other (non-planar) ring *via* four carbon atoms only.[408] (For the i.r. spectrum, see ref. 409.)

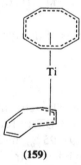

(159)

In solution, a fluxional process has been detected (by variable-temperature n.m.r. studies), in which the planar and non-planar rings exchange their respective conformations; if the planar ring is a COT dianion, the conformational change involves an intramolecular redox reaction, in which two electrons are transferred from one ligand to the other.[410]

(COT)²⁻Ti(COT) ⇌ (COT)Ti(COT)²⁻

The analogous zirconium complex **(160)** is formed (15%) by electrolytic reduction of zirconium(IV) chloride in the presence of COT when a magnesium anode is used, but with an aluminium anode the dichloride **(161)** is obtained

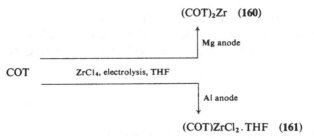

(COT)₂Zr (160)

$$(COT)_2Zr \quad (160)$$

(83%).[402,411] The structure of the tetrahydrofuranate of the zirconium complex (160) has been established as (160a), again featuring one planar and one non-planar COT ring;[412] the structure of the dichloride (161) is as shown.[413]

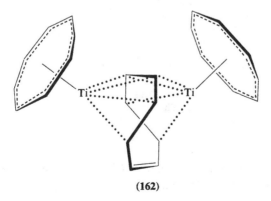

(160a)   (161)

The use of titanium(IV) butoxide in the electrosynthesis leads to another type of complex (162) (27%).[407]

COT $\xrightarrow[\text{Al anode, THF}]{\substack{\text{Ti(OBu)}_4, \\ \text{electrolysis}}}$ (COT)₃Ti₂ (162)

In this product (162), each titanium atom is bonded symmetrically to a planar eight-membered ring and to four carbon atoms of a central non-planar ring, in a double sandwich structure.[414,415] The same product (162) also results

(162)

(80%) when COT reacts with titanium(IV) butoxide in the presence of triethylaluminium; however, the use of a large excess of COT in this reaction produces $(COT)_2Ti$ **(159)** (80%), since $(COT)_3Ti_2$ reacts with COT.[416]

$$COT \xrightarrow[80°]{Ti(OBu)_4,\ Et_3Al} (COT)_3Ti_2 \xrightarrow[80°]{COT} (COT)_2Ti$$
$$\qquad\qquad\qquad\qquad\qquad (162)\qquad\qquad (159)$$

The zirconium complex **(160)** may be obtained (65%) by a similar process.[417]

$$COT \xrightarrow[\text{Toluene}, -45°\to\ 90°]{Zr(OBu)_4,\ Et_3Al} (COT)_2Zr \quad (160)$$

It should be noted, however, that the triethylaluminium used must be free from diethylaluminium hydride, which with zirconium(IV) butoxide and COT produces the complex hydride **(163)**.[417]

$$COT \xrightarrow{Zr(OBu)_4,\ Et_2AlH} (COT)ZrH_2 \quad (163)$$

Treatment of COT with tetra-allylzirconium also leads to **(160)** (98%).[417]

$$COT \xrightarrow[45-55°]{(CH_2=CHCH_2)_4Zr} (COT)_2Zr \quad (160)$$

Reaction of titanium(III) chloride with isopropylmagnesium bromide in the presence of COT and diphenylacetylene gives the complex **(164)** (11%).[418]

$$COT \xrightarrow[\substack{PhC\equiv CPh,\ Et_2O, \\ -78°\ \to\ r.-t.}]{TiCl_3,\ Pr^iMgBr} (COT)Ti(C_4Ph_4) \quad (164)$$

The structural evidence suggests that the product **(164)** is a sandwich compound containing a tetraphenylcyclobutadiene ligand.[418]

(164)

Electrolysis of $(CPD)TiCl_3$, or (better) $(CPD)_2TiCl_2$, in the presence of COT gives the 'mixed' complex **(165)** (25%).[407]

$$\text{COT} \xrightarrow[\substack{\text{Al anode, THF}}]{\substack{\text{e.g. (CPD)}_2\text{TiCl}_2, \\ \text{electrolysis}}} (\text{COT})\text{Ti(CPD)} \quad (\mathbf{165})$$

This is known to be a sandwich compound containing planar ligands.[419]

(**165**)

(For the e.p.r. spectrum, see ref. 420; for the photo-electron spectrum, see ref. 421.)

Group VA

Vanadium, niobium, tantalum. For ion-molecule reactions of (CPD)V(CO)$_4$ with COT, see ref. 422.

COT reacts with [Li(THF)$_4$]$^+$[Ph$_6$Ta]$^-$ to give the ionic complex (**166**) (23%), readily convertible to (**167**) (80%).[423]

$$\text{COT} \xrightarrow[\text{THF, reflux}]{[\text{Li(THF)}_4]^+[\text{Ph}_6\text{Ta}]^-} [\text{Li(THF)}_4]^+[(\text{COT})_3\text{Ta}]^- \quad (\mathbf{166})$$

$$[\text{Ph}_4\text{As}]^+\text{Cl}^- \downarrow \text{EtOH, r.-t.}$$

$$[\text{Ph}_4\text{As}]^+[(\text{COT})_3\text{Ta}]^- \quad (\mathbf{167})$$

Group VIA

Chromium, molybdenum, tungsten. COT displaces acetonitrile from (MeCN)$_3$Cr(CO)$_3$ to give the carbonyl complex (**168a**) (9%);[424] the analogous molybdenum and tungsten compounds (**168b**)[424] and (**168c**)[425,426] may be obtained in a similar manner (37% and 28% respectively). In related processes,

$$\text{COT} \xrightarrow[\text{Hexane, reflux}]{(\text{MeCN})_3\text{M(CO)}_3} (\text{COT})\text{M(CO)}_3 \quad (\mathbf{168})$$

$$a: \text{M} = \text{Cr}$$
$$b: \text{M} = \text{Mo}$$
$$c: \text{M} = \text{W}$$

COT reacts with (NH$_3$)$_3$Cr(CO)$_3$ to give (**168a**) (20%),[427] and with (diglyme)Mo(CO)$_3$ to yield (**168b**) (60–70%).[214] A low yield (5%) of the chro-

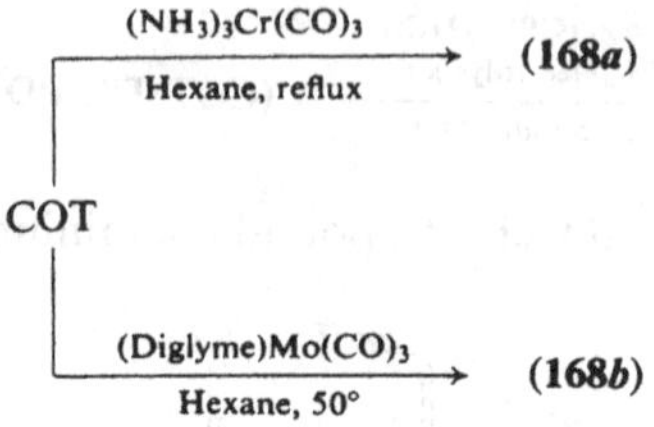
(NH₃)₃Cr(CO)₃
Hexane, reflux
(168a)
COT
(Diglyme)Mo(CO)₃
Hexane, 50°
(168b)

mium complex (**168*a***) also results from the reaction of COT with $Cr(CO)_6$ in acetonitrile.[424]

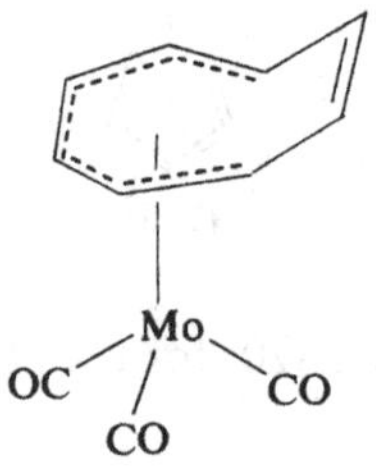
Mo
OC
CO
CO

(**168*b***)

Scheme 20

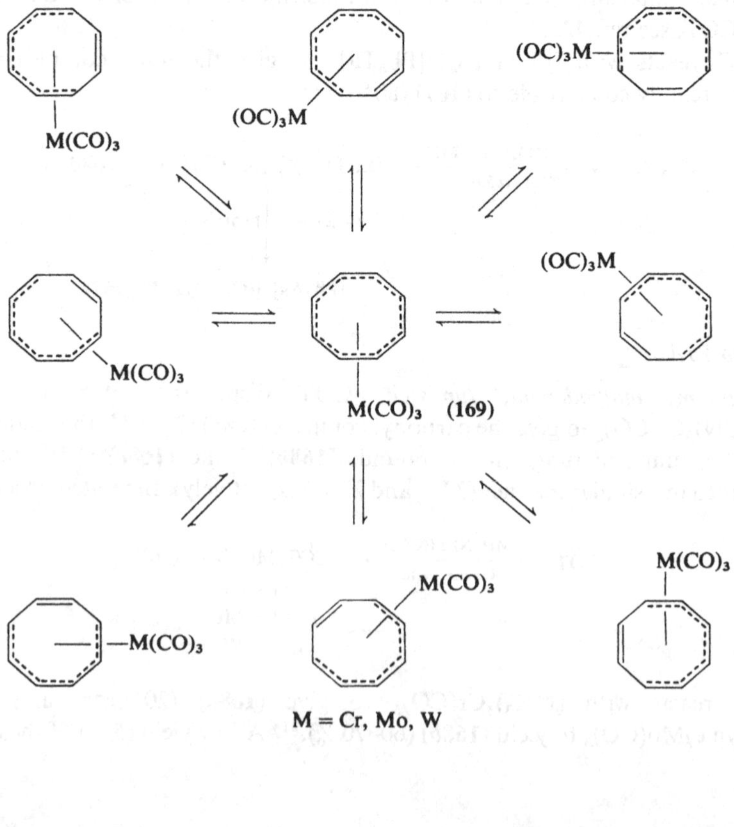
M(CO)₃
(OC)₃M
(OC)₃M—
M(CO)₃
(OC)₃M
M(CO)₃ (169)
—M(CO)₃
M(CO)₃
M(CO)₃

M = Cr, Mo, W

In (COT)Mo(CO)₃ (**168b**) the COT ring is partly flattened, the molybdenum atom being associated with six approximately coplanar carbon atoms (although not equidistant from each).[428]

The complexes (**168**) are fluxional in solution, the metal atom apparently moving around the ring (these shifts must necessarily involve ring-deformations).[214,424,427,429,430] The results of variable-temperature ^{13}C n.m.r. studies on the complexes (**168**) suggest that the mechanism of the fluxional process may involve random shifts in a symmetrical intermediate (**169**)[431,432] (scheme 20).

For the mass spectrum of (COT)Cr(CO)₃ (**168a**), see ref. 433.

Treatment of (CPD)CrCl₂.THF with isopropylmagnesium bromide in the presence of COT, followed by u.v. irradiation, affords the complex (**170**) (68 %).[434]

$$\text{COT} \xrightarrow[\text{(ii) } h\nu]{\substack{\text{(i) (CPD)CrCl}_2.\text{THF,} \\ \text{Pr}^i\text{MgBr, Et}_2\text{O, } -50° \to \text{r.-t.}}} \text{(COT)Cr(CPD)} \quad (\textbf{170})$$

The probable instantaneous structure of this fluxional molecule is as shown.

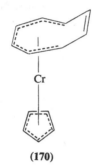

(**170**)

Treatment of COT with Mo(CO)₆ gives a product (**171**) (*ca.* 3 %), believed to incorporate a COT dimer.[435,436]

$$\text{COT} \xrightarrow[\text{Petroleum (b.p. 100–120°), reflux}]{\text{Mo(CO)}_6} \text{(C}_{16}\text{H}_{16}\text{)Mo(CO)}_4 \quad (\textbf{171})$$

Reaction between COT and molybdenum(VI) chloride oxide leads to the product (**172**)[437] (*cf.* cerium (p. 53) and titanium, zirconium and hafnium (p. 53)).

$$\text{COT} \xrightarrow[\text{C}_6\text{H}_6, \text{ reflux}]{\text{MoOCl}_4} \text{(C}_8\text{H}_7\text{)}_2\text{MoOCl}_2 \quad (\textbf{172})$$

Group VIIA

Manganese. Reaction of COT with Mn₂(CO)₁₀ affords (in low yield) the complex (**173**), containing a C₈H₉ ligand.[438]

$$\text{COT} \xrightarrow[150°]{\text{Mn}_2(\text{CO})_{10}} (\text{C}_8\text{H}_9)\text{Mn}(\text{CO})_3 \quad (173)$$

The structure of the product incorporates a bicyclo[3.3.0]octadienyl system[438] (*cf.* ruthenium (pp. 66, 69)).

Mn(CO)₃

(173)

With [HMn(CO)₄]₃, COT gives the complexes (174) (1.6%) and (175) (6.4%).[439]

$$\text{COT} \xrightarrow[\text{Hexane, reflux}]{[\text{HMn}(\text{CO})_4]_3} (\text{COT})\text{Mn}_2(\text{CO})_6 \quad (174)$$
$$+ \quad (\text{C}_8\text{H}_9)\text{Mn}(\text{CO})_3 \quad (175)$$

The first of these products (174) is apparently fluxional, and has the structure shown below[440]; the C₈H₉ ligand in (175) contains a free olefinic bond.[439]

(OC)₃Mn————Mn(CO)₃

Mn(CO)₃

(174)   **(175)**

For ion-molecule reactions of (CPD)Mn(CO)₃ with COT, see ref. 422.

Group VIII

Iron, ruthenium, osmium. Treatment of COT with iron(III) chloride in the presence of isopropylmagnesium chloride gives the product (176) (20%).[441]

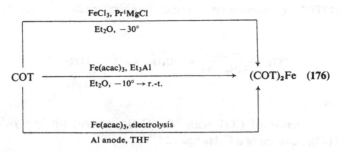

$$\text{COT} \begin{array}{c} \xrightarrow[\text{Et}_2\text{O, } -30°]{\text{FeCl}_3, \text{Pr}^i\text{MgCl}} \\ \xrightarrow[\text{Et}_2\text{O, } -10° \to \text{r.-t.}]{\text{Fe(acac)}_3, \text{Et}_3\text{Al}} \\ \xrightarrow[\text{Al anode, THF}]{\text{Fe(acac)}_3, \text{electrolysis}} \end{array} (\text{COT})_2\text{Fe} \quad (176)$$

A higher yield (*ca.* 75%) of this product results from a reaction employing Fe(acac)$_3$ and triethylaluminium.[442,443] Electrolysis of Fe(acac)$_3$ in the presence of COT, using an aluminium anode, also gives (**176**).[411] In the crystal, (COT)$_2$Fe (**176**) possesses a structure in which one of the COT rings uses a 1,3,5-triene system for coordination, while the other ring is attached *via* a 1,3-diene system (this is, of course, in accordance with EAN considerations).[444] The conformations adopted by the two ligands resemble those found in (COT)Mo(CO)$_3$ (**168b**) and (COT)Fe(CO)$_3$ (**177**) (see p. 62) respectively. In

(176)

solution, (COT)$_2$Fe (**176**) exhibits fluxional behaviour, of two kinds.[445] In the first process, the two ligands exchange roles, the changes of coordination being accompanied by changes of conformation. At −84° the 'triene' ligand is

'frozen', but the protons of the 'diene' ring continue to appear equivalent in the p.m.r. spectrum;[445] rapid 1,2-shifts of the iron atom are probably involved. (For 'broad-line' n.m.r. studies on the solid, see refs. 446, 447; for the ^{57}Fe Mössbauer spectrum, see ref. 448.)

COT reacts with Fe(CO)$_5$ under the influence of heat[449-451] or u.v. radiation[452,453] to afford (COT)Fe(CO)$_3$ (**177**) (up to 72%) and a binuclear compound '*trans*'-(COT)Fe$_2$(CO)$_6$ (**178**) (up to 31%).

$$COT \xrightarrow[\substack{\text{Ethylcyclohexane, reflux} \\ \text{or hexane, } h\nu}]{Fe(CO)_5} (COT)Fe(CO)_3 \quad (177)$$
$$+ \textit{'trans'}\text{-}(COT)Fe_2(CO)_6 \quad (178)$$

A second binuclear complex has been found amongst the products of the thermal reaction;[450] it was originally formulated as $(COT)Fe_2(CO)_7$, but is now known to be $(COT)Fe_2(CO)_5$[454,455] (see p. 199).

A very efficient conversion (up to 80%) of COT into $(COT)Fe(CO)_3$ (177) results from the reaction with $Fe(CO)_5$ in the presence of trimethylamine N-oxide.[456]

$$COT \xrightarrow[\substack{Me_3NO, \text{ e.g. } C_6H_6, \\ 0°, \text{ then reflux}}]{Fe(CO)_5} (COT)Fe(CO)_3 \quad (177)$$

The structure of the tricarbonyl complex (177), obtained by X-ray studies, features 1,3-diene-to-metal bonding.[457,458] The molecule is fluxional in

(177)

solution, the iron atom undergoing a sequence of extremely rapid 1,2-shifts.[459]

This fluxional process is sufficiently fast, even at $-150°$, to preclude n.m.r. observation of the protons in their instantaneous environments; 'wide-line' n.m.r. studies indicate that the molecule possesses considerable motional freedom even in the crystal.[455,460] The limiting ^{13}C n.m.r. spectrum, however, is observable below $-120°$,[461] and the nature of the rearrangement process has been elucidated by examination of the analogous ruthenium complex (see p. 65). (For other spectral data, see refs. 99 (u.v., i.r. and p.m.r.); 462, 463 (i.r. and Raman); 433, 464 (mass spectrum); 465 (chemical ionisation mass spectrum); 466, 467 (^{57}Fe Mössbauer).)

In '*trans*'-(COT)Fe$_2$(CO)$_6$ (**178**), the COT ligand exists in a 'chair' form with the Fe(CO)$_3$ groups attached on opposite sides of the ring;[457,458,468] the molecule is non-fluxional. (For the u.v., i.r. and p.m.r. spectra, see ref. 99; for the mass spectrum, see ref. 464; for [57]Fe Mössbauer spectral data, see ref. 466.)

(**178**)

Prolonged irradiation of Fe(CO)$_5$ in the presence of an excess of COT leads to two complexes in which the ligands are COT dimers.[469,470] Complex (**179**), m.p. 118°, is the chief product (58%) when the radiation is mainly in the visible region; when shorter wavelengths are used, complex (**180**), m.p. 172°, predominates (27%).[470]

$$COT \xrightarrow[\substack{C_6H_6, \\ hv, \text{ reflux}}]{Fe(CO)_5} (C_{16}H_{16})Fe(CO)_3 \quad (\mathbf{179}) \quad \text{m.p. } 118°$$
$$+ (C_{16}H_{16})Fe(CO)_3 \quad (\mathbf{180}) \quad \text{m.p. } 172°$$

Structure (**179**) may be assigned to the product of m.p. 118°,[470] while the higher-melting isomer has been shown to possess structure (**180**).[470,471]

(**179**)   (**180**)

(Stereochemistry assumed)

These products may also be obtained from COT and (COT)Fe(CO)$_3$ (see p. 64), and from the appropriate COT dimers (see pp. 367–8, 371).

The reaction of COT with Fe$_2$(CO)$_9$ gives not only (COT)Fe(CO)$_3$ (**177**) and '*trans*'-(COT)Fe$_2$(CO)$_6$ (**178**) but also '*cis*'-(COT)Fe$_2$(CO)$_6$ (**181**) (and allegedly a third isomer of formula (COT)Fe$_2$(CO)$_6$).[454]

$$COT \xrightarrow{Fe_2(CO)_9} (COT)Fe(CO)_3 \quad (\mathbf{177})$$
$$+ \text{'}\textit{trans}\text{'-}(COT)Fe_2(CO)_6 \quad (\mathbf{178})$$
$$+ \text{'}\textit{cis}\text{'-}(COT)Fe_2(CO)_6 \quad (\mathbf{181})$$

The complex (**181**) appears to be isostructural with the ruthenium analogue[472,473] (see p. 65).

The thermal reaction of COT with $Fe_3(CO)_{12}$ affords the tricarbonyl complex (177) (up to 88%).[474,475]

$$COT \xrightarrow[\substack{\text{Petroleum (b.p. 120–130°),} \\ \text{reflux}}]{Fe_3(CO)_{12}} (COT)Fe(CO)_3 \quad (177)$$

The same product (177) is also formed from COT and the cycloheptatriene complex $(CHT)Fe(CO)_3$.[450]

$$COT \xrightarrow[\text{Ethylcyclohexane, reflux}]{(CHT)Fe(CO)_3} (COT)Fe(CO)_3 \quad (177)$$

U.v. irradiation of COT in the presence of $(COT)Fe(CO)_3$ leads to the complexes of dimeric COT (179) and (180)[469,470,476] (see p. 63).

$$COT \xrightarrow[C_6H_6, h\nu]{(COT)Fe(CO)_3} (179) + (180)$$

COT displaces cyclohepta-1,4-diene from the complex (cyclohepta-1,4-diene)Fe(CO)(cyclohexa-1,3-diene) to give the product (182).[477]

$$(COT)Fe(CO)(cyclohexa-1,3-diene) \quad (182)$$

For the reaction of COT with ^{103}Ru atoms, see ref. 478.

With ruthenium(III) halides, COT gives polymeric complexes (183).[435]

$$COT \xrightarrow[\text{EtOH}]{RuX_3} [(COT)RuX_2]_n \quad (183) \qquad X = Cl, Br, I$$

COT reacts thermally or photochemically with $Ru_3(CO)_{12}$. When u.v. light is used to promote the reaction, the main product is the tricarbonyl complex (184) (60%), which is accompanied by the binuclear pentacarbonyl complex (185).[479]

$$COT \xrightarrow[\text{Cyclohexane, }h\nu]{Ru_3(CO)_{12}} (COT)Ru(CO)_3 \quad (184)$$
$$+ (COT)Ru_2(CO)_5 \quad (185)$$

In refluxing heptane, the products include the tricarbonyl (184) (12%), the pentacarbonyl (185) (34%), the '*cis*'-hexacarbonyl (186) (32%), and possibly the '*trans*'-hexacarbonyl (187);[480] additionally, the pentalene complex (188),[481] a complex (189) containing two (different) C_8H_9 ligands,[482] and two complexes (190) and (191) of a COT dimer,[483,484] have been isolated. Compounds (185) (60%) and (186) (30%) are obtained in refluxing xylene,[479] but reaction in

$$COT \xrightarrow[\text{Heptane, reflux}]{Ru_3(CO)_{12}} (COT)Ru(CO)_3 \quad \textbf{(184)}$$

$$+ (COT)Ru_2(CO)_5 \quad \textbf{(185)}$$

$$+ \text{'}cis\text{'-}(COT)Ru_2(CO)_6 \quad \textbf{(186)}$$

$$+ \text{'}trans\text{'-}(COT)Ru_2(CO)_6 \quad \textbf{(187)}$$

$$+ (Pentalene)Ru_3(CO)_8 \quad \textbf{(188)}$$

$$+ (C_8H_9)_2Ru_3(CO)_6 \quad \textbf{(189)}$$

$$+ (C_{16}H_{16})Ru(CO)_3 \quad \textbf{(190)}$$

$$+ (C_{16}H_{16})Ru_2(CO)_5 \quad \textbf{(191)}$$

refluxing octane affords yet another complex (**192**) as the principal product (87%).[480]

$$COT \xrightarrow[\text{Octane, reflux}]{Ru_3(CO)_{12}} (COT)_2Ru_3(CO)_4 \quad \textbf{(192)}$$

The tricarbonyl compound (**184**) is isostructural with $(COT)Fe(CO)_3$ (**177**),[485] and undergoes a similar fluxional process[472, 479, 486] (see p. 62). In this case the limiting p.m.r. spectrum is observable (at $-147°$), and the rearrangement mechanism has been established by computer-simulation of the spectrum.[487] (For the mass spectrum of $(COT)Ru(CO)_3$ (**184**), see ref. 464.)

Similarly, $(COT)Ru_2(CO)_5$ (**185**) appears to be isostructural with the fluxional $(COT)Fe_2(CO)_5$ (see p. 199),[472, 480] and '*trans*'-$(COT)Ru_2(CO)_6$ (**187**) with the non-fluxional '*trans*'-$(COT)Fe_2(CO)_6$ (**178**) (see p. 63).[472] (For the mass spectrum of $(COT)Ru_2(CO)_5$ (**185**), see ref. 464.)

The structure of crystalline '*cis*'-$(COT)Ru_2(CO)_6$ (**186**) features, in addition to a metal–metal bond, a π-allyl system, an olefin–metal bond, and a carbon–metal σ-bond (at least as rough approximations).[473]

$$(OC)_3Ru \longrightarrow Ru(CO)_3$$

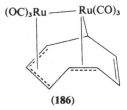

(**186**)

The observed p.m.r. spectrum of this compound[472] and that of the apparently isostructural '*cis*'-$(COT)Fe_2(CO)_6$ (**181**)[454] indicate that these molecules must either rearrange in solution to a more symmetrical structure, or that they are fluxional by virtue of a process such as the following.[473, 488]

$$(OC)_3M \longrightarrow M(CO)_3 \qquad (OC)_3M \longrightarrow M(CO)_3$$

$$\rightleftharpoons$$

$$M = Fe, Ru$$

(For the mass spectrum of '*cis*'-$(COT)Ru_2(CO)_6$, see ref. 464.)

The pentalene complex (188), the formation of which involves dehydrogenative transannular ring-closure of COT, possesses an instantaneous structure which may be represented as shown.[481] The molecule is fluxional,

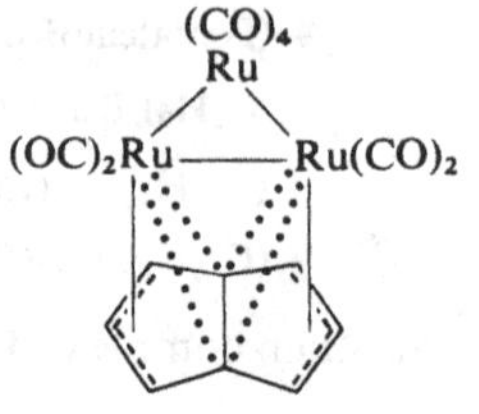

(188)

and the following oscillatory process is suggested to account for the observed p.m.r. spectrum.[481]

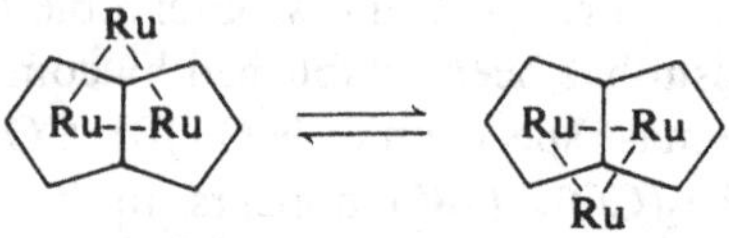

(CO groups omitted)

An X-ray structure determination has shown that the complex $(C_8H_9)_2$-$Ru_3(CO)_6$ (189) incorporates two different C_8H_9 ligands[489] (*cf.* the structures of $(C_8H_8.SiMe_3)Ru_2(SiMe_3)(CO)_5$ and $(C_8H_9)Ru(GeMe_3)(CO)_2$, p. 69).

(189)

The complex $(C_{16}H_{16})Ru(CO)_3$ (190) contains the illustrated dimeric-COT ligand,[483] a valence isomer of which is contained in the diruthenium complex

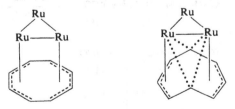

(190)   **(191)**

(191).[484] The fluxional process occurring in $(C_{16}H_{16})Ru_2(CO)_5$ **(191)** appears to involve a rearrangement of the bonding in the unbridged C_8 ring, similar to that postulated for '*cis*'-(COT)Ru$_2$(CO)$_6$ **(186)** (see p. 65), with a concomitant 1,3-shift within the other metal-bonded ring.[484]

The fluxional $(COT)_2Ru_3(CO)_4$ **(192)** is somewhat inadequately represented by the structure below;[490] the 'wide-line' p.m.r. spectrum of the solid is temperature-dependent.[491] Each COT ring is attached to two ruthenium

(192)

atoms, but the precise nature of the ring-to-metal bonding is not clear; it seems to be intermediate between the arrangements illustrated below.[490]

(One C_8 ring and CO groups omitted)

Treatment of COT with the hydridocarbonylruthenium cluster $H_4Ru_4(CO)_{12}$ affords, in addition to complexes of COT and of cyclo-octatriene, the free

COT dimer (**193**) together with the ruthenium carbonyl complexes (**190**) and (**191**).[483] The suggestion that (**193**) and (**190**) may be formed *via* (COT)Ru(CO)₃

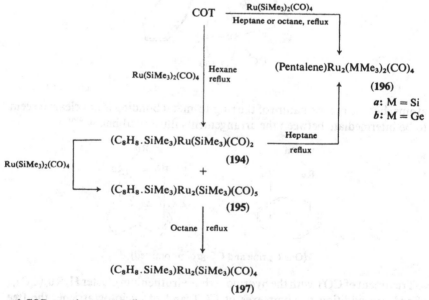

$$\text{COT} \xrightarrow[\text{Heptane, reflux}]{H_4Ru_4(CO)_{12}} \qquad + (\mathbf{190})$$
$$+ (\mathbf{191})$$
$$(\mathbf{193})$$

(**184**) is supported by the formation of the complex (**190**) from COT and (COT)Ru(CO)₃.[483]

$$\text{COT} \xrightarrow{(COT)Ru(CO)_3}$$

(**190**)

Displacement of cyclo-octa-1,5-diene from (COD)Ru(CO)₃* with COT provides a high-yield preparation of the tricarbonyl complex (**184**).[492]

$$\text{COT} \xrightarrow[\text{C}_6\text{H}_6\text{, reflux}]{(COD)Ru(CO)_3} (COT)Ru(CO)_3 \quad (\mathbf{184})$$

With Ru(SiMe₃)₂(CO)₄ in refluxing hexane, COT affords the complex (**194**);[482] a minor product is (**195**), which is best obtained from (**194**).[493] At

$$\text{COT} \xrightarrow[\text{Heptane or octane, reflux}]{Ru(SiMe_3)_2(CO)_4}$$

Ru(SiMe₃)₂(CO)₄ | Hexane reflux

(Pentalene)Ru₂(MMe₃)₂(CO)₄

(**196**)

a: M = Si
b: M = Ge

Ru(SiMe₃)₂(CO)₄

(C₈H₈·SiMe₃)Ru(SiMe₃)(CO)₂

(**194**)

+

(C₈H₈·SiMe₃)Ru₂(SiMe₃)(CO)₅

(**195**)

Heptane reflux

Octane | reflux

(C₈H₈·SiMe₃)Ru₂(SiMe₃)(CO)₄

(**197**)

* COD = cyclo-octa-1,5-diene.

higher temperatures (194) gives the pentalene complex (196*a*),[482] while (195) loses carbon monoxide to yield (197).[493]

Similar reactions occur with $Ru(GeMe_3)_2(CO)_4$, but in this case the pentalene complex (196*b*) is accompanied by several by-products,[494] two of which have been identified as (198*a*) and (199) respectively.[482]

$$(COT)Ru_2(MMe_3)_2(CO)_4 \quad \textbf{(198)} \quad a: M = Ge;$$
$$b: M = Si$$

$$(C_8H_9)Ru(GeMe_3)(CO)_2 \quad \textbf{(199)}$$

The structures of the products (194),[482] (195).[493] (196*b*),[494] (197),[493] (198*a*)[482] and (199)[482] are illustrated below. Compound (194) is fluxional, the n.m.r.

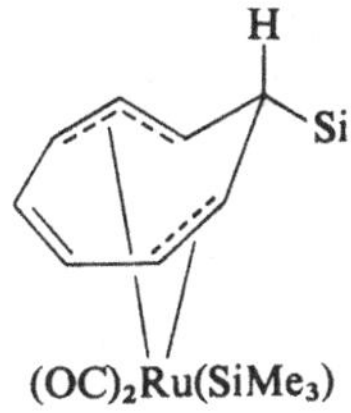

(194)

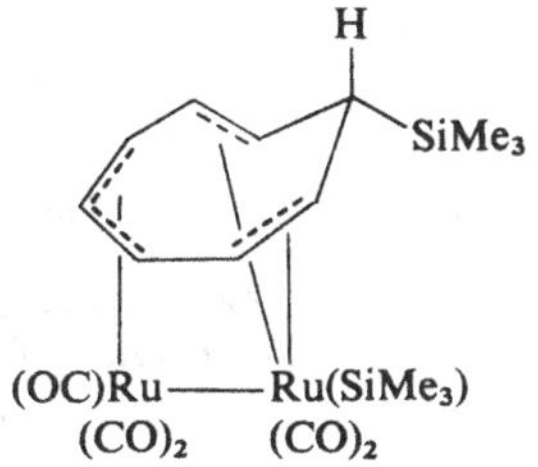

(195)

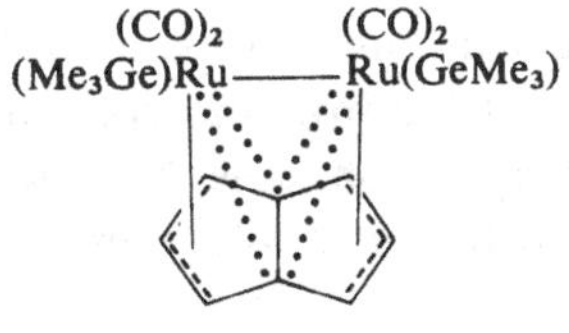

(196*b*)

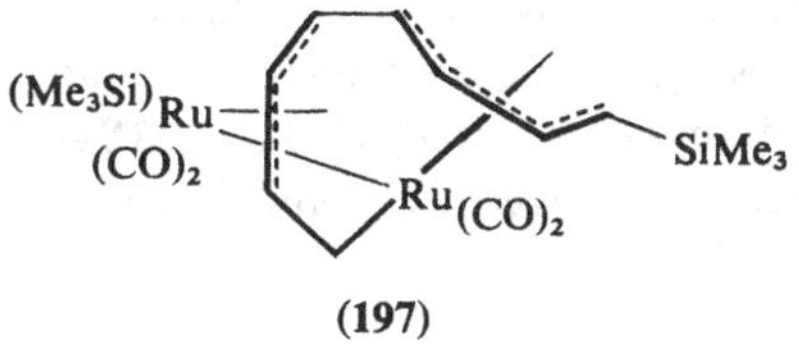

(197)

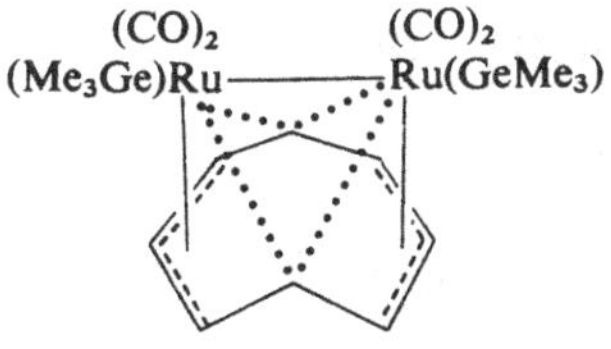

(198*a*)

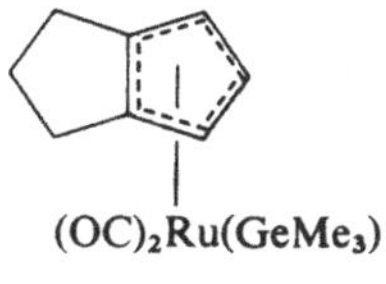

(199)

spectra being consistent with the operation of an oscillatory process as shown.[482]

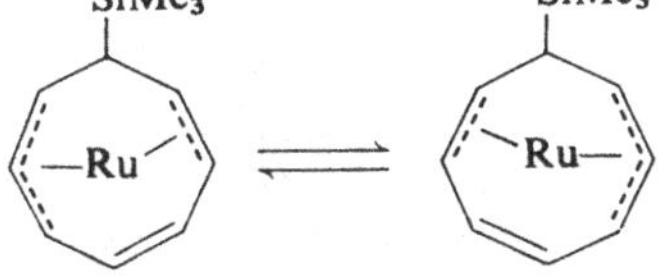

(SiMe₃ and CO groups on metal omitted)

The structures (195), (196), (198) and (199) may be compared with those of complexes (189) (see p. 66), (188) (see p. 66), (COT)Fe$_2$(CO)$_5$ (see p. 199) and (173) (see p. 60) severally. In complex (197), the C$_8$ ring has opened to form a chain, carrying an SiMe$_3$ group on one end and a σ-bonded Ru atom on the other; the octatetraene chain takes up a conformation such that each of the metal atoms is η^4-bonded.[493]

The reaction of COT with [Ru(SiMe$_3$)(CO)$_4$]$_2$ yields the complexes (188)[481] and (198b).[482]

Among the products from COT and Me$_2$Si.[CH$_2$]$_2$.SiMe$_2$.Ru(CO)$_4$ is the complex (200).[493] In the formation of compound (200), one SiMe$_2$ group

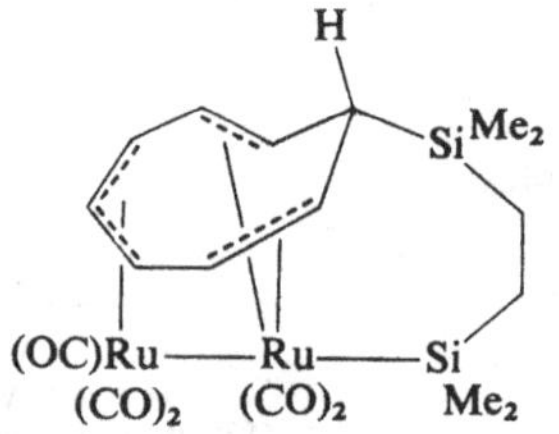

(200)

migrates to carbon while the other remains bonded to ruthenium; the metal-to-ring bonding is similar to that in complex (195) (see p. 69).

COT reacts with Os$_3$(CO)$_{12}$ under the influence of u.v. light to give a product (COT)Os(CO)$_3$ (201), which is *not* an analogue of the iron and ruthenium tricarbonyl complexes (177) and (184); in refluxing cyclohexane, however, it isomerises to give a compound (202) which is isostructural with these complexes.[495,496]

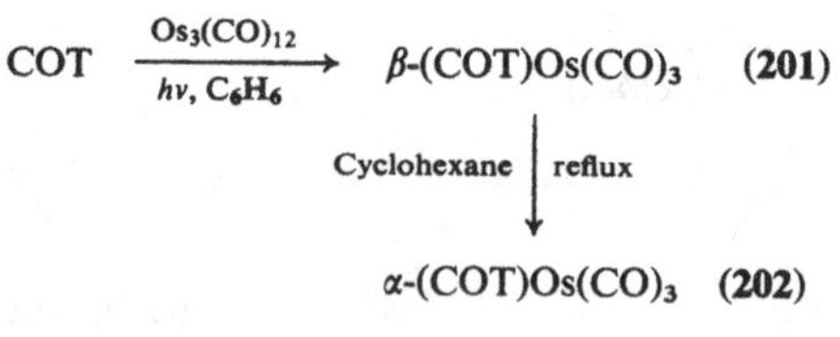

$$\text{COT} \xrightarrow[\text{hv, C}_6\text{H}_6]{\text{Os}_3\text{(CO)}_{12}} \beta\text{-(COT)Os(CO)}_3 \quad (201)$$

Cyclohexane | reflux

$$\alpha\text{-(COT)Os(CO)}_3 \quad (202)$$

The following structure has been suggested for the thermally labile β-isomer (201).[495,496]

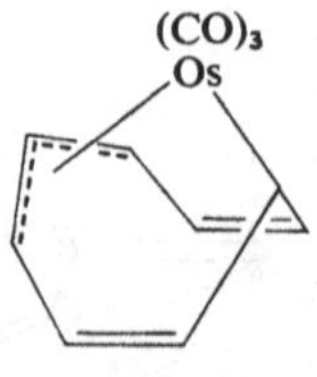

(201)

The p.m.r. spectrum of the stable isomer (202) is fully resolved at a higher temperature than for the analogous iron and ruthenium complexes (177) and (184)[496, 497] (see pp. 62, 65). Thus for the series $(COT)M(CO)_3$ (M = Fe, Ru, Os) the ease of the fluxional process, occurring *via* 1,2-shifts of the metal atom, decreases with increasing atomic number of the metal.

Cobalt, rhodium, iridium. Reduction of cobalt(II) chloride with sodium borohydride in the presence of COT gives a small yield (4 %) of the complex (203).[498]

$$COT \xrightarrow[\text{EtOH, } -30° \to \text{r.-t.}]{\text{CoCl}_2, \text{ NaBH}_4} \text{(Cyclo-octatrienyl)Co(COT)} \quad (203)$$

On the basis of the ^{1}H n.m.r. spectrum, the crude structures (203a) and (203b) both appear to be possibilities, although (203a) is favoured; the molecule is seemingly fluxional[498] (for a revised structure, see appendix).

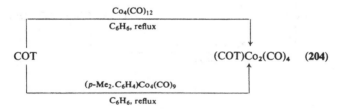

(203a) (203b)

With $Co_4(CO)_{12}$ a binuclear, apparently fluxional complex (204) is produced (2 %); a better yield (26 %) results from the use of $(p\text{-Me}_2.C_6H_4)Co_4(CO)_9$.[499]

$$COT \xrightarrow[\text{C}_6\text{H}_6, \text{ reflux}]{\begin{array}{c}\text{Co}_4\text{(CO)}_{12}\\ \\ (p\text{-Me}_2.\text{C}_6\text{H}_4)\text{Co}_4\text{(CO)}_9\\ \text{C}_6\text{H}_6, \text{ reflux}\end{array}} \text{(COT)Co}_2\text{(CO)}_4 \quad (204)$$

COT displaces carbon monoxide from $(CPD)Co(CO)_2$ to afford the product (205) (up to 30 %).[474, 500, 501]

$$COT \xrightarrow[ca. 140°]{\text{(CPD)Co(CO)}_2} \text{(CPD)Co(COT)} \quad (205)$$

The 'mixed' complex (205) is obtained in better yield by using u.v. irradiation,[99, 502] and under these conditions the binuclear complex (206) is also formed.[502]

$$\text{COT} \xrightarrow[h\nu]{\text{(CPD)Co(CO)}_2} \text{(205)} + \text{(CPD)}_2\text{Co}_2\text{(COT)}$$
$$\text{(206)}$$

Complexes (205) and (206) are formulated as shown below; structure (206) has been confirmed by X-ray studies.[503] (For u.v., i.r. and p.m.r. spectra of

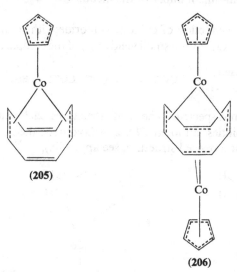

(205)

(206)

(205) and (206), see ref. 99.)

For ion-molecule reactions of COT with (CPD)Co(CO)$_2$, see ref. 422.

With PhC . Co$_3$(CO)$_9$, COT displaces carbon monoxide to give the fluxional complex (207) (9%); the same product is formed (16%) from PhC . Co$_3$(CO)$_6$-

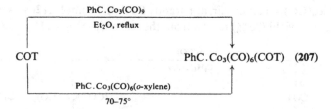

(207)

(*o*-xylene) by displacement of the xylene.[504] In the product (**207**), all three metal atoms in the metal cluster are bonded to the COT ligand.[504, 505]

COT reacts with hydrated rhodium(III) chloride to form the dimeric rhodium(I) complex (**208**).[339, 506]

$$\text{COT} \xrightarrow[\text{EtOH}]{\text{RhCl}_3.3\text{H}_2\text{O}} [(\text{COT})\text{RhCl}]_2 \quad (\textbf{208})$$

The same product is obtained in high yields (85–100 %) by displacement reactions using $[(\text{cyclo-octene})_2\text{RhCl}]_2$[507] or $(\text{butadiene})_2\text{RhCl}$.[508]

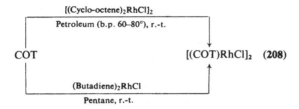

On the basis of the spectral evidence, the structure of the dimeric complex (**208**) is believed to be as shown.[507]

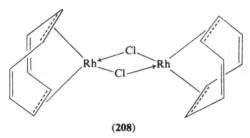

(**208**)

Treatment of $[(\text{cyclo-octene})_2\text{RhCl}]_2$ with thallium(I) acetylacetonate, followed by reaction with COT, affords the complex (**209**) (83 %);[509] the same product results (87 %) from the treatment of COT with $\text{Rh}(\text{acac})(\text{CO}_2)$.[510]

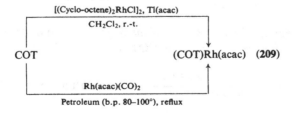

With $\text{Rh}_4(\text{CO})_{12}$, COT forms the complex (**210**) (77 %).[499]

$$\text{COT} \xrightarrow[\text{C}_6\text{H}_6,\ \text{reflux}]{\text{Rh}_4(\text{CO})_{12}} (\text{COT})_2\text{Rh}_4(\text{CO})_8 \quad (\textbf{210})$$

Reaction of hydrated rhodium(III) chloride with sodium tetraphenylborate in the presence of COT gives the neutral complex (211) (70–80%).[511]

$$\text{COT} \xrightarrow[\text{MeOH(aq.), r.-t. or } 50°]{\text{RhCl}_3.3\text{H}_2\text{O, NaBPh}_4} \text{(COT)Rh.BPh}_4 \quad \text{(211)}$$

In this product (211), the tetraphenylborate anion is co-ordinated to Rh⁺ *via* one of the phenyl rings, and the suggested structure is as follows.[511]

(211)

As with analogous cobalt compound, COT displaces carbon monoxide from (CPD)Rh(CO)₂. The thermal reaction affords the product (212) (11%),[474] while a photochemical procedure results in the binuclear compound (213) (62%).[512]

$$\text{COT} \begin{cases} \xrightarrow[\text{Petroleum (b.p. 120--160°), reflux}]{\text{(CPD)Rh(CO)}_2} \text{(CPD)Rh(COT)} \quad \text{(212)} \\ \\ \xrightarrow[hv,\ \text{decane, reflux}]{\text{(CPD)Rh(CO)}_2} \text{(COT)[(CPD)Rh]}_2 \quad \text{(213)} \end{cases}$$

The displacement of cyclo-octa-1,5-diene from the perchlorate [(COD)Rh. Bipy]⁺ ClO₄⁻ yields the corresponding COT complex (214).[513]

$$\text{COT} \xrightarrow[\text{MeOH, CH}_2\text{Cl}_2, \text{ r.-t.}]{\text{[(COD)Rh.Bipy]}^+ \text{ClO}_4^-} \text{[(COT)Rh.Bipy]}^+ \text{ClO}_4^- \quad \text{(214)}$$

Bipy = 2,2′-bipyridyl

COT reacts with [(cyclo-octene)₂IrCl]₂ or [(ethylene)₂IrCl]₂ to afford the COT complex (215) (63%), which is probably polymeric; in pyridine, however, the monomeric pyridinate (216) is formed.[514]

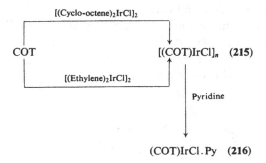

$$(COT)IrCl.Py \quad (216)$$

The acid H_3IrCl_6 apparently reacts with COT to give the hydrido-iridium(III) complex (217) (see ref. 515).

$$COT \xrightarrow[EtOH]{H_3IrCl_6} [(COT)IrHCl_2]_n \quad (217)$$

Nickel, palladium, platinum. Electrolysis of $Ni(acac)_2$ in the presence of COT, using a nickel or an aluminium anode, produces cyclo-octatetraenenickel (218) in high yields.[411,516,517] By working at $-60°$, the reaction can be made to give the bis(cyclo-octatetraene) complex (219), which decomposes above *ca.* $-40°$ to yield the mono(cyclo-octatetraene) compound (218).

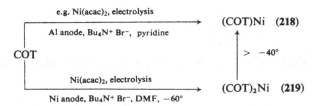

The reaction of $Ni(acac)_2$ with triethylaluminium in the presence of COT affords an alternative route to these products (218) (95%) and (219) (78.5%).[518]

COT

Ni(acac)₂, Et₃Al / C₆H₆, 0° → (COT)Ni (218)

Ni(acac)₂, Et₃Al / Toluene, −60° → (COT)₂Ni (219)

COT

(COD)₂Ni / C₆H₆, r.-t. → (COT)Ni (218)

(all-E-CDT)Ni / Et₂O, −40° → (COT)₂Ni (219)

The same products (218) and (219) are also formed in high yields by displacement reactions, using the cyclo-octa-1,5-diene or all-*E*-cyclododeca-1,5,9-triene complex of nickel(0).[518]

In the solid state, (COT)Ni (218) appears to be polymeric[518, 519] (see appendix); (COT)$_2$Ni (219) is monomeric.[518]

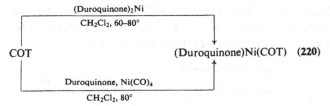

(219)

The action of COT on the duroquinone complex of nickel(0) yields the 'mixed' product (220) (40%),[520] and this is also formed (65%) by reaction of COT with duroquinone and Ni(CO)$_4$.[520] The proposed structure of (duro-

```
              (Duroquinone)₂Ni
              CH₂Cl₂, 60–80°
        ┌─────────────────────────────┐
        │                             │
        │                             ↓
COT     │                   (Duroquinone)Ni(COT)   (220)
        │                             ↑
        │                             │
        └─────────────────────────────┘
              Duroquinone, Ni(CO)₄
              CH₂Cl₂, 80°
```

quinone)Ni(COT) (220) is shown below; for a theoretical discussion, see ref. 521.

(220)

In the presence of COT, the mass spectrum of (CPD)Ni(NO) shows that [(CPD)Ni(COT)]$^+$ ions are formed.[522]

COT reacts with (F$_3$C.C≡C.CF$_3$)$_3$Ni$_4$(CO)$_4$ to give the complex (221).[523]

$$\text{COT} \xrightarrow{\;(F_3C.C{\equiv}C.CF_3)_3Ni_4(CO)_4\;} (COT)(F_3C.C{\equiv}C.CF_3)Ni_3(CO)_3 \quad (221)$$

The structure of this product, which is fluxional, incorporates a nickel cluster sandwiched between a planar COT ligand and the hexafluoro but-2-yne ligand, but the nature of the metal-to-ring bonding is not clear.[523]

(221)

COT displaces benzonitrile from palladium complexes of the type $(PhCN)_2PdX_2$ to give the products (222)[99, 524, 525] (table 15). (For heats of

$$COT \xrightarrow[C_6H_6, \ r.-t.]{(PhCN)_2PdX_2} (COT)PdX_2 \quad (222)$$

Table 15

X	Yield (%)	Ref.
Cl	86	524
Br	ca. 10	525

reaction, see ref. 526.) A square planar geometry for $(COT)PdCl_2$ (red) has been established; a (yellow) dimeric form also exists.[527]

In the presence of methanol, COT reacts with disodium tetrachloropalladate(II) to afford the dimeric complex (223), which apparently contains a 7-methoxycyclo-octa-1,3,5-triene ligand.[524]

$$COT \xrightarrow[MeOH, \ r.-t.]{Na_2PdCl_4.4H_2O} [(C_8H_9.OMe)PdCl]_2 \quad (223)$$

Reaction of COT with the cationic π-allyl complex (224) gives the product (225).[528]

(224)   (225)

The platinum(II) complex (226a) results from the reaction of COT with dipotassium tetrachloroplatinate(II); in the presence of potassium iodide, the corresponding di-iodide (226b) is produced (93%).[529] (For spectra of complexes of the type (226), see refs. 99, 530; for ^{81}Br n.q.r. measurements on $(COT)PtBr_2$, see ref. 531.)

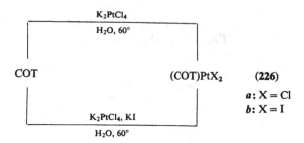

2
Substituted cyclo-octatetraenes

1. Monosubstituted derivatives

Formation

(a) *'Mixed' cyclotetramerisation of acetylenes.* The cyclotetramerisation of acetylene in the presence of nickel catalysts (see p. 2) may be adapted to produce substituted COTs, monosubstituted derivatives (**227**) being formed by the co-reaction of three molecules of acetylene and one molecule of a substituted acetylene; not unexpectedly, the yields are low (table 16).

(**227**)

Table 16

R	Yield (%)	Ref.
Me	16	(532), 533
CH=CH$_2$	20	13
Pr^n	25	533
Bu^n	16	(532), 533
Ph	17	(532), 533
CH$_2$OH	17	534
[CH$_2$]$_2$OH	24	534
[CH$_2$]$_3$OH	8	535
CH(OH)Me	11	535
C(OH)Me$_2$	13	535
CO.Me	1.5	535
[CH$_2$]$_2$NMe$_2$	8	534
[CH$_2$]$_3$CN	11	535

Vinylcyclo-octatetraene is a by-product of the Reppe synthesis of COT.[24, 25]

(b) *Photochemical cycloadditions of monosubstituted acetylenes to benzene.* The u.v. irradiation of suitable monosubstituted acetylenes in benzene leads to monosubstituted COTs (table 17), presumably *via* intermediate bicyclo[4.2.0]-octatrienes.

R	Yield (%)	Ref.
Bu^n	—	38
Bu^t	—	38
Ph	0.6	536
CO$_2$Me	8	536

Although the yields are very low, this reaction provides a useful one-step synthesis, from hexadeuteriobenzene, of the deuteriated products (228; R = Ph,[49] CO$_2$Me,[49] CO$_2$Et,[123] CN[49]).

(228)

(*c*) *Syntheses using organo-lithiums and analogous reagents.* Alkylation or arylation of COT may be effected with organo-lithiums, monosubstituted COTs being produced together with cyclo-octatrienes and their substituted derivatives (see p. 32); the yields are low (table 18). Phenylcyclo-octatetraene may also be prepared (22%) from COT using phenylsodium in refluxing benzene.[285]

(227)

Table 18

R	Yield (%)	Ref.
Et	29	286
Bu^n	14.5	286
[a] Ph	25	285, (538)
p-Me.C$_6$H$_4$	29	539
p-Bu^t.C$_6$H$_4$	22	539
p-Ph.C$_6$H$_4$	16	539
2,4,6-Me$_3$.C$_6$H$_2$	6 (crude)	539
p-Me$_2$N.C$_6$H$_4$	27, [21]	285, (538), [539]

[a] For the deuteriated derivative (227; R = C$_6$D$_5$), see ref. 537.

Superior yields of alkyl and aryl derivatives of COT result from the coupling reaction between bromocyclo-octatetraene and lithium 'organocuprates' (table 19); this is undoubtedly the method of choice for the preparation of monoalkyl- and monoarylcyclo-octatetraenes.

$$\xrightarrow[\text{Et}_2\text{O, e.g. } -70^\circ \rightarrow \text{r.-t.}]{\text{LiCuR}_2}$$

(227)

Table 19

R	Yield (%)	Ref.
Me	93[a]	540
Cyclopropyl	95, (90)	541, (542)
CH=CH₂	88	541
Ph	58	540
p-MeO.C₆H₄	80	541

[a] A lower yield (73%) results from the use of methyl-lithium.[540]

Metal–halogen exchange between n-butyl-lithium and bromocyclo-octatetraene affords cyclo-octatetraenyl-lithium, and this product may be used *in situ* for the preparation of a wide variety of COT derivatives (table 20). Similar

Table 20 (refers to the equation at the top of p. 82)

Reagent	R	Yield (%)	Ref.
D₂O	D	—	250
MeI	Me	(24), 81	(540), 543
CoCl₂ (catalyst)	Cyclo-octatetraenyl	24	544
Ethylene oxide	[CH₂]₂OH	[60], 70	(545), [546], 547
PhCHO	CH(OH)Ph	47	548
CH₂=CH.CHO	CH(OH).CH=CH₂	43	548
CH₂=CH.CO₂Et (CuCl)	[CH₂]₂.CO₂Et	5	547
HCO₂Me	CHO	79	541
Fe(CO)₅	CHO[a]	45	541
Me.CH=CH.CN	CO.CH=CHMe	75	548
PhCN	CO.Ph	(57), 63	(544), 548
CO₂	CO₂H	58.5	549
(CN)₂	CN	25	541
FClO₃	F	*ca.* 10	550
Me₃SiCl	SiMe₃	80	551
(Allyl)Me₂SiCl	SiMe₂(allyl)	50	551
(CPD)FeCl(CO)₂	Fe(CO)₂(CPD)	10	551
Me₃GeBr	GeMe₃	55	551
Me₃SnCl	SnMe₃	55	551
I₂	I	50	540

[a] The product is the iron tricarbonyl complex of (227; R = CHO).

(227)

conversions have been effected using the Grignard reagent (229) generated from bromocyclo-octatetraene (table 21).

(229)   (227)

Table 21

Reagent	R	Yield (%)	Ref.
$CH_2=CH.CH_2Br$ (LiCl, CuCl)	$CH_2.CH=CH_2$	44	547
Oxetan	$[CH_2]_3OH$	51	547
Ethylene oxide	$[CH_2]_2OH$	21	547
$CH_2=CH.CO_2Et$ (CuCl)	$[CH_2]_2CO_2Et$	20	547
CO_2	CO_2H	59	160
$(CN)_2$	CN	Up to 40	541

A coupling reaction between bromocyclo-octatetraene and methylmagnesium iodide (in the presence of cobalt(II) chloride) yields methylcyclo-octatetraene (36 %).[160]

(*d*) *Syntheses using the tricarbonyliron complex of COT.* It is possible to effect electrophilic substitution of the COT ring using the tricarbonyliron complex (177). Thus Vilsmeier-Haack formylation yields the aldehyde complex (230) (60 %), the functional group of which may be modified, if required, to give e.g. the methoxymethyl-compound (231); decomposition of these complexes with cerium(IV) ions occurs in high yield to afford the monosubstituted COTs (232) and (233)[552] (scheme 21).

Scheme 21

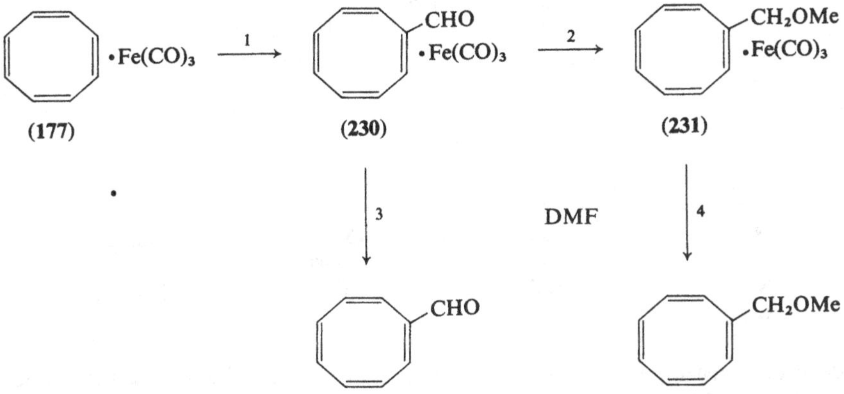

(177) (230) (231)

(232) (233)

Reaction	Reagents and conditions	Yield (%)
1	$POCl_3$, $HCONMe_2$, 4°	60
2	(i) $NaBH_4$, EtOH, 0° (ii) 64% HPF_6(aq.), 20° (iii) MeOH, r.-t.	74
3	$Ce(NO_3)_4$.$2NH_4NO_3$, EtOH	80
4	$Ce(NO_3)_4$.$2NH_4NO_3$, EtOH	90

(*e*) *Syntheses involving dehydrohalogenation.* The base-induced elimination of hydrogen halide from the appropriate 7,8-dihalocyclo-octa-1,3,5-triene (234) (see pp. 23–4) may be used to prepare the chloro- and bromo-derivatives of COT (table 22). Chloro- and bromocyclo-octatetraene are also formed

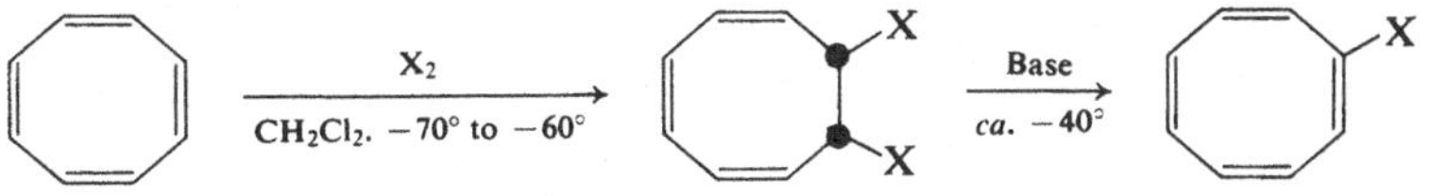

(234)

Table 22

X	Base	Yield (%)	Ref.
Cl	KOBu^t	74–83	540
Br	KOBu^t	85	540
Br	NaOMe	16	540

(26–50%) by the action of phenyl-lithium[553] or potassium t-butoxide[554] on the bicyclic dihalides (38/39) and (44) (pp. 23, 24).

Further treatment of bromocyclo-octatetraene with sodium or potassium alkoxides, or phenoxide, produces alkoxy- or aryloxy-derivatives of COT (235) (table 23). The intermediacy of 1,2-didehydrocyclo-octatetraene (236) in the reaction of bromocyclo-octatetraene with potassium t-butoxide has

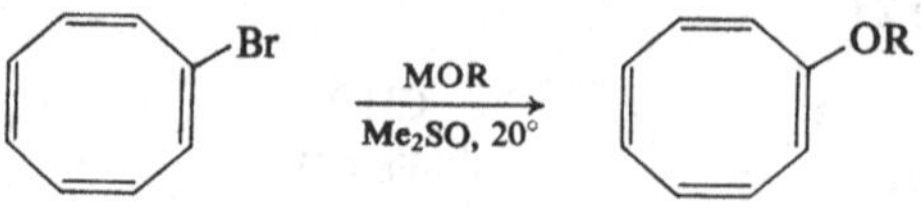

(235)

Table 23

M	R	Yield (%)	Ref.
Na	Me	75, (69)	540, (554)
K	Et	70–80	554
K	Bu^t	70–80	554, (555)
K	Ph	73	540

been demonstrated by trapping experiments with phenyl azide, tetracyclone, etc. (see p. 164), and generation of the 'yne' (236) in the presence of secondary amines leads to the dialkylamino-derivatives of COT (237)[541] (table 24).

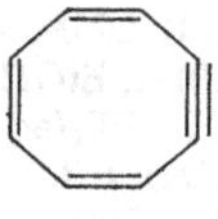

(236)

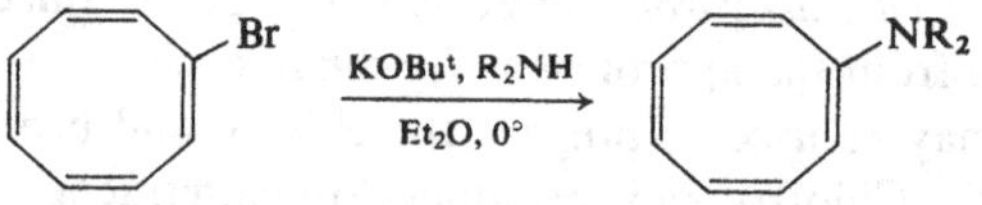

(237)

Table 24

R	Yield (%)
Me	79
Et	—

(*f*) *Syntheses involving skeletal transformations.* Methylcyclo-octatetraene is generated (33%), together with other hydrocarbons from which it is not completely separable, by treatment of the dibromocarbene-adduct (238) with potassium t-butoxide.[556]

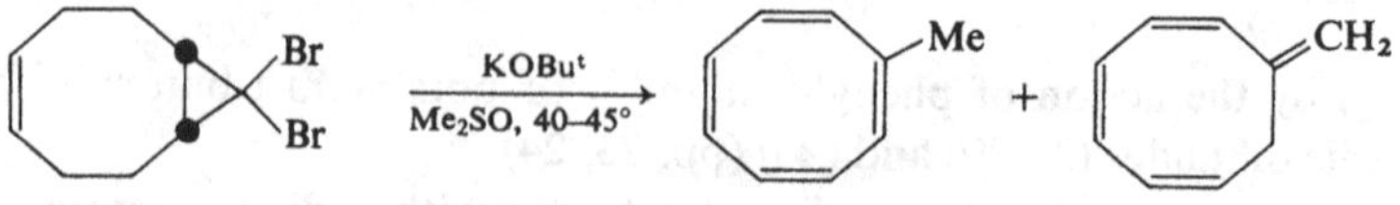

(238)   + bicyclic hydrocarbons

The proposed intermediate in this formation of methylcyclo-octatetraene is bicyclo[6.1.0]nona-2,4,6-triene (82*a*), which on similar treatment with base gives the same mixture of products (47% yield of methylcyclo-octatetraene)[556] (scheme 22).

Scheme 22

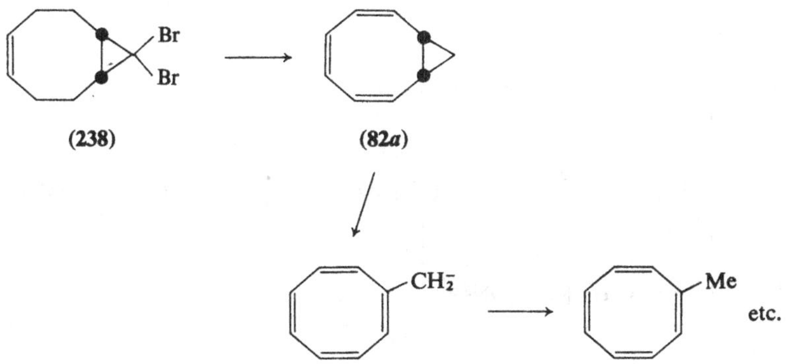

(238)　　　　　　(82*a*)

etc.

Flash thermolysis of the tricyclic diene (239), or of its initial thermolysis product (240), gives isopropylcyclo-octatetraene (29% or 40% respectively).[557]

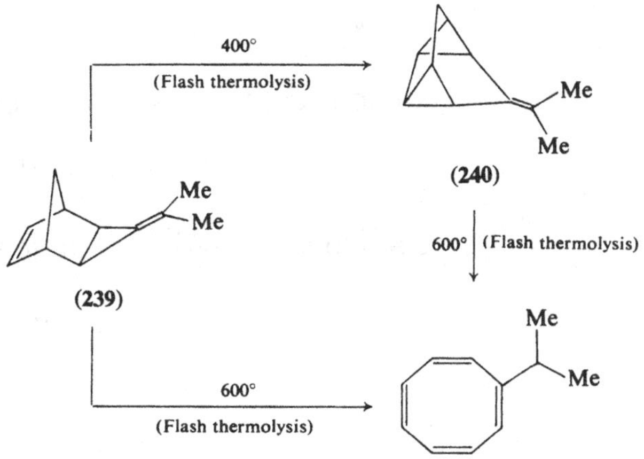

Cyanocyclo-octatetraene is formed (57%) by treatment of the oxime (241) with toluene-*p*-sulphonyl chloride[52] (see p. 297).

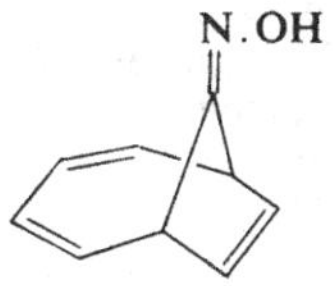

(241)

(*g*) *Functional group transformations.* The transformations outlined in schemes 23–36 lead to a further variety of monosubstituted COTs.

Scheme 23

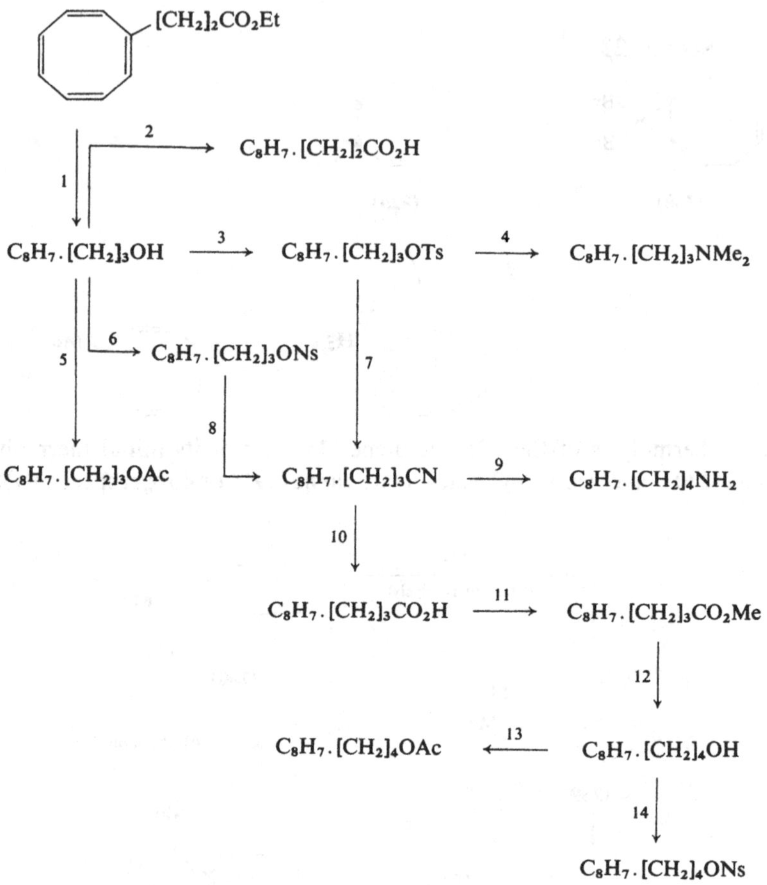

$C_8H_7.[CH_2]_2CO_2Et$

$$\xrightarrow{2} C_8H_7.[CH_2]_2CO_2H$$

$C_8H_7.[CH_2]_3OH \xrightarrow{3} C_8H_7.[CH_2]_3OTs \xrightarrow{4} C_8H_7.[CH_2]_3NMe_2$

$C_8H_7.[CH_2]_3ONs$

$C_8H_7.[CH_2]_3OAc$

$C_8H_7.[CH_2]_3CN \xrightarrow{9} C_8H_7.[CH_2]_4NH_2$

$C_8H_7.[CH_2]_3CO_2H \xrightarrow{11} C_8H_7.[CH_2]_3CO_2Me$

$C_8H_7.[CH_2]_4OAc \xleftarrow{13} C_8H_7.[CH_2]_4OH$

$C_8H_7.[CH_2]_4ONs$

Reaction	Reagents and conditions	Yield (%)	Ref.
1	LiAlH$_4$, Et$_2$O, r.-t.	90	547
2	CrO$_3$, H$_2$SO$_4$(aq.), Me$_2$CO, 25°	37	535
3	p-Me.C$_6$H$_4$.SO$_2$Cl, pyridine, 0°	96	535
4	Me$_2$NH, C$_6$H$_6$, r.-t.	54	535
5	Ac$_2$O, pyridine, 0° → r.-t.	—	547
6	p-NO$_2$.C$_6$H$_4$.SO$_2$Cl, pyridine, 0°	—	547
7	KCN, EtOH(aq.), reflux	74	535
8	NaCN, DMF, r.-t.	100	547
9	LiAlH$_4$, Et$_2$O, reflux	39	535
10	NaOH(aq.), reflux	69, (67)	535, (547)
11	CH$_2$N$_2$, Et$_2$O	100	547
12	LiAlH$_4$, Et$_2$O, r.-t.	91	547
13	Ac$_2$O, pyridine, 0° → r.-t.	—	547
14	p-NO$_2$.C$_6$H$_4$.SO$_2$Cl, pyridine, 0°	—	547

Scheme 24

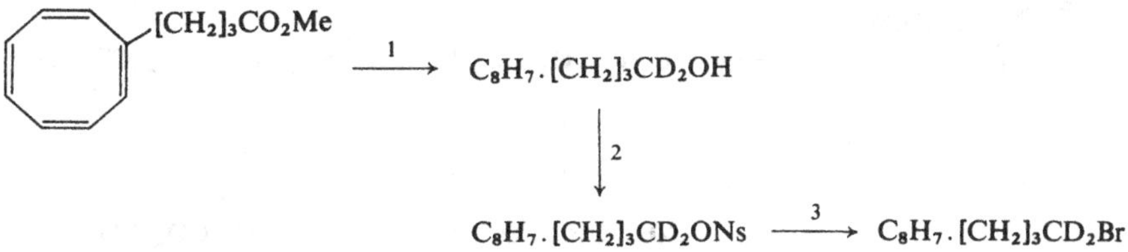

C_8H_7 $[CH_2]_3CO_2Me$ $\xrightarrow{1}$ $C_8H_7.[CH_2]_3CD_2OH$

$\downarrow 2$

$C_8H_7.[CH_2]_3CD_2ONs$ $\xrightarrow{3}$ $C_8H_7.[CH_2]_3CD_2Br$

Reaction	Reagents and conditions	Yield (%)	Ref.
1	LiAlD$_4$, Et$_2$O	91	558
2	p-NO$_2$.C$_6$H$_4$.SO$_2$Cl	—	558
3	PyH$^+$ Br$^-$, DMF	93	558

Scheme 25

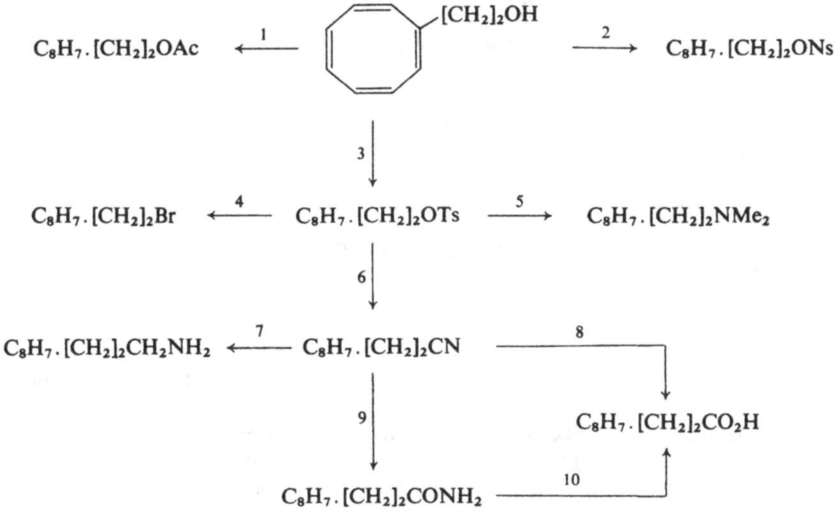

$C_8H_7.[CH_2]_2OAc$ $\xleftarrow{1}$ C_8H_7 $[CH_2]_2OH$ $\xrightarrow{2}$ $C_8H_7.[CH_2]_2ONs$

$\downarrow 3$

$C_8H_7.[CH_2]_2Br$ $\xleftarrow{4}$ $C_8H_7.[CH_2]_2OTs$ $\xrightarrow{5}$ $C_8H_7.[CH_2]_2NMe_2$

$\downarrow 6$

$C_8H_7.[CH_2]_2CH_2NH_2$ $\xleftarrow{7}$ $C_8H_7.[CH_2]_2CN$ $\xrightarrow{8}$

$\downarrow 9$ $C_8H_7.[CH_2]_2CO_2H$

$C_8H_7.[CH_2]_2CONH_2$ $\xrightarrow{10}$

Reaction	Reagents and conditions	Yield (%)	Ref.
1	Ac$_2$O, *ca.* 100°	91	534
	or Ac$_2$O, pyridine, r.-t.	95 (crude)	546, (547)
2	p-NO$_2$.C$_6$H$_4$.SO$_2$Cl, pyridine, 0°	82	547
3	p-Me.C$_6$H$_4$.SO$_2$Cl, pyridine	92	534
4	CaBr$_2$, MeO[CH$_2$]$_2$OH, 50°	72	534
5	Me$_2$NH, C$_6$H$_6$, r.-t.	53	534
6	KCN, EtOH(aq.), reflux	78	534
7	LiAlH$_4$, Et$_2$O, reflux	56	534
8	NaOH(aq.), reflux	90	534
9	H$_2$O$_2$, Na$_2$CO$_3$, Me$_2$CO(aq.), r.-t.	36	559
10	NaNO$_2$, HCl, dioxan(aq.), r.-t.	35	559

Scheme 26

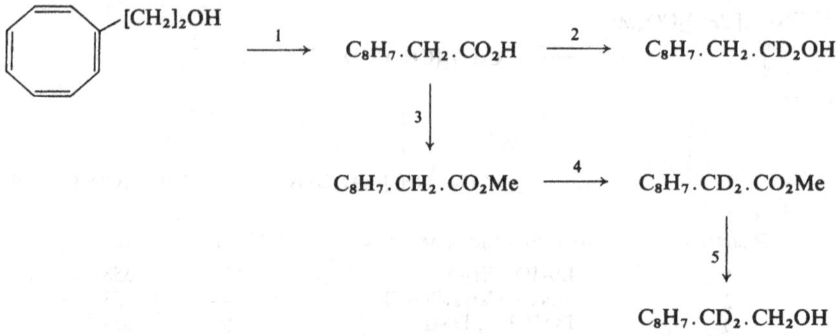

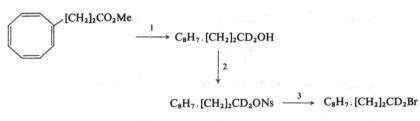

[CH₂]₂OH
1
C₈H₇.CH₂.CO₂H
2
C₈H₇.CH₂.CD₂OH
3
C₈H₇.CH₂.CO₂Me
4
C₈H₇.CD₂.CO₂Me
5
C₈H₇.CD₂.CH₂OH

Reaction	Reagents and conditions	Yield (%)	Ref.
1	CrO_3, H_2SO_4(aq.), Me_2CO	61	546
2	$LiAlD_4$, Et_2O, reflux	61	546
3	CH_2N_2, Et_2O, $-5° \rightarrow$ r.-t.	81 (crude)	546
4	Na, MeOD, 50°	—	546
5	$LiAlH_4$, Et_2O, reflux	—	546

Scheme 27

[CH₂]₂CO₂Me
1
C₈H₇.[CH₂]₂CD₂OH
2
C₈H₇.[CH₂]₂CD₂ONs
3
C₈H₇.[CH₂]₂CD₂Br

Reaction	Reagents and conditions	Yield (%)	Ref.
1	$LiAlD_4$	90	558
2	*p*-$NO_2.C_6H_4.SO_2Cl$	87	558
3	$PyH^+ Br^-$, DMF	70	558

Scheme 28

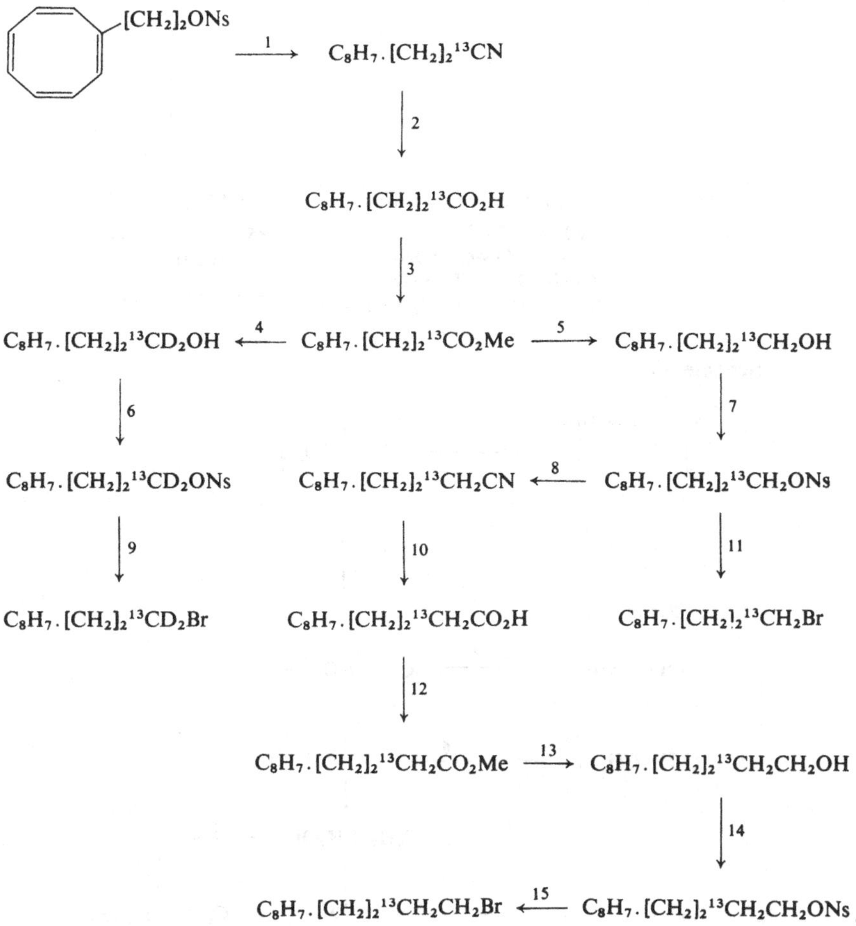

$C_8H_7 \cdot [CH_2]_2ONs \xrightarrow{\;1\;} C_8H_7 \cdot [CH_2]_2{}^{13}CN$

(depicted with cyclooctatetraene ring bearing $[CH_2]_2ONs$)

$\Big\downarrow 2$

$C_8H_7 \cdot [CH_2]_2{}^{13}CO_2H$

$\Big\downarrow 3$

$C_8H_7 \cdot [CH_2]_2{}^{13}CD_2OH \xleftarrow{\;4\;} C_8H_7 \cdot [CH_2]_2{}^{13}CO_2Me \xrightarrow{\;5\;} C_8H_7 \cdot [CH_2]_2{}^{13}CH_2OH$

$\Big\downarrow 6 \qquad\qquad\qquad\qquad\qquad\qquad\qquad\qquad\qquad\qquad\qquad \Big\downarrow 7$

$C_8H_7 \cdot [CH_2]_2{}^{13}CD_2ONs \qquad C_8H_7 \cdot [CH_2]_2{}^{13}CH_2CN \xleftarrow{\;8\;} C_8H_7 \cdot [CH_2]_2{}^{13}CH_2ONs$

$\Big\downarrow 9 \qquad\qquad\qquad\qquad\qquad \Big\downarrow 10 \qquad\qquad\qquad\qquad\qquad\qquad \Big\downarrow 11$

$C_8H_7 \cdot [CH_2]_2{}^{13}CD_2Br \qquad C_8H_7 \cdot [CH_2]_2{}^{13}CH_2CO_2H \qquad C_8H_7 \cdot [CH_2]_2{}^{13}CH_2Br$

$\Big\downarrow 12$

$C_8H_7 \cdot [CH_2]_2{}^{13}CH_2CO_2Me \xrightarrow{\;13\;} C_8H_7 \cdot [CH_2]_2{}^{13}CH_2CH_2OH$

$\Big\downarrow 14$

$C_8H_7 \cdot [CH_2]_2{}^{13}CH_2CH_2Br \xleftarrow{\;15\;} C_8H_7 \cdot [CH_2]_2{}^{13}CH_2CH_2ONs$

Reaction	Reagents and conditions	Yield (%)	Ref.
1	Na^{13}CN, DMF	87	558
2	NaOH(aq.)	87	558
3	CH$_2$N$_2$	95	558
4	LiAlD$_4$	95	558
5	LiAlH$_4$	95	558
6	p-NO$_2$.C$_6$H$_4$.SO$_2$Cl	—	558
7	p-NO$_2$.C$_6$H$_4$.SO$_2$Cl	80	558
8	NaCN, DMF	—	558
9	PyH$^+$ Br$^-$, DMF	97	558
10	NaOH(aq.)	—	558
11	PyH$^+$ Br$^-$, DMF	91	558
12	CH$_2$N$_2$	—	558
13	LiAlH$_4$	—	558
14	p-NO$_2$.C$_6$H$_4$.SO$_2$Cl	—	558
15	PyH$^+$ Br$^-$, DMF	—	558

Scheme 29

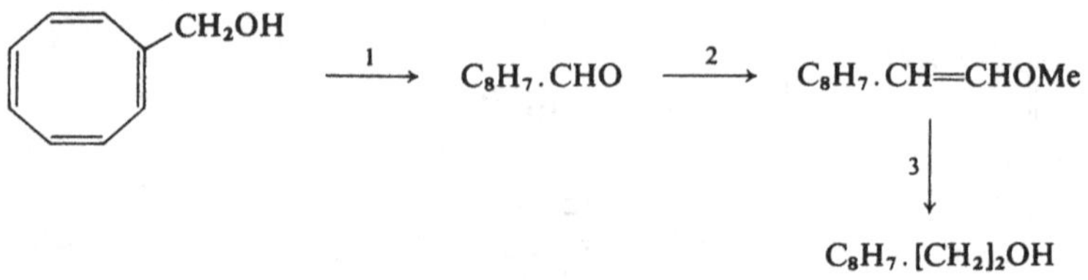

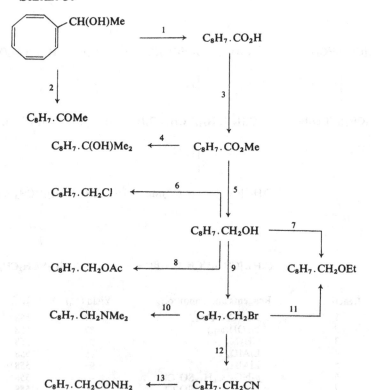

$$\text{CH}_2\text{OH} \xrightarrow{1} \text{C}_8\text{H}_7.\text{CHO} \xrightarrow{2} \text{C}_8\text{H}_7.\text{CH}{=}\text{CHOMe}$$

$$\downarrow 3$$

$$\text{C}_8\text{H}_7.[\text{CH}_2]_2\text{OH}$$

Reaction	Reagents and conditions	Yield (%)	Ref.
1	MnO$_2$–C, CHCl$_3$, r.-t.	85	546
2	Ph$_3$P=CHOMe, EtOH, r.-t.	94 (crude)	546
3	(i) Hg(OAc)$_2$, THF(aq.), r.-t. (ii) NaBH$_4$, NaOH(aq.), r.-t.	91 (crude)	546

Scheme 30

$$\text{CH(OH)Me} \xrightarrow{1} \text{C}_8\text{H}_7.\text{CO}_2\text{H}$$

C$_8$H$_7$.COMe

C$_8$H$_7$.C(OH)Me$_2$ ← C$_8$H$_7$.CO$_2$Me

C$_8$H$_7$.CH$_2$Cl ← C$_8$H$_7$.CH$_2$OH → C$_8$H$_7$.CH$_2$OEt

C$_8$H$_7$.CH$_2$OAc ← C$_8$H$_7$.CH$_2$NMe$_2$ ← C$_8$H$_7$.CH$_2$Br

C$_8$H$_7$.CH$_2$CONH$_2$ ← C$_8$H$_7$.CH$_2$CN

Reaction	Reagents and conditions	Yield (%)	Ref.
1	NaOBr(aq.), 10° → r.-t.	37	535
2	CrO$_3$, H$_2$SO$_4$(aq.), Me$_2$CO, 0°	53	535
3	CH$_2$N$_2$, Et$_2$O, r.-t.	89	549
4	MeMgI, Et$_2$O	16	535
5	LiAlH$_4$, Et$_2$O, r.-t.	99	210, (560)

6	{ SOCl$_2$, pyridine, CHCl$_3$, r.-t.	33	210, (560)
	{ or *N*-chlorosuccinimide, Me$_2$S, CH$_2$Cl$_2$,		
	{ $-30^\circ \to 0^\circ$	77	210
7	KOBu^t, EtBr, THF, r.-t.	75	210
8	—	—	561
9	PBr$_3$, pyridine, hexane, $0^\circ \to$ r.-t.	65	559
10	Me$_2$NH, C$_6$H$_6$, r.-t.	43	559
11	NaOEt, EtOH, r.-t.	95	210
12	KI, KCN, THF(aq.), reflux	44	559
13	H$_2$O$_2$, Na$_2$CO$_3$, Me$_2$CO(aq.), r.-t.	74	559

Scheme 31

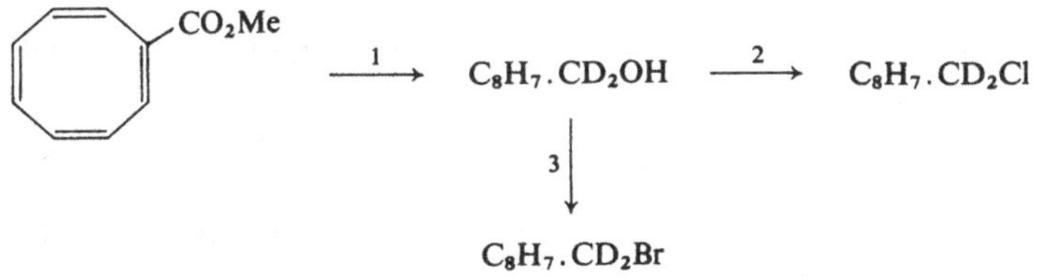

$$\text{(cyclooctatetraene)—CO}_2\text{Me} \xrightarrow{1} \text{C}_8\text{H}_7 . \text{CD}_2\text{OH} \xrightarrow{2} \text{C}_8\text{H}_7 . \text{CD}_2\text{Cl}$$

$$\xrightarrow{3} \text{C}_8\text{H}_7 . \text{CD}_2\text{Br}$$

Reaction	Reagents and conditions	Yield (%)	Ref.
1	LiAlD$_4$, Et$_2$O, r.-t.	76	210, (558), (560)
2	SOCl$_2$, pyridine, CHCl$_3$, r.-t.	35	210, 560
3	PBr$_3$	94[a]	558

[a] Overall yield from the methyl ester.

Scheme 32

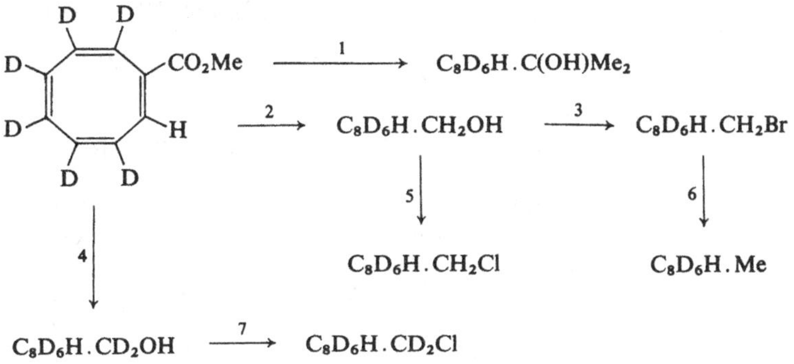

Reaction	Reagents and conditions	Yield (%)	Ref.
1	MeMgI	—	123
2	LiAlH$_4$, Et$_2$O, $0^\circ \to 25^\circ$	64	49
3	Ph$_3$PBr$_2$, DMF, 25°	84	49
4	LiAlD$_4$, Et$_2$O, r.-t.	95	210
5	*N*-Chlorosuccinimide, Me$_2$S, CH$_2$Cl$_2$, $-30^\circ \to 0^\circ$	60	210
6	LiAlH$_4$, Et$_2$O, reflux	67	49
7	*N*-Chlorosuccinimide, Me$_2$S, CH$_2$Cl$_2$, $-30^\circ \to 0^\circ$	65	210

Scheme 33

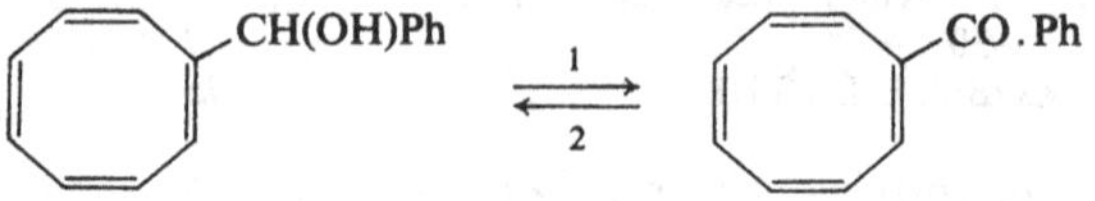

Reaction	Reagents and conditions	Yield (%)	Ref.
1	CrO_3-pyridine, C_6H_6, r.-t.	66	548
2	$LiAlH_4$, Et_2O, reflux	89	544

Scheme 34

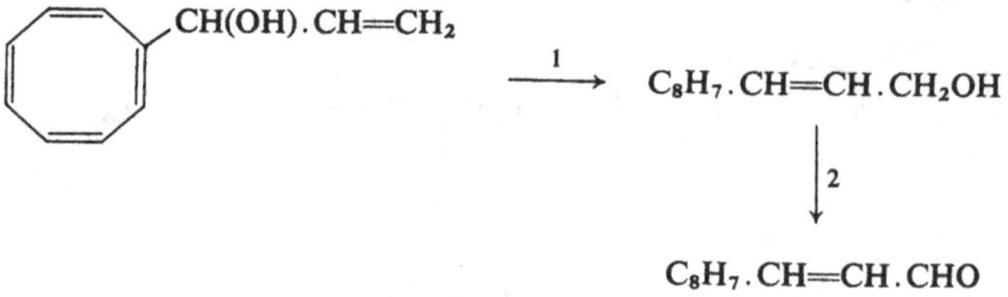

Reaction	Reagents and conditions	Yield (%)	Ref.
1	HCl, Me_2CO(aq.), r.-t.	56	548
2	CrO_3-pyridine, C_6H_6, r.-t.	40	548

Scheme 35

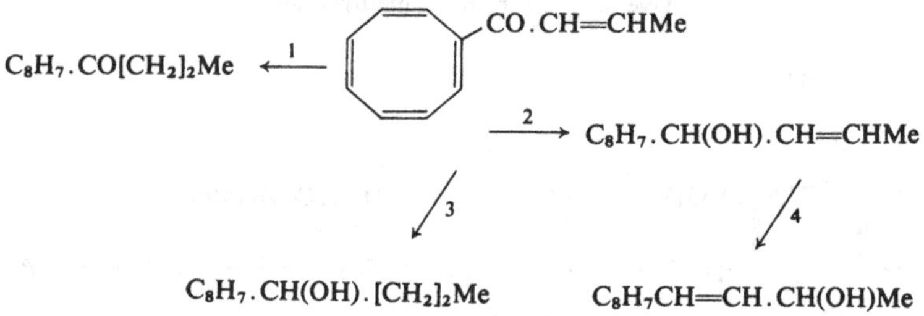

Reaction	Reagents and conditions	Yield (%)	Ref.
1	$LiAlH_4$, Et_2O, 0°	67	548
2	$Al(OPr^i)_3$, toluene, reflux	8	548
3	KBH_4, MeOH(aq.), reflux	85	548
4	HCl, Me_2CO(aq.), r.-t.	87	548

Scheme 36

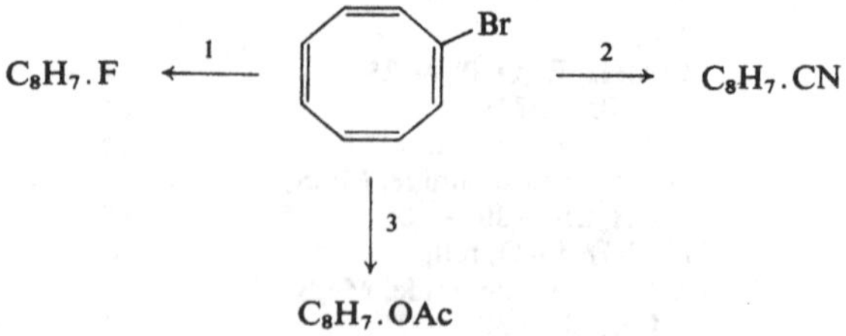

Reaction	Reagents and conditions	Yield (%)	Ref.
1	AgF, pyridine, r.-t.	67	562
2	$K_4Ni_2(CN)_6$, KCN, MeOH, r.-t.	8	541
3	AgOAc, HOAc, 80°	39	540

Structure

The bond lengths and angles in the eight-membered ring of cyclo-octatetraene-carboxylic acid (obtained by X-ray crystallography[563]) are tabulated (table 25).

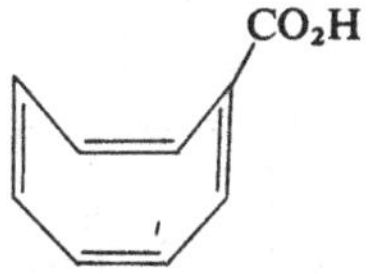

Table 25

	Bond	Length (Å)
	C=C	1.322
	C—C	1.470
Angle C(=C)—C		126°
Torsion angle around =C—C=		57°

A monosubstituted COT possessing a *rigid* non-planar eight-membered ring should be capable of existing in two enantiomeric forms, (**227*a***) and (**227*b***), and early attempts were made to effect the separation of such forms (see e.g. refs. 285, 549). When rapid ring-inversion and/or bond-shift processes are

Scheme 37

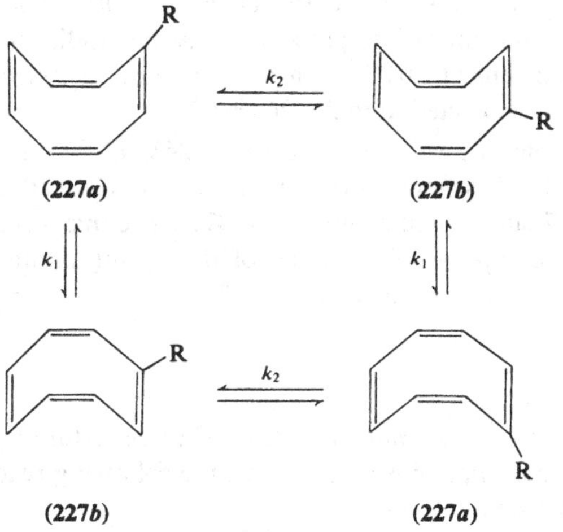

Table 26. Ring-inversion

R	Temp. (°C)	k_1 (s^{-1})	$\Delta G^{\pm}$(kcal mol^{-1})	Ref.
CHMe$_2$	−25	—	14.8	564
C(OH)Me$_2$[a]	−2	7.8	14.7	123
OEt	0	604	12.5	104
OCHMe$_2$	0	361	12.7	104

[a] Substituent group in the hexadeuteriocyclo-octatetraene (228).

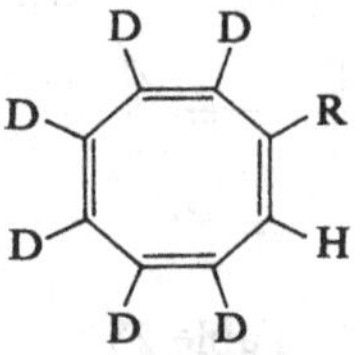

(228)

Table 27. Bond-shift

R	Temp. (°C)	k_2 (s^{-1})	$\Delta G^{\pm}$(kcal mol^{-1})	Ref.
C(OH)Me$_2$[a]	−2	*ca.* 0.1	*ca.* 17.1	123
C(OH)Me$_2$[a]	41	5.4	17.4	123
CO$_2$CH$_2$Me[a]	40	126	15.3	123
OMe	0	(0.45), 0.61	16.2	(554), 104
OEt	0	0.88	16.0	104
OCHMe$_2$	0	1.86	15.6	104
OCMe$_3$	0	(6), 7.29	14.9	(554), 104
F	−33	*ca.* 44	*ca.* 12	550

[a] Substituent group in the hexadeuteriocyclo-octatetraene (228).

operating (see scheme 37), such attempts have no chance of success.

Ring-inversion and bond-shift processes may be studied in suitable substituted COTs by means of variable-temperature n.m.r. spectroscopy, and the available results are collected in tables 26 and 27.

Four different bicyclic valence tautomers, (242)–(245), are possible for a monosubstituted COT (scheme 38); however, Diels–Alder adducts are mostly derived from the 7-substituted compound (242), since this reacts more rapidly than the others (see pp. 101–2). (Values of the Eyring parameters for (227; R = Ph) → (242; R = Ph) are given in ref. 135.)

Reactions

(a) *Functional group transformations.* Most of these naturally lead to other substituted COTs (see pp. 86–92); however, the following reactions result in derivatives which are not COTs.

Scheme 38

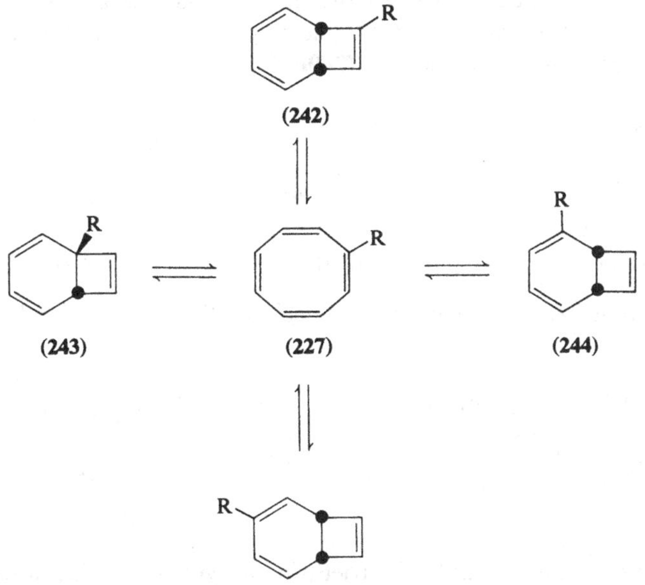

(242)

(243) (227) (244)

(245)

Hydrolysis of the chloromethyl-derivative (246) in buffered aqueous ethanol leads to a mixture of the methylenecyclo-octatrienes (247) (*ca.* 66%) and (248) (*ca.* 22%); these products are extremely acid-sensitive, rearranging to *o*-tolylacetaldehyde.[210]

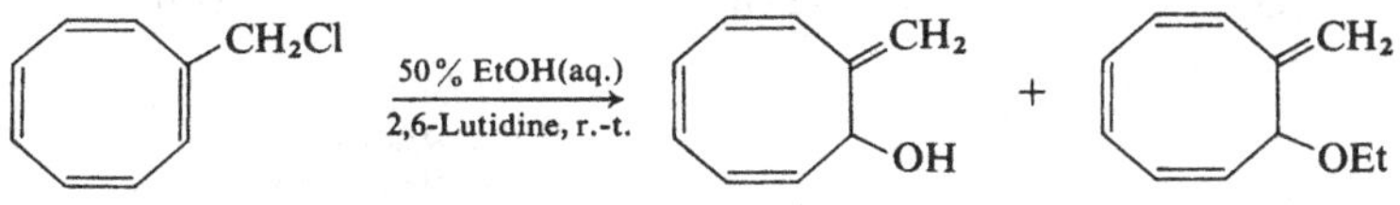

(246) $\xrightarrow[\text{2,6-Lutidine, r.-t.}]{\text{50% EtOH(aq.)}}$ (247) + (248)

The operation of π-bond participation in solvolytic reactions of suitable monosubstituted COTs is revealed by studies using the sulphonate esters (249).

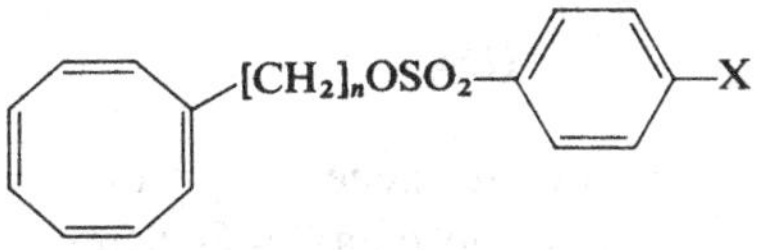

(249)

a: $n = 2$, X = Br
b: $n = 2$, X = NO$_2$
c: $n = 3$, X = NO$_2$
d: $n = 4$, X = NO$_2$

The acetolysis (NaOAc–HOAc) of (**249a**) or (**249b**) gives rise to varying amounts of the tetrahydroazulene (**250**), the dihydronaphthalene (**251**), and other products,[545-547] and that of (**249d**) to the bicyclic tetraene (**252**) etc.[547] (very little rearrangement is observed with (**249c**)[547]).

Acid-catalysed rearrangement of the alcohol (**253**) gives the product (**254**) (83%).[548]

Treatment of the nitrile (**255**) with triethylamine yields a product formulated as either (**256**) or (**257**) (87%), which may be converted into the amide (**258**) or (**259**) (32%).[559]

N,N-Dimethylaminocyclo-octatetraene is hydrolysed extremely readily in the presence of acids, forming cyclo-octa-2,4,6-trienone.[541]

With phenylmagnesium bromide, the benzoyl-derivative (260) undergoes 1,4-addition and subsequent ring-opening, with the formation of the acyclic ketone (261) (20%).[544]

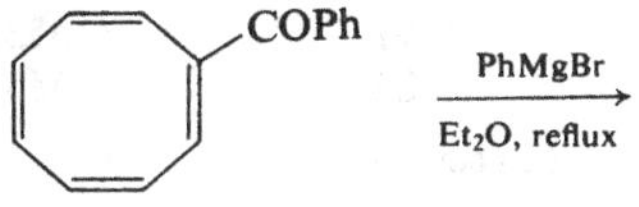

(260) → [PhMgBr / Et₂O, reflux] → Ph[CH=CH]₄COPh

(261)

Treatment of chlorocyclo-octatetraene with sodium iodide gives *E-β*-iodo-styrene (262) (72%)[553] (*cf.* the thermal rearrangement of halocyclo-octa-tetraenes, below).

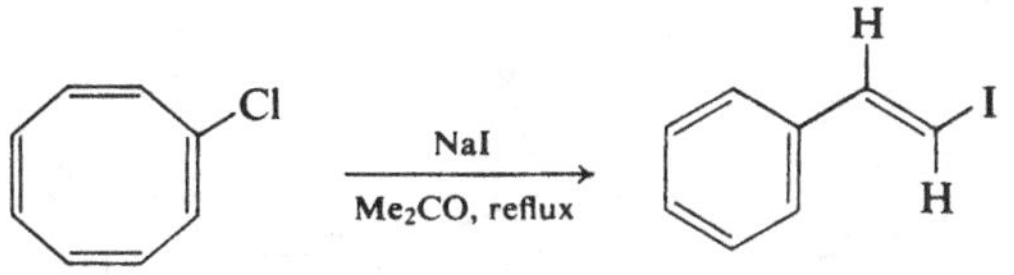

(262)

(*b*) *Thermolysis.* The operation of a degenerate rearrangement in certain mono-substituted COTs when subjected to a temperature of *ca.* 550° (flow system) is revealed by protium scrambling in the hexadeuterio-derivatives (228)[49] (see p. 80). (For carbon and hydrogen scrambling in the mass spectrometry of monosubstituted COTs, see ref. 558.)

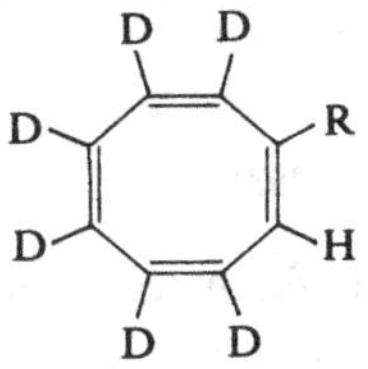

(228)

R = Me, Ph, CO₂Me

The acetoxy-, fluoro-, chloro- and bromo-derivatives of COT undergo thermal rearrangement to give *E-β*-substituted styrenes (263) (table 28); the isomerisation is accelerated by acids.[566] The mechanism of the rearrangement

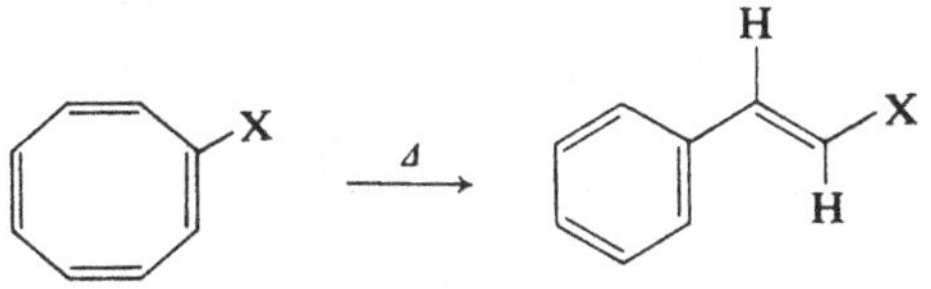

(263)

Table 28

X	Temp. (°C)	Yield (%)	Ref.
OAc	120[a]	100	565
F	100	12	562
Cl	200–210	91	553, (565)
Br	90–103	84	553, (566)

[a] In the presence of acetic acid.

has been elucidated,[566, 567] and is illustrated in scheme 39. Ionisation of the valence tautomer (264) generates a cation which may be formulated as a homocyclopropenium ion (265).* Ion-recombination can give the strained

Scheme 39

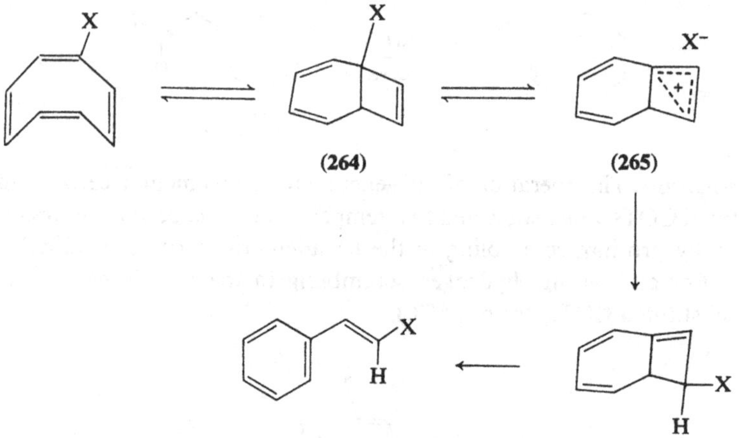

(264)   (265)

(263)   ←   (267)

cyclobutene derivative (267), in which conrotatory ring-opening leads to the *E*-styrene (263) (very little of the *Z*-isomer is produced).

In contrast, *N,N*-diethylaminocyclo-octatetraene rearranges to give the α-substituted styrene (268).[541]

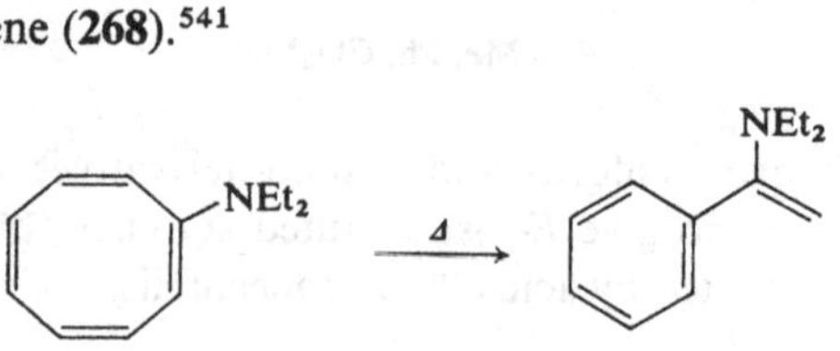

(268)

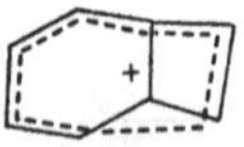

(266)

* Or, alternatively, as the homotropylium ion (266).[562]

(c) *Reactions with electrophiles.* Protonation of methyl- or phenylcyclo-octatetraene occurs preferentially on the 'inside' of the 'tub' structure (*cf.* COT, p. 22), and affords only one isomer (**269**) of the six possible homotropyl-ium ions.[215]

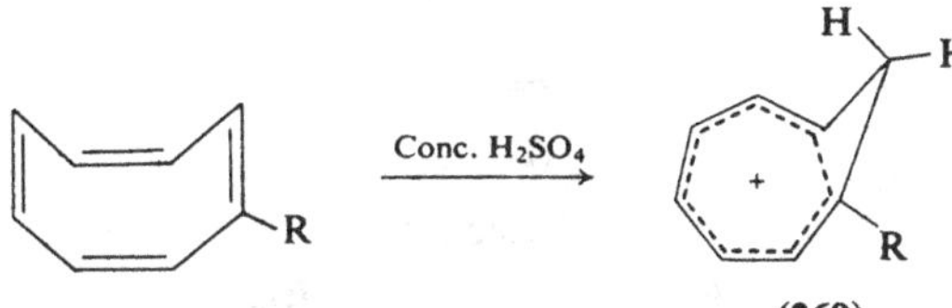

(**269**)

R = Me, Ph

It is proposed that the cation (**270**), the most stable of the possible classical structures, is an intermediate in this conversion (or alternatively, that the transition state resembles (**270**)).[215] Similarly, protonation of methoxycyclo-

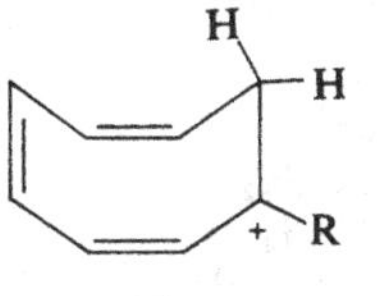

(**270**)

octatetraene gives the homotropylium ion (**271**), but this rearranges at 51° to (**272**), which on hydrolysis yields acetophenone.[568] In FSO_2OD, *ca.* 75%

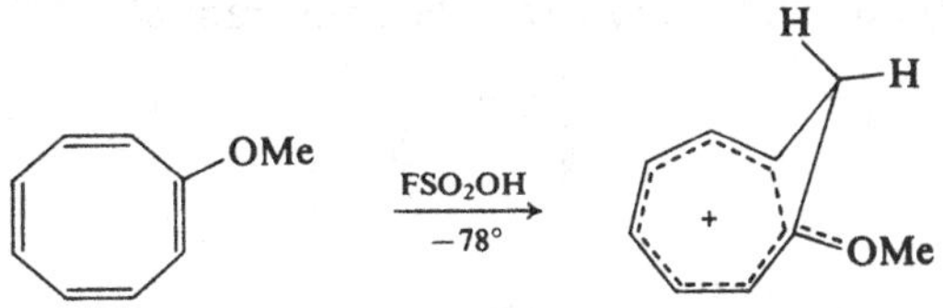

(**271**)

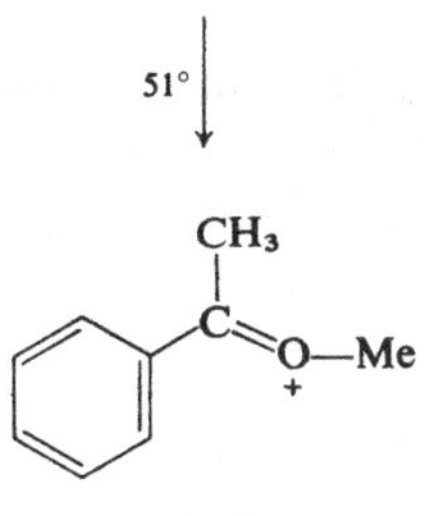

51°

(**272**)

of the resulting homotropylium ions contain deuterium in the *endo*-position.[568]

For the addition of bromine to monosubstituted COTs, see ref. 569 (for the low-temperature bromination of bromocyclo-octatetraene, see refs. 43, 565).

(*d*) *Reduction.* The anion radicals (**273**) derived from monosubstituted COTs have been studied by means of e.s.r. spectroscopy (table 29). For theoretical studies, see also refs. 258, 575.

(**273**)

Table 29

R	Ref.
D	250
Me	247, 570
Et	247, 537, 570, 571
Pr^n	247, 570
Bu^n	247, 570
[a]Ph	80, 572–574
p-Me.C_6H_4	539
p-Bu^t.C_6H_4	539
p-Ph.C_6H_4	539
2,4,6-Me_3.C_6H_2	539
p-Me_2N.C_6H_4	539
Cyclo-octatetraenyl	574
OBu^t	537, 571

[a] For the deuteriated species (**273**; R = C_6D_5), see ref. 537.

Bromocyclo-octatetraene reacts with potassium in tetrahydrofuran-hexamethylphosphoramide to generate the anion radical (**274**).[576]

(**274**)

For the polarographic reduction of methoxycyclo-octatetraene, see refs. 75, 577.

The reported monosubstituted COT^{2-} species (**275**) are listed in table 30.

(**275**)

Table 30

R	Ref.
Me	578
Ph	80
OMe	578
$OCHMe_2$	579

Examples of the catalytic hydrogenation of monosubstituted COTs to give cyclo-octane derivatives (276) are listed in table 31; in some instances the intermediate cyclo-octene (277) may be isolated (see p. 30).

$$\text{(227)} \quad \xrightarrow[\text{Catalyst}]{H_2} \quad \text{(277)} \quad \longrightarrow \quad \text{(276)}$$

Table 31

R	Catalyst and solvent	Yield (%)	Ref.
Me	PtO_2,[a] HOAc	53	533
Et	PtO_2,[a] HOAc	69	286
Pr^n	PtO_2,[a] HOAc	64	533
Bu^n	PtO_2,[a] HOAc	60	(286), 533
$CH{=}CH_2$	PtO_2,[a] HOAc	86, (64)[b]	24, (25)
Cyclo-octatetraenyl	PtO_2,[a] HOAc	67	544
Ph	10% Pd–C, MeOH	80	285
$p\text{-}Me_2N.C_6H_4$	10% Pd–C, MeOH	61	285
$[CH_2]_2OH$	PtO_2,[a] EtOH	84	534
CH_2OH	PtO_2,[a] EtOH	94	534
$CH(OH)Ph$	10% Pd–C, EtOH	67[c]	548, (549)
COPh	10% Pd–C, EtOH	38	549
COPh	1% Pd–$CaCO_3$, EtOH	66[d]	549
CO_2H	10% Pd–C, EtOH	97.5[e]	549
Cl	PtO_2,[a] HOAc, NaOAc	72[f]	553
Br	PtO_2,[a] HOAc, NaOAc	—[f]	553

[a] Pre-reduced.   [b] Product (276; R = Et).   [c] Product (276; R = CH_2Ph).   [d] Product (277; R = COPh).   [e] Product (277; R = CO_2H).   [f] Product (276; R = H).

(e) *Cycloadditions.* The [2 + 2] cycloaddition of certain monosubstituted COTs to COT itself produces very low yields of the 'mixed' dimers (278) (table 32), convertible into monosubstituted [16]annulenes (see p. 359).[160, 580]

The dimerisation of (227; R = CO_2Me, OMe) occurs at 100–150°, but the products have not been characterised.[160]

In principle, monosubstituted COTs could undergo [4 + 2] cycloadditions *via* any of the four possible bicyclic valence tautomers (242)–(245) (see p. 95). In most cases, however, the adducts are predominantly those of the 7-substituted bicyclo[4.2.0]octa-2,4,7-triene (242), steric factors presumably resulting

(227)   (278)

Table 32

R	Temp. (°C)
Ph	100
CO$_2$Me	100
F	70
Cl	100

in a slower rate of addition for the 1-, 2- and 3-substituted compounds. Thus with e.g. maleic anhydride, monosubstituted COTs afford adducts of structure (279). The known reactions of this general type are listed in table 33. (For secondary deuterium isotope effects in Diels–Alder reactions of the hexa-

Table 33 (refers to the equation at the top of p. 103)

R	Dienophile	Conditions	Yield (%)	Ref.
Me	Maleic anhydride	{ Benzene, reflux	55	543
		{ Toluene, reflux	30	533
Me	Tetracyanoethylene	EtOAc, reflux	61	49
Et	Maleic anhydride	165–170°	60	533
Et	Tetracyanoethylene	—	—	567
Cyclopropyl	Maleic anhydride	Toluene, reflux	—	542
Ph	Fumaroyl chloride	EtOAc or THF, 70°	—	135
Ph	Maleic anhydride	Benzene, reflux	92, (91), [69]	135, (285), [542]
Ph	Tetracyanoethylene	{ EtOAc or THF, 70°	—	135
		{ EtOAc, reflux	58	49
Ph	Dicyanomaleimide	EtOAc or THF, 70°	—	135
Ph	*p*-Benzoquinone	Benzene, reflux	(76), 81	(285), [536], 581
CO.CH═CHMe	Maleic anhydride	*o*-Cl$_2$.C$_6$H$_4$, reflux	41	548
CO.Ph	Maleic anhydride	*o*-Cl$_2$.C$_6$H$_4$, reflux	70	548
CO$_2$Me	Maleic anhydride	Xylene, reflux	64 (crude)	542
CO$_2$Me	Tetracyanoethylene	Dioxan, reflux	62	49, (567)
OMe	Maleic anhydride	{ Benzene, 70°	99	582
		{ r.-t.	12[a]	542
OAc	Maleic anhydride	Benzene, reflux	59	543
OAc	Tetracyanoethylene	—	—	567
F	Tetracyanoethylene	EtOAc, reflux	48	562
Cl	Tetracyanoethylene	—	—	567
Br	Tetracyanoethylene	—	—	567

[a] The isolated product was the keto-anhydride (280).

(242)   (279)

deuterio-compounds (228; R = Me, Ph, CO_2Me, CN), see ref. 583.) Vinyl-cyclo-octatetraene forms a 1:2 adduct with maleic anhydride at 140–160°

(280)   (228)

(yield 23%),[26] but the structure of the product is unknown. Phenylcyclo-octatetraene is reported to react with dimethyl acetylenedicarboxylate at 130°.[581]

(281)   (282)

(283)   (284)

104   *Substituted cyclo-octatetraenes*

Table 34

Yield (%)

R	(282)	(283)	(284)
Me	—	18	78
Ph	74	—	17

In contrast with the above reactions which proceed *via* a bicyclic valence tautomer, methoxy- and phenoxycyclo-octatetraene react directly with tetracyanoethylene (at room-temperature) *via* the homotropylium zwitterion (**281**) (which presumably gains additional stabilisation from the OR group), affording 1,2- and/or 1,4-addition products (**282**)–(**284**)[582] (table 34). (For photolysis of (**284**; R = Me), see ref. 582.)

With dicyanoacetylene, two different types of [4 + 2] adduct, (**285**) and (**286**) have been obtained[562] (table 35).

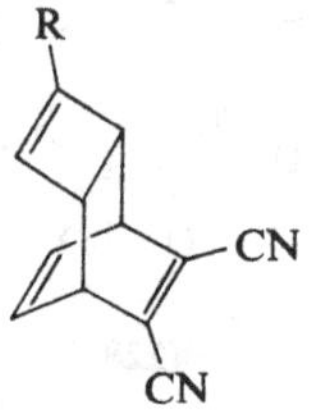

(285)

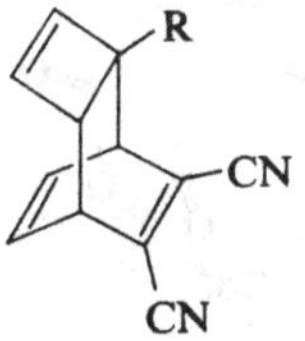

(286)

Table 35

Yield (%)

R	Conditions	(285)	(286)
F	Et₂O, r.-t.	—	0.9
Cl	90°	9	—

With 1,2,4-triazoline-3,5-diones, the isomers (**287**)–(**290**) may be produced, in some cases accompanied by products of 1,4-addition (**291**) (*cf.* pp. 44–6) (table 36).

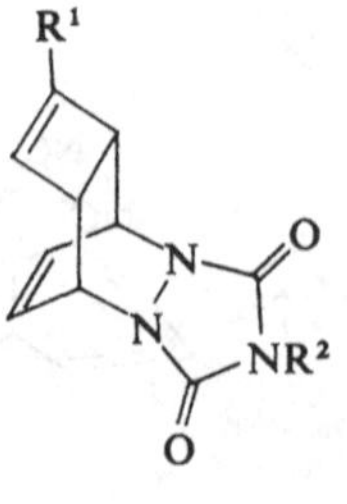

(287)

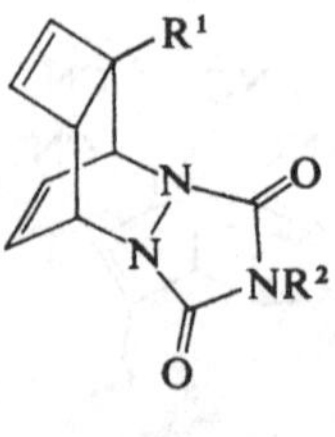

(288)

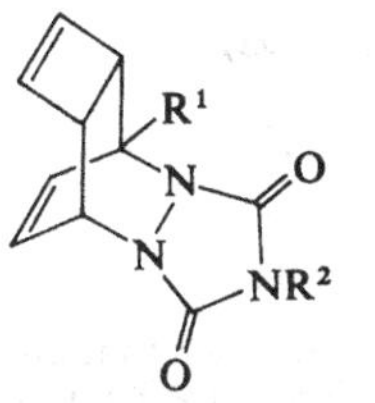

(289) 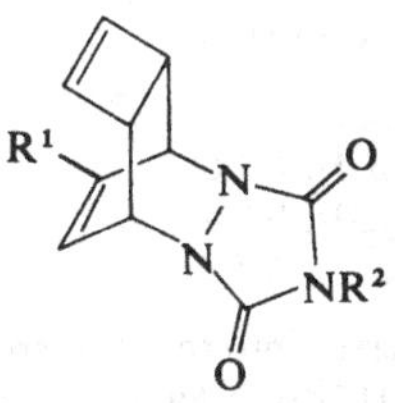   (290) 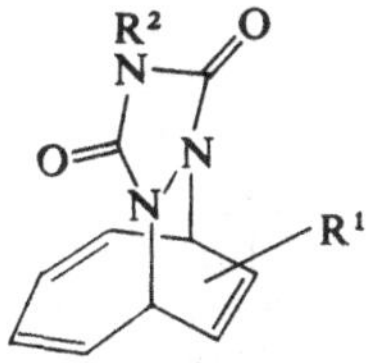   (291)

Table 36

| | | | Yield (%) | | | | | |
R¹	R²	Conditions	(287)	(288)	(289)	(290)	(291)	Ref.
Me	Ph	EtOAc, reflux	11	—	—	—	11ª	569
Ph	Me	EtOAc, reflux	84	—	1	—	—	569
CH₂OMe	Ph	EtOAc, reflux	42	14	7.5	—	17ª	569
CH₂OAc	Ph	EtOAc, reflux	32	—	—	—	10ᵇ	569
CN	Ph	EtOAc, reflux	26	25	31	2–4	—	569
CO₂Me	Ph	EtOAc, reflux	8	—	27	2	10ᶜ	569
F	Me	EtOAc, 65°	—	62	—	—	—	562
F	Ph	EtOAc, reflux	—	95	—	—	—	562
Cl	Ph	EtOAc, 60°	27	30	—	—	—	567
Br	Ph	EtOAc, reflux	12	25	—	—	10ᶜ	567
I	Ph	EtOAc, 80°	13	—	—	—	8.5	364
I	Ph	EtOAc, 20°	—	—	—	—	12	364

ª Two isomers.   ᵇ Three isomers.   ᶜ Only one isomer detected.

1,4-Addition occurs with chlorosulphonyl isocyanate (*cf.* COT, pp. 45–6), with the formation of the isomeric products (292) and (293)[367] (table 37).

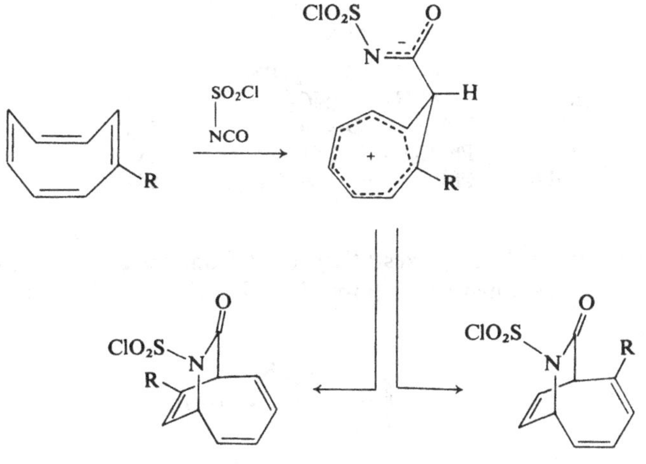

(292)   (293)

Table 37

R	Conditions	Product ratio[a] (292) : (293)	
Me	CH_2Cl_2, 0° → 20°	65	35
Ph	CH_2Cl_2, reflux	80	20
p-MeO.C_6H_4	CH_2Cl_2, 0° → r.-t.	45	55
OMe	CH_2Cl_2, −78°	17	83

[a] The initial products were converted into the corresponding lactams, which were characterized as the imino-ethers (294) and (295); when R = OMe, two additional products were the ketones (296) and (297).

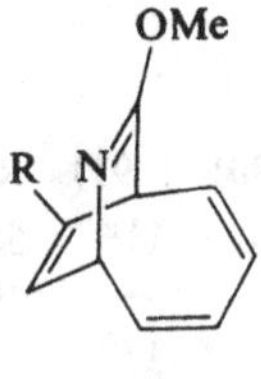

(294) 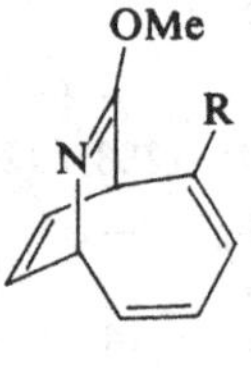   (295) 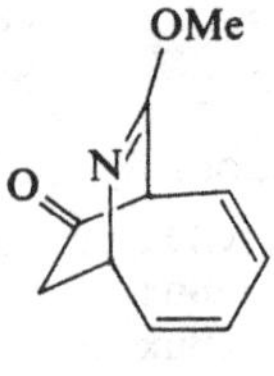   (296) 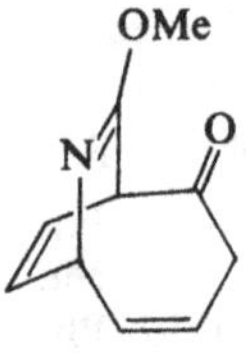   (297)

(*f*) *Formation of metal derivatives.* Silver nitrate complexes of monosubstituted COTs (227) are readily formed in boiling ethanol (see p. 51) (table 38).

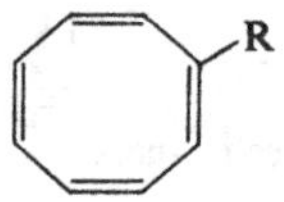

(227)

Table 38

R	Complex	Ref.
Me	2(Me.C_8H_7).3AgNO$_3$	533
Et	Et.C_8H_7.2AgNO$_3$	533
Pr^n	Pr^n.C_8H_7.2AgNO$_3$	533
Ph	Ph.C_8H_7.AgNO$_3$	285
COPh	PhCO.C_8H_7.AgNO$_3$	544

The iron tricarbonyl complexes (298) result from the action of iron carbonyls on monosubstituted COTs (table 39). N.m.r. studies indicate that for

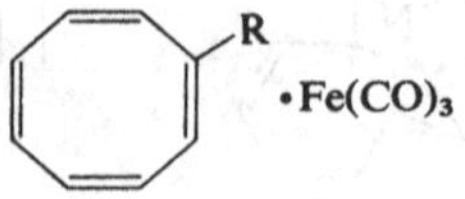

(298)

Table 39

R	Reagents and conditions	Yield (%)	Ref.
Me	Fe$_2$(CO)$_9$, hexane, reflux	49	584, (585)
CH$_2$D[a]	Fe$_2$(CO)$_9$, hexane, reflux	—	586
Et	—	—	585
CPh$_3$	—	—	585, 587
Ph	{ Fe$_2$(CO)$_9$, hexane, reflux	50	588
	{ Fe$_3$(CO)$_{12}$, benzene, reflux[b]	60	(585), 590
CO$_2$Me	Fe$_2$(CO)$_9$, hexane, reflux	80	(585), 588, (591)
OMe	Fe$_2$(CO)$_9$, Et$_2$O, reflux	32[c]	588
Br	—	—	587
SiMe$_3$	Fe$_2$(CO)$_9$, heptane	40[d]	551
SiMe$_2$(allyl)	Fe$_2$(CO)$_9$, heptane	60	551
GeMe$_3$	Fe$_2$(CO)$_9$, heptane	60	551
SnMe$_3$	Fe$_2$(CO)$_9$, heptane	10[d]	551
Fe(CO)$_2$(CPD)	Fe$_2$(CO)$_9$, heptane	7	551

[a] Substituent group in (228).
[b] Photoactivation of the reaction with Fe$_2$(CO)$_9$ or Fe$_3$(CO)$_{12}$ in benzene may also be used; with Fe(CO)$_5$ (no solvent) a binuclear complex (299) is an additional product.[589]
[c] The cyclo-octatrienone complexes (300) (10%) and (301) (5%) are also formed.
[d] A binuclear complex (299) (*ca.* 2%) is also formed.[551]

(228)

(299)

(300)

(301)

(302*a*)

(302*b*)

the complex (298; R = Me), in the most favourable arrangement of the co-ordinated diene system the methyl group is attached to one of its internal carbon atoms, and that the valence tautomerism (302a) ⇌ (302b) is rapid at room temperature.[592, 593] Examination of the deuteriated derivative (303a ⇌ 303b) has provided confirmation.[586] In contrast, electron-withdrawing groups tend to repel the $Fe(CO)_3$ group.[593]

(For mass spectra of iron tricarbonyl complexes (298), see ref. 594.)

(303a)   (303b)

($\Delta G^{\ddagger} = 7.5$ kcal mol)

Amongst the identified products resulting from the reactions of methyl- or phenylcyclo-octatetraene with $Ru_3(CO)_{12}$ or $[Ru(SiMe_3)(CO)_4]_2$ are the pentalene complexes (304), of which the 2-substituted compounds (304a) are fluxional but the 1-substituted isomers (304b) are not. [595]

(304)

a: $R^1 = H$; $R^2 = Me$, Ph
b: $R^1 = Me$, Ph; $R^2 = H$

The reaction of bicyclo-octatetraenyl with $Ru_3(CO)_{12}$ gives the complex (305) (see appendix).

$\xrightarrow[\text{Toluene, reflux}]{Ru_3(CO)_{12}}$ (Bicyclo-octatetraenyl)$Ru_3(CO)_6$

(305)

Palladium(II) and platinum(II) complexes of type (306) are formed by treatment of phenylcyclo-octatetraene with sodium tetrachloropalladate or

(306)

Table 40

M	Yield (%)
Pd	41
Pt	24

tetrachloroplatinate[596] (table 40). (Similar complexes have been obtained from the salts Na_2MBr_4, but they have not been characterised.)

2. Disubstituted derivatives

Formation

(a) *'Mixed' cyclotetramerisation of acetylenes.* Co-reaction of acetylene and disubstituted acetylenes may be effected in the presence of nickel catalysts to afford 1,2-disubstituted COTs (307) in low yield (table 41).

(307)*

Table 41

R^1	R^2	Yield (%)	Ref.
Me	Me	19	533
Ph	Ph	14	532
CH_2OAc	CH_2OAc	22[a]	597
CO_2Me	CO_2Me	10[a]	597

[a] The initial product was hydrolysed before purification.

A similar reaction involving two molecules of acetylene and two of phenyl-acetylene gives 1,4- and 1,5-diphenylcyclo-octatetraene (together with other isomers).[598]

(b) *Photochemical cycloadditions of disubstituted acetylenes to benzene.* The u.v. irradiation of certain disubstituted acetylenes in benzene produces 1,2-disubstituted COTs (table 42); in one case, evidence for the intermediacy of a bicyclo[4.2.0]octatriene has been provided.[38]

* The existence of bond-shift isomerism in 1,2-, 1,3- and 1,4-disubstituted COTs complicates their structural representation (see pp. 115–17) and their nomenclature.

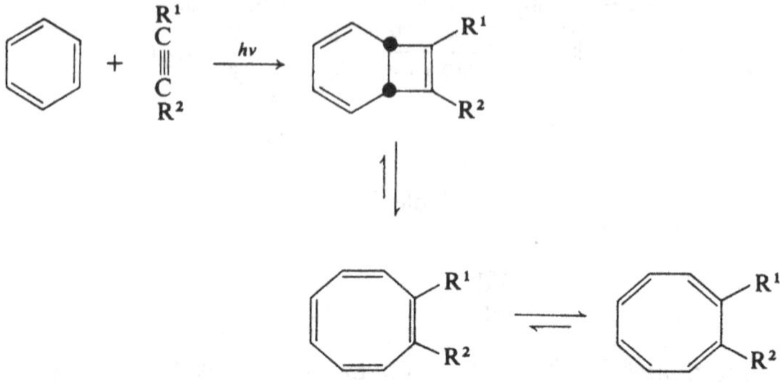

(307)

Table 42

R¹	R²	Yield (%)	Ref.
Me	CO₂Me	(45), 50	(546), 599
Ph	CO₂Me	—	38
CF₃	CF₃	(Very low)	600
CO₂Me	CO₂Me	32, (15)	536, (601, 602)

(c) *Syntheses using organo-lithium reagents.* The reaction of phenyl-lithium with phenylcyclo-octatetraene yields all the four possible positional isomers of the disubstituted derivative; the resulting mixture of 1,2-, 1,3-, 1,4- and 1,5-diphenylcyclo-octatetraene is separable by countercurrent distribution.[598]

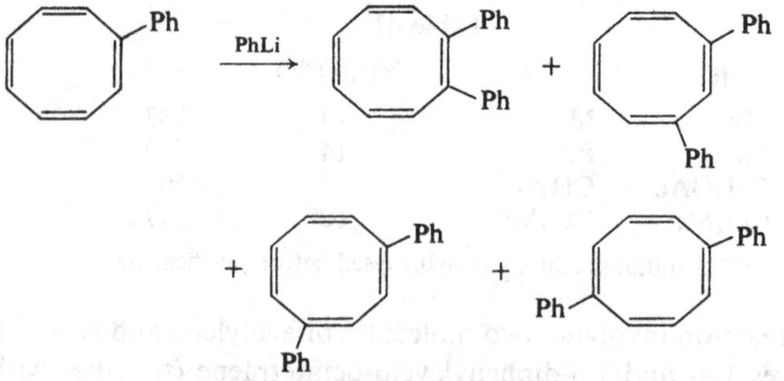

Using a mixture of 1-bromo-4-methyl- and 1-bromo-5-methylcyclo-octatetraene (from methylcyclo-octatetraene, see p. 99), halogen–metal exchange followed by reaction with ethylene oxide leads to the 1,4- and 1,5-disubstituted products (308) and (309) (overall yield from methylcyclo-octatetraene 19%; isomer ratio 3:1).[43,545,546]

1,4-Dimethylcyclo-octatetraene is produced (95%) from the 1,4-dibromo-compound and lithium 'dimethylcuprate'[565] (but see ref. 43).

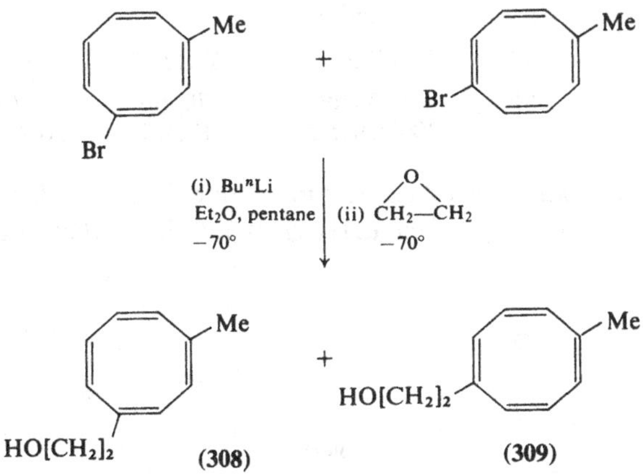

(308) (309)

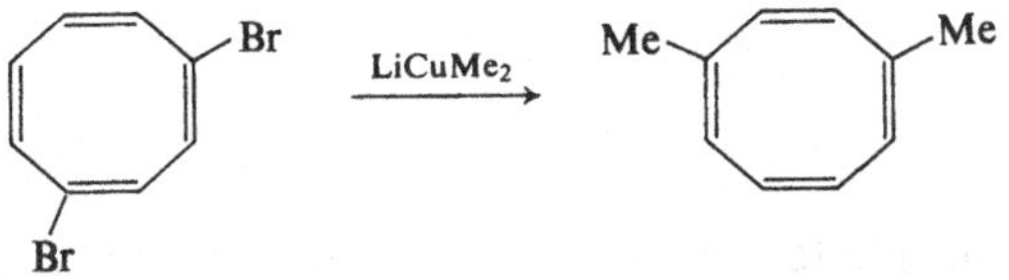

(*d*) *Syntheses involving dehydrohalogenation.* 1,4-Dibromocyclo-octatetraene may be prepared from the monobromo-compound by low-temperature bromination and subsequent dehydrobromination;[565] the product is, however, contaminated with the 1,5-dibromo-compound.[43]

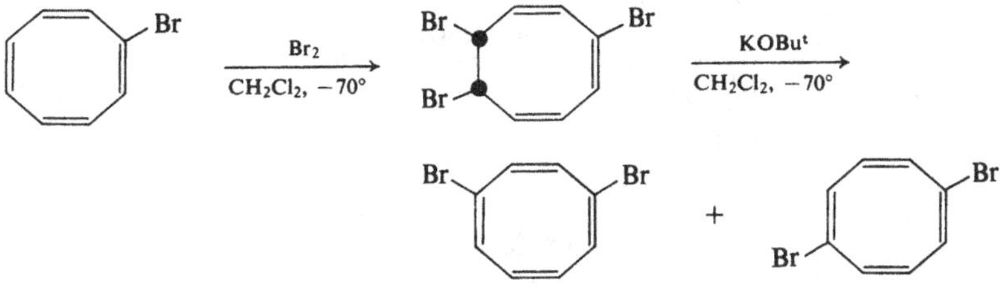

(*e*) *Syntheses involving skeletal transformations.* Disubstituted COTs result from the gas-phase thermolysis of disubstituted semibullvalenes, probably *via* a diradical intermediate; thus 1,3- and 1,5-dimethylcyclo-octatetraene are obtained from the appropriate dimethylsemibullvalenes (**310**) (table 43). (For thermolytic interconversions of disubstituted COTs, see p. 118.)

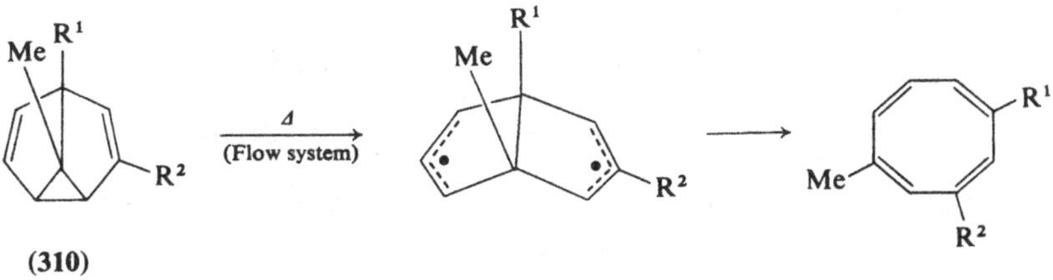

(310)

Table 43

	R^1	R^2	Conditions	Yield (%)	Ref.
a	H	Me	445°/35 mm	78, (69)	43, (603)
b	Me	H	390°/30 mm	(71), 76	(42), 43

The photolytic extrusion of carbon monoxide from the disubstituted 9-oxobicyclo[4.2.1]nona-2,4,7-triene (311) gives 1,2-dimethylcyclo-octatetraene (85%).[604]

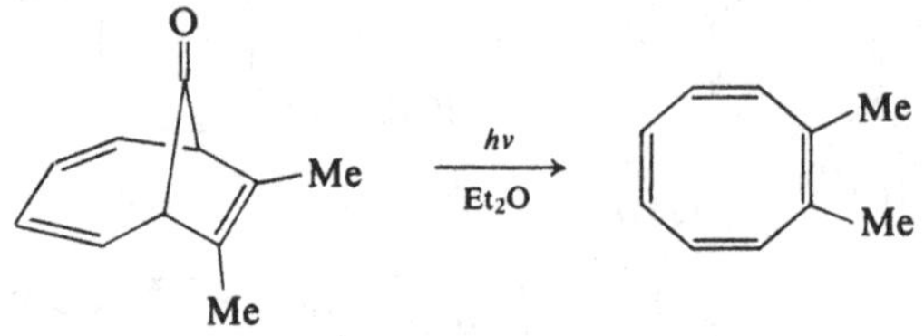

(311)

Similarly, the photolysis of disubstituted 9-thiabicyclo[4.2.1]nona-2,4,7-triene 9,9-dioxides (312) leads to 1,2- or 1,4-dimethylcyclo-octatetraene (table 44).

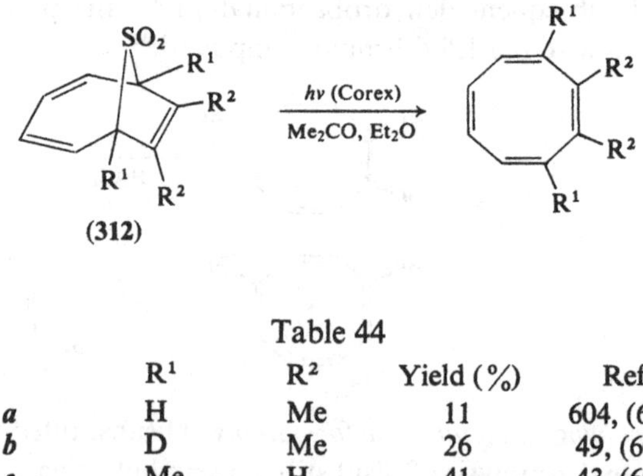

(312)

Table 44

	R^1	R^2	Yield (%)	Ref.
a	H	Me	11	604, (605)
b	D	Me	26	49, (606)
c	Me	H	41	43, (606)

(*f*) *Functional group transformations.* The transformations outlined in schemes 40–43 result in other disubstituted COTs.

Scheme 40

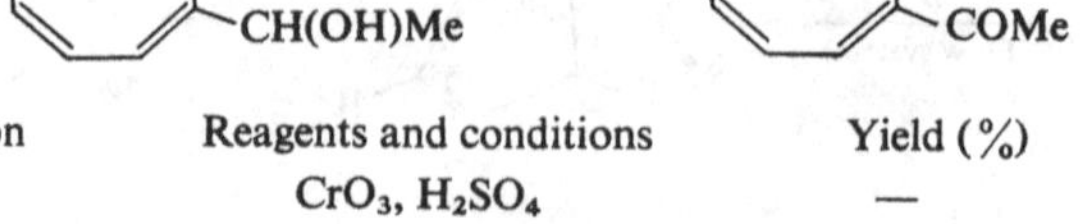

Reaction	Reagents and conditions	Yield (%)	Ref.
1	CrO_3, H_2SO_4	—	607

Scheme 41

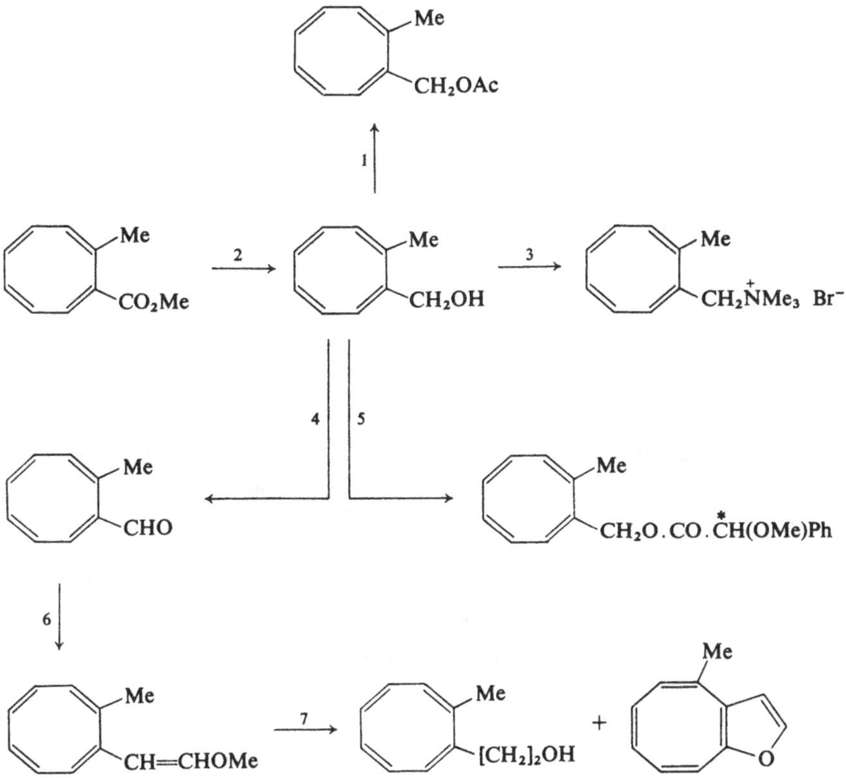

Me

CH₂OAc

1

Me
CO₂Me

2 →

Me
CH₂OH

3 →

Me
CH₂ṄMe₃ Br⁻

4 | 5

Me
CHO

Me
CH₂O.CO.ĊH(OMe)Ph

6 ↓

Me
CH=CHOMe

7 →

Me
[CH₂]₂OH

+

Me

O

Reaction	Reagents and conditions	Yield (%)	Ref.
1	Ac₂O, *ca.* 100°	87	542
2	LiAlH₄, Et₂O, r.-t. or reflux	96–100, (72)	542, 546, (599)
3	{ (i) PBr₃, pyridine, hexane } { (ii) Me₃N, MeOH }	40 (overall)	599
4	MnO₂–C, CHCl₃, r.-t.	93 (crude)	546
5	(−)-O-Methylmandelyl chloride	—	608
6	Ph₃P=CHOMe, Et₂O, reflux	82 (crude)	546
7	{ (i) Hg(OAc)₂, THF(aq.), r.-t. } { (ii) NaBH₄, NaOH(aq.), r.-t. }	28 + 56	546

Structure

A structural peculiarity of 1,2-, 1,3- and 1,4-disubstituted COTs is that bond-shift leads to a non-enantiomeric structure (bond-shift isomer).

For a 1,2-disubstituted COT, bond-shift and ring-inversion processes result in the equilibration of two pairs of enantiomers, (313*a*) and (313*b*), and (314*a*) and (314*b*) (scheme 44). The 'skew' structure (314) is generally favoured,

Scheme 42

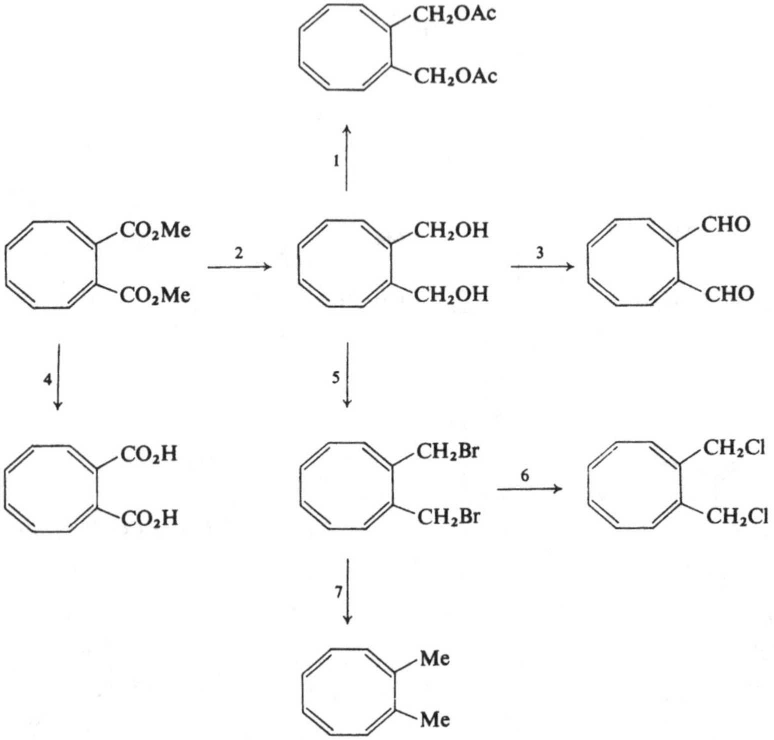

Reaction	Reagents and conditions	Yield (%)	Ref.
1	Ac$_2$O, *ca.* 100°	75	542
2	LiAlH$_4$, AlCl$_3$, r.-t.	82 (crude)	43, (609)
3	MnO$_2$, CHCl$_3$, r.-t.	15	610
	or NiO$_2$	Up to 40	611
4	KOH, EtOH	—	602
5	PBr$_3$, pyridine, Et$_2$O, 0–20°	53	612
	or Ph$_3$PBr$_2$, DMF, r.-t.	78	43
6	LiCl, DMF, r.-t.	90	612
7	LiAlH$_4$, Et$_2$O, reflux	87	43

Scheme 43

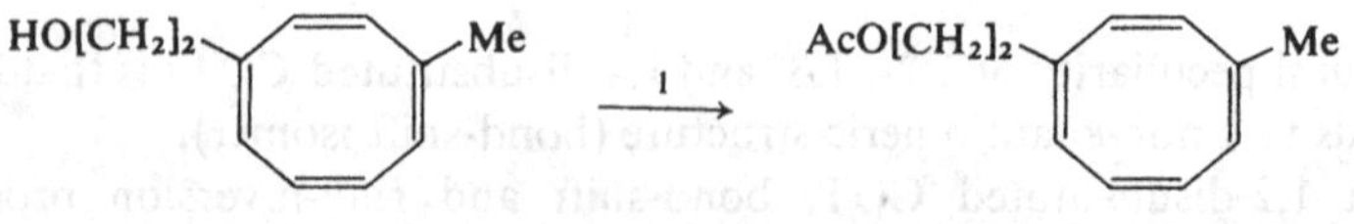

Reaction	Reagents and conditions	Yield (%)	Ref.
1	Ac$_2$O, pyridine, r.-t.	87 (crude)	546

Scheme 44

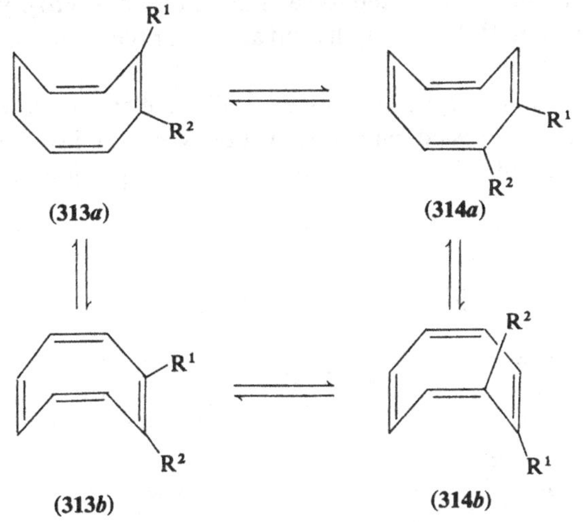

(313*a*) (314*a*)

(313*b*) (314*b*)

(When $R^1 = R^2$, (313*a*) ≡ (313*b*))

presumably to minimise repulsion between the substituent groups. The n.m.r. spectra of compounds (307; $R^1 = Me$, $R^2 = CH_2O.CO.CH(OMe)Ph$),[608] (307; $R^1 = Me$, $R^2 = CO_2Me$),[608] (307; $R^1 = Ph$, $R^2 = CO_2Me$),[613] (307; $R^1 = R^2 = CHO$),[610] (307; $R^1 = R^2 = CO_2H$)[613] and (307; $R^1 = R^2 = CO_2Me$)[602,613] reveal that in each case the 'skew' form predominates in

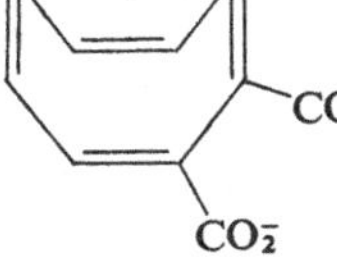

(307)

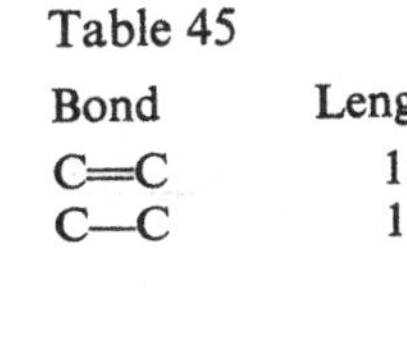

Table 45

Bond	Length (Å)
C=C	1.335
C—C	1.464

Angle   126°

Torsion angle around   57°

solution. The dicarboxylic acid (**307**; $R^1 = R^2 = CO_2H$) forms a cyclic an-hydride only with great difficulty,[613] and its calcium salt is known to crystallise in the 'skew' configuration[614] (bond lengths and angles for the eight-membered ring are given in table 45[614]).

The rate of ring-inversion in 1,2-disubstituted COTs is relatively slow,[599, 608] in accordance with the suggested planar transition state for inversion (see p. 10), and the rate of bond-shift is even slower.[608] The proportion of the less

Scheme 45

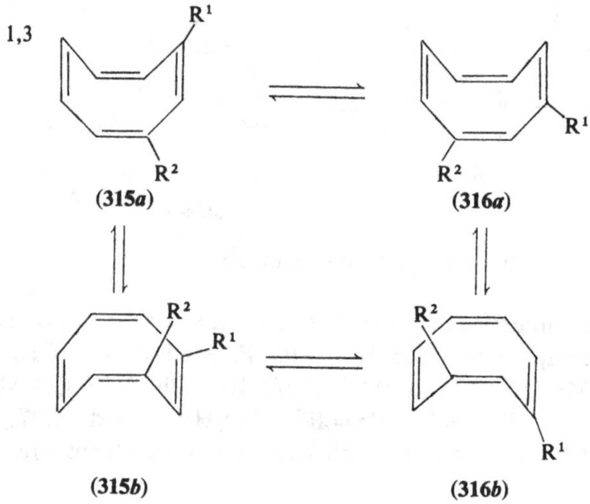

Scheme 46

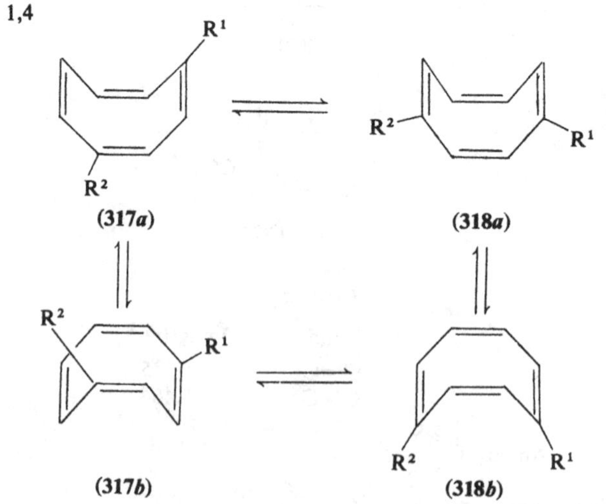

(When $R^1 = R^2$, (**318a**) ≡ (**318b**))

stable form (313) can be increased by low-temperature u.v. irradiation; e.g. for the compound (307; $R^1 = Me$, $R^2 = CO_2Me$) the ratio of the bond-shift isomers (313):(314) in deuteriochloroform at 27° is *ca.* 1:17, but after u.v. irradiation at −50° to −30°, the ratio is *ca.* 1:1 (measured at −42°).[608]

Bond-shift and ring-inversion in 1,3- and 1,4-disubstituted COTs is illustrated (schemes 45 and 46).

The analogous processes in 1,5-disubstituted COTs, however, involve only the enantiomeric forms (319*a*) and (319*b*) (scheme 47).

Scheme 47

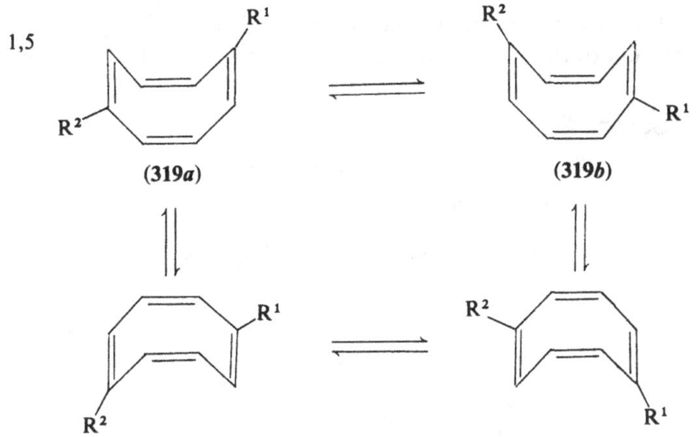

(319*a*)　　　　　　(319*b*)

(319*b*)　　　　　　(319*a*)

Reactions

(*a*) *Functional group transformations.* While most of these transformations give rise to other disubstituted COTs (see pp. 112–14) or annulated COTs (see pp. 123–4), the following reactions lead to a different system.

Treatment of 1,2-di(halomethyl)-compounds (320) with potassium t-butoxide yields the 1,4-dehydrohalogenated products (321)[612] (table 46).

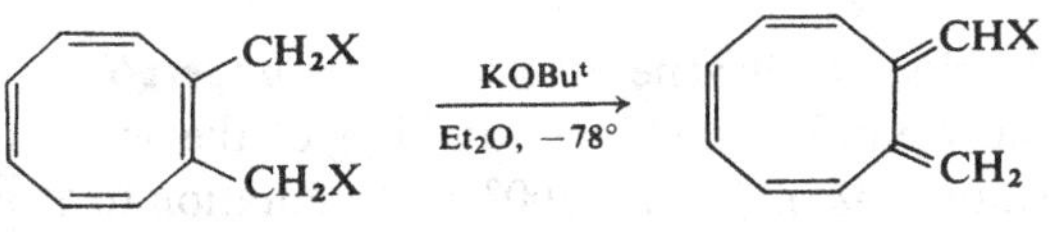

(320)　　　　　　(321)

Table 46

	X	Yield (%)
a	Cl	97
b	Br	83

Thermolysis of the quaternary ammonium hydroxide (322) gives the dimethylenecyclo-octatriene (321c) (35%).[599]

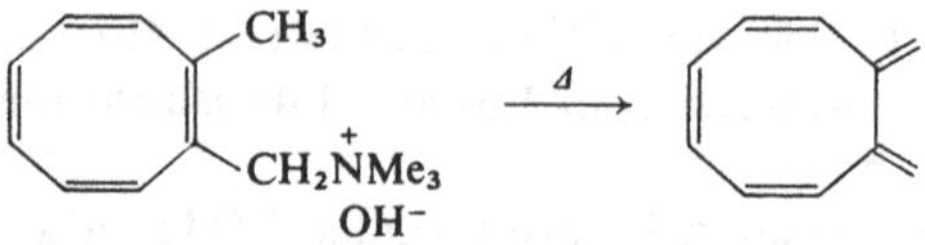

(322)   (321c)

(*b*) *Thermolysis.* At temperatures *ca.* 500° the dimethylcyclo-octatetraenes interconvert (scheme 48); 1,4-dimethylcyclo-octatetraene is the *initial* product from the 1,2- or the 1,5-dimethyl-derivative, and is kinetically dominant in the rearrangement of the 1,3-dimethyl-compound.[49, 603, 606] The main mechanism

Scheme 48

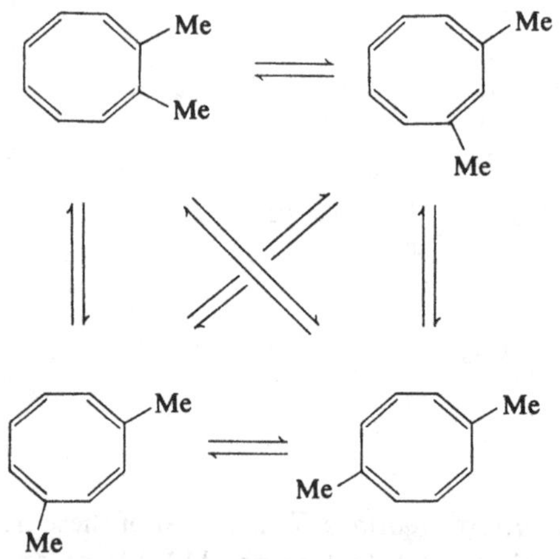

for these rearrangements is illustrated (scheme 49) for 1,5-dimethylcyclo-octatetraene; intramolecular Diels–Alder reactions in the two possible valence tautomers (323) and (324) being followed by *retro*-processes in the alternative mode.[49, 603]

1,2-Dimethylcyclo-octatetraene is known to undergo an additional *degenerate* rearrangement, revealed by scrambling of the deuterium label in the dideuterio-derivative (325).[49] At *ca.* 600°, fragmentation to aromatic products predominates.[49]

Thermal isomerisation of 1,4-dibromocyclo-octatetraene gives the dibromostyrene (326) (up to 92%)[565] (see pp. 97–8).

(*c*) *Bromination.* 1,4-Dinitrocyclo-octatetraene is reported to afford the dibromo-derivative (327).[615]

Scheme 49

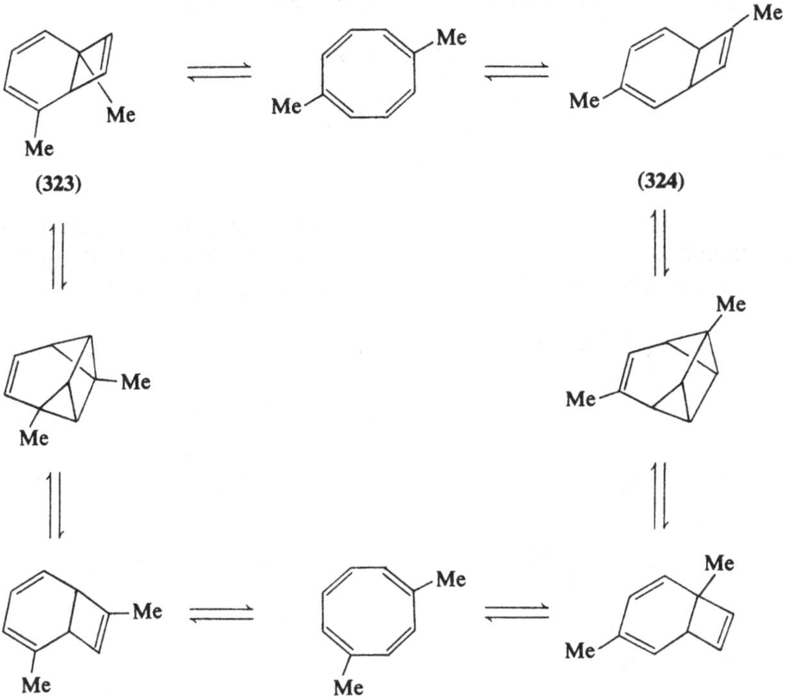

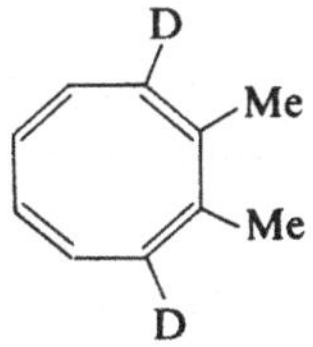

(323)

(324)

(325)

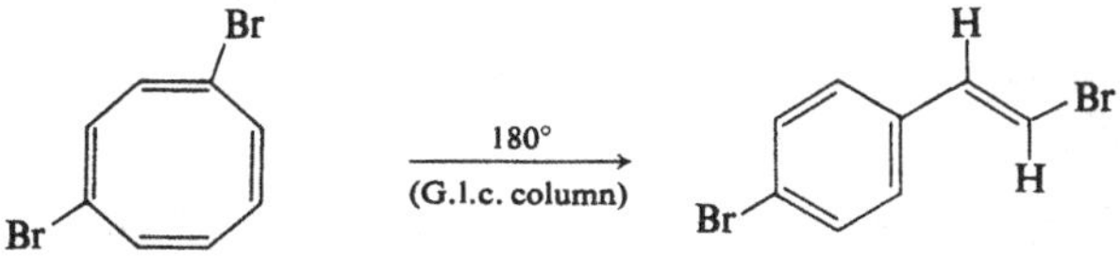

(326)

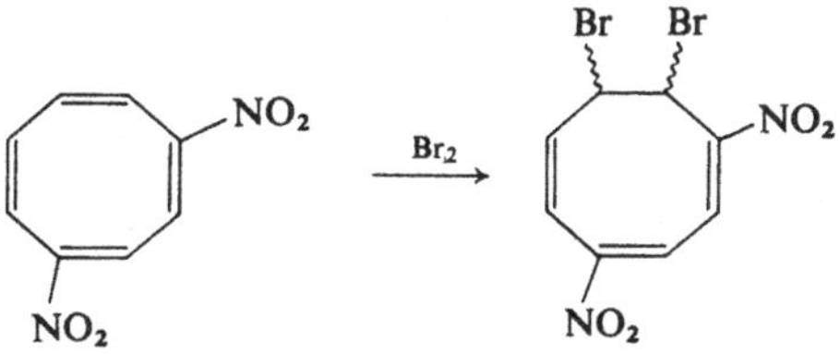

(327)

(*d*) *Reduction.* For alkali metal and electrolytic reduction of 1,2-di(methoxy-carbonyl)cyclo-octatetraene to the anion radical (**328**), see ref. 616.

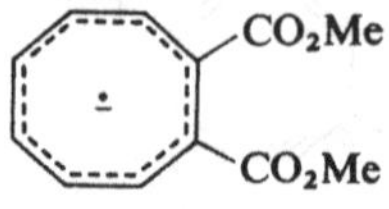

(**328**)

The dianions (**329**)–(**332**) are produced from the parent dimethylcyclo-octatetraenes by the action of potassium in liquid ammonia[43] (the polaro-graphic reduction of these hydrocarbons is also described in ref. 43).

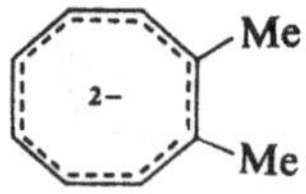

(**329**)

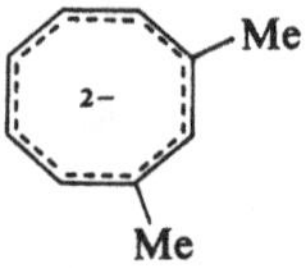

(**330**)

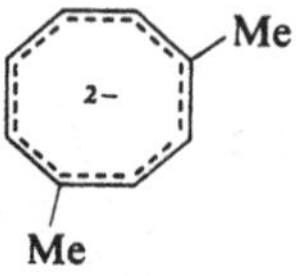

(**331**)

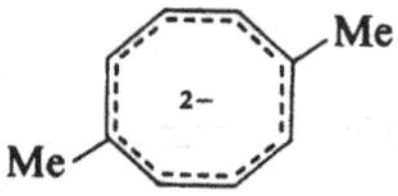

(**332**)

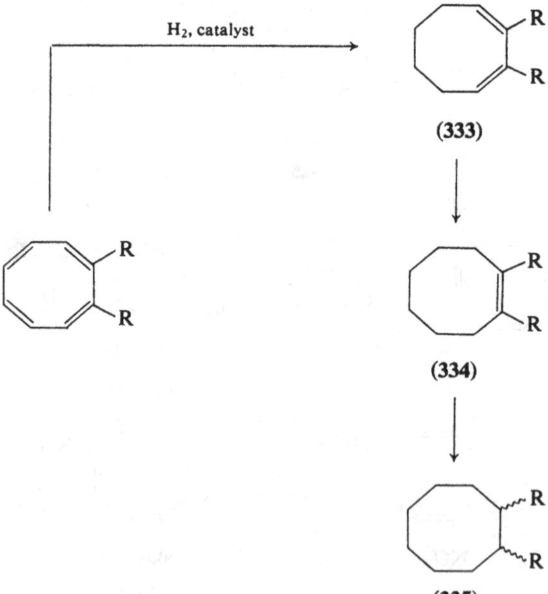

(**333**)

(**334**)

(**335**)

Table 47

Product	R	Catalyst	Solvent	Yield (%)	Ref.
(333)	Ph	1% Pd–CaCO₃, then 10% Pd–C	EtOAc	17	617
(333)	CF₃	Pd–C	—	—	600
(333)	CO₂H	10% Pd–C	MeOH(aq.)	32 (crude)	536
(334)	Me	1% Pd–CaCO₃	MeOH	66	533
(334)	Ph	10% Pd–C	EtOAc	52	617
(334)	CO₂H	Pd–C	MeOH(aq.)	—	536
(335)	CH₂OH	PtO₂[a]	HOAc	58[b]	597

[a] Pre-reduced. [b] The product has R = Me.

The catalytic hydrogenation of 1,2-disubstituted COTs sometimes allows the isolation of 2,3-disubstituted cyclo-octa-1,3-dienes (333) or cyclo-octenes (334) (table 47).

The hydrogenation of 1,3-, 1,4- or 1,5-diphenylcyclo-octatetraene over 10% Pd–C in ethanol gives the corresponding cyclo-octane (mixture of *cis*- and *trans*-isomers).[598]

(*e*) *Cycloadditions.* 1,2-Dicyanocyclo-octatetraene reacts with cyclohepta-trienylidene (generated from the sodium salt of the tosylhydrazone of tropone) to afford a low yield (3%) of the product (336).[618]

(336)

1,2-Disubstituted COTs undergo Diels–Alder addition with maleic an-hydride to give adducts of structure (337); tetracyanoethylene reacts similarly (table 48). (For the tetracyanoethylene adduct of the 3,8-dideuterio-derivative of 1,2-dimethylcyclo-octatetraene (325), see ref. 606.)

(337)

Table 48

R^1	R_2	Dienophile	Conditions	Yield (%)	Ref.
Me	Me	Maleic anhydride	165°	68	533
			C_6H_6, reflux	43	542
Me	Me	Tetracyanoethylene	EtOAc, reflux	52	49
Me	CH_2OAc	Maleic anhydride	C_6H_6, reflux	69	542
CH_2OAc	CH_2OAc	Maleic anhydride	C_6H_6, reflux	90	542

1,5-Dimethylcyclo-octatetraene gives a tetracyanoethylene adduct of structure (338) (55%).[43]

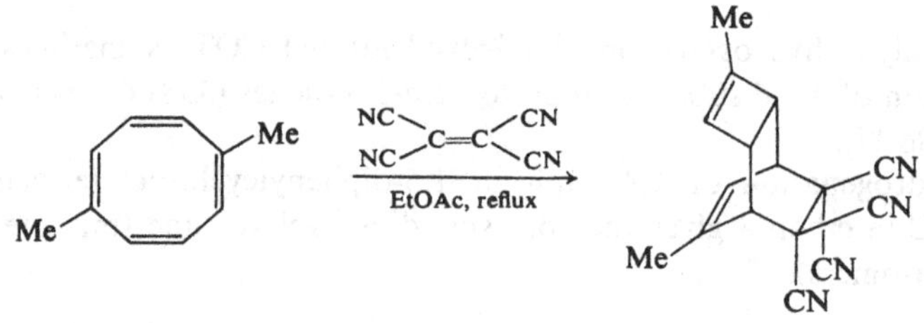

(338)

(*f*) *Formation of metal derivatives.* 1,2-Dimethylcyclo-octatetraene forms a silver nitrate complex (339) in boiling ethanol.[533]

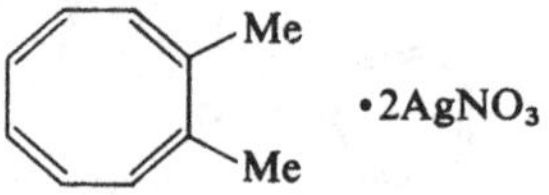

(339)

The 1,2-dialdehyde (340) reacts with $Fe(CO)_5$ to give the lactone complex (341) (8%).[619]

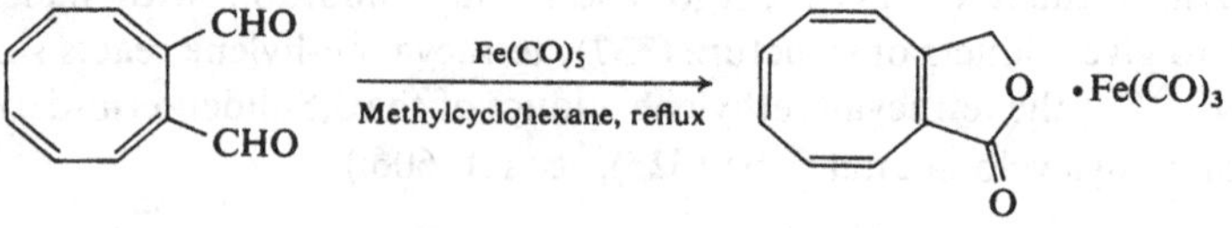

(340)   (341)

Treatment of the 1,2-dicarboxylic ester (342) with $Fe_3(CO)_{12}$ gives a low yield (12%) of the complex (343).[619]

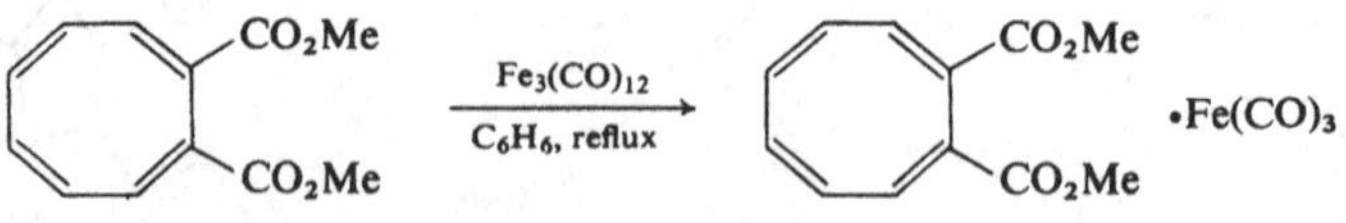

(342)   (343)

The 1,4-disubstituted COT (**344**) reacts with $Fe_2(CO)_9$ to afford a complex (**345**) (crude yield 28.5%), which has been subjected (as the *p*-nitrobenzoate) to X-ray analysis.[545, 546]

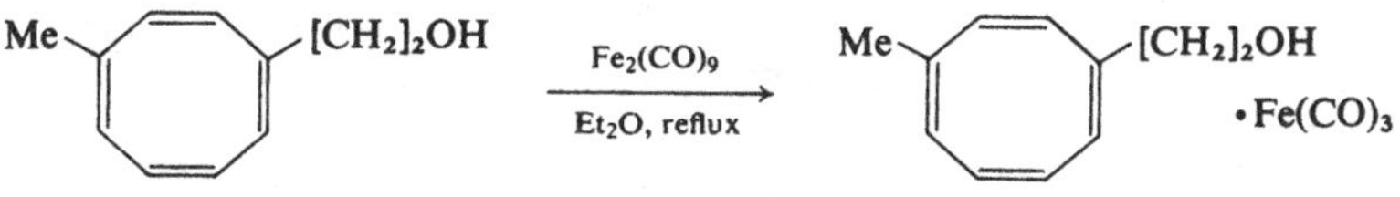

(**344**)   (**345**)

3. Annulated and bridged derivatives

Formation

(*a*) *Cyclisation reactions of 1,2-disubstituted COTs.* A variety of annulated COT derivatives may be prepared from suitable 1,2-disubstituted COTs (schemes 50–52) (see also p. 113).

Scheme 50

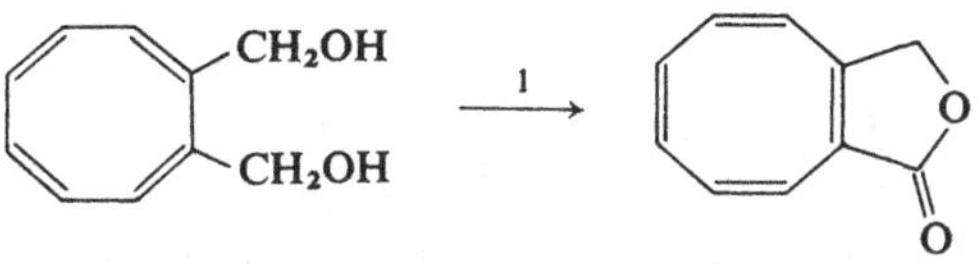

Reaction	Reagent and conditions	Yield (%)	Ref.
1	MnO_2, $CHCl_3$, r.-t.	39	610

Scheme 51

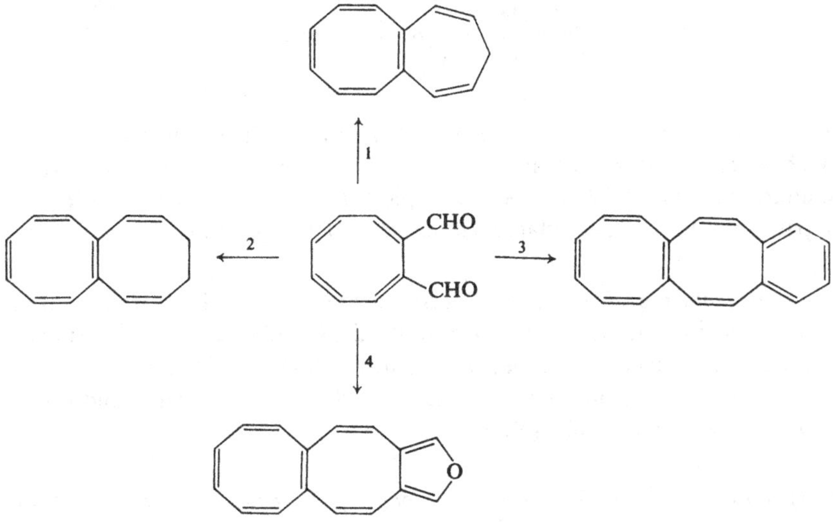

Reaction	Reagents and conditions	Yield (%)	Ref.
1	$Ph_3P{=}CH.CH_2.CH{=}PPh_3$	15	611
2	$Ph_3P{=}CH[CH_2]_2CH{=}PPh_3$	5–10	611
3	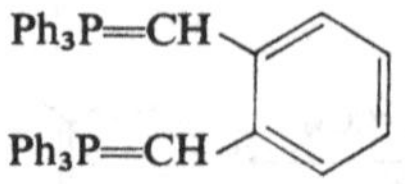 $Ph_3P{=}CH$ ... $Ph_3P{=}CH$	1–2	611
4	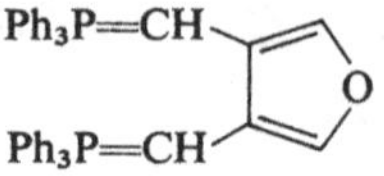 $Ph_3P{=}CH$ ... O, DMF, 90° $Ph_3P{=}CH$	15	610, (620)

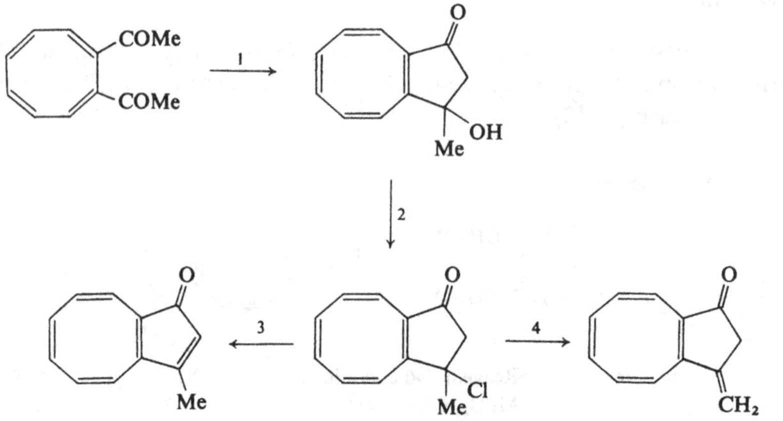

Scheme 52

Reaction	Reagents and conditions	Yield (%)	Ref.
1	NaOMe	—	607
2	$SOCl_2$, pyridine, Et_2O	—	607
3	Chromatography on neutral Al_2O_3	—	607
4	Chromatography on SiO_2	—	607

(b) *Cycloadditions to 1,2-didehydrocyclo-octatetraenes.* The 'yne' species (236), which is generated by the action of potassium t-butoxide on bromocyclo-octatetraene (see pp. 83–4), may be employed *in situ* as a dienophile for the preparation of certain annulated derivatives of COT (scheme 53).

(c) *Photochemical cycloadditions of acetylenes to benzenes.* U.v. irradiation of cyclo-octyne in benzene affords the annulated COT (346) (56%).[625] A similar reaction with toluene gives a methyl-derivative (347) (25–30%).[625]

An intramolecular photoaddition may be effected in the compound (348), leading to the product (349) (5%).[626]

(d) *Syntheses from 7,8-dimethylenecyclo-octa-1,3,5-trienes.* U.v. irradiation of

Scheme 53

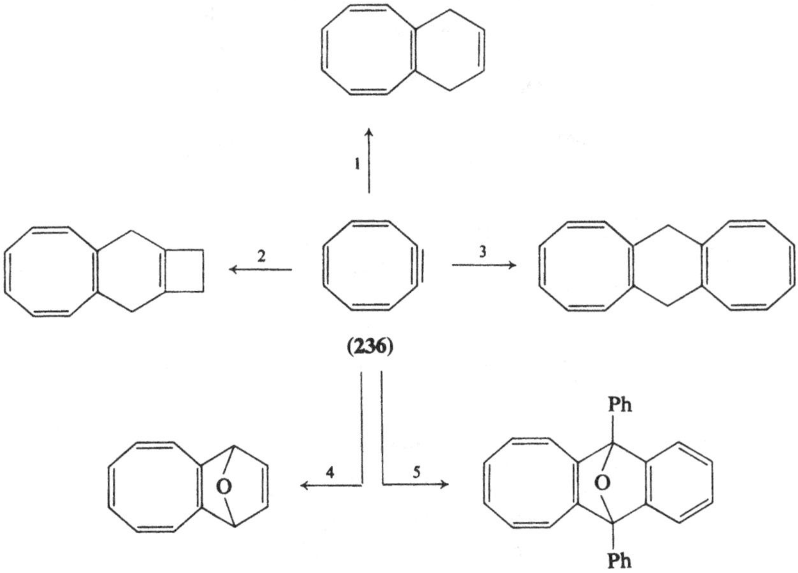

(236)

Reaction	Diene	Yield (%)	Ref.
1	Buta-1,3-diene	83[a]	621
2	1,2-Dimethylenecyclobutane	13[a]	621
3	7,8-Dimethylenecyclo-octa-1,3,5-triene	22	612, (622)
4	Furan	48	623
5	1,3-Diphenylbenzo[c]furan	87	624

[a] Based on the bromocyclo-octatetraene consumed.

either of the dimethylenecyclo-octatrienes (321*b*) or (321*c*) affords the cyclised product (350)[627, 628] (table 49). (For the irradiation of (321*a*), see p. 237.)

The addition of dienophiles to the dimethylene compound (321*c*) also

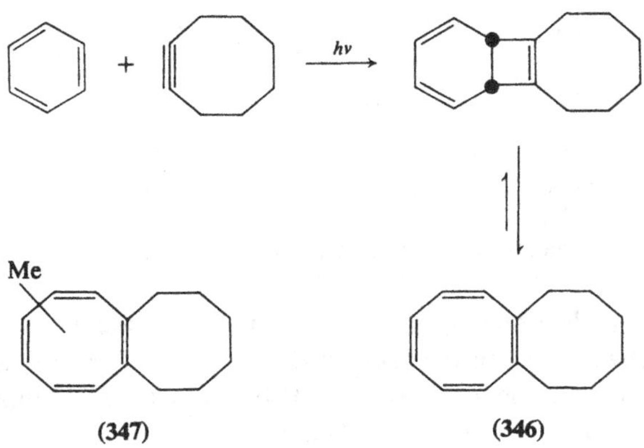

(347)   (346)

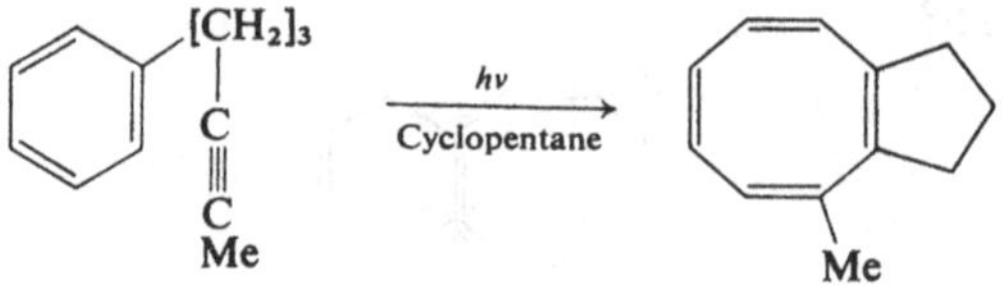

(348)　　　　　　　　　　　(349)

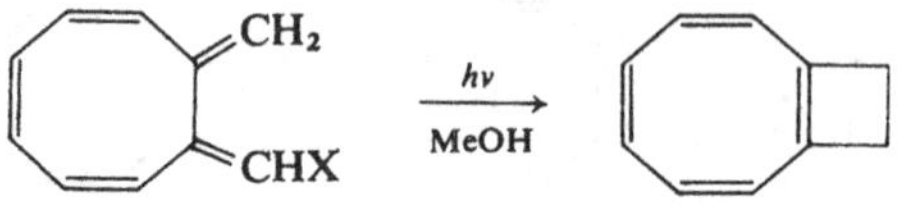

(321)　　　　　　　　　　　(350)

Table 49

	X	Yield (%)
b	Br	*ca.* 7
c	H	*ca.* 10

leads to annulated COTs; thus maleic anhydride gives the adduct (351), and similar reactions occur (under the same conditions) with other dienophiles[612] (table 50). The addition of 1,2-didehydrocyclo-octatetraene (236) to the dimethylene compound (321*c*) has already been mentioned (see p. 125).

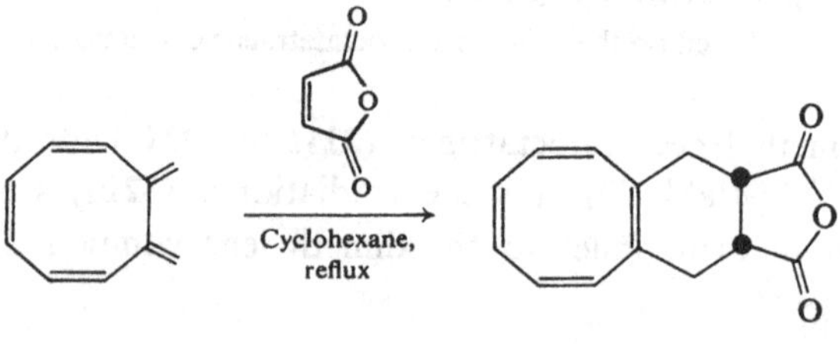

(321*c*)　　　　　　　　　　　(351)

Table 50

Dienophile	Yield (%)
Maleic anhydride	37
p-Benzoquinone	34
Dimethyl acetylenedicarboxylate	52

Addition of maleic anhydride to the monohalodimethylene compounds (321*a*) and (321*b*) results in the product (352) (45–47%); presumably loss of hydrogen halide from the initial adduct (353) to give (354) is followed by the addition of a second molecule of the dienophile.[612]

(e) *Syntheses involving skeletal transformations.* Photolysis of the carbonyl-bridged compound (355) gives the product (356) (30 %).[604]

Table 51

n	R	Conditions	Yield (%)	Ref.
2	H	400°/10 mm	—	605
3	H	hv(Corex), Me₂CO	—	605
4	H	400°/10 mm	92	604
		hv (Corex), Me₂CO	75	604
4	D	hv(Vycor), Me₂CO, Et₂O	25	629
5	H	hv (Corex), Me₂CO	37	604

Thermolytic or photolytic extrusion of sulphur dioxide from the system (357) similarly leads to annulated COTs (358) (table 51).

Gas-phase thermolysis of the propellatriene system (359) also yields annulated COTs (360) (table 52). The rearrangement mechanism is indicated in scheme 54.[629, 630]

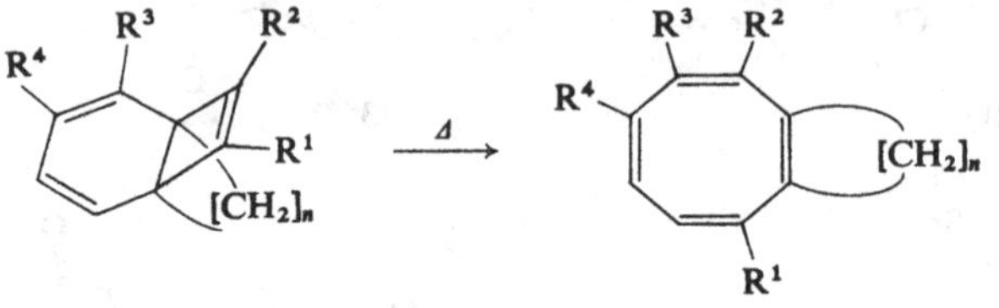

(359)   (360)

Table 52

n	R^1	R^2	R^3	R^4	Conditions	Yield (%)	Ref.
3	H	H	H	H	615°/5 mm	—[a]	629, (630)
4	H	H	H	H	430°/2.5 mm	60	629, (630)
4	D	D	H	H	380°/2.5 mm	30	629, (630)
4	Me	D	H	H	480°	—	631
4	H	H	Me	H	—	—[b]	631
4	H	H	H	Me	—	—	631
4	Me	Me	H	H	500°/2.5 mm	76	629, (630)
5[c]	H	H	H	H	485°/2.5 mm	47	629, (630)

[a] The main product (by a factor of 2) is indane.
[b] The product is contaminated with an exomethylene isomer.
[c] The system (361) exists almost exclusively as the 1,4-bridged COT (362).[632]

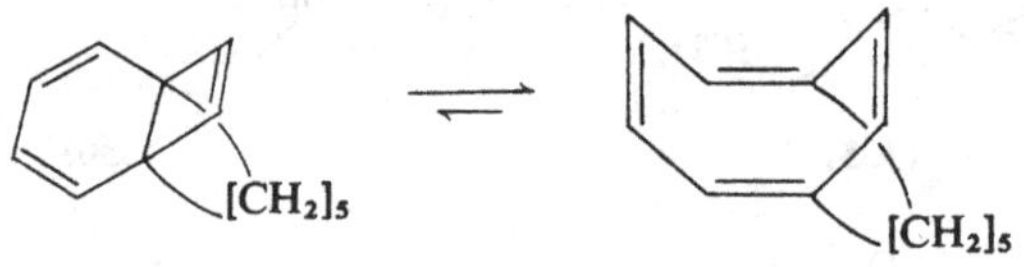

(361)   (362)

Scheme 54

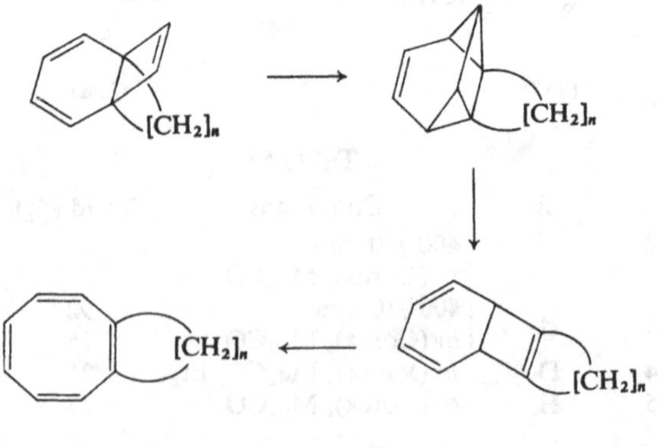

Skeletal rearrangement of the propellatriene system (363) to the bicyclic system (364) is catalysed by $Mo(CO)_6$ (table 53). The simplest mechanistic

(363)　　　　　　　　(364)

Table 53

n	R^1	R^2	R^3	R^4	R^5	R^6	Yield (%)	Ref.
4	H	H	H	H	H	H	41	633
4	D	D	H	H	H	H	—	633
4	H	H	D	H	H	D	—	631
4	H	Me	H	H	H	H	50	631
4	D	Me	H	H	H	H	—	631
4	H	H	Me	H	H	H	87[a]	631
4	H	H	H	Me	H	H	—[b]	631
4	Me	Me	H	H	H	H	*ca.* 36[c]	634
4	H	H	H	Me	Me	H	—	634
5	H	H	H	H	H	H	15[d]	633
5	D	D	H	H	H	H	—	633

[a] Initial reaction in refluxing benzene, followed by further heating at 125°.
[b] Mixture of (365) and (366) obtained.
[c] Mixture of the double bond isomers (367) (*ca.* 36%) and (368) (*ca.* 40%) obtained, together with (369) (*ca.* 24%).
[d] Reaction in refluxing toluene.

(365)　　　　　　　　(366)

(367)　　　　　　(368)　　　　　　(369)

interpretation of these results involves a [1,5] sigmatropic shift of a trigonal cyclobutene carbon atom[631, 634] (scheme 55).

Scheme 55

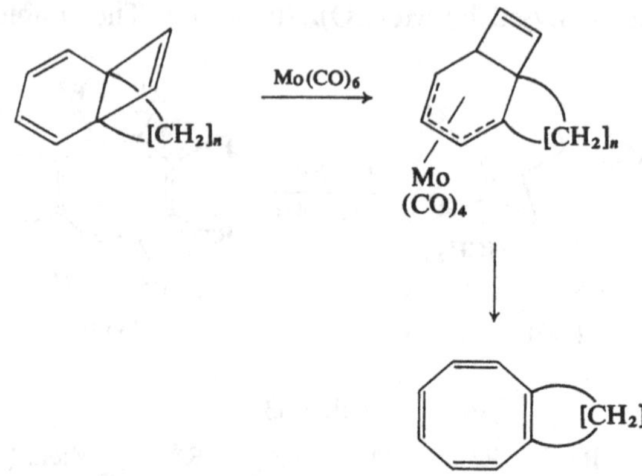

The bridged semibullvalene (370) rearranges at 460–500° to give the annulated COT (356) (69%); results obtained with deuterium-labelled starting material suggest the illustrated reaction path[42, 635] (scheme 56).

Scheme 56

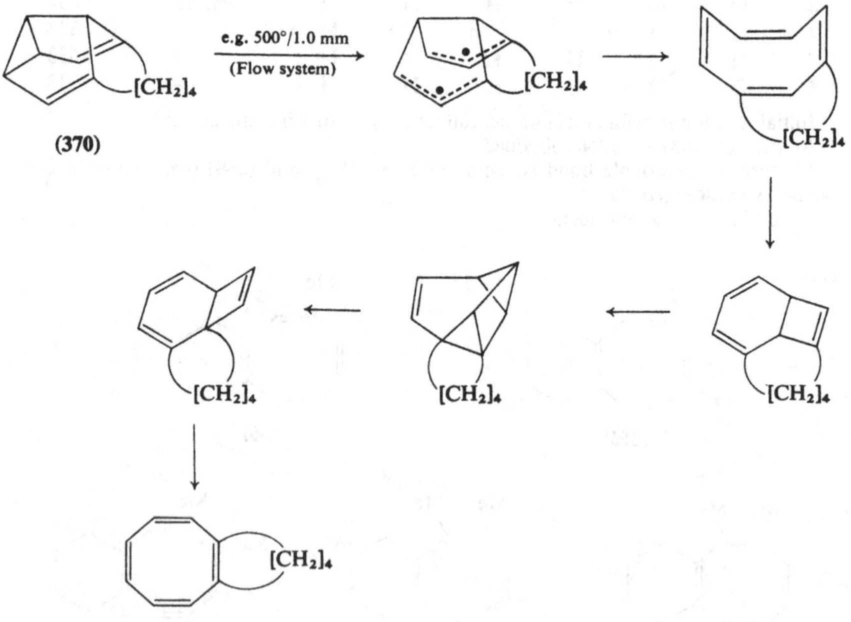

(370)

(356)

Structure

1,2-Annulated COTs presumably resist the bond shift process (371a) → (371b) when $n < 4$; when $n \geqslant 4$ very little torsional strain is imposed on structure (371b) by the ring-fusion (as judged from molecular models).

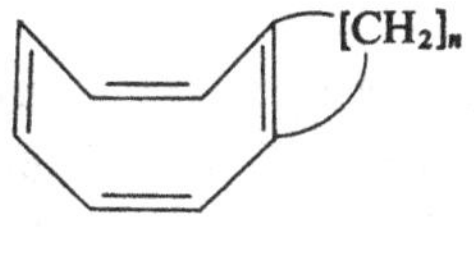

(371*a*)

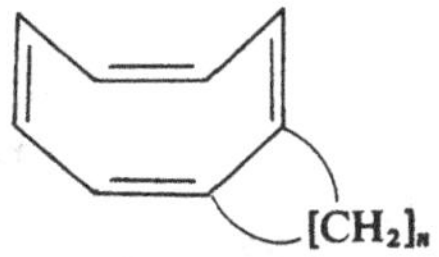

(371*b*)

The strained 1,3-bridged COT (**372**) has been invoked as an intermediate in the thermal rearrangement of the bridged semibullvalene (**370**) (see p. 130); the 1,4-bridged COT (**362**) is more stable than its valence tautomer (**361**) (see p. 128).

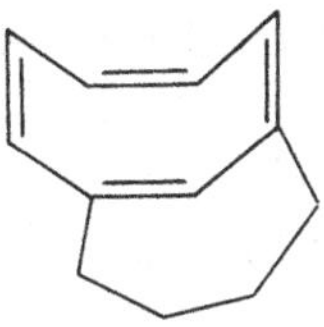

(**372**)

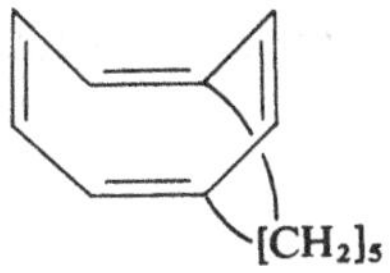

(**362**)

Reactions

(*a*) *Thermolysis.* Thermolysis of the annulated COT (**356**) does not result in the formation of structural isomers (*cf.* disubstituted COTs, pp. 118, 119), but at *ca.* 600° (flow system) a degenerate rearrangement occurs (see p. 17); using the 3,8-dideuterio-derivative (**373**), the stepwise shift of deuterium first to the 4,7-positions and then to the 5,6-positions may be detected.[49]

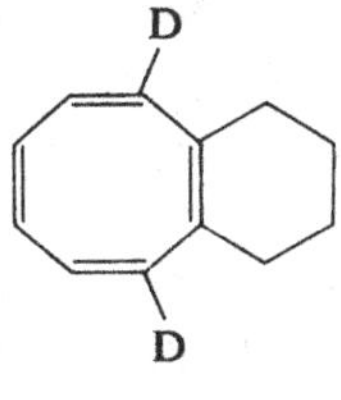

(**373**)

(*b*) *Dehydrogenation.* Treatment of compounds (**374**) with 2,3-dichloro-5,6-dicyano-*p*-benzoquinone gives the corresponding benzo-derivatives (**375**) (table 54).

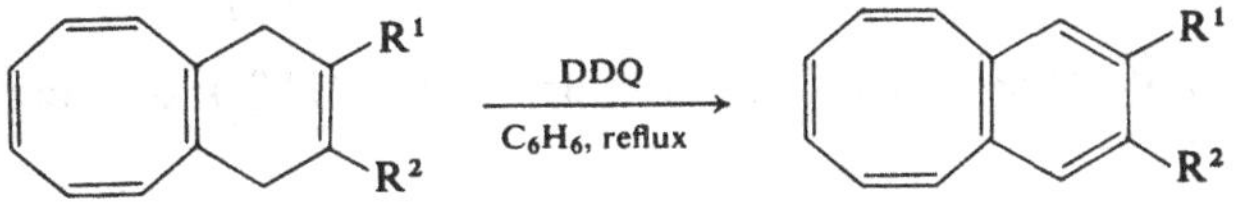

(**374**)   (**375**)

Table 54

R¹	R²	Yield (%)	Ref.
H	H	68ᵃ	621
CO₂Me	CO₂Me	81	612
—[CH₂]₂—		99	621
—[CH$\overset{z}{=}$CH]₃—		47ᵇ	612

ᵃ The final stage in an efficient synthesis of benzocyclo-octatetraene from COT.
ᵇ Using Pd–C in refluxing decalin, the yield is 36%.

(*c*) *Reduction.* For the radical anion of (**350**), see appendix.

(*d*) *Cycloadditions.* Reaction of the annulated COT (**356**) with maleic anhydride gives the adduct (**376**); analogous products are formed from other dienophiles (table 55). For tetracyanoethylene adducts of deuterio-derivatives of (**356**) etc.,

(**356**)   (**376**)

Table 55

Dienophile	Conditions	Yield (%)	Ref.
Maleic anhydride	Toluene, 110°	35	542
Tetracyanoethylene	EtOAc, reflux	65	629, (630)
N-Phenyltriazolinedione	EtOAc, r.-t. reflux	83	629

see refs. 49, 606, 629, 630, 631, 633, 635. For N-phenyltriazolinedione adducts of methyl-substituted derivatives (**365**)–(**369**), see refs. 629, 631, 634.

(*e*) *Formation of metal derivatives.* The 1,4-bridged COT (**362**) reacts with Mo(CO)₆ to give the complex (**377**) (87.5%).[633]

(**362**)   (**377**)

The lactone (**378**) forms the complex (**379**) (11%) on treatment with Fe(CO)₅.[619]

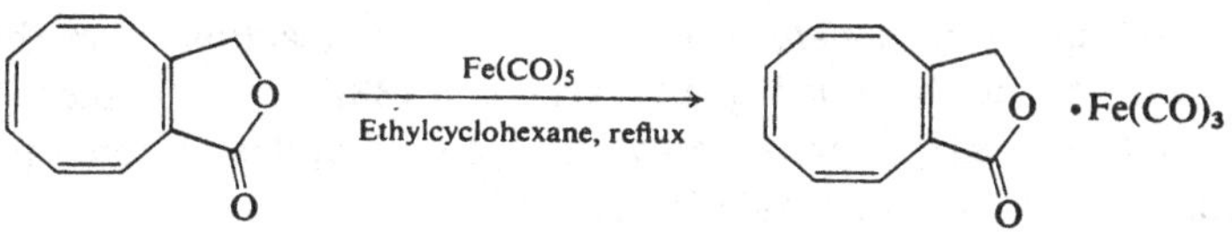

(378)   (379)

4. Trisubstituted derivatives

Formation

(a) *Photochemical cycloadditions of disubstituted acetylenes to monosubstituted benzenes.* The photoaddition of disubstituted acetylenes to benzonitrile affords the 1,2,3-trisubstituted COTs (380) in very low yield[636] (table 56).

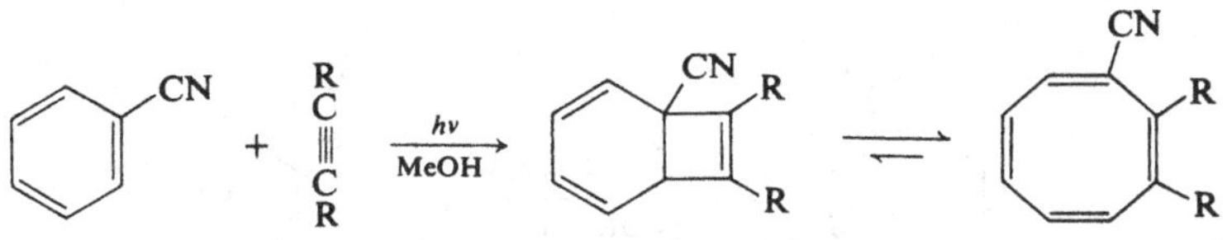

(380)

Table 56

	R	Yield (%)
a	Et	8
b	Bu^n	—

Scheme 57

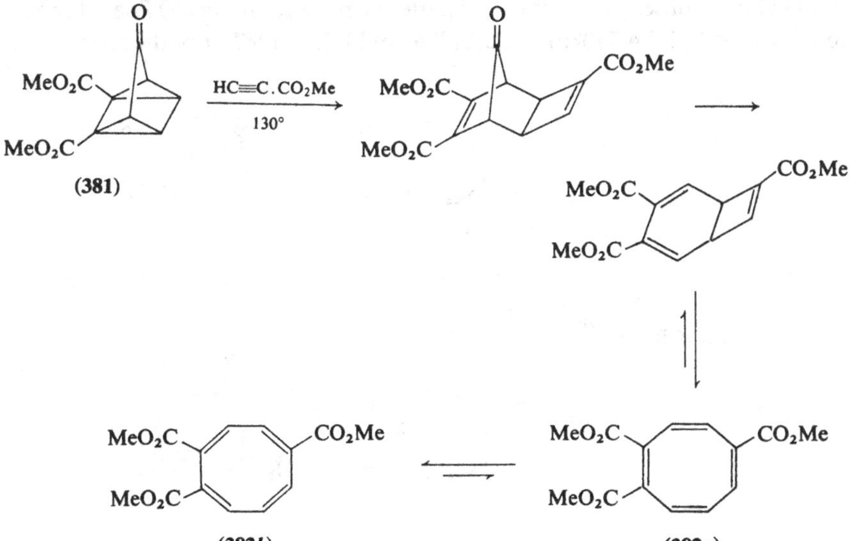

(381)

(382b)   (382a)

(*b*) *Synthesis involving skeletal transformation.* The reaction of methyl propiolate with the disubstituted quadricyclanone (**381**) leads to the 1,2,5-trisubstituted COT (**382**);[637] presumably $[_\pi 2 + _\sigma 2 + _\sigma 2]$ cycloaddition is followed by extrusion of carbon monoxide (scheme 57).

Structure

The n.m.r. spectrum of compound (**382**) indicates that it exists almost exclusively in the form (**382*b***).[637]

Reactions

Hydrogenation of the trisubstituted COT (**380*a***) over Pd–SrCO$_3$ affords, as the major product, the 1,3-diene (**383**) (*ca.* 67%).[636]

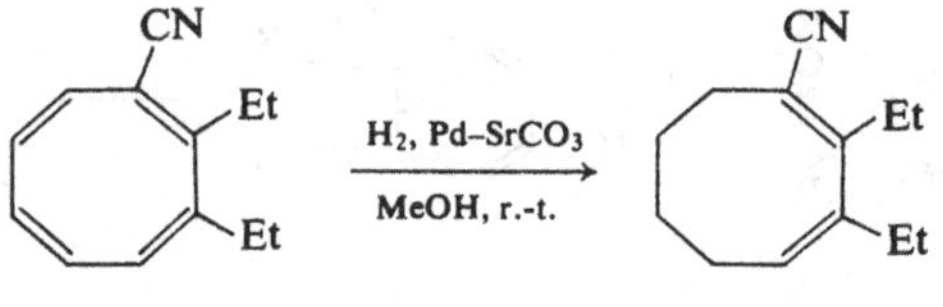

(**380*a***)   (**383**)

5. Tetrasubstituted derivatives

Formation

(*a*) *Cyclotetramerisation of monosubstituted acetylenes.* The cyclotetramerisation of monosubstituted acetylenes may be effected by a variety of catalysts (table 57); the observed substitution patterns in the resulting COTs are limited to 1,2,4,6 (**384**), 1,2,4,7 (**385**), 1,2,5,6 (**386**) and 1,3,5,7 (**387**). For the production

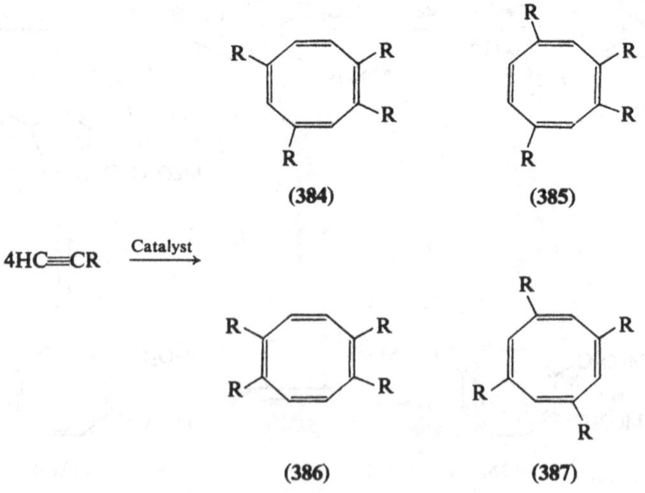

(**384**)   (**385**)

$$4HC{\equiv}CR \xrightarrow{\text{Catalyst}}$$

(**386**)   (**387**)

Table 57

R	Catalyst and conditions	Product	Yield (%)[a]	Ref.
CH_2OH	$Ni(acac)_2$, C_6H_6, reflux	(387)	Up to 71	638
$C(OH)Me_2$	e.g. $Ni(acac)_2$,[b] 90–120°	(384) + (385)	40	639
CO_2Me	$Ni(PCl_3)_4$, cyclohexane, 22–55°	(384)	83	640
CO_2Me	$PdCl_2$, NaOAc, HOAc, 80°	(386)	(Very low)	641
CO_2Et	$Ni(PCl_3)_4$, cyclohexane, 25–46°	Mainly (384)	30	640
OEt	$NaNH_2$, NH_3(liq.), $-70°$	(387)	10–24	642
OPr^n	$NaNH_2$, NH_3(liq.), $-70°$	(387)	9.5	642

[a] Based on reacted monomer.
[b] The reaction is also catalysed by $Ni(CO)_4$, $Ni(COD)_2$, $Ni(CN)_2$, $Ni(OEt)_2$ and $NiCl_2$–$NaBH_4$.[639] For other studies using various nickel chelate 'ato' complexes, see refs. 643–645.

of (387; R = Ph) (6%) by the self-condensation of $PhC(OEt)_2Me$, see ref. 646.

(b) *Syntheses involving skeletal transformations.* Photo-dimerisation of 4,6-dimethyl-2-pyrone (388) leads to a mixture of the dimers (389), (390) and (391), which is readily converted to 1,3,5,7-tetramethylcyclo-octatetraene (26%, based on unrecovered starting material) by brief heating above the melting-point.[647, 648]

(388)

hv
(Pyrex) | C_6H_6

(389)   +   (390)   +   (391)

Δ

Me

U.v. irradiation of 4,5-diphenyl-2-pyrone (**392**) gives 1,2,4,7-tetraphenyl-cyclo-octatetraene (up to 24%) (the product is itself photolabile (see p. 145)); the reaction sequence of scheme 58 is postulated.[649]

Scheme 58

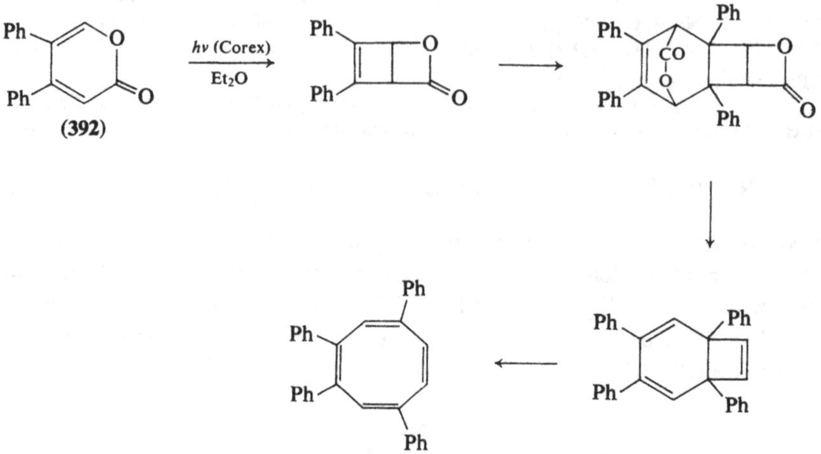

The disubstituted quadricyclanone (**381**) reacts with dimethyl acetylene-dicarboxylate to afford 1,2,5,6-tetramethoxycarbonylcyclo-octatetraene[637] (*cf.* p. 133), presumably *via* a [$_\pi 2 + _\sigma 2 + _\sigma 2$] cycloaddition and subsequent loss of carbon monoxide (scheme 59).

Scheme 59

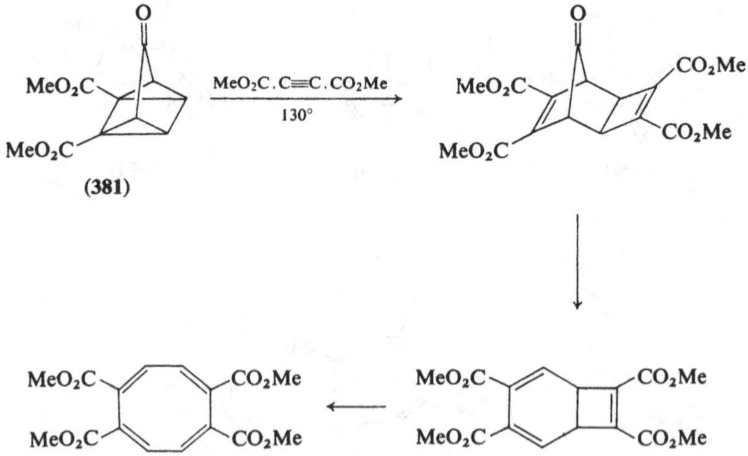

(393) (394)

(395)

Photodecarbonylation of the norbornenone (393) gives the bicyclo[4.2.0]-octa-2,4,7-triene (394),[650] which is slowly transformed at 65° (half-life *ca.* 90 min.) to an equilibrium mixture of 1,6-dimethyl-7,8-diphenylcyclo-octatetraene and the valence tautomer (395)[651] (see p. 143).

The generation of disubstituted cyclobutadienes leads to tetrasubstituted dimers possessing the tricyclo[4.2.0.0^{2,5}]octa-3,7-diene skeleton; thermolysis of this system then affords tetrasubstituted COTs (see p. 3). Thus 1,2,4,5-, 1,2,4,7- and 1,2,5,6-tetramethylcyclo-octatetraene may be prepared by dechlorination of a mixture of the dichlorodimethylcyclobutenes (396), (397) and (398), followed by thermolysis of the resulting dimethylcyclobutadiene dimers (399), (400) and (401).[652] Thermolytic ring-opening of (399)–(401) is almost quantitative.

The same three tetramethylcyclo-octatetraenes result (total yield 8.8%) from photolysis of the dimethylcyclobutenedicarboxylic anhydride (402)[653] (see p. 159).

1,2,4,7- and 1,3,5,7-tetraphenylcyclo-octatetraene may be obtained (24% and 34% respectively) from the quaternary ammonium salt (403) by treatment with potassium t-butoxide, without isolation of the intermediate diphenylcyclobutadiene dimers (404) and (405).[654]

1,3,5,7-Tetramethylcyclo-octatetraene is produced (58%) by u.v. irradiation of the tetramethylsemibullvalene (406).[183]

Bromination–dehydrobromination of the bicyclic diene (408) yields 1,2,3,8-tetramethylcyclo-octatetraene (98%), which resists bond-shift isomerisation to the 1,2,3,4-tetramethyl-derivative[655] (see p. 142).

1,2,3,4-Tetramethylcyclo-octatetraene is available from the sulphone (409) (70–80%).[655]

(396)   (397)   (398)

Na–Hg | Et$_2$O, r.-t.

(399)   (400)   (401)

175°   175°

Product	Yield (%)
(399)	24
(400) + (401)	7.5

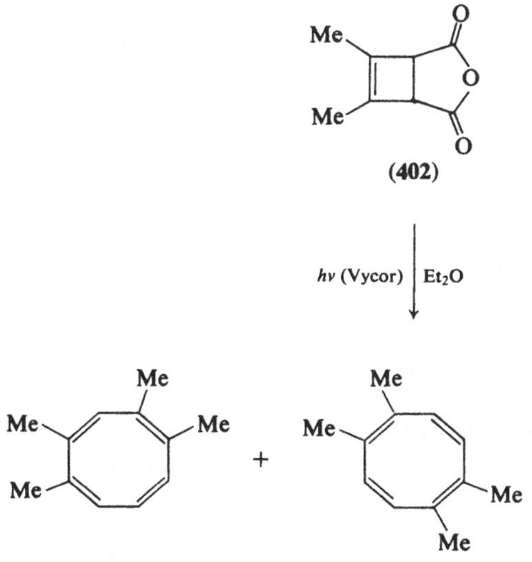

(402)

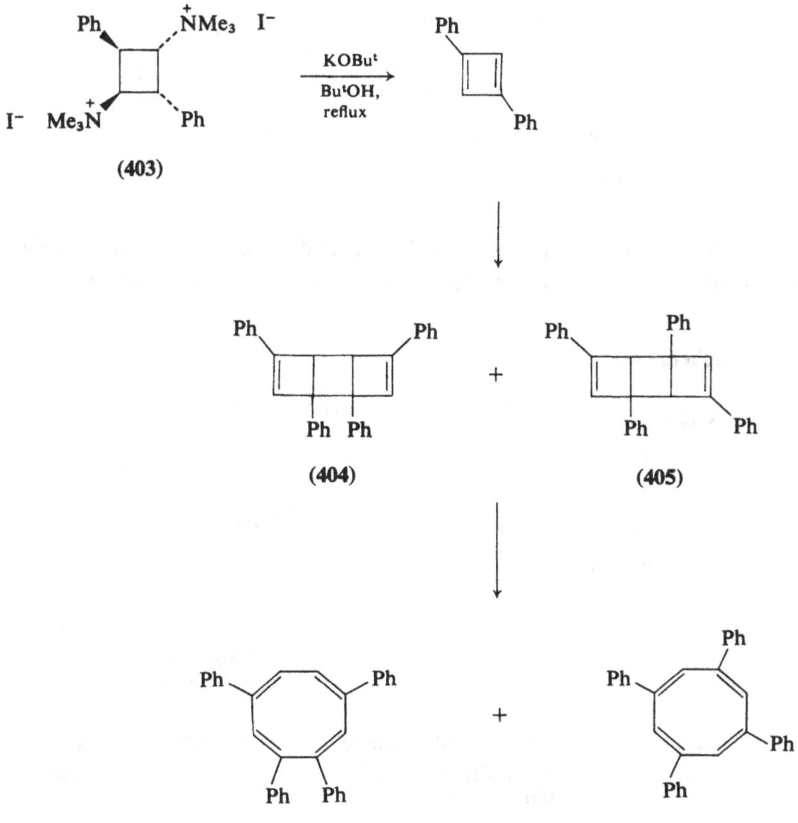

hv (Vycor) | Et₂O

(403)

KOBuᵗ
BuᵗOH,
reflux

(404)   (405)

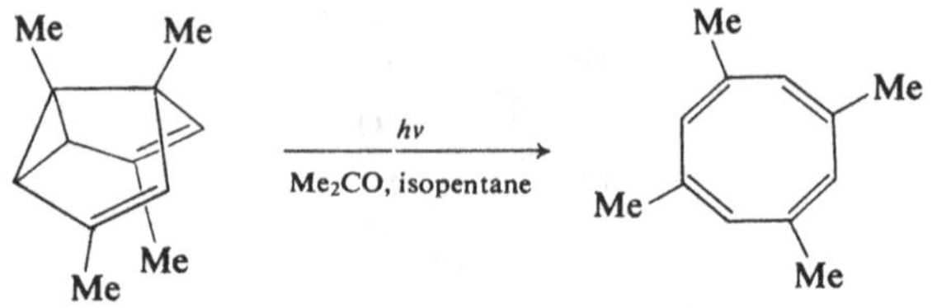

Me   Me

$\xrightarrow[\text{Me}_2\text{CO, isopentane}]{h\nu}$

Me
Me

Me

Me

Me

Me

(406)

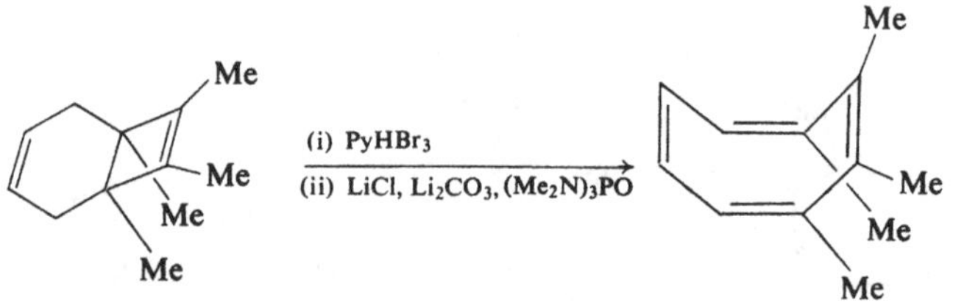

Me

Me

(i) PyHBr$_3$
$\xrightarrow[\text{(ii) LiCl, Li}_2\text{CO}_3\text{, (Me}_2\text{N)}_3\text{PO}]{}$

Me
Me

Me

Me

Me

Me
Me

(408)

SO$_2$

(i) Bu^{n}Li
(ii) MeI
$\xrightarrow[\text{(iv) LiAlH}_4]{\text{(iii) Bu}^n\text{Li}}$

Me

Me
Me

Me

Me
Me

Me

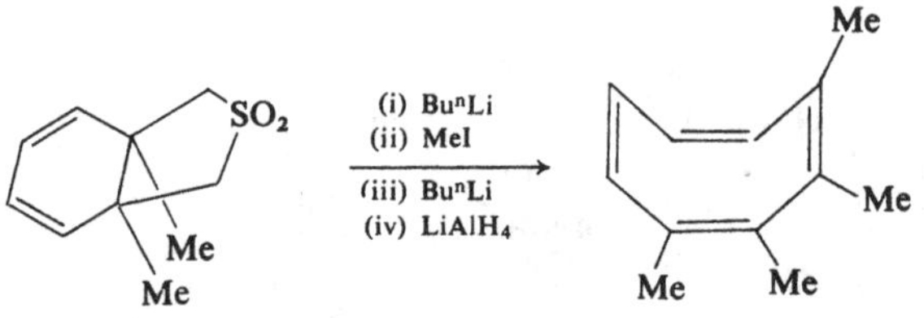

(409)

(*c*) *Functional group transformations.* The transformations in schemes 60–62 lead to other tetrasubstituted COTs. Acid treatment (as in scheme 62) of

Scheme 60

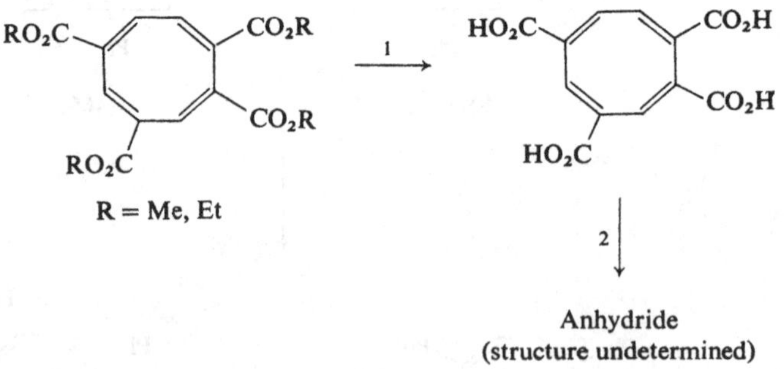

RO$_2$C

CO$_2$R

$\xrightarrow{1}$

HO$_2$C

CO$_2$H

CO$_2$R

CO$_2$H

RO$_2$C

HO$_2$C

R = Me, Et

2

Anhydride
(structure undetermined)

Reaction	Reagents and conditions	Yield (%)	Ref.
1	KOH, EtOH, reflux	100	640
2	100°, dioxan	—	640

Scheme 61

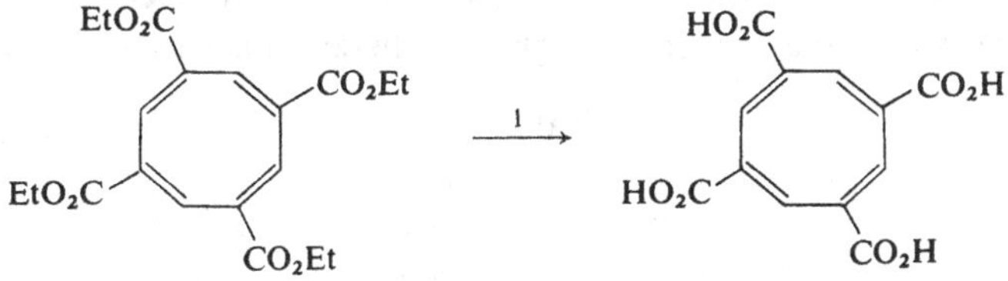
EtO2C
CO2Et
EtO2C
CO2Et
1
HO2C
CO2H
HO2C
CO2H

Reaction	Reagents and conditions	Yield (%)	Ref.
1	KOH, EtOH, reflux	100	640

Scheme 62

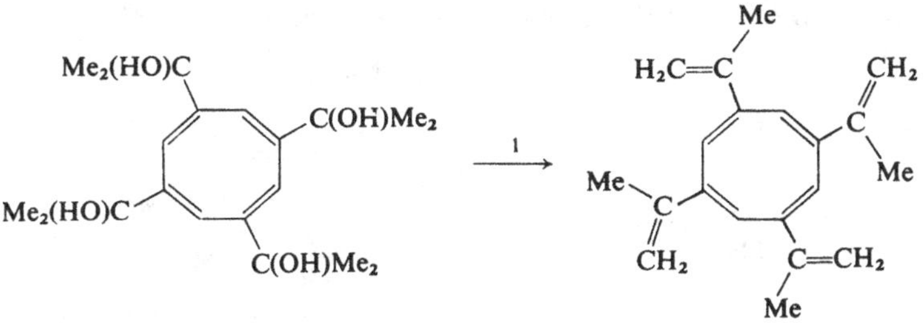
Me2(HO)C
C(OH)Me2
Me2(HO)C
C(OH)Me2
1
Me
H2C=C
CH2
C
Me
Me
C
CH2
C=CH2
Me

Reaction	Reagents and conditions	Yield (%)	Ref.
1	*p*-Me.C_6H_4.SO_2OH, toluene, reflux	88	639

(**410**) and (**411**) results in the cyclised products (**412**) (97%) and (**413**) (93%) respectively.[639]

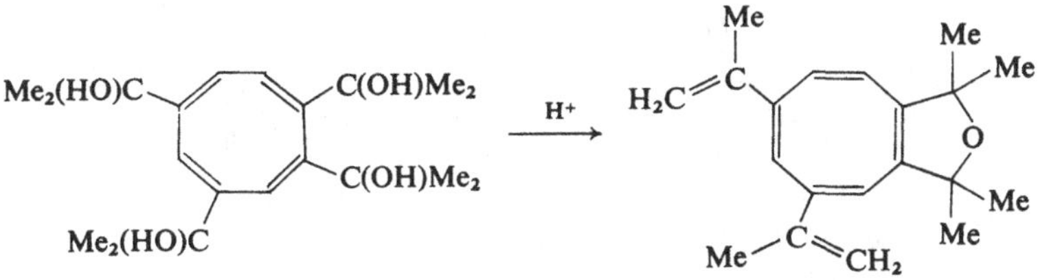
Me2(HO)C
C(OH)Me2
C(OH)Me2
Me2(HO)C
H+
Me
H2C=C
Me
Me
O
Me
Me
Me
C
CH2

(**410**) (**412**)

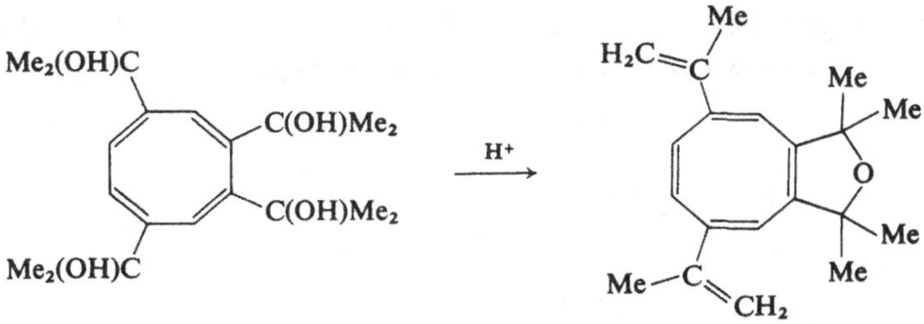
Me2(OH)C
C(OH)Me2
C(OH)Me2
Me2(OH)C
H+
Me
H2C=C
Me
Me
O
Me
Me
Me
C
CH2

(**411**) (**413**)

Structure

1,3,5,7-Tetramethylcyclo-octatetraene has been subjected to X-ray analysis;[656] the bond lengths and angles in the eight-membered ring are listed in table 58.

Table 58

Bond	Length (Å)
C=C	1.330
C—C	1.481

Angle C⟨C/C⟩   125°

For the p.m.r. spectrum of the 1,3,5,7-tetramethyl-compound, see refs. 657–659; for the calculated barrier to ring-inversion, see refs. 126, 127. The bond-shift process (**414a**) ⇌ (**414b**) becomes mobile only above 120°.[657]

(**414a**)   (**414b**)

($\Delta G^* = ca.$ 22.5 kcal mol^{-1})

The bond-shift isomers (**415a**) and (**415b**) are actually isolable, the attainment of planarity by the C$_8$-ring being inhibited by severe substituent interactions.[655] (Interconversion can be achieved thermally or photochemically,

(**415a**)   (**415b**)

giving mixtures in which (**415b**) predominates[655].)

P.m.r. spectral evidence indicates that the 1,2,5,6-tetramethyl ester (**416**) exists (at 30°) almost exclusively in the form (**416a**).[637]

(**416a**)   (**416b**)

The 1,2,3,4-tetramethyl-derivative (**415a**) contains 25% of the bicyclic valence tautomer (**417**) at room-temperature;[655] ((**415b**) is apparently homogeneous.)

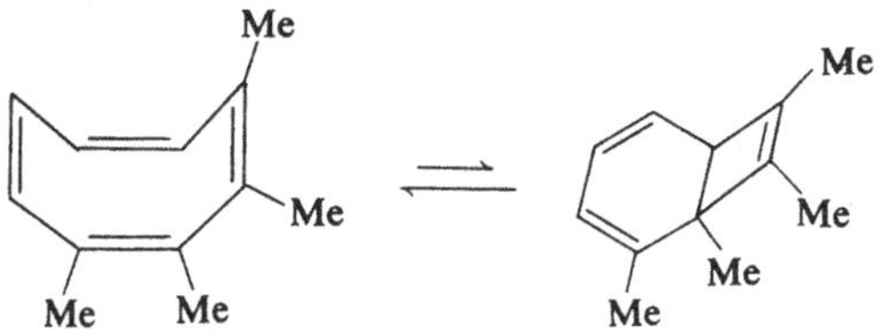

(**415a**)　　　　　(**417**)

The tetrasubstituted compound (**418**) exists in equilibrium with the bicyclic isomer (**395**) (which is crystallisable), the position of the equilibrium changing very little between 0° and 100°; at 120°, however, the bond-shift isomer (**419**) emerges, together with a second bicyclic valence tautomer (**420**)[651].

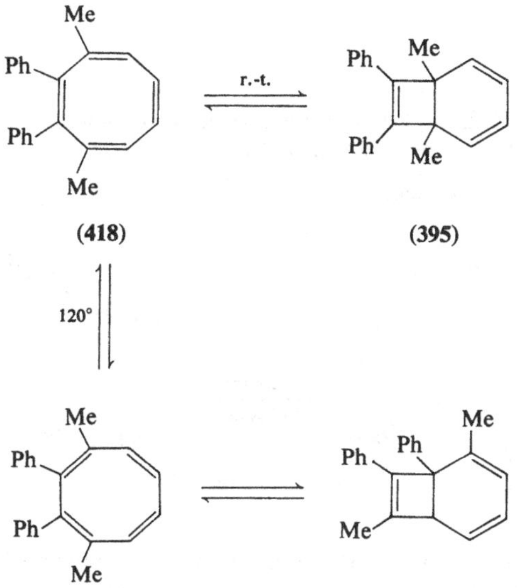

(**418**)　　　　　(**395**)

120°

(**419**)　　　　　(**420**)

Reactions

(*a*) *Functional group transformations.* Transformations leading to other COTs have already been mentioned (see pp. 140–1).

1,3,5,7-Tetraethoxycyclo-octatetraene is hydrolysed quantitatively to the cyclo-octatrienone (**421**) on a silica column, while treatment with alcohols in the presence of hydrochloric acid gives the tetra-alkoxydioxa-adamantanes (**422**) (up to 60%).[642]

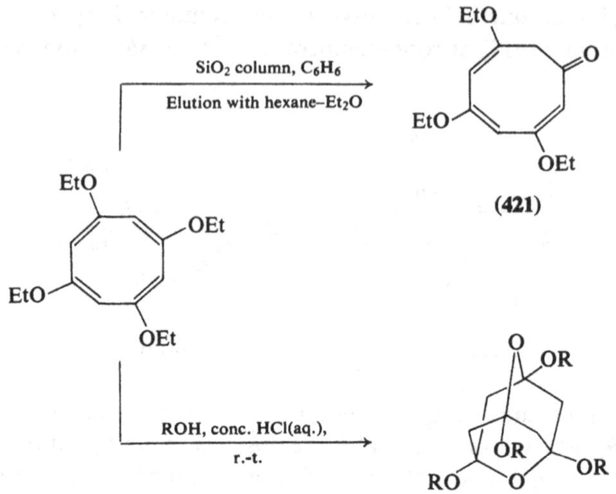

(421)

(422)

R = Me, Et

(*b*) *Thermolysis.* 1,2,4,6-Tetraethoxycarbonylcyclo-octatetraene undergoes fragmentation at 340–350° to form (as the main products) the ethyl esters of propiolic and trimellitic acids.[640]

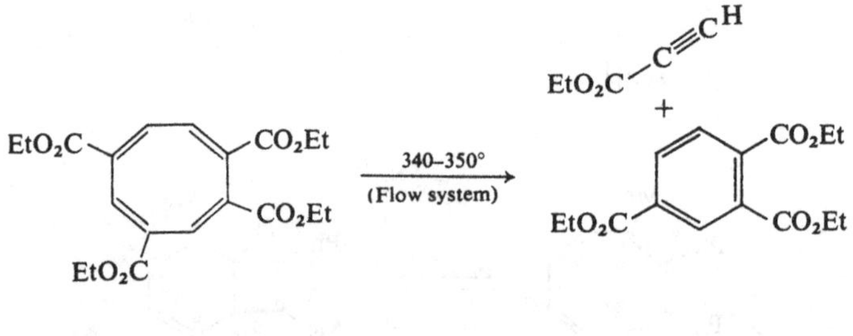

1,3,5,7-Tetramethylcyclo-octatetraene is partly converted (*ca.* 10%) at 250° into the semibullvalene (**406**)[183] (*cf.* semibullvalene, p. 169).

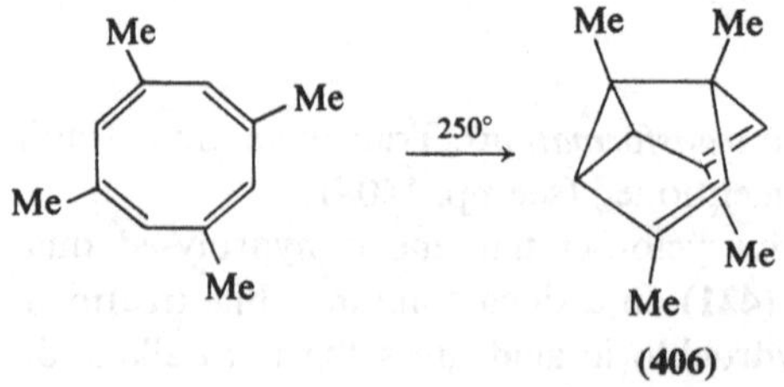

(406)

(c) *Photolysis.* U.v. irradiation of 1,3,6,8-tetraphenylcyclo-octatetraene in dilute solution effects fragmentation to *p*-terphenyl (25%) and diphenyl-acetylene (20%).[660, 661] In concentrated solution however, a photo-isomer formulated as the *Z,Z,Z,E*-cyclo-octatetraene (**423**) accumulates; this is surprisingly stable, reverting to the starting material at 60° with a half-life of *ca.* 1 h[182, 660] (but see appendix).

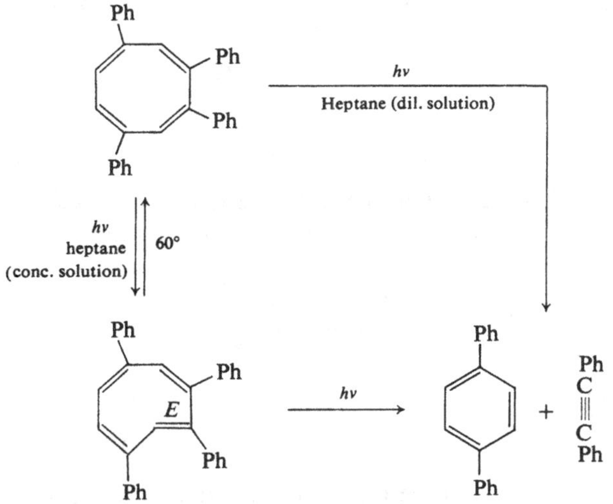

(**423**)

(d) *Reduction.* Polarographic studies of the reduction of the stable bond-shift isomers 1,2,3,4- and 1,2,3,8-tetramethylcyclo-octatetraene indicate the difficulty experienced by these systems in attaining a planar conformation, but they both afford the dianion (**424**) with potassium in ND_3.[655]

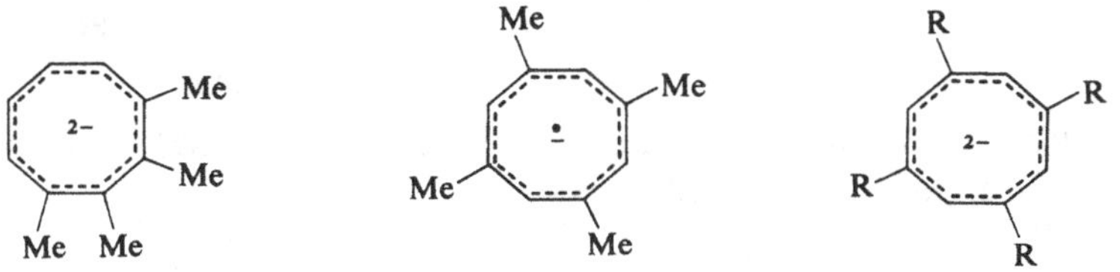

(**424**)   (**425**)   (**426**)

a: R = Me
b: R = Ph

The anion radical (**425**) of 1,3,5,7-tetramethylcyclo-octatetraene may be generated by the action of sodium or potassium in hexamethylphosphor-amide,[537, 571, 662] or electrolytically.[662] For the p.m.r. spectrum of the dianion (**426a**), see refs. 658, 659.

For alkali metal and electrochemical reductions of 1,3,5,7- and 1,3,6,8-tetraphenylcyclo-octatetraene, see refs. 80, 573; for the preparation of the dianion (426b), see ref. 663.

The catalytic hydrogenation of 1,3,5,7-tetramethylcyclo-octatetraene affords four stereoisomers of the cyclo-octane (427).[664]

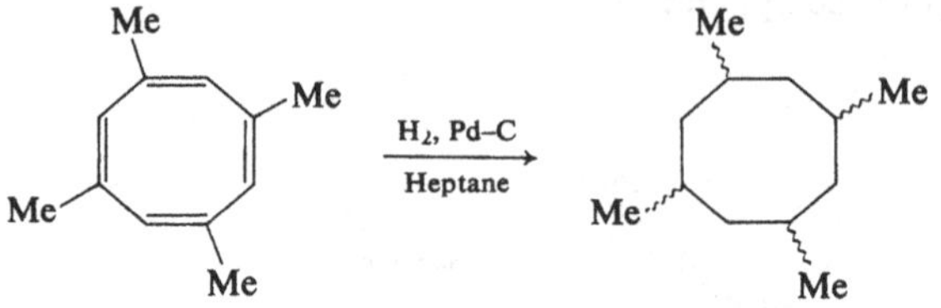

(427)

The catalytic hydrogenation of the 1,3,5,8-tetra-esters (428) may be controlled to yield cyclo-octenes, formulated as (429).[640] For this substitution

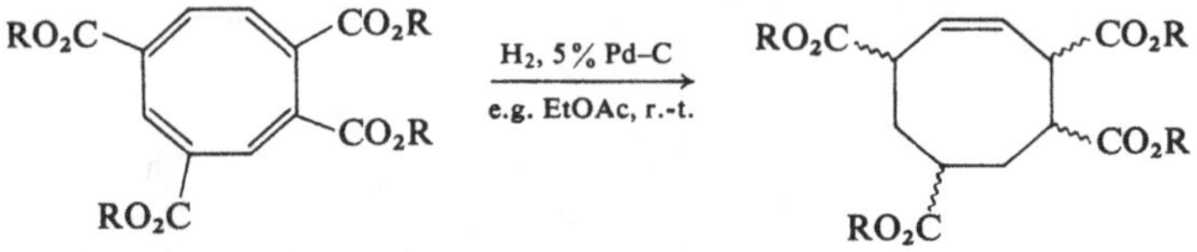

(428)

R = Me, Et

(429)

pattern, total saturation of the ring with this catalyst system requires a prolonged reaction time; in contrast, the complete hydrogenation of 1,3,5,7-tetraethoxycarbonylcyclo-octatetraene to (430) is rapid.[640]

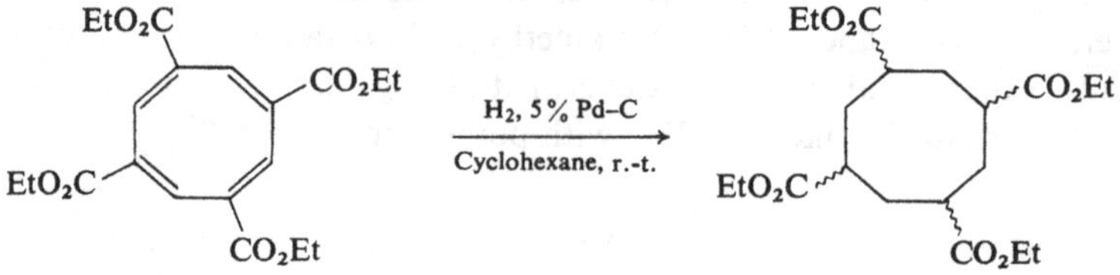

(430)

(e) *Cycloadditions.* The equilibrium mixture of (418) and (395) reacts with maleic anhydride to give the adduct (431).[651]

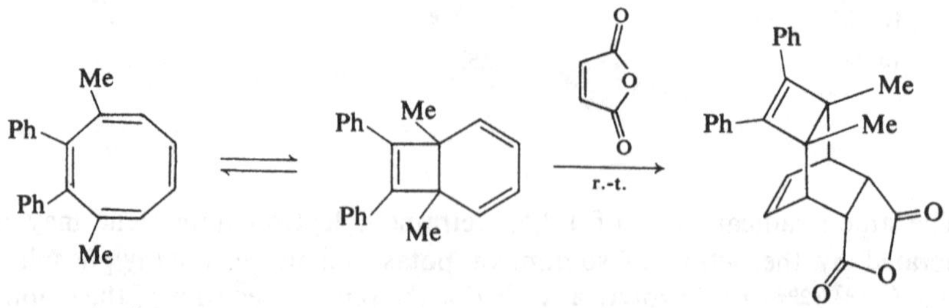

(418)   (395)   (431)

A similar adduct is obtained with *N*-methyltriazolinedione;[651] this also reacts with the equilibrium mixture of (419) and (420) to afford the adduct (432).[651]

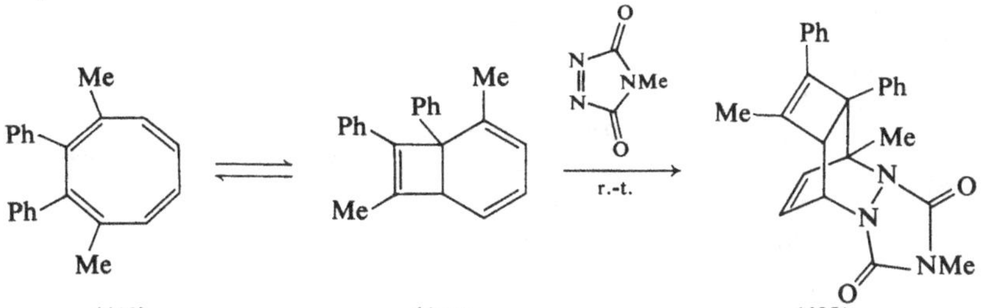

(419) (420) (432)

With *N*-phenyltriazolinedione, the 1,2,3,4- and 1,2,3,8-tetramethyl-compounds yield the adducts (433) and (434) respectively.[655]

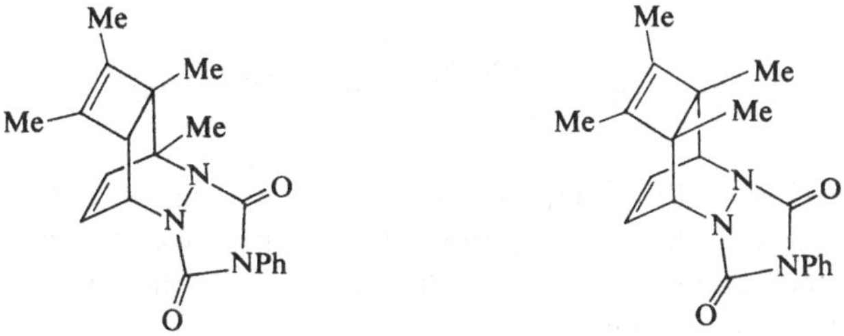

(433) (434)

For the tetracyanoethylene adducts of 1,2,4,5-, 1,2,4,7- and 1,2,5,6-tetramethylcyclo-octatetraene, see ref. 652.

(*f*) *Formation of metal derivatives.* The 1,3,5,7-tetramethylcyclo-octatetraenide (435) incorporates a planar dianion, as shown by X-ray crystallography.[665]

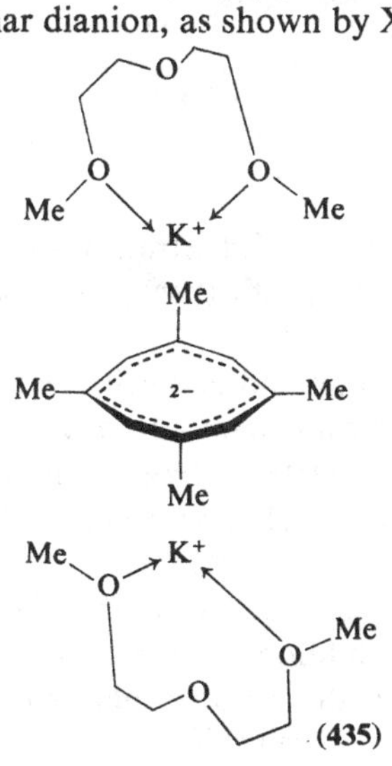

(435)

1,3,5,7-Tetramethylcyclo-octatetraene effects the following displacement reactions to afford the chromium (**436a**) (60%),[648] molybdenum (**436b**) (51%)[648, 666] and tungsten (**436c**) (*ca.* 64%)[648] tricarbonyl complexes. The

structure of (TMCOT)Cr(CO)₃ (**436a**)[667] resembles that of (COT)Mo(CO)₃ (**168b**) (see p. 58). The ¹H [648, 666] and ¹³C [432] spectra of these compounds reveal that they undergo a complex fluxional process in solution, involving 1,2-shifts of the metal atom.

(**436a**)

The equilibrium mixture of 1,2,3,4-tetramethylcyclo-octatetraene and its bicyclic valence tautomer reacts with (benzylidene–acetone)Fe(CO)₃ to give the complex (**437**) (79%).[655]

The 1,2,3,8-tetramethyl-compound is much less reactive, but with a large excess of the reagent yields the analogous product (**438**).[655]

1,3,5,7-Tetramethylcyclo-octatetraene reacts with iron carbonyls to form a variety of complexes, some of which have no parallel amongst the complexes of COT itself.

With Fe₃(CO)₁₂, the main product is the tricarbonyl complex (**439**), containing a bicyclic ligand; in addition, the binuclear pentacarbonyl complex (**440**) is produced, together with an isomer (**441**) in which the ligand has apparently experienced a 1,3-hydrogen shift.[668]

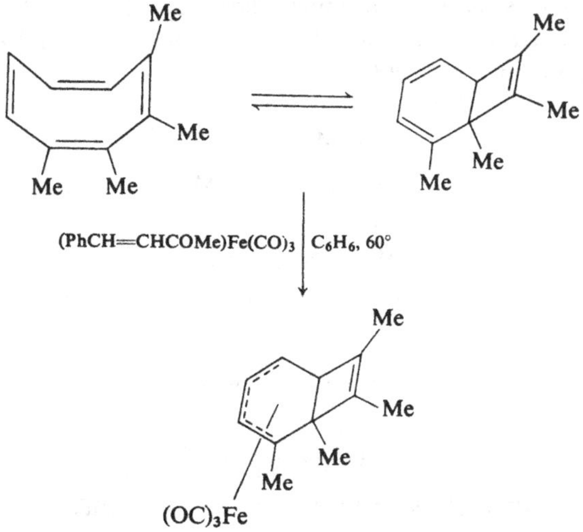
Me
Me
Me
Me
Me
Me
Me
Me
(PhCH=CHCOMe)Fe(CO)3
C6H6, 60°
Me
Me
Me
Me
(OC)3Fe

(437)

(Stereochemistry not certain)

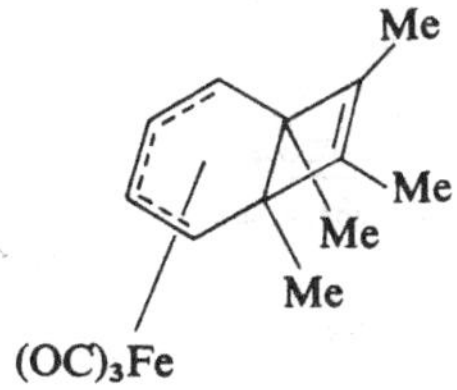
Me
Me
Me
Me
(OC)3Fe

(438)

(Stereochemistry not certain)

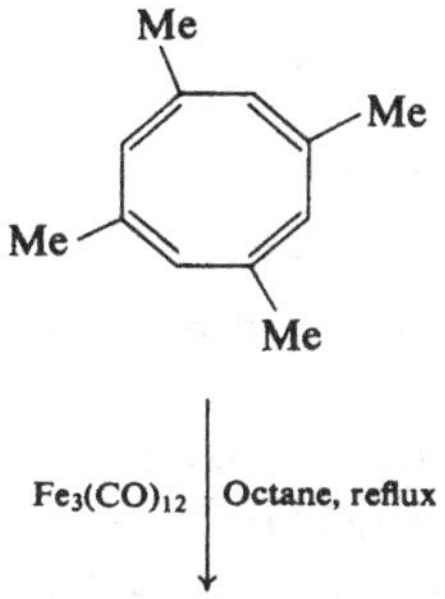
Me
Me
Me
Me
Fe3(CO)12
Octane, reflux

(1,3,5,7-Tetramethylbicyclo[4.2.0]octa-2,4,7-triene)$Fe(CO)_3$ **(439)**

\+ $(TMCOT)Fe_2(CO)_5$ **(440)**

\+ (1,3,5-Trimethyl-7-methylenecyclo-octa-1,3,5-triene)$Fe_2(CO)_5$ **(441)**

Complex (440) is the main product when $Fe_2(CO)_9$ in refluxing octane is used, but the reaction also yields the isomer (441) and the '*trans*'-binuclear hexacarbonyl complex (442).[668]

$Fe_2(CO)_9$ | Octane, reflux

(TMCOT)Fe₂(CO)₅ (440)

+ (1,3,5-Trimethyl-7-methylenecyclo-octa-1,3,5-triene)Fe₂(CO)₅ (441)

+ '*trans*'-(TMCOT)Fe₂(CO)₆ (442)

In refluxing hexane, however, $Fe_2(CO)_9$ gives a minute yield of the '*cis*'-binuclear hexacarbonyl complex (443).[668]

$\xrightarrow[\text{Hexane, reflux}]{Fe_2(CO)_9}$ '*cis*'-(TMCOT)Fe₂(CO)₆ (443)

Structure (439) is in accord with the spectral evidence.[668] The complex

(OC)₃Fe

(439)

(Stereochemistry not certain)

(TMCOT)Fe₂(CO)₅ (440), which is fluxional, possesses a crystal structure analogous with that of (COT)Fe₂(CO)₅[669] (see p. 199). In compound (441), which contains an Fe–Fe bond, an Fe(CO)₃ grouping is bonded to a π-allyl system in the ring, while the remaining Fe(CO)₂ is attached to another allylic system involving an exocyclic methylene group, and also to an endocyclic double bond.[670] The complexes (442) and (443) are apparently analogues of

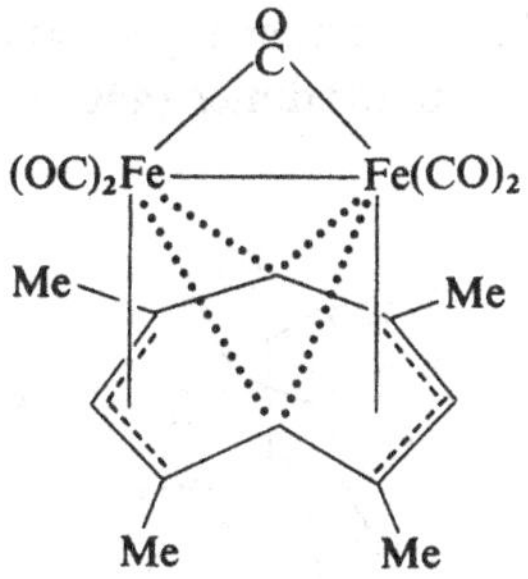

(440)

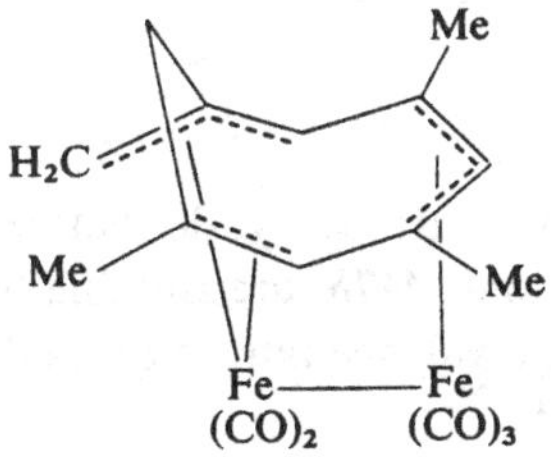

(441)

'*trans*'- and '*cis*'-(COT)Fe$_2$(CO)$_6$ respectively (see pp. 63, 65).[668]

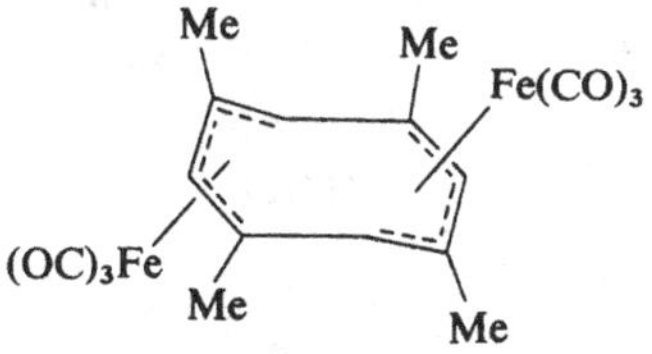

(442)

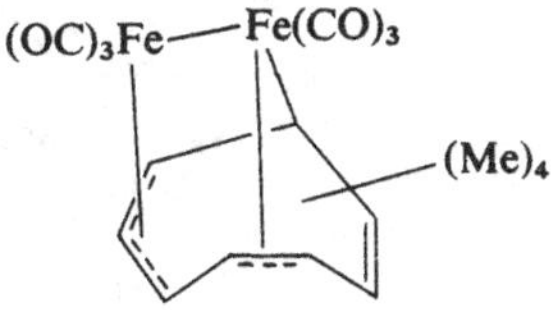

(443)

6. Pentasubstituted derivatives

Formation

1,2,3,4,6- and 1,2,3,5,8-pentamethylcyclo-octatetraene are formed from the sulphone (444) and the diene (445) respectively, using reaction sequences similar to those employed for the 1,2,3,4- and 1,2,3,8-tetramethyl-compounds[671] (see p. 140).

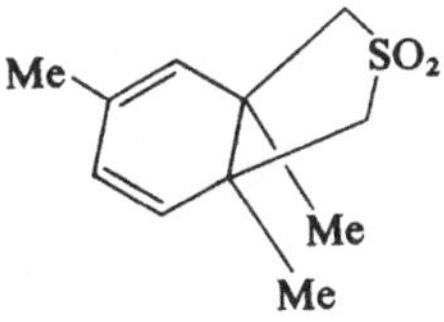

(444)

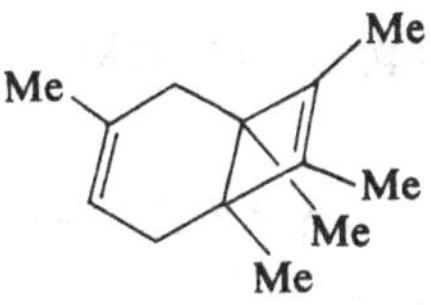

(445)

2,4,6,8-Tetraphenylcyclo-octatetraene-1-carboxylic acid is obtained (50%) when 4,6-diphenyl-2-pyrone photodimer (446) is treated with trifluoroacetic acid.[672]

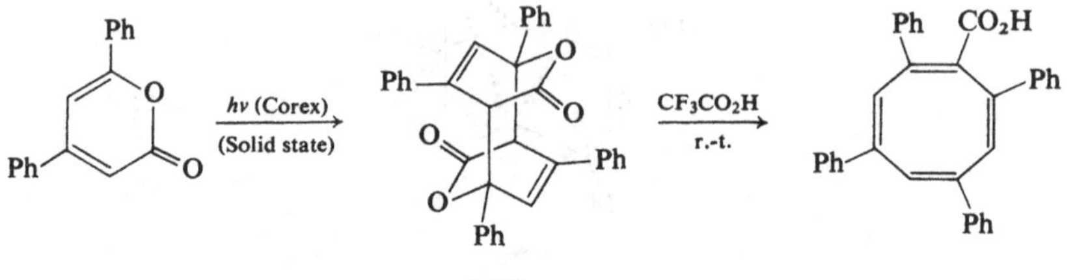

(446)

Structure

Like the 1,2,3,4- and 1,2,3,8-tetramethyl-derivatives (see p. 142), the pentamethyl-compounds (447*a*) and (447*b*) are isolable bond-shift isomers; (447*a*) equilibrates with the bicyclic valence tautomers (448) and (449), but (447*b*) is seemingly homogeneous.[671]

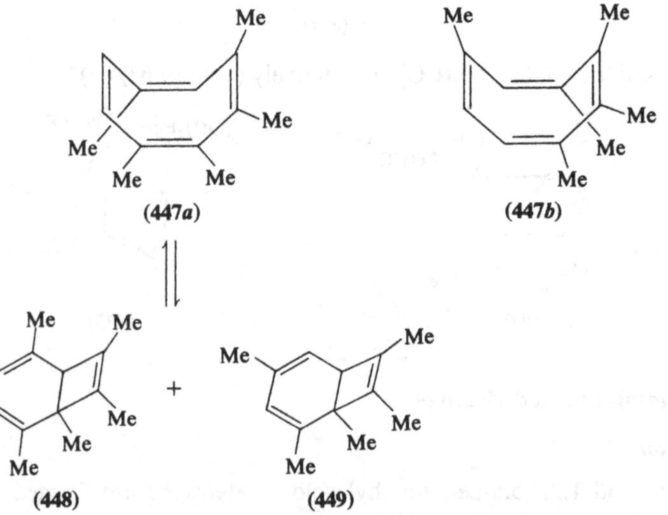

(447*a*)

(447*b*)

(448)   +   (449)

7. Hexa- and heptasubstituted derivatives

These are apparently unknown.

8. Octasubstituted derivatives

Formation

(*a*) *Cyclotetramerisation of disubstituted acetylenes.* The cyclotetramerisation of diphenylacetylene may be achieved (up to 7.5% yield) in the presence of certain ether-free organomagnesium compounds, of which phenylmagnesium bromide is the most efficient.[673, 674]

PhC≡CPh

$$\begin{matrix} Ph \\ C \\ \parallel \\ C \\ Ph \end{matrix} \quad + \quad \begin{matrix} Ph \\ C \\ \parallel \\ C \\ Ph \end{matrix} \quad \xrightarrow[\text{Xylene, reflux}]{PhMgBr}$$

PhC≡CPh

A minute yield (0.06%) of octaphenylcyclo-octatetraene is formed by u.v. irradiation of diphenylacetylene.[675]

(*b*) *Syntheses involving skeletal transformations.* The dimerisation of tetrasubstituted cyclobutadienes results in octasubstituted tricyclo[4.2.0.0²,⁵]octa-3,7-dienes, which undergo thermal rearrangement to yield octasubstituted COTs (see pp. 137–9); in some cases the cyclobutadiene dimers rearrange under the reaction conditions of their formation.

The tetramethylcyclobutadiene dimer (**450**) (shown to possess *anti*-stereochemistry) may be obtained by heating the nickel(ii) chloride complex (**451**) at 120°; at higher temperatures octamethylcyclo-octatetraene is formed (scheme 63). The low yields of octamethylcyclo-octatetraene obtained by these procedures result from its thermal rearrangement (see p. 160).

Scheme 63

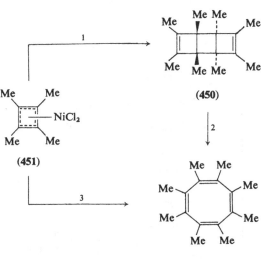

(**450**)

(**451**)

Reaction	Conditions	Yield (%)	Ref.
1	120°, H₂O	63–65	676
2	300°/3 mm	10	220
3	185–200°/0.001 mm	7–21	677

The cyclobutene (452), on treatment with phenyl-lithium, forms a debrominated product which may be the cyclobutadiene dimer (453) (80%); the same compound results (40%) from a reaction between the dilithio- and di-iodobutadienes (454) and (455).[678] This product slowly rearranges to octaphenylcyclo-octatetraene (even in the solid state) at room temperature.[678] Octaphenylcyclo-octatetraene is also formed by thermolysis of the di-iodobutadiene (455) (5%), and by treatment of the dilithio-compound (454) with copper(II) bromide (15%).[679] Homolytic extrusion of dimethyltin bromide from the compound (456) affords octaphenylcyclo-octatetraene in high yield (85%);[680] thermolysis of tetraphenylmercurole (457) gives the same product.[679]

(456)

(457)

The dimeric (tetra-arylcyclobutadiene)palladium(II) chloride complexes (458) form octa-arylcyclo-octatetraenes (459) on treatment with triphenyl-phosphine or (preferably[681]) tributylphosphine (table 59).

(458)   (459)

Table 59

Ar^1	Ar^2	n	R	Yield (%)	Ref.
Ph	Ph	1	Ph	*ca.* 70	682, (683)
p-Me.C_6H_4	p-Me.C_6H_4	3	Bu^n	38	681
p-Me.C_6H_4	p-MeO.C_6H_4	1	Ph	60	684
p-MeO.C_6H_4	p-MeO.C_6H_4	3	Bu^n	94	681
p-Cl.C_6H_4	p-Cl.C_6H_4	2	Bu^n	73	685

Treatment of the cyclobutene (460) with methyl-lithium gives a mixture of products, chiefly (461) (37%), (462) (22%) and (463) (36%); isomerisation of (462) to (461) and of (463) to (464) is effected by caesium fluoride, and at 300° these products rearrange to octakis(trifluoromethyl)cyclo-octatetraene in quantitative yield.[686]

The isolable tetrasubstituted cyclobutadiene (465), when heated at 110°, yields the octasubstituted COT (466) (70%).[687]

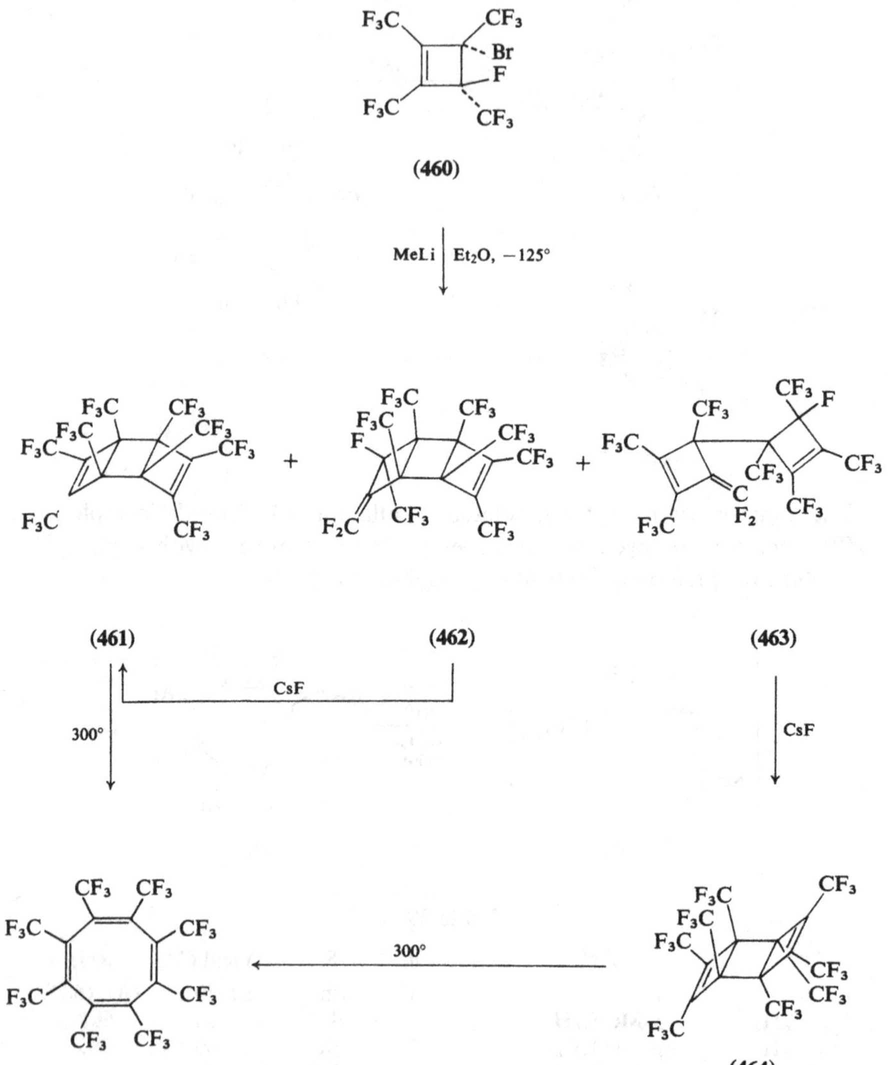

(460)

MeLi | Et₂O, −125°

+   +

(461)   (462)   (463)

CsF

300°   CsF

← 300° ←

(464)

110° →

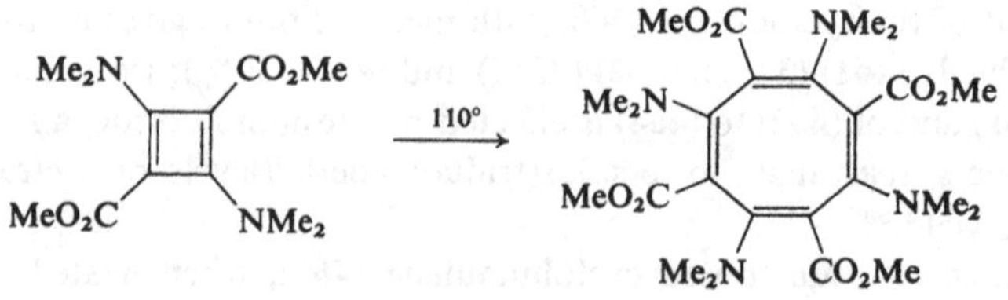

(465)   (466)

Dehydrobromination of the cyclobutene (467) gives the octasubstituted COT (468) in low yield (5%).[688]

Dehalogenation of the cyclobutenes (469) affords the cyclobutadiene dimers (470), which are readily converted into the related octahalocyclo-octatetraenes (471) (scheme 64).[689] The octachloro-diene (470; X = Cl) is also formed

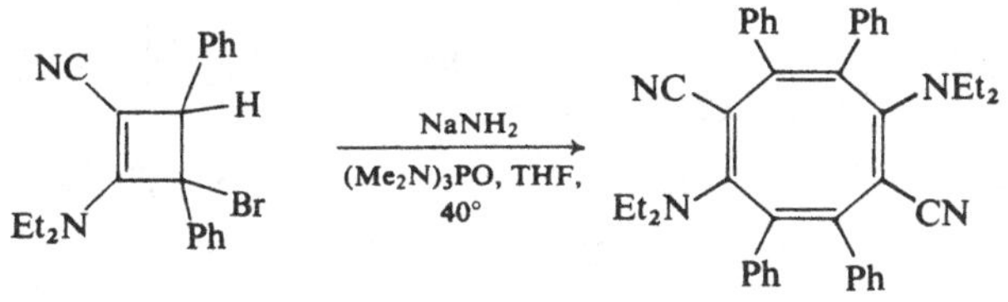

(467)   (468)

on photolysis of the ozonide (472) derived from the 'Dewar-benzene' (473) (*ca.* 22% overall yield).[689]

Scheme 64

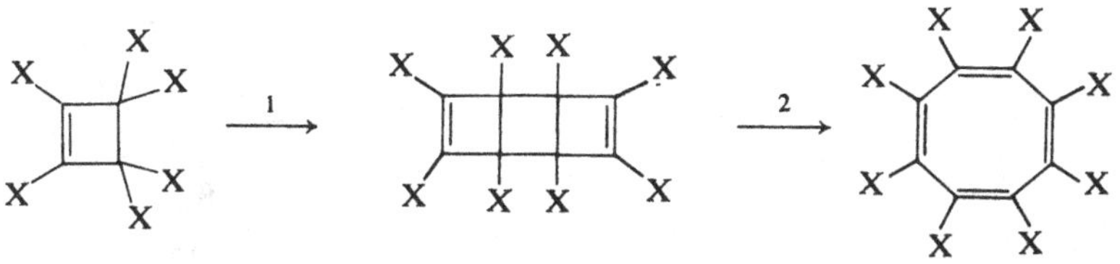

(469)   (470)   (471)

X = Cl, Br

Reaction	X	Conditions	Yield (%)
1	Cl	Li–Hg, Et$_2$O, r.-t.	14–22
1	Br	Li–Hg, Et$_2$O, r.-t.	27–40
2	Cl[a]	Cl$_2$C=CCl$_2$, reflux	75
2	Br	*o*-Dichlorobenzene, reflux	58

[a] See also ref. 690.

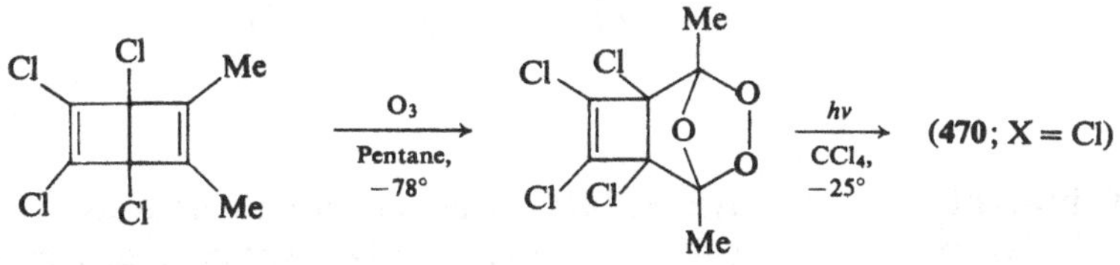

(473)   (472)

A special preparative method for halocyclo-octatetraenes involves thermal dimerisation of the enynes (474), which results in dimers of structure (475); acid-catalysed rearrangement then leads to the octasubstituted COTs (476) (table 60).[691-694]

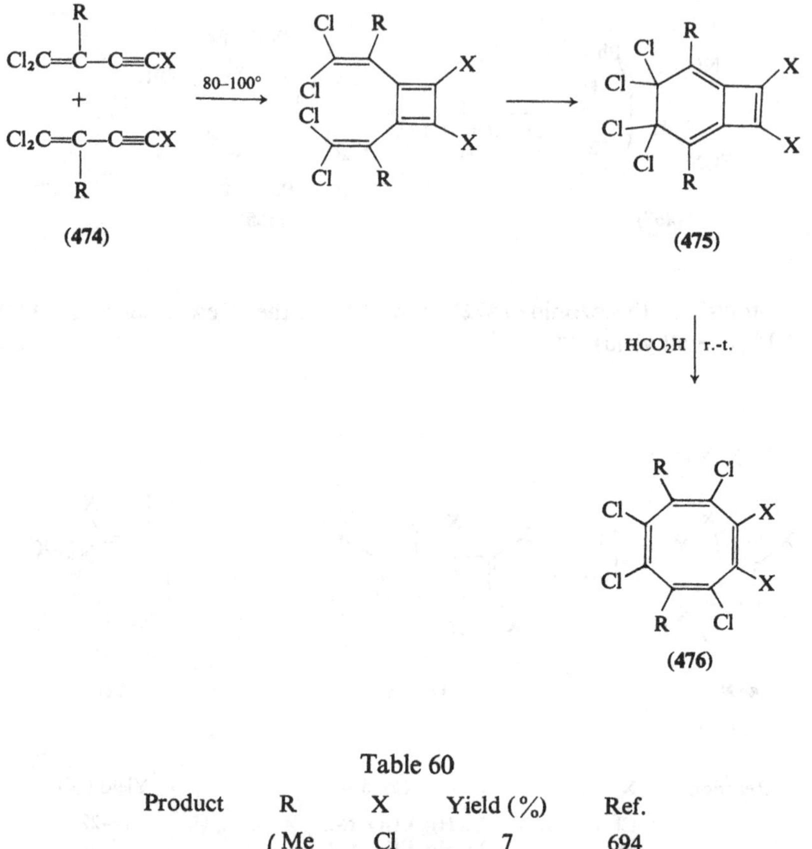

(474)   (475)

$$\text{HCO}_2\text{H} \quad \text{r.-t.}$$

(476)

Table 60

Product	R	X	Yield (%)	Ref.
(475)	Me	Cl	7	694
	Cl	Cl	35	695
	Cl	Br	33–37	693
	Br	Cl	39	694
(476)	Me	Cl	60 (crude)	694
	Cl	Cl	35	695
	Cl	Br	83	693, (696)
	Br	Cl	33	694

U.v. irradiation of the anhydride (477) gives multifarious products, one of which is octamethylcyclo-octatetraene (5.3%); the reaction sequence involved in its formation may be as shown.[697]

The sensitised photolysis of the heterocycle (478) affords octamethylcyclo-octatetraene almost exclusively.[698]

Structure

The crystal structures of octamethyl- and octaphenylcyclo-octatetraene* have been determined, with the results (for the COT ring) listed in table 61.

* The structure of the compound now known[699-701] to be octaphenylcyclo-octatetraene was for some time in doubt.[678, 680, 683, 702]

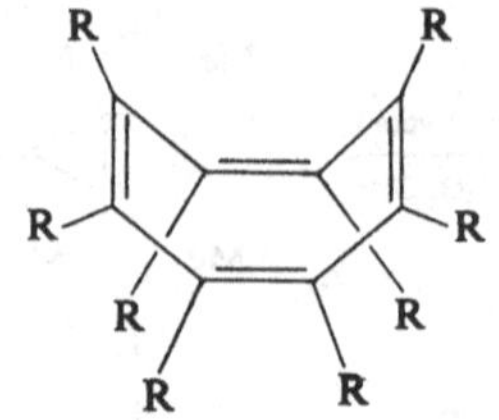

Table 61

R	Bond	Length (Å)	Ref.
Me	C=C	1.326	116
Ph	C=C	(1.35), 1.342	(700), 701
Me	C—C	1.483	116
Ph	C—C	(1.51), 1.493	(700), 701
Me	Angle	122°	116
Ph	Angle	(120°), 122°	(700), 701
Me	Torsion angle around	67°	116

For X-ray confirmation of the structure of octachlorocyclo-octatetraene, see ref. 703.

It is evident that the presence of eight substituent groups produces a significant flattening of the COT ring (see pp. 9, 93, 115, 142).

For the calculated barrier to ring-inversion of octamethylcyclo-octatetraene, see ref. 127.

Reactions

(a) *Thermal isomerisation.* Octamethylcyclo-octatetraene rearranges at 200–210°,[677] giving a product now formulated as (479).[220] (Thus thermolysis of the

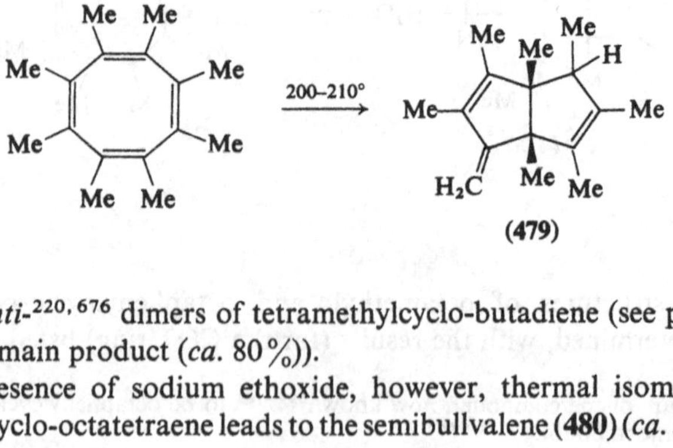

(479)

syn-[704] or anti-[220, 676] dimers of tetramethylcyclo-butadiene (see p. 153) gives (479) as the main product (ca. 80%)).

In the presence of sodium ethoxide, however, thermal isomerisation of octamethylcyclo-octatetraene leads to the semibullvalene (480) (ca. 37%).[705, 706]

Me Me Me Me
Me⟍ ⟋Me 1% NaOEt   Me⟍Me⟍ ⟋Me
Me⟍ ⟋Me EtOH, 240° Me⟍ ⟍Me
Me Me Me Me

(480)

Octachlorocyclo-octatetraene, when kept at 180° for a short time, affords the bicyclic isomer **(481)** (41 %); prolonged heating effects further rearrangement to give a second isomer **(482)** (58 %).[707]

Cl Cl   Cl Cl Cl   Cl Cl Cl
Cl⟍ ⟋Cl 180° Cl⟍ ⟋Cl ⟶ Cl⟍ ⟋Cl
Cl⟍ ⟋Cl   Cl⟍ ⟋Cl   Cl⟍ ⟋Cl
Cl Cl   Cl Cl   Cl Cl

(481)   **(482)**

(b) *Photo-isomerisation.* Light-induced isomerisation of octakis(trifluoro-methyl)cyclo-octatetraene leads to the formation of the cubane **(483)** and cuneane **(484)** as end-products.[686]

F_3C CF_3 F_3C CF_3 F_3C F_3C CF_3
F_3C⟍ ⟋CF_3 hν F_3C⟍ ⟋CF_3 + F_3C⟍ ⟋CF_3
F_3C⟍ ⟋CF_3 F_3C⟍ ⟋CF_3 F_3C⟍ ⟋CF_3
F_3C CF_3 F_3C CF_3 F_3C CF_3

(483)   **(484)**

(c) *Oxidation, and reactions with electrophiles.* With perbenzoic acid, octamethylcyclo-octatetraene gives a product formulated as **(485)** (96 %);[708] this structure may need correction (see refs. 220, 706).

Treatment of octamethylcyclo-octatetraene with fluorosulphonic acid–antimony(III) fluoride produces a species formulated as the bicyclic dication **(486)**, probably *via* protonation followed by hydride abstraction; proton abstraction by base then affords the $C_{16}H_{22}$ hydrocarbon **(487)**.[220]

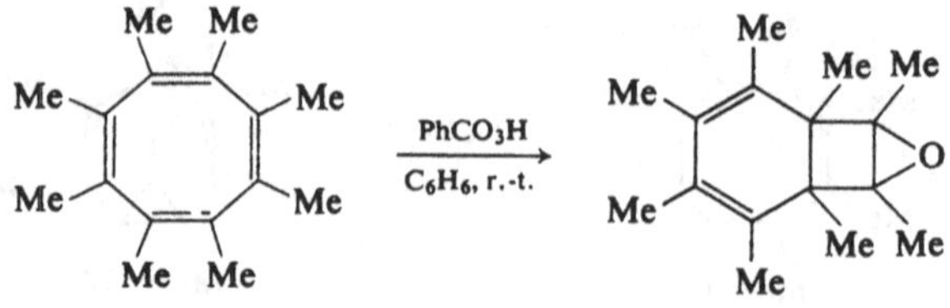

PhCO₃H, C₆H₆, r.-t.

(485)

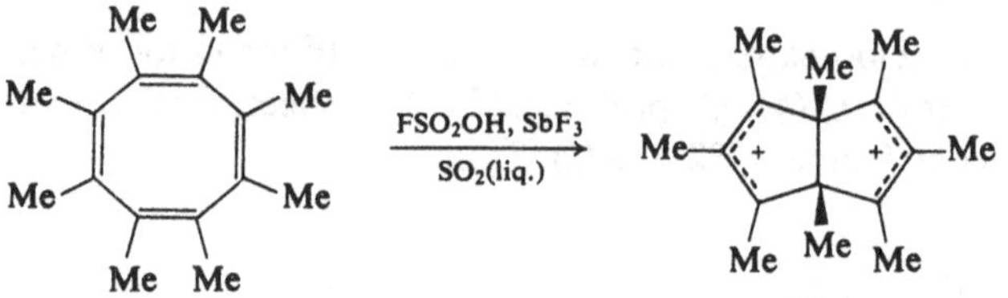

FSO₂OH, SbF₃, SO₂(liq.)

(486)

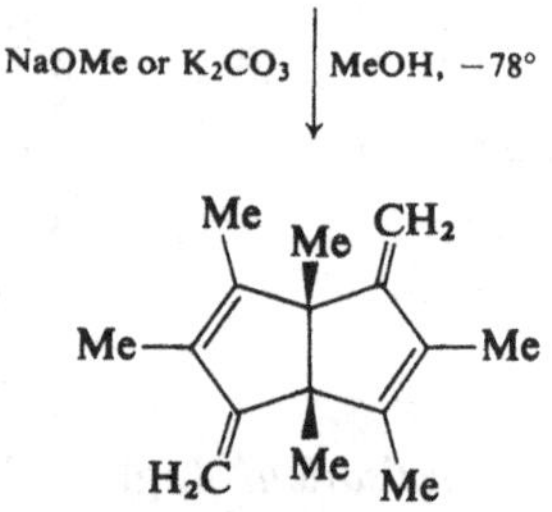

NaOMe or K₂CO₃ | MeOH, −78°

(487)

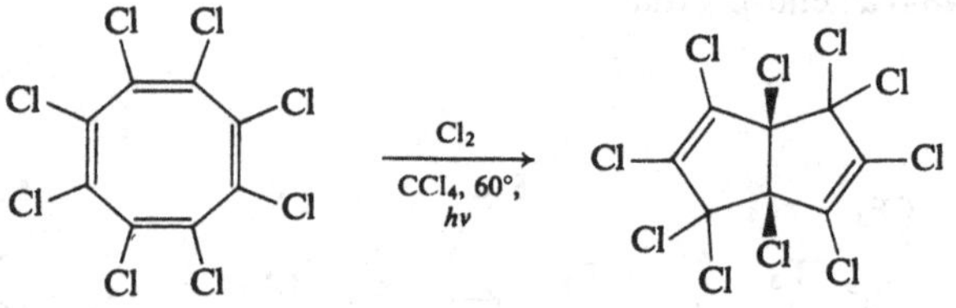

Cl₂, CCl₄, 60°, hv

(488)

AlX₃, CS₂, reflux

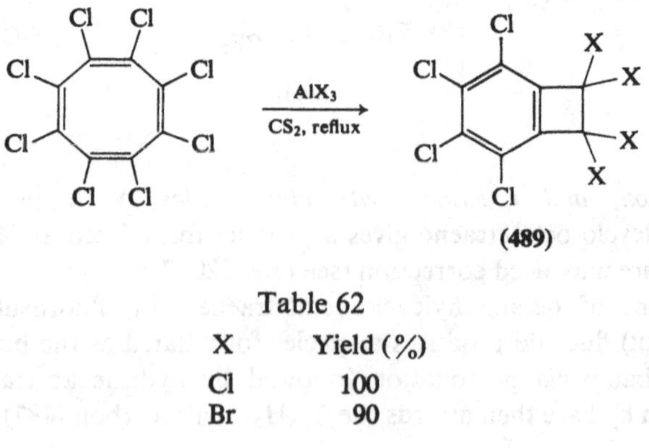

(489)

Table 62

X	Yield (%)
Cl	100
Br	90

Chlorination of octachlorocyclo-octatetraene yields the bicyclic decachloro-derivative (**488**) (81 %).[707]

With aluminium halides, octachlorocyclo-octatetraene affords the benzo-cyclobutenes (**489**)[709] (table 62).

(*d*) *Cycloadditions.* See appendix.

3

Further reactions of compounds derived from cyclo-octatetraenes

1. 1,2-Didehydrocyclo-octatetraene

1,2-Didehydrocyclo-octatetraene (**236**) may be generated by dehydrobromination of bromocyclo-octatetraene (see pp. 83–4).

Reactions. The preparation of some monosubstituted COTs and annulated derivatives from (**236**) has already been described (see pp. 83–4, 124–5); other reactions are outlined in scheme 65.

Scheme 65

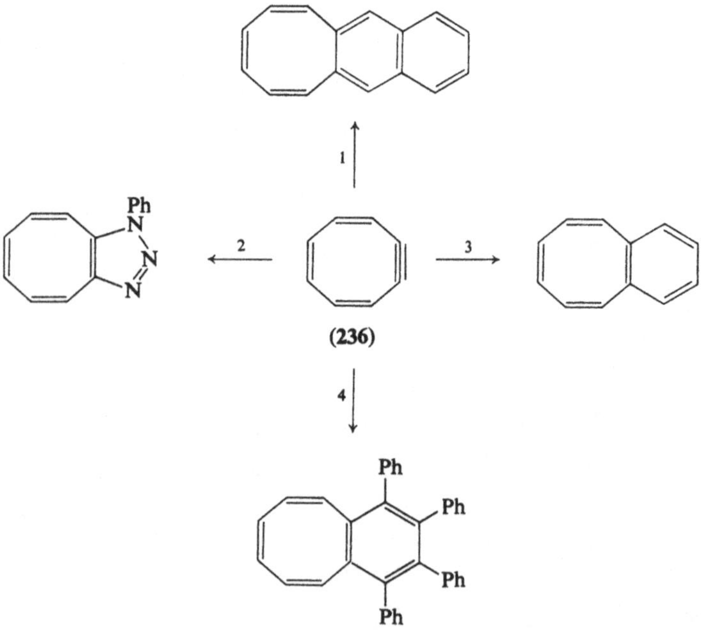

(**236**)

Reaction	Reagents and conditions	Yield (%)	Ref.
1	THF, r.-t.	10	624
2	Phenyl azide, Et$_2$O, r.-t.	38	624
3	1-(*N*,*N*-Diethylamino)buta-1,3-diene, Et$_2$O, r.-t.	5	624
4	Tetracyclone, Et$_2$O, r.-t.	(72), 91	(624), 710

2. Bicyclo[4.2.0]octa-2,4,7-trienes

These valence tautomers of COTs have been discussed in the sections dealing with the parent compounds.

For the (low-temperature) preparation of bicyclo[4.2.0]octa-2,4,7-triene, see pp. 245, 354.

3. Bicyclo[3.3.0]octa-2,6-dienes, bicyclo[3.3.0]octa-1,3,6-trienes and bicyclo[3.3.0]octa-1,3,7-trienes

The perchloro-compounds (481), (482) and (488) are available from octa-chlorocyclo-octatetraene (see pp. 161–3).

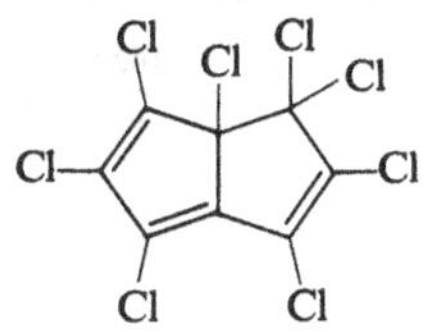

(481) 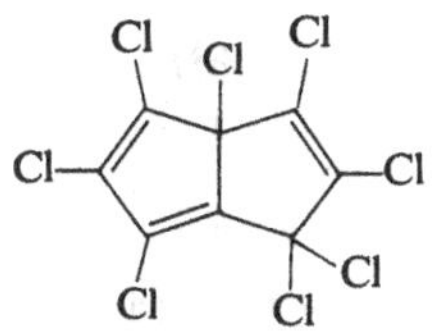   (482) 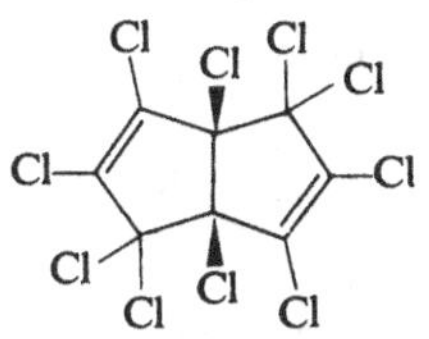   (488)

Reactions. Most of the known[707] chemistry of these systems is summarised in schemes 66 and 67. In addition, dechlorination of compounds (481) or (482) with tin(II) chloride affords a dimer of hexachloropentalene (490) (78–86 %).[707]

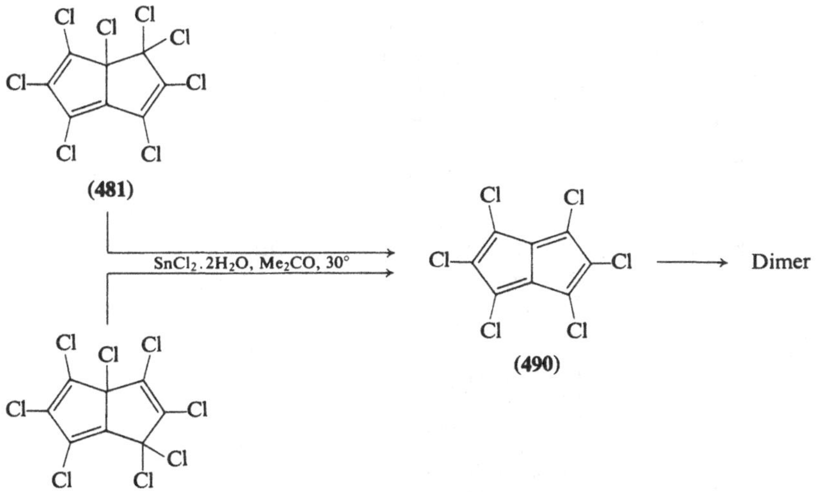

(481)

SnCl$_2$.2H$_2$O, Me$_2$CO, 30°

(490)   Dimer

(482)

Scheme 66

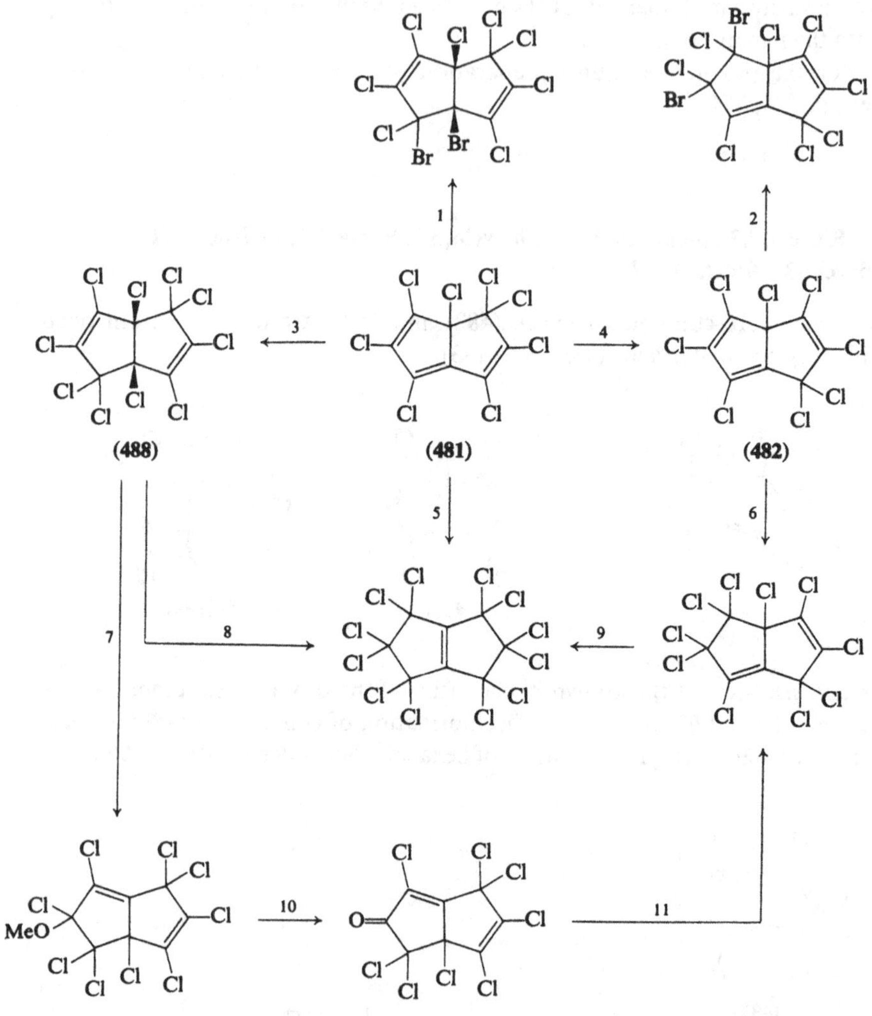

Reaction	Conditions	Yield (%)
1	Br$_2$, r.-t.	77
2	Br$_2$, r.-t.	88
3	Cl$_2$, CCl$_4$, 40°, hv	68
4	AlCl$_3$, CS$_2$, r.-t.	100
5	Cl$_2$, CCl$_4$, 60°, hv	88
6	Cl$_2$, CCl$_4$, 40°, hv	27
7	NaOH, MeOH, reflux	60
8	Cl$_2$, CCl$_4$, 60°, hv	88–95
9	Cl$_2$, CCl$_4$, 60°, hv	88–95
10	Conc. H$_2$SO$_4$, r.-t.	72
11	PCl$_5$, 180°	78

Scheme 67

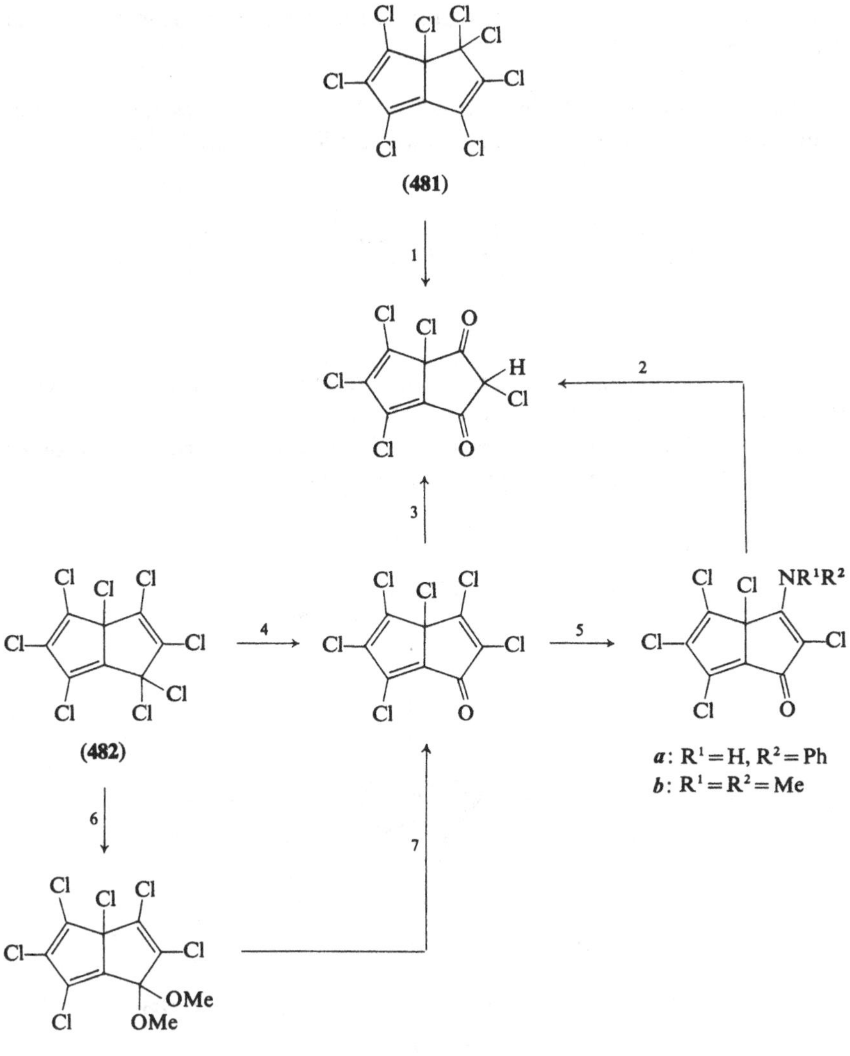
(481)
1
2
3
4
5
(482)
6
7
a: R¹ = H, R² = Ph
b: R¹ = R² = Me

Reaction	Conditions	Yield (%)
1	Conc. H_2SO_4, r.-t.	77
2	Conc. H_2SO_4, r.-t.	52–68
3	HCO_2H, 80°	72
4	HCO_2H, r.-t.	65
	or $Ag(OCOCF_3)$, C_6H_6, r.-t.	74
5	(*a*) $PhNH_2$, Et_2O, r.-t.	81
	(*b*) Me_2NH(aq.), Et_2O, r.-t.	76
6	MeOH, reflux	53
7	Conc. H_2SO_4, r.-t.	72

4. Bicyclo[3.3.0]octa-1,4,6-triene

The dihydropentalene (**24**) may be prepared by the thermal isomerisation of COT (see p. 18).

Reactions. Proton-abstraction with n-butyl-lithium generates the 'aromatic' pentalenyl dianion, isolable as the crystalline dilithium salt.[711, 712]

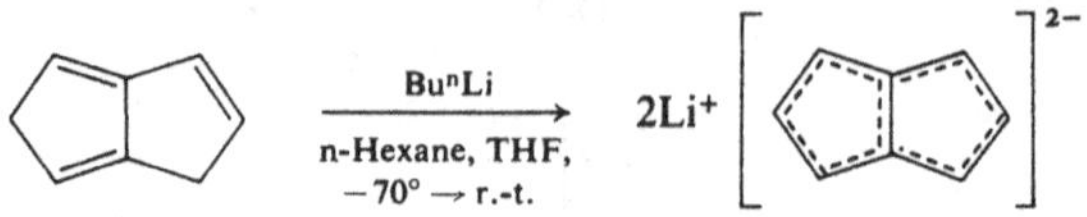

(**24**)

The dihydropentalene (**24**) reacts with maleic anhydride to form an adduct[712] (no structural evidence available).

The dihydropentalene (**24**) forms a thallium derivative from which other metal derivatives may be obtained[713] (scheme 68).

Scheme 68

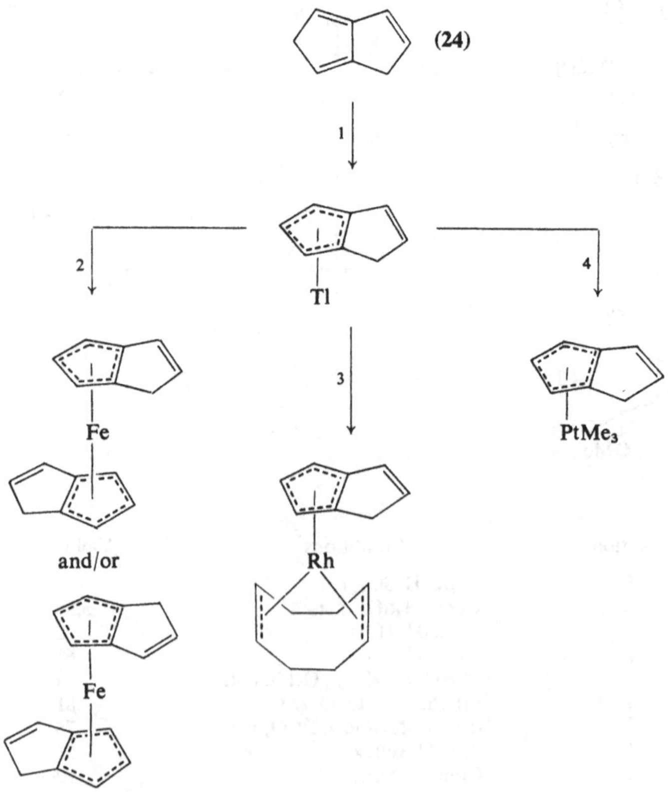

Reaction	Reagents and conditions	Yield (%)
1	Tl_2SO_4, NaOH(aq.), r.-t.	100
2	$FeCl_2$, THF, r.-t.	49
3	[(COD)RhCl]$_2$, THF, r.-t.	76
4	Me_3PtI, THF, r.-t.	43

5. Semibullvalenes

Semibullvalene (18) is formed directly from COT by u.v. irradiation (see p. 18), but a more satisfactory preparative procedure, which may also be used for substituted derivatives, is the indirect route *via* 9,10-diazatricyclo[4.2.2.0^{2,5}]-deca-3,7-dienes (see pp. 353, 355). The octamethyl-compound (480) is available from octamethylcyclo-octatetraene (see pp. 160–1).

(18)   (480)

The semibullvalene structure undergoes an extremely rapid fluxional process *via* a degenerate Cope rearrangement.[44,714]

For other leading references on semibullvalene, see ref. 635.
For studies with substituted semibullvalenes, see refs. 715, 716.

Reactions. The gas-phase thermolysis of semibullvalene leads to COT (see p. 4); u.v. irradiation also induces equilibration with COT (see p. 18).

Photo-isomerisation of octamethylsemibullvalene yields the cuneane (491) (10–20%).[706]

(480)   (491)

The octamethyl-compound readily absorbs oxygen to form the peroxide (492) (75%),[705, 706] and is easily dehydrogenated to give the dimethylene compound (487) (up to 80%)[705, 706] (for a possible mechanism, see ref. 635).

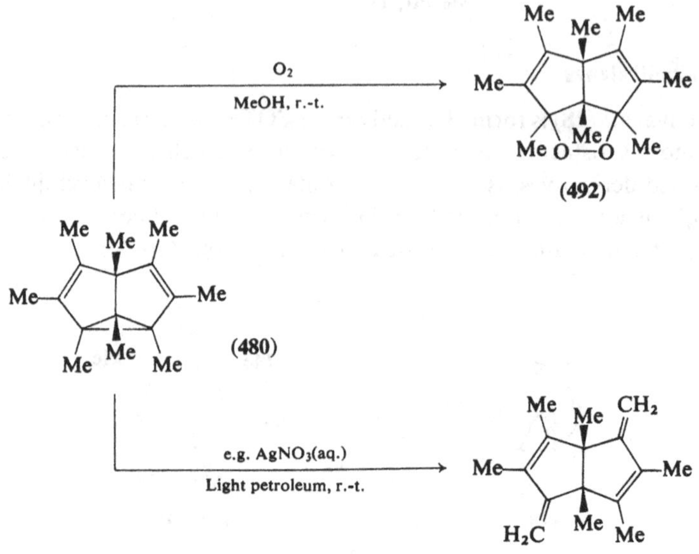

(492)

(480)

(487)

Semibullvalene adds bromine in a stereoselective manner to afford the *cis*, *exo*-dibromide (493) (67%).[717]

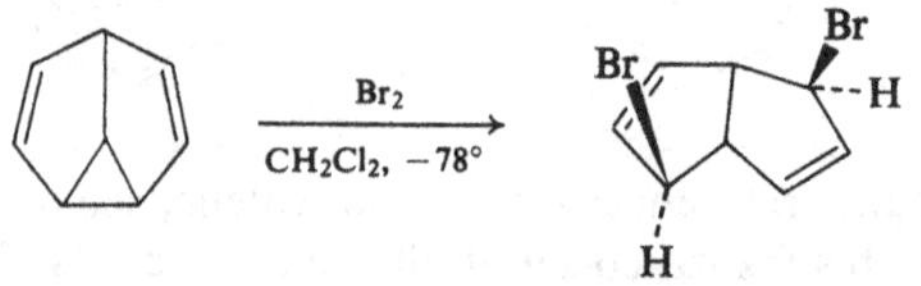

(493)

Catalytic hydrogenation of semibullvalene gives the tetrahydro-derivative (494).[714]

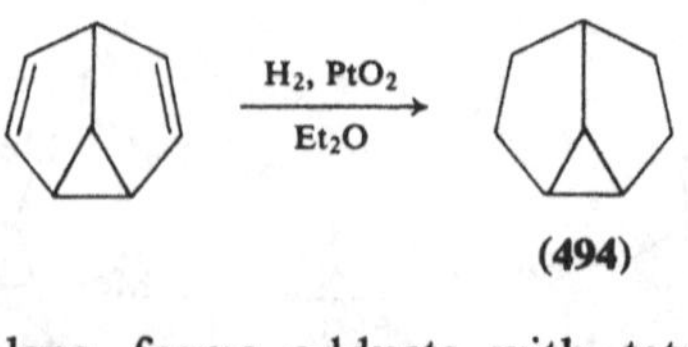

(494)

Octamethylsemibullvalene forms adducts with tetracyanoethylene and diethyl azodicarboxylate[705, 706] (see appendix).

Silver and tungsten complexes of semibullvalene are known[718] (for the silver-catalysed rearrangement of semibullvalene to COT, see p. 4).

U.v. irradiation at 25° in the presence of $Fe(CO)_5$ affords the complexes (495) and (496); at −50° the products are (496) and (497), the latter disproportionating at 45° to semibullvalene and (498).[719] Spectral evidence indicates

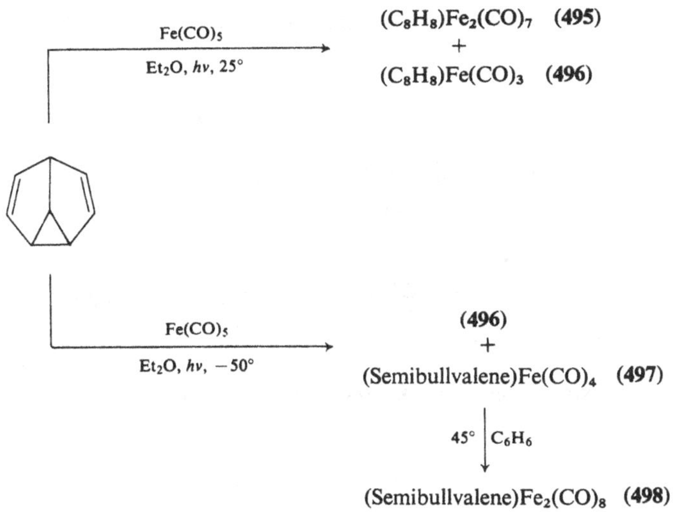

(C₈H₈)Fe₂(CO)₇ $(C_8H_8)Fe_2(CO)_7$ (495)

Fe(CO)₅, Et₂O, hv, 25° → $(C_8H_8)Fe_2(CO)_7$ (495) + $(C_8H_8)Fe(CO)_3$ (496)

Fe(CO)₅, Et₂O, hv, −50° → (496) + (Semibullvalene)Fe(CO)₄ (497)

45° | C₆H₆

(Semibullvalene)Fe₂(CO)₈ (498)

that the C_8H_8 ligand in the complexes (495) and (496) is bonded as shown.[720]

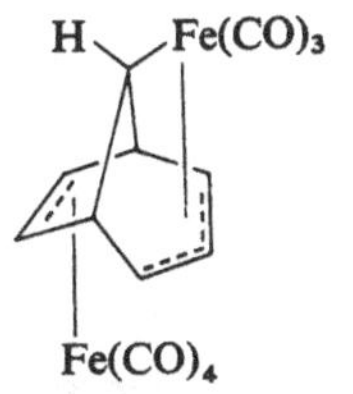

(495)

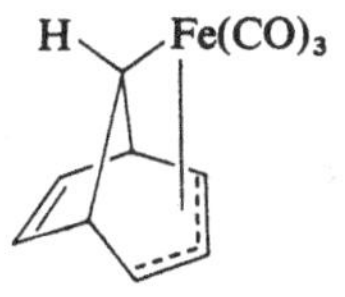

(496)

Thermal reactions occur with $Fe_2(CO)_9$; at room-temperature the main products are the binuclear species (495) and (498);[720] on heating, however, the

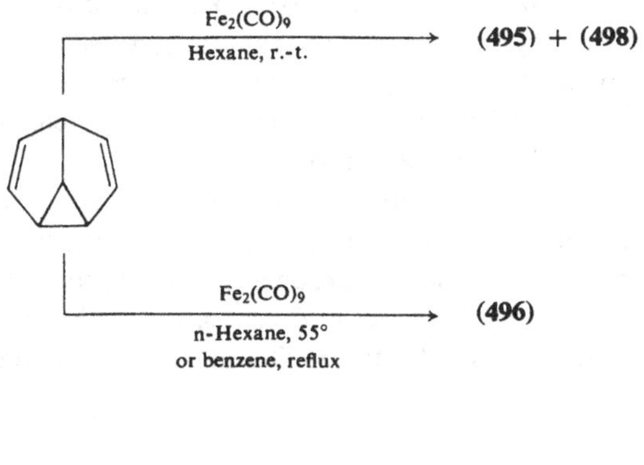

Fe₂(CO)₉, Hexane, r.-t. → (495) + (498)

Fe₂(CO)₉, n-Hexane, 55° or benzene, reflux → (496)

mononuclear tricarbonyl complex (496) is obtained (70%).[721,722] The complexes (497) and (498), containing the semibullvalene ligand, are formulated as follows.[719,720]

(497)   (498)

Semibullvalene reacts with methanolic potassium tetrachloropalladate, giving the dimeric complex (499) (75%).[723]

$$\xrightarrow[\text{MeOH}]{\text{K}_2\text{PdCl}_4}$$ [(MeO.C_8H_8)PdCl]$_2$ (499)

The suggested structure of the product is given below.

(499)

6. Metal derivatives

(a) Lithium, sodium and potassium derivatives

Donation of two electrons from an alkali metal to the COT ring leads to the dianion (60) (see pp. 28, 48–9).

Reactions. The cyclo-octatetraenyl dianion (60) is capable of transferring electrons to other species in several different ways.

Anion radicals may be generated from neutral substrates, e.g. COT (see p. 28), 6,6-diphenylfulvene,[724] phenanthraquinone,[725] dicyanodiphenyl-ethylenes,[724] nitrosobenzene,[726] nitrobenzene,[726] *m*-dinitrobenzene,[727] nitro-stilbenes,[728,729] benzofurazan,[730] benzothiadiazoles,[731] etc. With benzonitrile, electron-transfer induces trimerisation.[378]

The ability of COT^{2-} to function as a reducing agent is illustrated by its reactions with benzil,[725] azobenzene,[732] azoxybenzene,[732] *N*-nitrosodiphenyl-amine,[378] sulphur dichloride,[380] benzenesulphenate esters,[733] metal halides[378] (see also below), and by its reductive dimerisation of the tropylium ion,[734] the 2,4,6-trimethylpyrylium ion[735] and *p*-nitroso-*N,N*-dimethylaniline.[378]

The dianion (60) naturally acts as a proton-acceptor, and with water or alcohols yields mixtures of cyclo-octa-1,3,5-triene and cyclo-octa-1,3,6-triene (see p. 28). The use of deuterium oxide leads to the dideuterio-derivatives (500) (80%) and (501) (20%).[736]

(60)　　　　　　(500)　　　　　　(501)

The basicity of COT^{2-} is enhanced in the excited state, and u.v. irradiation in the presence of weakly acidic proton-donors, e.g. amines, acetylenes, gives the cyclo-octatrienyl anion (502); this is eventually protonated further to yield cyclo-octatrienes (which then give their photo-products (see p. 216)], or is deprotonated to regenerate COT^{2-}.[737]

(502)

The most important synthetic applications of COT^{2-} result from its reactivity as a dicarbanion.

With alkyl halides, alkylation of COT^{2-} occurs with the formation of dialkylcyclo-octatrienes.[738, 739] The products formed from methyl iodide apparently consist of a *ca.* 4:1 mixture of the *cis*- and *trans*-dimethylcyclo-octa-1,3,5-trienes, (503) and (504), together with the *trans*-dimethylcyclo-octa-1,3,6-triene (505) (total yield 84%).[379]

(503)　　　　　　(504)

(505)

Silylation with chlorotrimethylsilane gives the cyclo-octa-1,3,6-triene-derivative (506)[740] (*cf.* ref. 741); chlorodimethylsilane, however, yields a mixture of (507) and (508) (*ca.* 2:1) (35%).[742]

$2Li^+$ ⬡ $2-$

Me_3SiCl | Et_2O, $-50° \rightarrow$ r.-t.

Me_2SiHCl | Et_2O, $-50° \rightarrow$ r.-t.

SiMe₃ ... SiMe₃

SiHMe₂ ... SiHMe₂

+

SiHMe₂ ... SiHMe₂

(506)   (507)   (508)

Reaction of COT²⁻ with *gem*-dihalides affords the bicyclo[6.1.0]nona-2,4,6-triene system (82), possibly by generation of a carbenoid which then adds to COT.[307, 743] The known reactions of this type are listed in table 63 (the use of dilithium cyclo-octatetraenide in liquid ammonia is recommended[744]). For

$2M^+$ ⬡ $2-$ $R^1{>}CX_2$ / R^2 Low temp. $\rightarrow$ r.-t.   → structure (82) with R^1, R^2

(82)

Table 63

M	R¹	R²	X	Solvent	Yield (%)	Ref.
K	H	H	Cl	THF	45	743, (745)
Li	H	Me	Cl	THF	20	307
—	H	Buᵗ	Cl	NH₃	—	746
Li	H	OMe	Cl	Me₂O	—ᵃ	747
—	H	CN	Br	THF	—	748
K	H	Br	Br	THF	*ca.* 15	749
K	OMe	H	Cl	THF	Up to 49	743, 745
—	CN	H	Br	THF	—	748
Li	F	H	Cl	Et₂O	—ᵇ	747
Li or K	Cl	H	Cl	THF	Up to 52	743, 745
Li	Me	Me	Cl	NH₃	88	744
K	Clᶜ	Meᶜ	Cl	THF	(Very low)	307
K	Cl	Cl	Cl	THF	19	743

ᵃ At −80°, the *syn*-isomer is the predominant product, but epimerises (90%) even at room-temperature to give the *anti*-isomer.
ᵇ At −78°, the product is a mixture of *syn*- and *anti*-isomers (25:75); at 0° the ratio is 13:87.
ᶜ Stereochemistry uncertain.

similar reactions with monosubstituted COT^{2-} species, leading to the products **(509)–(512)**, see ref. 578.

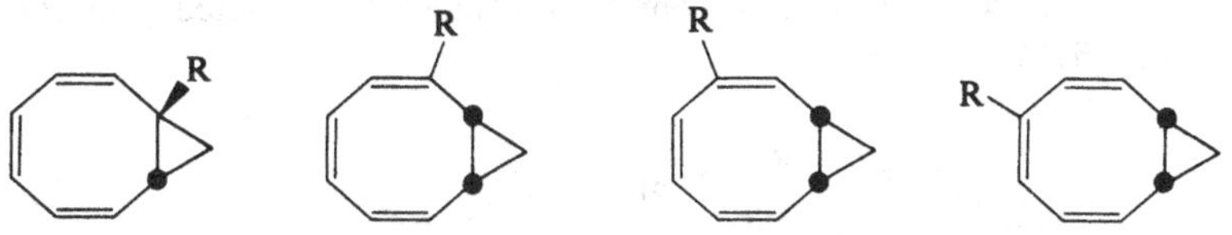

| **(509)** | **(510)** | **(511)** | **(512)** |

$R = Me, OMe*$

Dilithium 1,3,5,7-tetramethylcyclo-octatetraenide reacts with dichloromethane to afford the product **(513)** (83 %).[578]

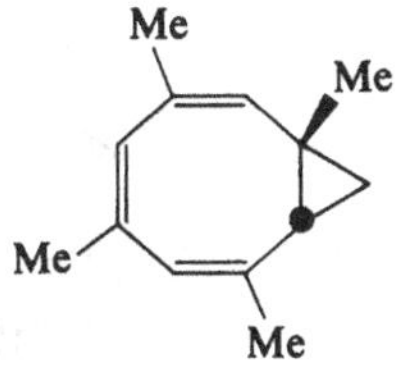

(513)

With dichlorophenylphosphine, COT^{2-} gives the 9-phosphabicyclo[6.1.0]-nona-2,4,6-triene **(514)** (64 %).[380, 750]

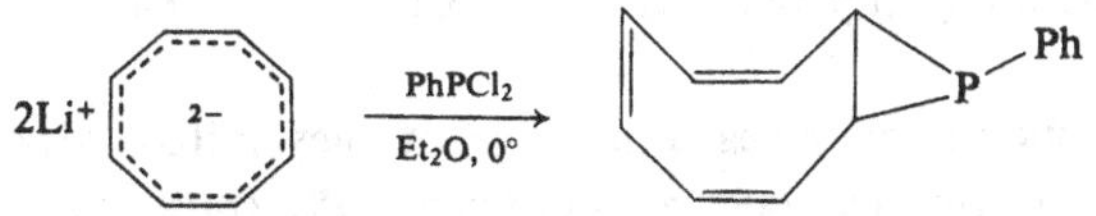

(514)

Treatment of COT^{2-} with 1,2-dibromoethane yields bicyclo[6.2.0]deca-2,4,6-triene **(515a)**, and an analogous reaction occurs with 1,2,3-trichloropropane (table 64).

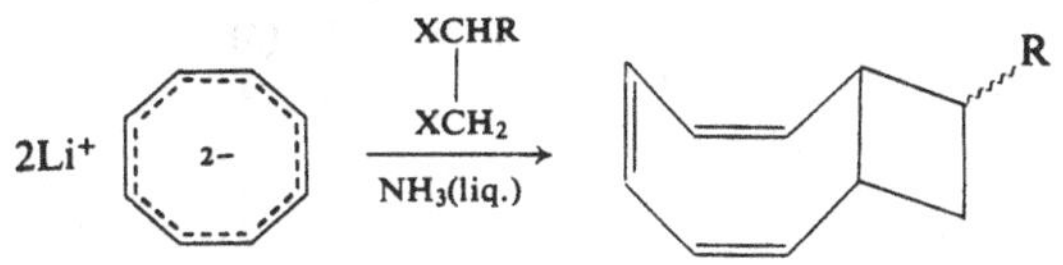

(515)

Table 64

	R	X	Yield (%)	Ref.
a	H	Br	52	751
b	CH_2Cl	Cl	54	382

* Only one isomer (structure **(512**; R = OMe)) isolated in a pure state.

Reaction with 1,3-dibromopropane affords a mixture of bicyclo[6.3.0]-undeca-2,4,6-triene (**516a**) and its tricyclic valence isomer (**517a**) (ratio *ca.* 3:7); with 1,3-dichloro-2-tetrahydropyranyloxypropane, the isolated product has structure (**517b**) (table 65).

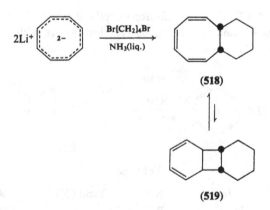

(516)

(517)

Table 65

	R	X	Yield (%)	Ref.
a	H	Br	48.5	(752), 753
b	2-Tetrahydropyranyloxy	Cl	64	382

1,4-Dibromobutane gives a mixture of bicyclo[6.4.0]dodeca-2,4,6-triene (**518**) and its tricyclic valence tautomer (**519**) (ratio *ca.* 7:3, total yield 35%).[752, 753]

(518)

(519)

The acylation of COT^{2-} with acid chlorides may lead to various products including bicyclo[4.2.1]nonatrienes (**520**) and (**521**), bicyclo[6.1.0]nonatrienes (**522**), and acyclic tetraenes (**523**) (table 66) (*cf.* refs. 738, 739, 754). With

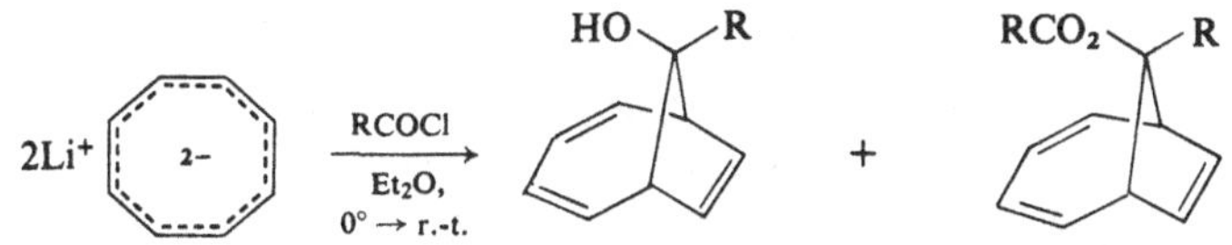

(520)   (521)

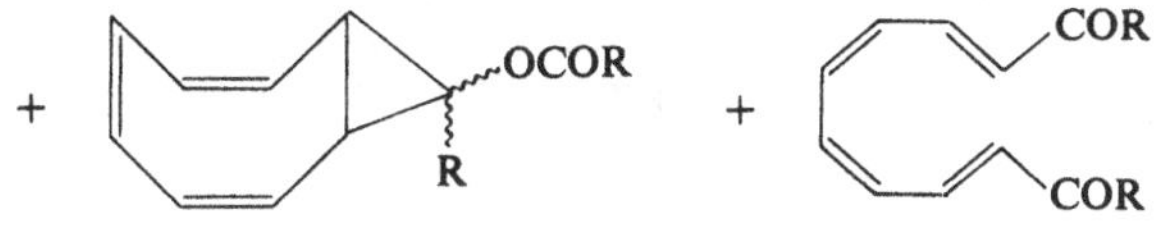

(522)   (523)

Table 66

Yield (%)

R	(520)	(521)	(522)	(523)	Ref.
Me[a]	18–19	12–13	19–22	7–8	377
PhCH$_2$	60	—	—	—	378
Ph[a, b]	—	58–62	—	6–7	377
p-Br.C$_6$H$_4$[a]	—	44	—	5	377

[a] 2Li$^+$COT^{2-} added to the acid chloride.
[b] Using 2K$^+$COT^{2-} in THF, the products are (521) (38%) and the all-E-pentaene (524) (0.3%).

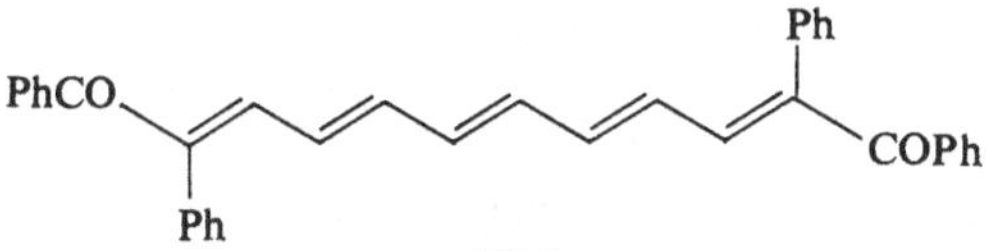

(524)

phthaloyl chloride, the bicyclo[6.1.0]nonatriene-lactone (525) is produced (28%).[377]

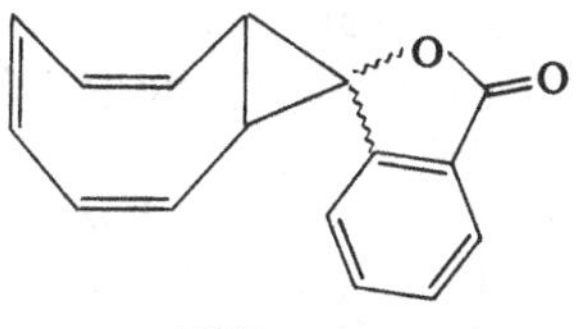

(525)

Analogous reactions[378] with acid anhydrides are listed in table 67. Phthalic

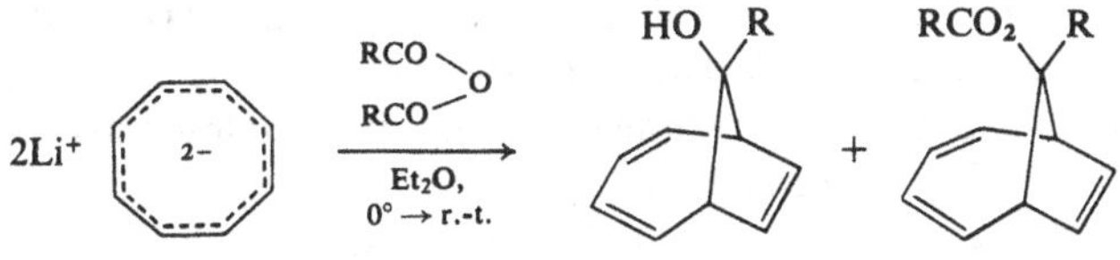

(520)   (521)

Table 67

Yield (%)

R	(520)	(521)
Me	22.5	39
Ph[a, b]	52	—

[a] If the reaction mixture is refluxed, the exclusive product is (521) (51 %).
[b] Reaction at −60° produces (520) (43 %), together with the tetraene (526; R = Ph) (4.2 %).

RCO⌁⌁⌁⌁COR

(526)

anhydride yields the bicyclo[4.2.1]nonatriene-lactone (527) (48 %).[378]

(527)

The manifold acylation products of COT^{2-} may arise as outlined in scheme 69. The initially formed acylcyclo-octatrienyl anion (528) may not only undergo further (vicinal) C-acylation to give the precursor (529) of acyclic tetraenes, but also give rise to the bicyclic [4.2.1] and [6.1.0] systems (530) and (531), each of which may be O-acylated.

Carboxylation of COT^{2-} with carbon dioxide gives mainly the acyclic tetraene dicarboxylic acid (532) (50–65%), together with small amounts (2–3 %) of the cyclic isomer (533)[755, 756] (*cf.* ref. 11) (but see appendix).

Aldehydes and ketones also yield, initially, mixtures of disubstituted cyclo-octa-1,3,5-trienes and cyclo-octa-1,3,6-trienes; however, the former products isomerise to bicyclo[4.2.0]octa-2,4-dienes or open-chain tetraenes[725] (table 68) (*cf.* refs. 259, 738, 739, 754). Phthalaldehyde gives a low yield (3.6%) of the tetracyclic diol (537).[725] Fluorenone affords the cyclo-octa-1,3,6-triene derivative (534; R^1R^2 = fluorenyl) (42%).[725]

In general, the reaction of esters with COT^{2-} furnishes the bicyclo[4.2.1]-nonatrienes (520)[378] (table 69). Ethyl formate (at −60°), however, affords the

Scheme 69

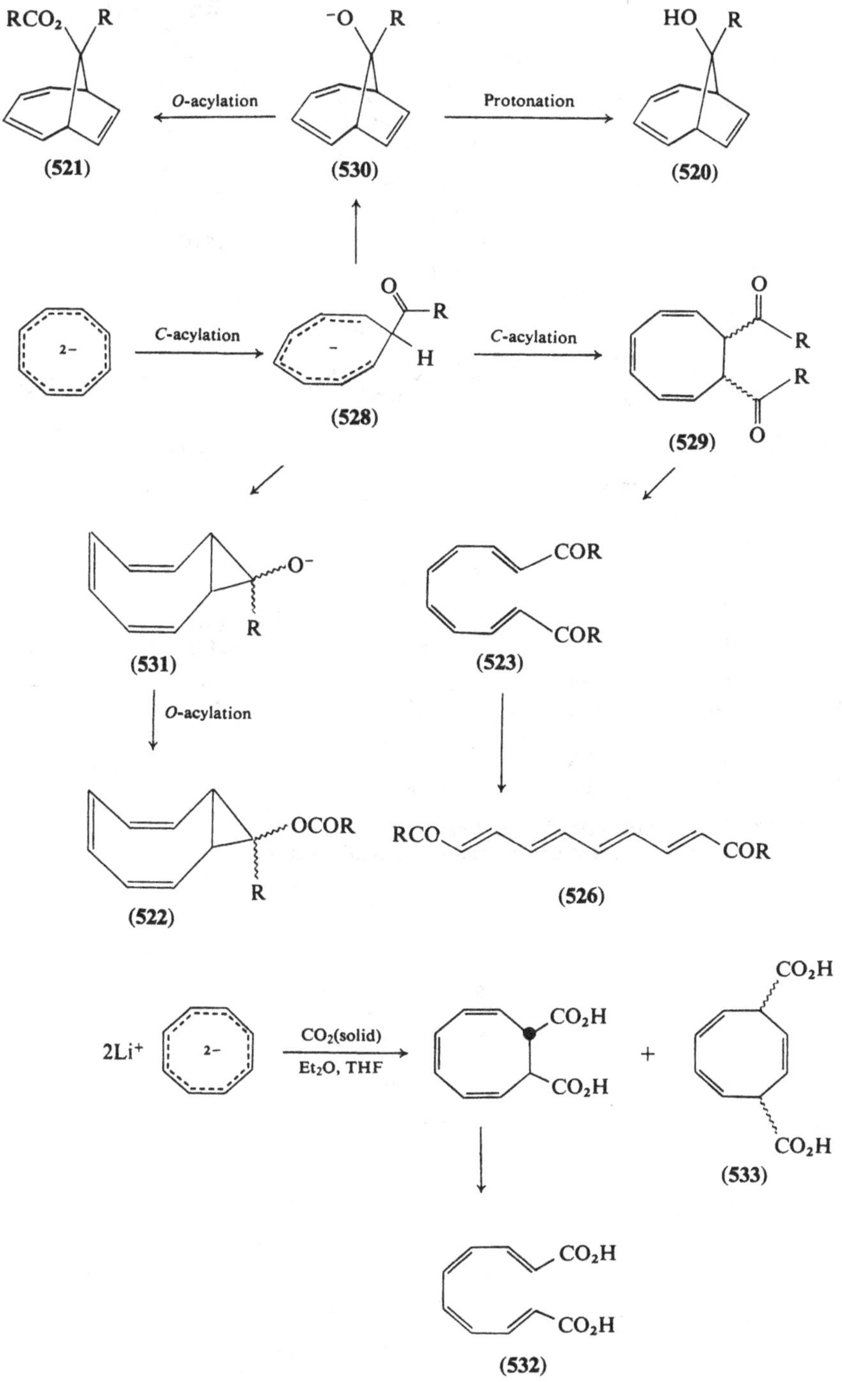

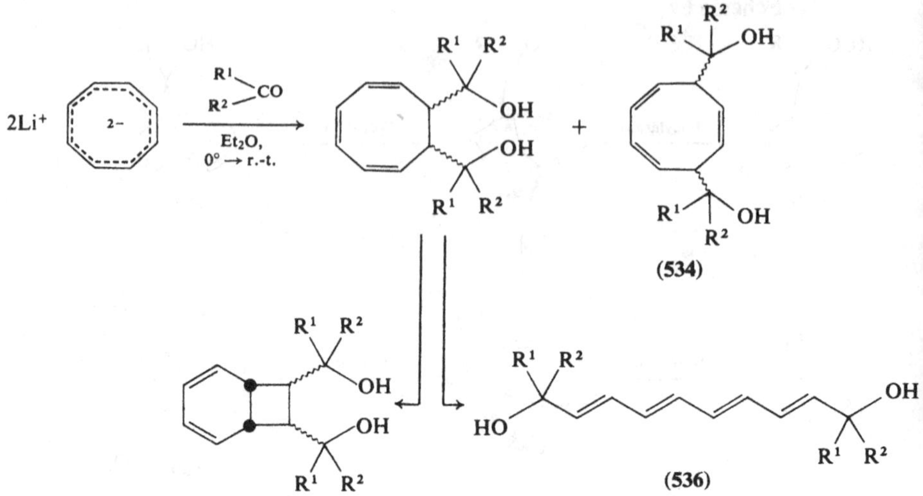

(534)

(535)   (536)

Table 68[a]

	R¹	R²	(534)	(535)	(536)
				Yield (%)	
a	Me	H	*ca.* 31	*ca.* 10	—
b	Ph	H	66	15	—
c	Me	Me		96[b]	—
d	Ph	Ph	64	—	28

[a] $2Li^+COT^{2-}$ added to aldehyde or ketone.
[b] Mixture of products not separated; ratio (534):(535) = 38:62.

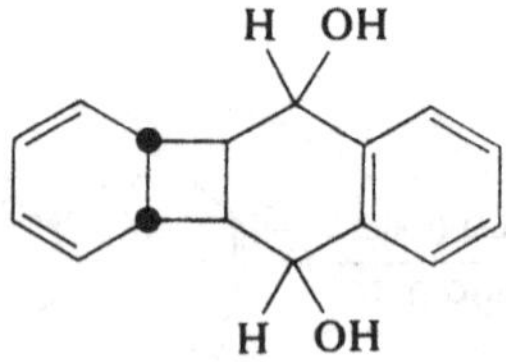

(537)

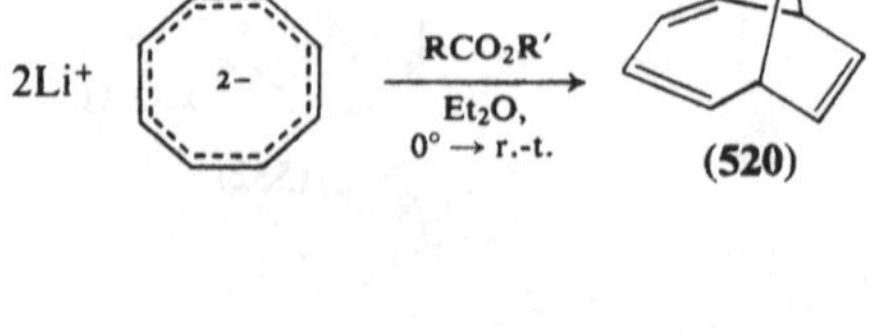

(520)

Table 69

R	R'	Yield (%)	Ref.
Me	Et	77[a]	378
Ph	Me	74, (46)	378, (757)
p-MeO.C$_6$H$_4$	Me	48	757

[a] Reduced to 31 % if 2Li$^+$COT^{2-} is added to the ester, rather than the reverse.

tetraene-dialdehyde (538) (mixture of geometrical isomers) (25%), and diethyl oxalate and methyl chloroformate give the tetraene-diesters (539a) (12%) and (539b) (28%) respectively; diethyl phthalate gives the bicyclo[6.1.0] nonatriene-lactone (525) (15%).[378]

OHC[CH=CH]$_4$CHO   RO$_2$C[CH=CH]$_4$CO$_2$R

(538)   (539)

a: R = Et
b: R = Me

Reaction with phosgene leads to a mixture of 9-oxobicyclo[4.2.1]nona-2,4,7-triene (540) (17–20%) and 3-chloroindene (22–25%),[758] or, under slightly different conditions, 1-chloroindene (33%).[759]

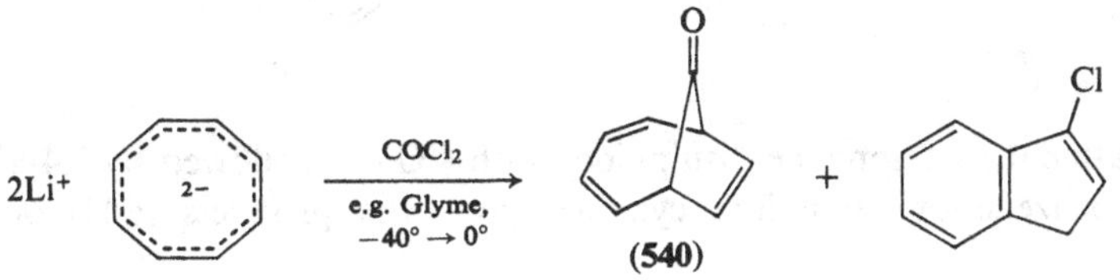

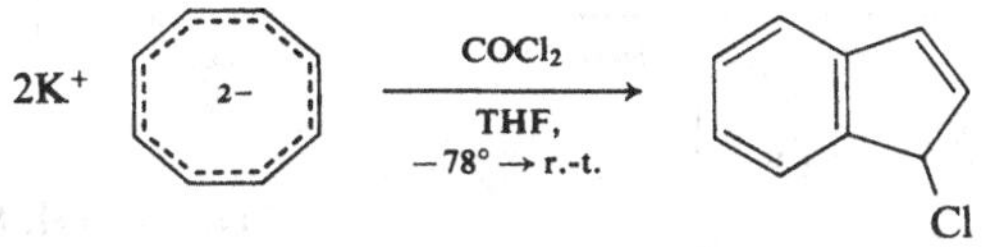

The reaction of COT^{2-} with dimethylcarbamoyl chloride provides a high-yield (up to 76%) synthesis of the carbonyl-bridged triene (540).[52,757]

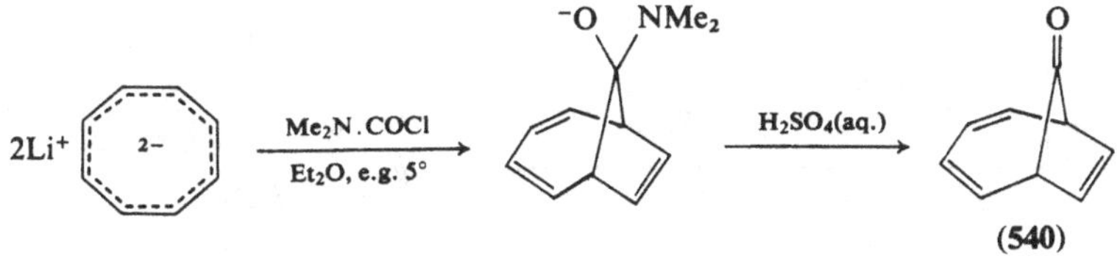

1,1-Dimethoxytrimethylamine gives a mixture of *syn-* and *anti*-9-dimethyl-aminobicyclo[6.1.0]nonatrienes, (541) and (542) (3:2)[747] (*cf. gem*-dihalides, p. 174):

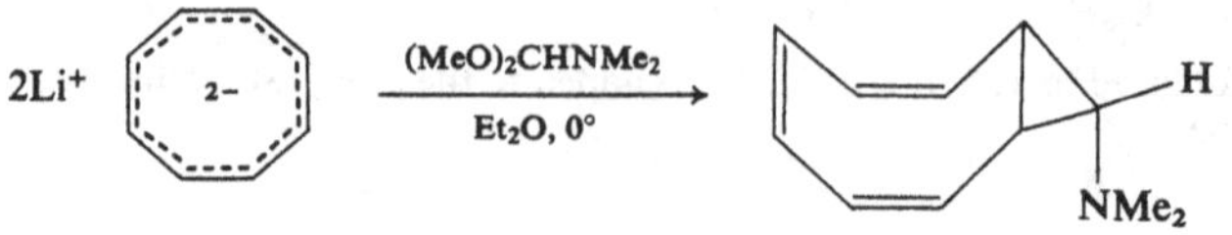

(541)

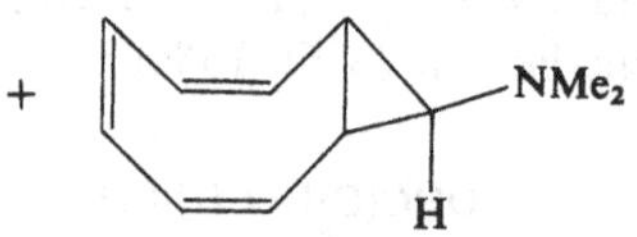

(542)

An efficient preparation (74%) of the 9-azabicyclo[4.2.1]nonatriene system (543) results from the use of isoamyl nitrite.[320]

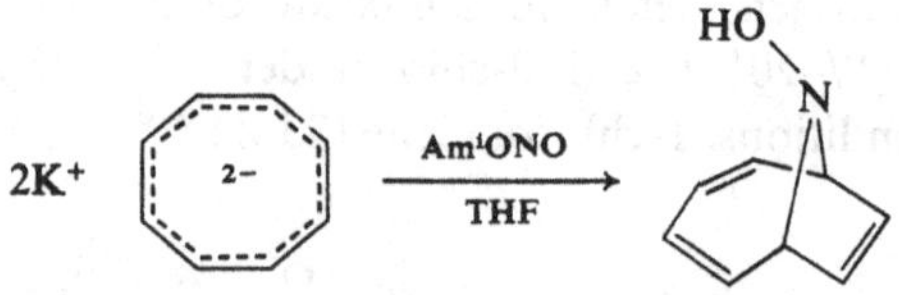

(543)

Treatment of thiophene disulphides with COT^{2-}, followed by 2,4-dinitrochlorobenzene or potassium cyanide, gives the products (544) or (545)

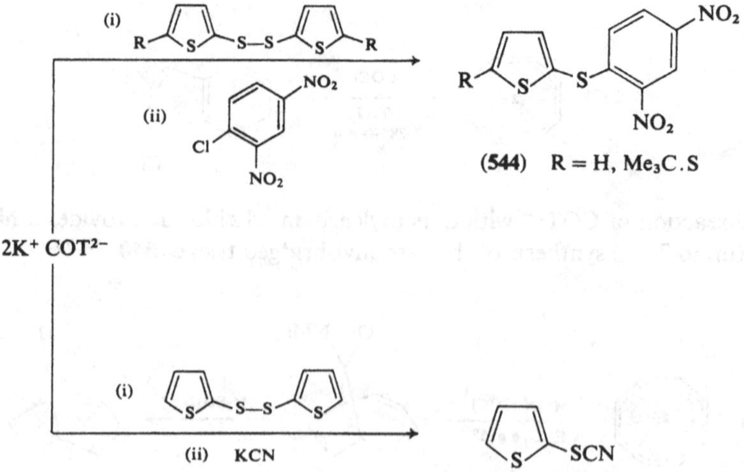

(544)   R = H, $Me_3C.S$

(545)

respectively.[760] 2,5-Dithiocyanatothiophene reacts with one equivalent of COT^{2-}, followed by benzoyl chloride, to give compound (546); using two equivalents of COT^{2-}, the product (547) is formed.[760] Reaction with COT^{2-} in the presence of acetic acid leads to the product (548).[760]

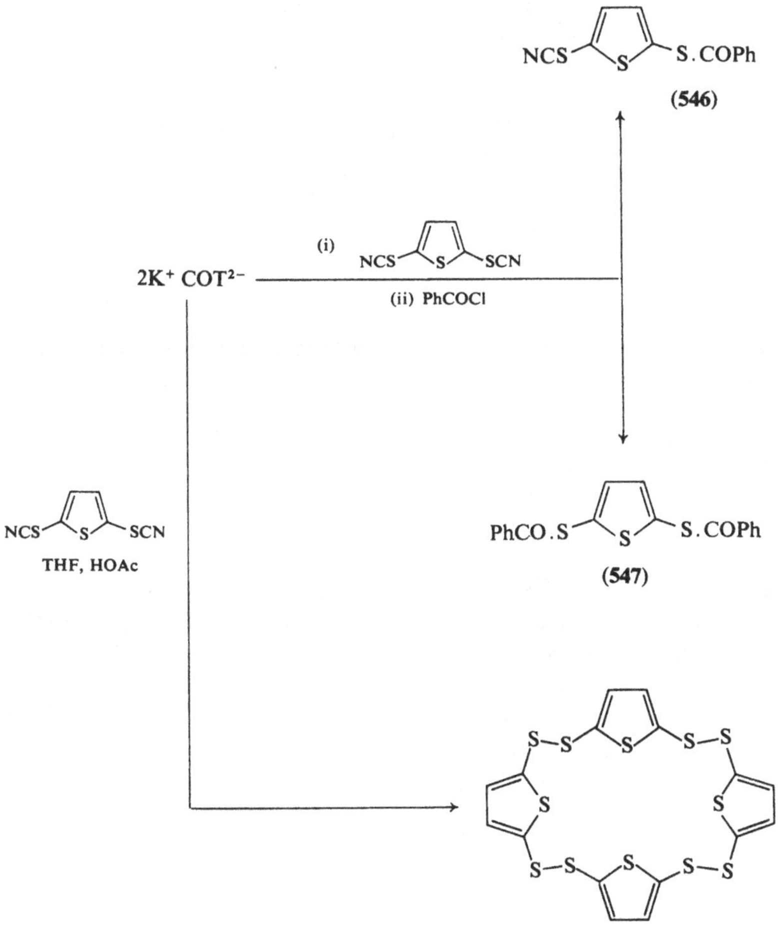

NCS⌒ᴥ⌒S.COPh

(546)

2K$^+$ COT^{2-} ⟶ (i) NCS⌒ᴥ⌒SCN / (ii) PhCOCl

NCS⌒ᴥ⌒SCN

THF, HOAc

PhCO.S⌒ᴥ⌒S.COPh

(547)

(548)

The cyclo-octatetraenyl dianion is the progenitor of numerous metal derivatives containing co-ordinated COT.

Reaction with various lanthanide metal(III) chlorides in tetrahydrofuran affords the solvated dimeric complexes (549) in high yield[761, 762] (the best yields are obtained by the addition of 2K$^+$COT^{2-} to the metal chloride[762]).

2K$^+$COT^{2-} $\xrightarrow[\substack{\text{THF,} \\ -10° \to \text{r.-t.}}]{\text{MCl}_3}$ [(COT)MCl.2THF]$_2$ (549)

M = Ce, Pr, Nd, Sm

(Complexes (COT)NdX.nTHF (X = Br, I) have been reported[763].) The crystal structure of the cerium compound (**549**; M = Ce) is as shown.[761,764]

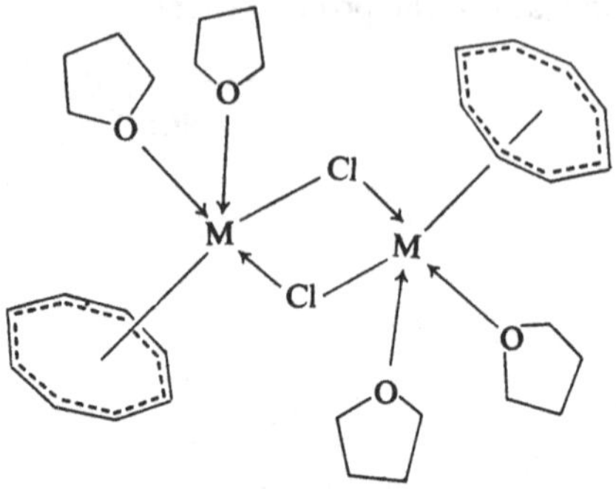

(**549**; M = Ce)

In the presence of COT, similar reactions produce the ionic complexes (**550**) (36–78 %).[762,765]

$$2K^+COT^{2-} \quad \xrightarrow[\substack{THF, \\ ca. -10° \to r.-t.}]{MCl_3, COT} \quad K^+[(COT)_2M]^- \quad (\mathbf{550})$$

M = Y, La, Ce, Pr, Nd, Sm, Gd, Tb

The cerium complex (**550**; M = Ce), on treatment with cerium(III) chloride in tetrahydrofuran, gives (**549**; M = Ce) quantitatively.[762] The diglyme-solvated cerium complex (**550**; M = Ce) possesses the structure shown.[766]

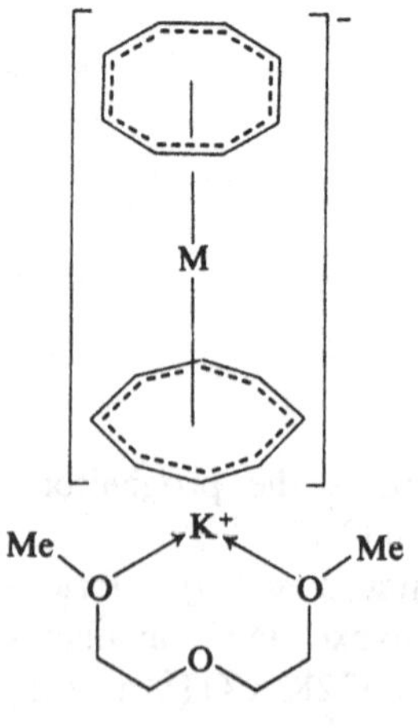

(**550**; M = Ce) (diglyme-solvated)

With dipyridinium hexachlorocerate(IV), COT^{2-} yields the complex (**551**) (64 %).[767]

$$2K^+COT^{2-} \quad \xrightarrow[\text{THF, r.-t.}]{(PyH^+)_2[CeCl_6]^{2-}} \quad (COT)_2Ce \quad (\mathbf{551})$$

(For the application of ligand field theory to this type of complex, see ref. 768.)
'Mixed' sandwich complexes of the type (552) result from the reaction of
COT^{2-} with yttrium or lanthanide complexes $(CPD)MCl_2 . 3THF$; the products
are desolvated by heating under reduced pressure, giving (553).[769]

$$2K^+COT^{2-} \xrightarrow[\text{THF}]{(CPD)MCl_2 . 3THF} (COT)M(CPD).THF \quad (552)$$

$$\downarrow 50°/10^{-3} \text{ mm}$$

$$M = Y, Ho, Er \qquad (COT)M(CPD) \quad (553)$$

Treatment of COT^{2-} with actinide metal(IV) chlorides gives complexes of
structure (155) (see p. 186) (table 70). For the i.r. spectra of (155; M = Th, U),

$$2K^+COT^{2-} \xrightarrow[\substack{\text{THF,} \\ \text{e.g.} -20° \to \text{r.-t.}}]{MCl_4} (COT)_2M \quad (155)$$

$$M = Th, Pa, U, Np, Pu$$

Table 70

M	Yield (%)	Ref.
Th	60	(770), 771
Pa	1	772, (773)
U	60–80	381, (774)
Np	—	775
Pu[a]	—	775

[a] Complex prepared using $(Et_4N)_2PuCl_6$.

see ref. 409. For a 'wide-line' n.m.r. study of (155; M = U), see ref. 776.
Analogous preparations using monosubstituted dianions (275) result in the

(275) (554)

Table 71

R	M	Yield (%)	Ref.
Et	U	92	777, (778)
Bu^n	U	90	777, (778)
Ph	U	87	777
$CH=CH_2$	U	97	777
Cyclopropyl	U	88	777
Et	Np	—	778
Bu^n	Np	—	778
Et	Pu[a]	—	778
Bu^n	Pu[a]	—	778

[a] Complex prepared using $(Et_4N)_2$ $PuCl_6$.

complexes (554) (table 71). The 1,3,5,7-tetrasubstituted compounds (555a) (53%),[777,779] (555b)[779] and (555c) (64%)[663] may be obtained in similar fashion, and likewise the octamethylcyclo-octatetraene complex (556).[780] The

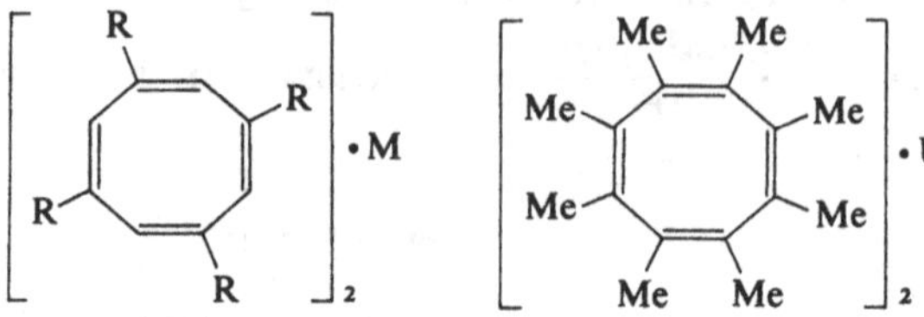

(555)   (556)

a: R = Me, M = U
b: R = Me, M = Np
c: R = Ph, M = U

complexes of this type are known to have sandwich structures, and have been dubbed 'uranocene' etc. The ligands in (COT)$_2$M (155; M = Th, U) are arranged as shown,[781,782] while in (TMCOT)$_2$U (555a) two crystallographic-ally independent molecules exist in the unit cell, one with eclipsed and the other with staggered methyl groups.[783,784] In these compounds the question

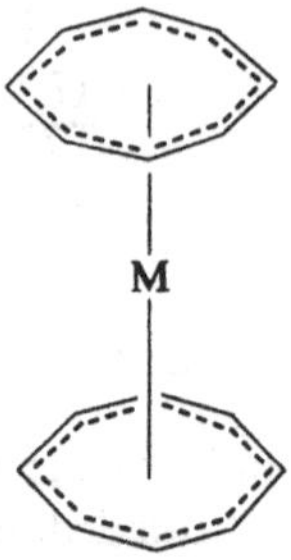

M   (155)

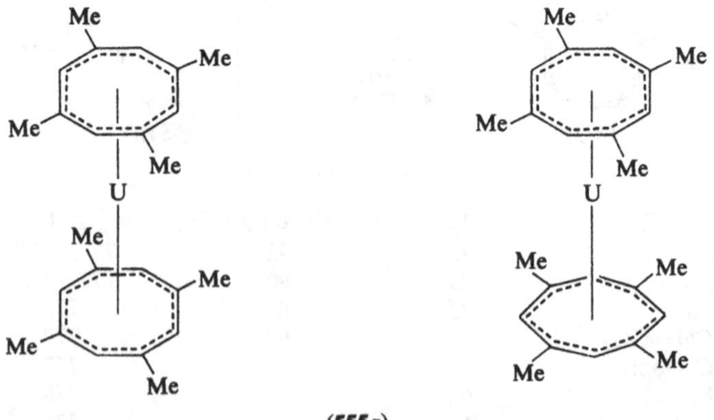

(555a)

of f-orbital participation in the metal–ligand bonding is of great interest (see e.g. refs. 381, 779, 780, 785; for a review, see ref. 786). For the application of ligand field theory, see ref. 768.

Complexes (557) may be obtained from actinide metal(III) bromides.[787]

$$2K^+COT^{2-} \xrightarrow[\text{THF, } -10° \text{ to } -20° \rightarrow \text{r.-t.}]{MBr_3} K^+[(COT)_2M]^-.THF \quad (557)$$

$$M = Np, Pu$$

Dipyridinium pentachlorothallate(III) reacts with COT^{2-} to afford the thallium(III) complex (558) (65%).[788]

$$2K^+COT^{2-} \xrightarrow[\text{THF, r.-t.} \rightarrow \text{reflux}]{(PyH^+)_2[TlCl_5]^{2-}} (COT)TlCl \quad (558)$$

Titanium(IV) chloride reacts with COT^{2-} to give the product (159) (see p. 54) (53%).[416]

$$2Na^+COT^{2-} \xrightarrow[\text{C}_6\text{H}_6, 70°]{TiCl_4} (COT)_2Ti \quad (159)$$

The 'mixed' sandwich complex (165) (see pp. 56–7) may be prepared (60%) by treatment of COT^{2-} with a mixture of titanium(IV) chloride and sodium cyclopentadienide.[789]

$$2K^+COT^{2-} \xrightarrow[\text{Toluene, reflux}]{TiCl_4, Na(CPD)} (COT)Ti(CPD) \quad (165)$$

The same product results (in lower yields) from the use of $(CPD)TiCl_3$ or $(CPD)_2TiCl_2$;[789] an analogous complex (559) (50–60%) is formed from the sodio-derivative of indene.[790]

$$(COT)Ti(indenyl) \quad (559)$$

With titanium(III) chloride there is obtained the dimeric complex (560) (*cf.* (157*a*), p. 54) (64%),[791] which in ether forms the tetrameric (561).[792]

$$2K^+COT^{2-} \xrightarrow[\text{THF, reflux}]{TiCl_3} [(COT)TiCl.THF]_2 \quad (560)$$

$$\text{Et}_2\text{O,} \quad \text{r.-t.} \downarrow$$

$$[(COT)TiCl]_4 \quad (561)$$

Reaction of COT^{2-} with hafnium(IV) chloride results in the complex (562) (70–80%).[793]

$$2Na^+COT^{2-} \xrightarrow[\text{Xylene, 120°}]{HfCl_4} (COT)_2Hf \quad (562)$$

The vanadium complex $(COT)_2V$ may be prepared from COT^{2-}.[409,416] (For the e.p.r. spectrum, see ref. 420.)

Niobium(v) chloride affords the ionic complex (**563a**) (70%), from which the other salts (**564**) (85%) and (**565**) are readily obtained[423] (see p. 57). In the

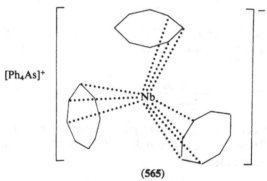

$$2K^+COT^{2-} \xrightarrow[\substack{\text{Toluene, THF,}\\ -78° \to \text{r.-t.}}]{\text{NbCl}_5} K^+ [(COT)_3Nb]^- \quad (563a)$$

$$[Li(THF)_4]^+ [(COT)_3Nb]^- \longleftarrow \begin{array}{c} \text{LiCl} \\ \text{THF,} \\ \text{r.-t.} \end{array} \quad \begin{array}{c} [Ph_4As]^+Cl^- \\ CH_2Cl_2, \\ \text{r.-t.} \end{array} \longrightarrow [Ph_4As]^+ [(COT)_3Nb]^-$$

$$(564) \qquad\qquad\qquad\qquad\qquad\qquad\qquad (565)$$

salt (**565**), which is fluxional, two of the COT rings are η^3-bonded to the niobium, while the third ring is apparently η^4-bonded.[423]

(**565**)

(The exact nature of the bonding within the rings is not clear)

Complexes $(COT)_3M_2$ (M = Cr, Mo, W) are obtainable from COT^{2-}.[416]

The 'mixed' ruthenium complex (**566**) is obtained (20–30%) by reaction of COT^{2-} with the polymeric norbornadiene complex $[(NBD)RuCl_2]_n$; the cyclo-octa-1,3,5-triene complex (**567**) is a co-product (4%).[794]

$$2K^+COT^{2-} \xrightarrow[\text{THF}]{[(NBD)RuCl_2]_n} (COT)Ru(NBD) \quad (566)$$

$$+ \text{(Cyclo-octa-1,3,5-triene)Ru(NBD)} \quad (567)$$

If a similar reaction is carried out using potassium dissolving in COT, a low yield of the binuclear product (**568**) is obtained in place of complex (**566**).[794]

$$(COT)Ru_2(NBD)_2 \quad (568)$$

Reaction of COT^{2-} with the cyclo-octa-1,5-diene complex $[(COD)RuCl_2]_n$

gives an ill-defined product of formula $C_{16}H_{20}Ru$ (10%), the cyclo-octa-1,3,5-triene complex (569) (10%), and a small amount (*ca.* 1%) of a binuclear product $C_{24}H_{32}Ru_2$.[794]

$$2K^+COT^{2-} \xrightarrow{[(COD)RuCl]_n} C_{16}H_{20}Ru$$

$$+ (\text{Cyclo-octa-1,3,5-triene})Ru(COD) \quad (569)$$

$$+ C_{24}H_{32}Ru_2$$

With the reagents $(PPh_3)_4RuCl_3$ and $[(PPhMe_2)_6Ru_2Cl_3]^+Cl^-$, COT^{2-} affords the complexes (570) (50%) and (571) (20%) respectively.[794]

$$\xrightarrow{(PPh_3)_4RuCl_3} (COT)Ru(PPh_3)_2 \quad (570)$$

$$2K^+COT^{2-}$$

$$\xrightarrow{[(PPhMe_2)_6Ru_2Cl_3)^+Cl^-} (COT)Ru(PPhMe_2)_3 \quad (571)$$

Treatment of COT^{2-} with the osmium complex $[(COD)OsCl_2]_n$ results in the 'mixed' products (572) (*ca.* 25%) and (573) (1–2%).[794]

$$2K^+COT^{2-} \xrightarrow{[(COD)OsCl_2]_n} (COT)Os(COD) \quad (572)$$

$$+ (\text{Cyclo-octa-1,3,5-triene})Os(COD) \quad (573)$$

The complexes $(COT)M$ (M = Co, Ni) may be obtained from COT^{2-}.[416] Reaction of COT^{2-} with $(COD)PtCl_2$ in the presence of cyclo-octa-1,5-diene affords the product (574) (30–50%).[795]

$$2Li^+COT^{2-} \xrightarrow[\text{Et}_2\text{O, } -20°]{(COD)PtCl_2, \text{ COD}} (COD)_2Pt \quad (574)$$

(*b*) *Silver derivatives*

For preparation from COT, see pp. 51–2.

Reactions. Treatment of the silver complex (147*b*) (see p. 51) with bromine at low temperature affords the salt (575*a*), containing the *endo*-bromohomotropylium ion; at 20° in sulphur dioxide conversion to the *exo*-form (575*b*) occurs.[230]

[(COT)Ag]⁺BF₄⁻ → (575a) — rendered below:

$$[(COT)Ag]^+BF_4^- \xrightarrow[\substack{ClCH_2CH_2Cl, \\ -35°}]{Br_2}$$

(147b)

(575a)

BF_4^-

SO_2(liq.) | 20°

(575b)

BF_4^-

(c) Magnesium derivative

For preparation from COT, see p. 52.

Reactions. Treatment of COT in hexamethylphosphoramide with dichloro-dimethylsilane in the presence of magnesium gives the 9-silabicyclo[4.2.1]nona-triene (576) (*ca.* 20%);[796] the reaction may involve COT²⁻ (see p. 174, but *cf.* ref. 797).

$$COT \xrightarrow[Mg, (Me_2N)_3PO]{Me_2SiCl_2}$$

(576)

The thorium and uranium complexes (155; M = Th, U) (see pp. 185–6) are produced (in very low yield) by heating magnesium cyclo-octatetraenide with the metal(IV) fluoride.[773]

(d) Yttrium and lanthanide derivatives

For preparation from COT, see pp. 52–3; from COT²⁻, see pp. 183–5.

Reactions. The lanthanide complexes [(COT)MCl.2THF]₂ (see p. 183) react with sodium cyclopentadienide to give the 'mixed' sandwich complexes (552) (see p. 185).[769]

$$[(COT)MCl.2THF]_2 \xrightarrow[THF]{(CPD)Na} (COT)M(CPD).THF$$

(549)   (552)

M = Y, Nd

(e) Actinide derivatives

For preparation from COT, see p. 53; from COT^{2-}, see pp. 185–7.

Reactions. The 'uranocene' (577) undergoes catalytic hydrogenation, and cyclopropanation, affording the ethyl- and cyclopropyl-derivatives, (578) and (579).[777]

(f) Thallium derivatives

For preparation from COT^{2-}, see p. 187.

Reactions. Treatment of the thallium(III) complex (558) (see p. 187) with potassium yields the thallium(I) complex (580) (46 %);[788] reaction with sodium–naphthalene, however, gives the product (581) (40 %).[798]

(g) Titanium and hafnium derivatives

For preparation from COT, see pp. 53–7; from COT^{2-}, see p. 187.

Reactions. The titanium complex (560) (see p. 187) reacts with allyl Grignard reagents in ether to form the π-allyl complexes (582); in tetrahydrofuran, the isolated products are the solvated derivatives (583) (table 72). With hydrogen chloride, these compounds give the tetrameric complex (561)[799] (see p. 187).

[(COT)TiCl.THF]₂   (560)

X = Cl, Br

$R^1CH{=}C.CH_2MgX$ $\begin{matrix}R^2\\|\end{matrix}$ Et₂O or THF

(COT)Ti—⟩—R²   (582)   or   (COT)Ti—⟩—R².THF   (583)

with R¹ substituents

HCl | Et₂O

[(COT)TiCl]₄   (561)

Table 72

R¹	R²	Yield (%) (582)	(583)	Ref.
H	H	57	44	799
H	Me	40	63	799
Me	H	(50), 70	34	(791), 799

With sodium cyclopentadienides, products (584) are obtained[800] (table 73).

[(COT)TiCl.THF]₂   (560) $\xrightarrow[\text{Et}_2\text{O}]{\text{Na}^+ \text{C}_5\text{H}_4\text{-R}}$ (COT)Ti—⟨C₅H₄⟩—R   (584).

Table 73

R	Yield (%)
Me	60
Buᵗ	61
SiMe₃	63

Indenyl and fluorenyl derivatives may be prepared similarly (77% and 63% respectively).[800] (For e.p.r. spectral studies on the indenyl complex, see ref. 801.)

Reactions with the sodium salts of carborane dianions afford mixed-ligand titanacarboranes (585), which may be isolated as the tetra-ethylammonium salts, and oxidised to the neutral species (586).[802]

$$[(COT)TiCl]_2 \xrightarrow[\text{THF}]{Na_2C_2B_nH_{n+2}} Na^+[(COT)Ti(C_2B_nH_{n+2})]^- \quad (585)$$

(158)

$$\downarrow H_2O_2$$

$n = 9$ or 10

$$(COT)Ti(C_2B_nH_{n+2}) \quad (586)$$

The proposed structure for (586, $n = 9$) is as shown.

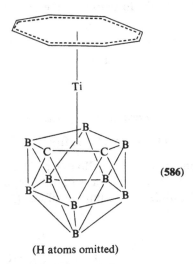

(586)

(H atoms omitted)

The titanium complex (165) (see pp. 56–7) reacts quantitatively with iodine to give either the mono- or the tri-iodide (587) or (588), depending on the relative proportions of the reactants.[803]

$$(COT)Ti(CPD) \xrightarrow[\text{Et}_2O, \text{ r.-t.}]{I_2} (COT)TiI(CPD) \quad \text{or} \quad (COT)TiI_3(CPD)$$

(165)   (587)   (588)

Treatment of the hafnium complex (562) (see p. 187) with hydrogen chloride in tetrahydrofuran affords the solvated product (589) (*ca.* 80%); removal of the solvent may be effected (quantitatively) to give complex (590), which with allylmagnesium halides yields the bis(π-allyl)-compound (591) (60%).[793] For

$$(COT)_2Hf \xrightarrow[\text{THF}]{HCl} (COT)HfCl_2 . THF \quad (589)$$

(562)

$$\downarrow 70° \text{ High vacuum}$$

$$(COT)Hf \xleftarrow[\text{Et}_2O, -35°]{CH_2=CH.CH_2MgX} (COT)HfCl_2 \quad (590)$$

(591)

analogous reactions of (COT)₂Zr (160) (see p. 55), and additional transformations, see ref. 417.

(h) Niobium and tantalum derivatives

For preparation from COT, see p. 57; from COT²⁻, see p. 188.

Reactions. The COT ligands in the complexes (563) (see p. 188) are displaced by carbon monoxide with the formation of the products (592).[423]

$$K^+ [(COT)_3M]^- \xrightarrow[\substack{THF, \\ 110°, 50 \text{ atm}}]{CO} K^+ [M(CO)_6]^-$$

(563)   (592)

$$M = Nb, Ta$$

(i) Chromium, molybdenum and tungsten derivatives

For preparation·from COT, see pp. 57–9; from COT²⁻, see p. 188.

Reactions. The chromium complex (170) (see p. 59) is protonated by HPF₆ to give the salt (593) (*ca.* 35%); reaction with lithium aluminium hydride then gives (594) (*ca.* 90%), while methyl-lithium affords (595).[434]

Protonation of the molybdenum tricarbonyl complex (168*b*) (see pp. 57–8) affords the homotropylium derivative (596), the incoming proton occupying the *exo*-position.[214] The tricarbonyl complex (168*b*) readily takes up carbon

monoxide to give the tetracarbonyl compound (597) (65%), which in solution rapidly forms the product (598); protonation of the tetracarbonyl (597) is accompanied by loss of carbon monoxide and the formation of the homotropylium tricarbonyl complex (596).[804]

Protonation of the tungsten tricarbonyl and tetracarbonyl complexes (168*c*) (see p. 57) and (599) is analogous to that of the molybdenum compounds, and the resulting cation has the homotropylium structure (600) (incoming proton *exo*).[805]

(168c)

(599)

(600)

(*j*) *Iron, ruthenium and osmium derivatives*

For preparation from COT, see pp. 60–71; from COT^{2-}, see pp. 188–9.

Reactions. The iron(0) complex $(COT)_2Fe$ (**176**) (see pp. 60–1) catalyses the cyclotrimerisation of disubstituted acetylenes,[806] and the reaction of mono- and disubstituted acetylenes with norbornadiene, which gives rise to cyclopentadiene and 1,3-disubstituted or 1,2,3,4-tetrasubstituted benzenes (20–40%).[807]

$R^1 = H$; $R^2 = Me$, Ph   or $R^1 = R^2 = Me$, Ph

The [4 + 2] cycloaddition of buta-1,3-diene to but-2-yne proceeds at 18° (55% yield) in the presence of $(COT)_2Fe$.[806] This also acts as a selective oligo-merisation catalyst for buta-1,3-diene, which affords the trimer (**601**) as the main product, and it promotes the reaction of ethylene with buta-1,3-diene to give Z-hexa-1,4-diene (**602**).[808]

$$CH_2{=}CH.CH{=}CH.CH_2.CH{=}CH.[CH_2]_2.CH{=}CH.CH_3 \quad (601)$$

$$CH_2{=}CH.CH_2.CH\overset{Z}{=}CH.CH_3 \quad (602)$$

Reaction of $(COT)_2Fe$ (**176**) at 0° with buta-1,3-diene followed by carbon monoxide gives the complex (**603**) (*ca.* 25%).[809]

$$(COT)_2Fe \xrightarrow[\text{(ii) CO, 0}^\circ\text{, 2 atm}]{\text{(i) Butadiene, hexane, 0}^\circ} (COT)(\text{butadiene})Fe(CO) \quad (603)$$

(176)

The structure of **(603)** is shown.[810] (For the ^{57}Fe Mössbauer spectrum, see ref. 448.)

(603)

With carbon monoxide alone, $(COT)_2Fe$ forms the tricarbonyl complex **(177)** (see p. 62) (50%).[808]

$$(COT)_2Fe \xrightarrow[20^\circ\text{, 1 atm}]{CO} (COT)Fe(CO)_3 \quad (177)$$

Reaction with hydrogen in the presence of phosphorus(III) ligands results in complexes of the type **(604)**[442] (table 74). A similar reaction with the bidentate

$$(COT)_2Fe \xrightarrow[\substack{H_2,\text{ pressure} \\ \text{toluene, } \Delta}]{\begin{array}{c} R^1 \\ P-R^2 \\ R^3 \end{array}} H_2Fe\left[P\begin{array}{c} R^1 \\ R^2 \\ R^3 \end{array} \right]_4 \quad (604)$$

Table 74

R^1	R^2	R^3	Pressure (atm)	Temp. (°C)
Ph	OMe	OMe	100	80
OPr^i	OPr^i	OPr^i	100	75
OPh	OPh	OPh	1000	70
O(*o*-tolyl)	O(*o*-tolyl)	O(*o*-tolyl)	1000	70
	[OCH$_2$]$_3$CEt		100	150
F	F	F	200	160[a]

[a] The initial product was $Fe(PF_3)_5$, which was hydrogenated in a second stage.

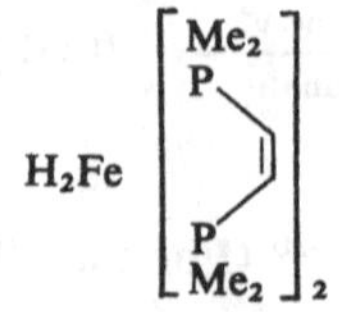

H_2Fe
Me_2
P
P
Me_2
2

(605)

The iron tricarbonyl complex (**177**) decomposes at *ca.* 160°, but the substituted COT ligands in (**606**)–(**608**) (see pp. 106–8, 204) rearrange to give complexes of bicyclo[4.2.0]octa-2,4,7-trienes (**609**)–(**611**).[811]

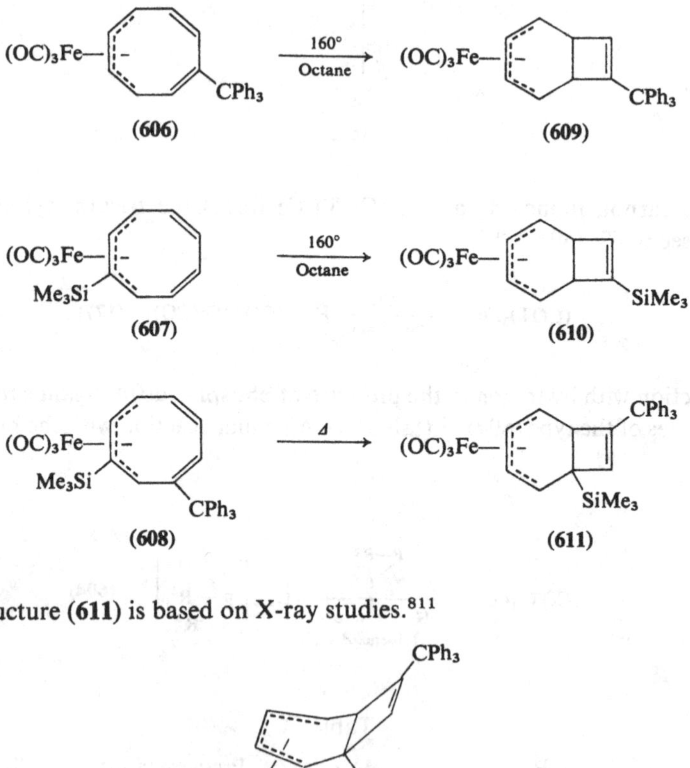

$(OC)_3Fe$
CPh_3
160°
Octane
$(OC)_3Fe$
CPh_3
(606)
(609)
$(OC)_3Fe$
Me_3Si
160°
Octane
$(OC)_3Fe$
$SiMe_3$
(607)
(610)
$(OC)_3Fe$
Me_3Si
CPh_3
Δ
$(OC)_3Fe$
CPh_3
$SiMe_3$
(608)
(611)

Structure (**611**) is based on X-ray studies.[811]

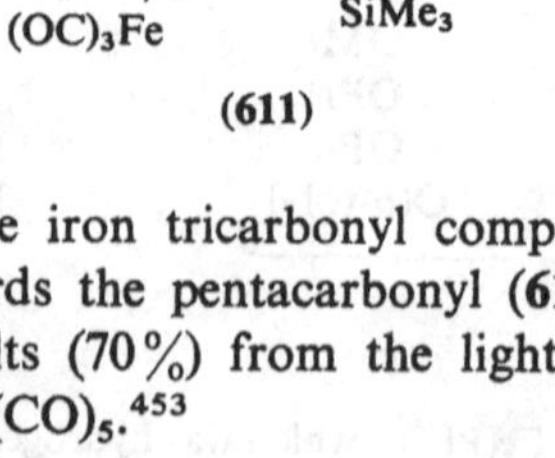

CPh_3
$(OC)_3Fe$
$SiMe_3$

(611)

U.v. irradiation of the iron tricarbonyl complex (**177**), or the '*trans*'-hexacarbonyl (**178**), affords the pentacarbonyl (**612**) in good yield;[812] the hexacarbonyl (**178**) results (70%) from the light-induced reaction of the tricarbonyl (**177**) with $Fe(CO)_5$.[453]

(COT)Fe(CO)₃ $\xrightarrow[\text{Hexane}]{hv}$

(177)

Fe(CO)₅ | e.g. Hexane, *hv*

(COT)Fe₂(CO)₅   (612)

'*trans*'-(COT)Fe₂(CO)₆ $\xrightarrow[\text{Hexane}]{hv}$

(178)

The molecule of (COT)Fe₂(CO)₅ (612) incorporates two π-allyl systems, a formal four-centre four-electron bonding system, and a bridging carbonyl group.[813] This structure is fluxional in solution,[454] and a rapid rotation of the

(612)

ring (relative to the Fe–Fe bond) is invoked as an explanation.[813] (For 'wide-line' n.m.r. studies on the solid, see refs. 455, 460.)

If (COT)Fe(CO)₃ is irradiated at low temperature, and then a disubstituted acetylene is added, the system (613) is produced[814] (table 75) (see p. 206).

(613)

Table 75

R¹	R²	Yield (%)
Ph	Ph	*ca.* 20
CO₂Me	CO₂Me	*ca.* 20

The iron tricarbonyl complex (177) undergoes protonation in e.g. concentrated sulphuric acid to give the bicyclic cation (614) (incoming proton *endo*).[474,815,816] However, using e.g. fluorosulphonic acid-sulphuryl fluoride at −120°, it may be shown that the initial product is the monocyclic species

(615), which at −60° is converted into the bicyclic product (614).[584,817] In 93 % aqueous sulphuric acid, cleavage of the cyclopropane ring occurs with the formation of a species tentatively formulated as (616).[584] Similar protonation

of the methylcyclo-octatetraene complex (302) (see p. 107) affords the initial products (617) and (618) (ratio *ca.* 2:1), which rearrange to give the species (619) and (620) respectively.[818]

The bicyclic cationic complex (614) may be isolated in the form of the crystal-line tetrafluoroborate (621);[474,815,816] treatment of the salt with sodium borohydride then gives the bicyclo[5.1.0]octa-2,4-diene complex **(622)** (16%), which may be reconverted into the original tetrafluoroborate by hydride-abstraction with triphenylmethyl tetrafluoroborate.[474,815,819]

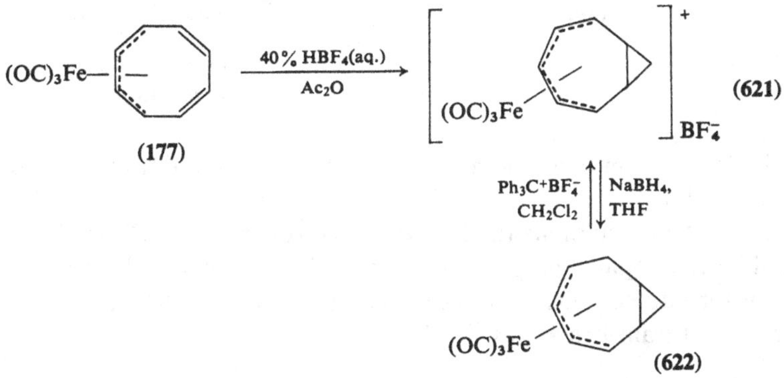

By the steps outlined in scheme 70, the tetrafluoroborate **(621)** may be transformed into 2,3-homotropone **(623)**.[820]

Scheme 70

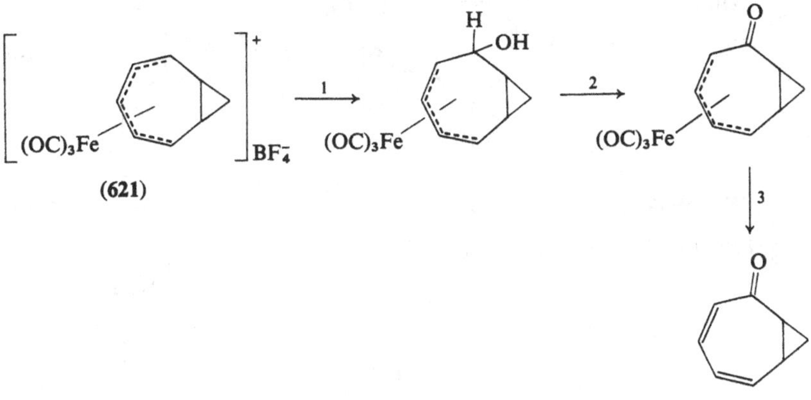

Reaction	Reagents and conditions
1	NaOH, Me$_2$CO(aq.), low temp.
2	CrO$_3$, pyridine
3	Ce(NO$_3$)$_4$.2NH$_4$NO$_3$

The iron tricarbonyl complex **(177)** reacts with aluminium chloride to give the carbonyl insertion product **(624a)** (*ca.* 40%), which on treatment with carbon monoxide liberates barbaralone (90%).[821]

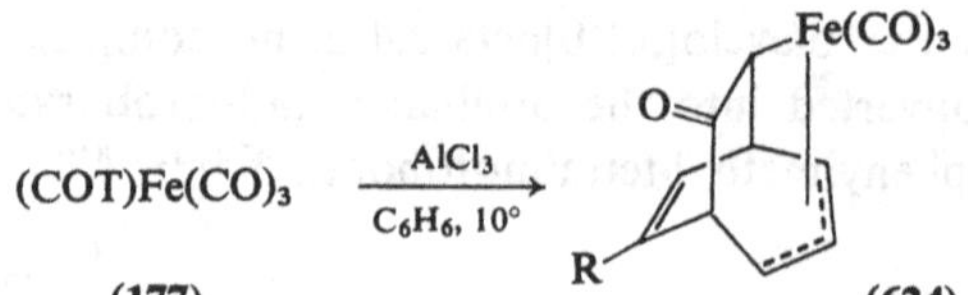

(COT)Fe(CO)₃ $\xrightarrow[C_6H_6,\ 10°]{AlCl_3}$

(177)   (624)

a: R = H
b: R = Me

A similar reaction with (methylcyclo-octatetraene)Fe(CO)₃ (302) gives the complex (624b).[821]

Under certain conditions the iron tricarbonyl complex (177) may be substituted by electrophilic reagents. The use of this property in the preparation of monosubstituted COTs has been described (see pp. 82–3); scheme 71 outlines other transformations.[552, 822]

Scheme 71

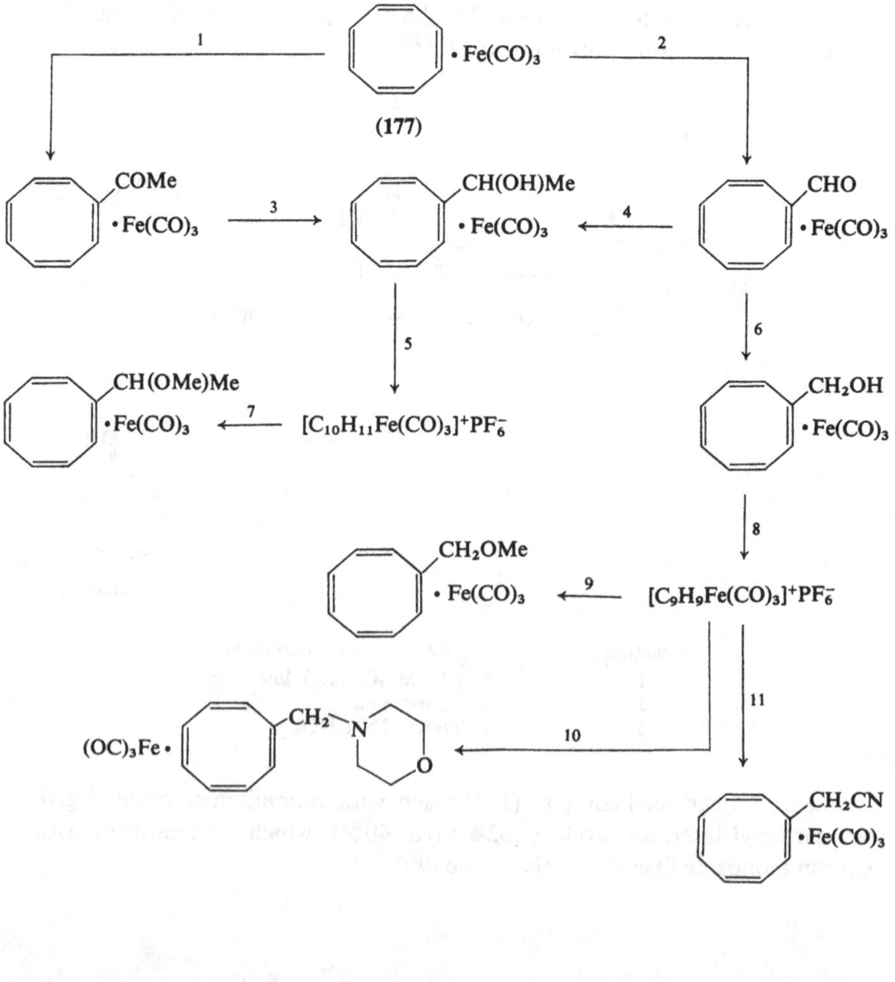

Reaction	Reagents and conditions	Yield (%)
1[a]	MeCOCl, AlCl$_3$, CH$_2$Cl$_2$, 0°	5
2	HCONMe$_2$, POCl$_3$, 4°	60
3	NaBH$_4$, EtOH, 0°	85
4	MeMgX	—
5	65 % HPF$_6$(aq.), Et$_2$O, 20°	90
6	NaBH$_4$, EtOH, 0°	93
7	MeOH, r.-t.	83
8	65 % HPF$_6$(aq.), Et$_2$O, 20°	89
9	MeOH, r.-t.	89
10	Morpholine, Et$_2$O, r.-t.	—
11	NaCN, Me$_2$CO(aq.), r.-t.	—

[a] A diacetyl-derivative may also be isolated (0.9 % yield). The main acetylation product appears to be the cation (625), isolable as the hexafluorophosphate (see below).

(625)

With acetyl chloride in the presence of aluminium chloride, followed by ammonium hexafluorophosphate, the iron tricarbonyl complex (177) affords the salt (626) (28 %), which with sodium methoxide gives the methoxy-derivative (627); removal of the metal then gives the free bicyclic diene (628).[552]

(177)

(626)

NaOMe | MeOH, r.-t.

(628)

Ce^{4+} / MeOH

(627)

Using the same reaction conditions as for (COT)Fe(CO)$_3$ (see above), the phenylcyclo-octatetraene complex (629) may be converted into the formyl-derivative (630) (60 %).[590]

(629)   (630)

The complexes (298) (p. 106) react with trityl tetrafluoroborate to give salts which on hydrolysis yield the products (631)[551] (table 76).

(298)   CPh₃   (631)

Table 76

R	Yield (%)
Ph	11[a]
SiMe₃	15
GeMe₃	12

[a] Position of trityl group uncertain.

The electrochemical reduction of (COT)Fe(CO)₃ is mentioned in ref. 823. Displacement of COT from the iron tricarbonyl complex (177) is effected by carbon monoxide at 100° and high pressure.[824] Triphenylphosphine in refluxing benzene[824] or ethylcyclohexane[450] also displaces COT, giving the product (632) (for the kinetics of this type of displacement, see ref. 825). Triphenylarsine and triphenylstibene, however, displace carbon monoxide to afford complexes of the type (633) (80% and 50% respectively.)[450]

The iron tricarbonyl complexes of COTs (298) undergo cycloaddition reactions with e.g. tetracyanoethylene, forming products now known to possess structure (634)[587, 826] (table 77). Demetallation of the adduct (634a; R = H)

(298)

(634)

$a: R^1 = R\ (R^2 = R^3 = R^4 = R^5 = H)$

$b: R^2 = R\ (R^1 = R^3 = R^4 = R^5 = H)$

$c: R^3 = R\ (R^1 = R^2 = R^4 = R^5 = H)$

$d: R^4 = R\ (R^1 = R^2 = R^3 = R^5 = H)$

$e: R^5 = R\ (R^1 = R^2 = R^3 = R^4 = H)$

Table 77

Yield (%)

R	(634a)	(634b)	(634c)	(634d)	(634e)	Ref.
H			96			(474), 826, 827
Me	71	21.5	—	—	—	(587), 588
Ph₃C			75ᵃ			587
Ph	16	23	—	39	—	(587), 588
CO₂Me	—	—	—	64	23	588
OMe	—	—	31.5	—	—	588
Br			81ᵃ			587

ᵃ Only one isomer reported.

gives the dihydroquinacene **(635)** (84 %),[826, 827] starting material for a projected synthesis of the dodecahedrane system.[828, 829] For demetallation of substituted derivatives **(634)**, see ref. 588.

(634a; R = H)

(635)

Similar cycloadditions occur with 1,1-dicyano-2,2-bis(trifluoromethyl)-ethylene (table 78); the structures of the products, originally formulated as 1,2-cycloadducts, are probably analogous to the tetracyanoethylene-derivatives **(634)**.

Table 78

	R	Conditions	Yield (%)	Ref.
(298)	H	Hexane, r.-t.	95	(830), 831
	Ph_3C	Hexane, r.-t.	25	587

Reaction of the iron tricarbonyl complex **(177)** with phenylacetylene affords a product formulated as **(636)**.[832]

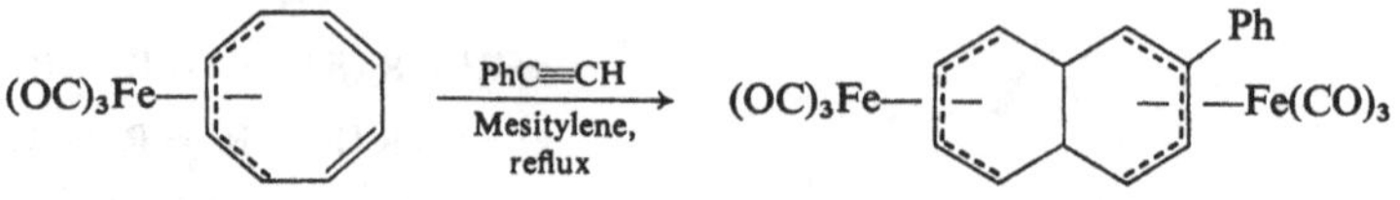

(177) **(636)**

With disubstituted acetylenes, however, the metal is displaced to give the bicyclo[4.2.2]decatetraenes **(613)**[832] (table 79) (see p. 199).

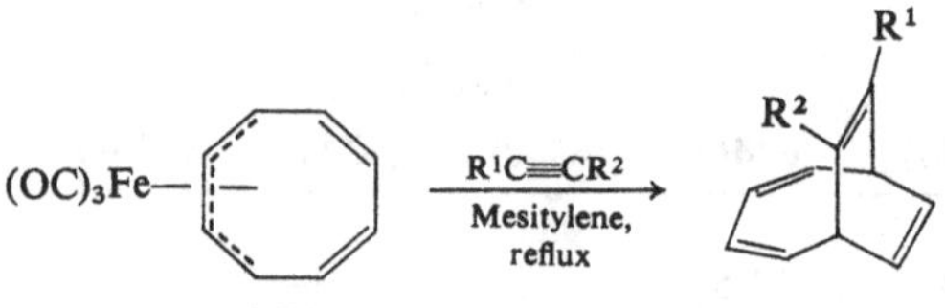

(177) **(613)**

Table 79

R^1	R^2	Yield (%)
Ph	Ph	Up to 35
Ph	CO_2Me	15
Ph	$SiMe_3$	*ca.* 10

Hexafluoroacetone reacts with the iron complex **(177)** to give '*trans*'-(COT)$Fe_2(CO)_6$ **(178)** and a metal-free product (12%), tentatively assigned structure **(637)**;[831] in view of the proved structure (**634*a***; R = H) for the tetracyanoethylene-adduct, it is likely that the hexafluoroacetone-adduct possesses structure **(638)** (*cf.* the reaction of tetracyanoethylene with (benzo-cyclo-octatetraene)$Fe(CO)_3$, followed by demetallation[588]).

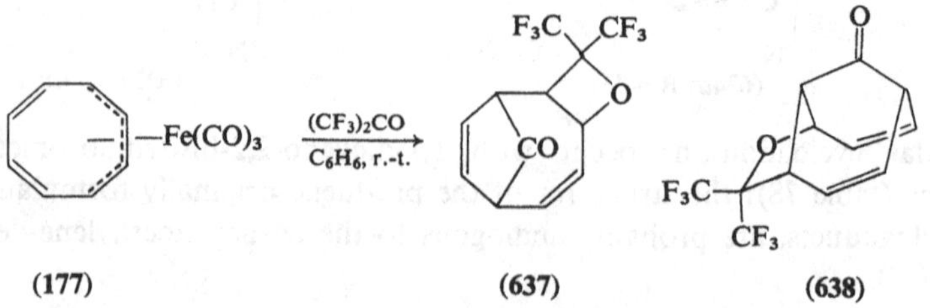

(177) **(637)** **(638)**

A similar structure (**639**) probably results from the reaction of the 1,1-dicyano-2,2-bis(trifluoromethyl)ethylene-adduct (presumably (**640**)) with carbon monoxide, nitric oxide, or 1,2-bis(diphenylphosphino)ethane[831] (table 80).

(**640**)   (**639**)

Table 80

Reagents and conditions	Yield (%)
CO (100 atm), CH_2Cl_2, 80°	53
or NO, CH_2Cl_2, r.-t.	46
or $Ph_2PCH_2 . CH_2PPh_2$, CH_2Cl_2, r.-t.	54

In contrast with the 1,3-cycloaddition of tetracyanoethylene etc., 4-phenyl-1,2,4-triazoline-3,5-dione reacts with $(COT)Fe(CO)_3$ to afford a product (**641**) (29%) resulting from 1,4-cycloaddition to the ligand.[831]

(**177**) $Fe(CO)_3$ (**641**)

(Stereochemistry not certain)

1,4-Attack on the complex (**177**) is also adopted by chlorosulphonyl isocyanate; the product obtained (80%) after dechlorosulphonylation has structure (**642**), and on demetallation yields the free ligand (**643**).[826]

Light-induced reactions of $(COT)Fe(CO)_3$ with COT, giving derivatives of COT dimers, have already been mentioned (see p. 63).

In the presence of tetrafluorobenzyne, $(COT)Fe(CO)_3$ is apparently converted into '*trans*'-$(COT)Fe_2(CO)_6$ (**178**) (25%).[833]

Highly efficient (95%) demetallation of $(COT)Fe(CO)_3$ is effected by trimethylamine N-oxide in refluxing benzene.[834]

(OC)₃Fe—(177) $\xrightarrow[\text{(ii) PhSH, pyridine, Me}_2\text{CO}]{\text{(i) ClSO}_2\cdot\text{NCO, CH}_2\text{Cl}_2, 0°}$ (642) [structure, Fe(CO)₃]

(177)

(642)

$\downarrow$ Ce⁴⁺

(643)

Reaction of (COT)Fe(CO)₃ with Ru₃(CO)₁₂ results in the 'mixed' binuclear complex (644).[835]

$$\text{(COT)Fe(CO)}_3 \xrightarrow[\text{Xylene, reflux}]{\text{Ru}_3\text{(CO)}_{12}} \text{(COT)FeRu(CO)}_5 \quad \text{(644)}$$

From the spectroscopic evidence, the structure of this fluxional product is as shown (cf. (612), p. 199).

(OC)₂Fe ⋯⋯ Ru(CO)₂

(644)

The ruthenium complex (645), in refluxing octane, is transformed into the binuclear compound (646), which at higher temperatures forms the pentalene complex (304) (mixture of isomers)[811] (see p. 108). Thermolysis of (trityl-cyclo-octatetraene)Ru(CO)₃ also affords a polynuclear species.[811]

Protonation of (COT)Ru(CO)₃ (184) (p. 65) initially leads to the bicyclic cation (647), isolable as the tetrafluoroborate; on standing in solution, how-ever, this bicyclic cation isomerises to a monocyclic species best formulated as (648)[836] (cf. the protonation of (COT)Fe(CO)₃ (177), pp. 199–200). Treatment

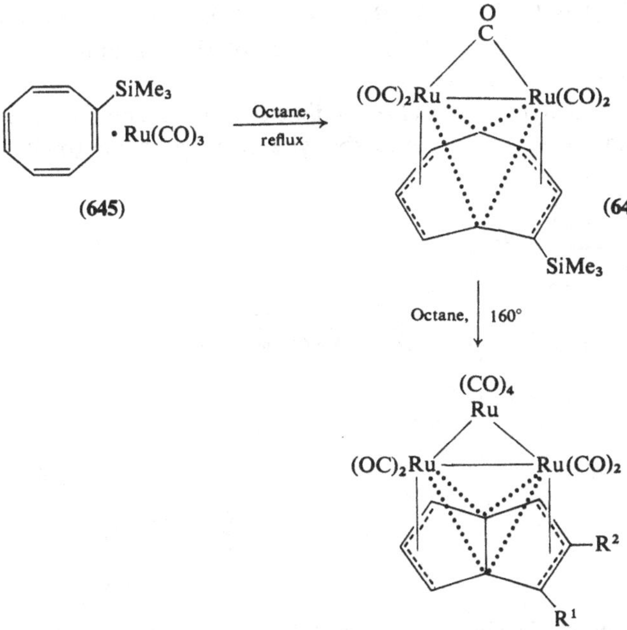

(645) → (646)

(304)

$R^1 = SiMe_3$, $R^2 = H$

$R^1 = H$, $R^2 = SiMe_3$

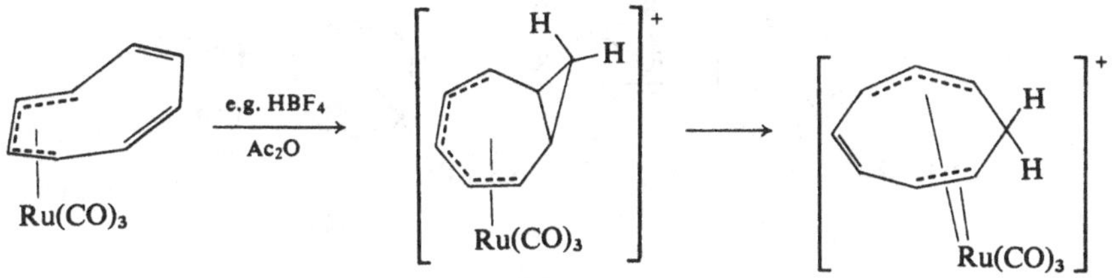

(184)   (647)   (648)

of the salts of (648) with cyanide ions, or nitromethane in the presence of sodium carbonate, gives the neutral complexes (649a) or (649b).[836]

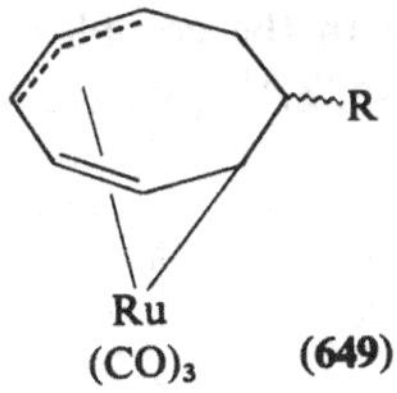

(649)

a: R = CN

b: R = CH$_2$NO$_2$

COT is displaced from (COT)Ru(CO)₃ (**184**) by e.g. triphenylphosphine, iodine, mercury(II) chloride;[479] for kinetics of the reactions with tertiary phosphines, see ref. 837.

The pentacarbonyl complex (**191**) of dimeric COT (see pp. 65–7) reacts with carbon monoxide to give a hexacarbonyl species, probably (**650**), which is fluxional.[484]

(191)

$$\xrightarrow[\text{Me}_2\text{CO, 40°, 10 atm}]{\text{CO}}$$

(650)

The ruthenium tricarbonyl complex (**184**) reacts with tetracyanoethylene and related compounds to form 1:1 adducts[831] (table 81); the original formulation of the products as 1,2-cycloadducts (**651**) is questionable (see the corresponding iron compounds, pp. 204–5, 207).

(184)   (651)

Table 81

R¹	R²	R³	R⁴	Conditions	Yield (%)
CN	CN	CN	CN	CH_2Cl_2, r.-t.	18
CN	CN	CF₃	CF₃	C_6H_6, r.-t.	16
CN	CF₃	CN	CF₃	CH_2Cl_2, r.-t.	44

Reaction of hexafluoroacetone with the complex (**184**) affords an adduct (14 %) with the suggested structure (**652**)[831] (but see p. 206).

(184)   (652)

The ruthenium complex (**183**) (see p. 64) reacts with allylmagnesium chloride to give the bis-allyl-derivative (**653**).[838]

$$[(COT)RuCl_2]_n \xrightarrow[\text{Et}_2\text{O, r.-t.}]{CH_2=CH.CH_2MgCl} (COT)Ru(allyl)_2 \quad (653)$$
$$(183)$$

Protonation of α-(COT)Os(CO)$_3$ (**202**) (see p. 70) is similar to that of (COT)Ru(CO)$_3$ (**184**)[836] (see pp. 208–9). For the displacement of COT by trimethyl phosphite, see ref. 496.

(k) Cobalt, rhodium and iridium derivatives

For catalytic and displacement reactions of complex (**203**) (see p. 71), see ref. 498.

The 'mixed' cobalt complex (**205**) (see pp. 71–2) is transformed by the action of heat into the product (**206**).[839]

$$(COT)Co(CPD) \xrightarrow{133°} (COT)Co_2(CPD)_2$$
$$(205) \qquad\qquad (206)$$

Protonation of the 'mixed' complexes (**654a**) and (**654b**) may be effected by trifluoroacetic acid to give, initially, the bicyclic cations (**655**), which subsequently isomerise to the monocyclic form (**656**)[840] (*cf.* ruthenium, pp. 208–9).

$$a: M = Co \qquad b: M = Rh \qquad c: M = Ir$$

With aqueous HPF$_6$, the cobalt and iridium complexes (**654a**) and (**654c**) afford the salts (**657**); the cobalt salt (**657**; M = Co) with sodium cyanide yields the cyano-derivative (**658**) (63%).[840] The rhodium complex (**654b**), however, reacts with HPF_6 to form a hexafluorophosphate salt of the bicyclic cation (655).[840]

At low temperatures, metal-protonation of the iridium complex (654c) occurs to give the hydrido-species (659), which at room-temperature affords a mixture of (655c) and (656c).[840]

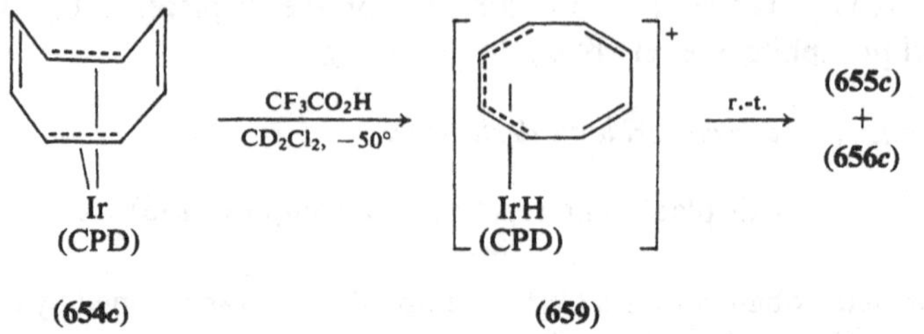

Some transformations of the dimeric rhodium complex (208) (see p. 73) are outlined in scheme 72. (For the dipole moments of (COT)Rh(acac) (209) and (COT)Rh₂(acac)₂ (660b), see ref. 843.)

Scheme 72

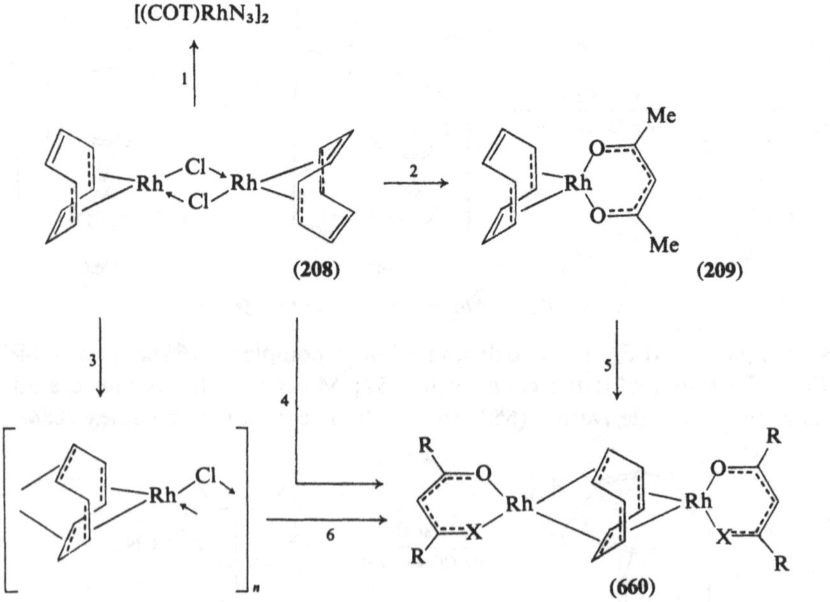

a: R = Ph, X = S
b: R = Me, X = O

Reaction	Reagents and conditions	Yield (%)	Ref.
1	NaN$_3$, MeOH, r.-t.	*ca.* 100	841
2	CH$_3$COCH$_2$COCH$_3$, K$_2$CO$_3$,	80	507
	light petroleum (b.p. 60–80°), r.-t.		
3	CHCl$_3$ or CS$_2$, r.-t.	*ca.* 65	507
4	PhCOCH$_2$CSPh, K$_2$CO$_3$,	48[a]	842
	petroleum, r.-t.		
5	130°	77[b]	510
6	(As for reaction 2)	65[b]	507

[a] Product (660*a*).   [b] Product (660*b*).

COT is displaced from the rhodium complex (208) by e.g. triphenyl phosphite; for heats of reaction, see ref. 509.

The 'mixed' complex (661) results from the reaction of [(COT)IrHCl$_2$]$_n$ (217) (see p. 75) with sodium cyclopentadienide; treatment with iodine then leads to (CPD)IrI$_2$ (662).[515]

$$[(COT)IrHCl_2]_n \xrightarrow[\text{THF}]{\text{(CPD)Na}} (COT)Ir(CPD) \quad (661)$$

(217)

$$I_2 \Big| \begin{array}{l} CH_2Cl_2, \\ \text{r.-t.} \end{array}$$

$$(CPD)IrI_2 \quad (662)$$

(*l*) Nickel and platinum derivatives

For preparation from COT, see pp. 75–8.

Reactions. COT is displaced from (COT)Ni(duroquinone) (220) (p. 76) by tertiary phosphines and phosphites; for kinetics, see ref. 844.

The nickel(0) complex (218) (pp. 75–6) reacts with 2,4,6-tri-t-butylphenoxy radicals to yield the product (663) (66%).[845]

The platinum complex (226*b*) (pp. 77–8) reacts with Grignard reagents to give products of the types (664)–(666)[846, 847] (table 82). (For the ^{1}H n.m.r. spectrum of (665; R = Me), see ref. 530).

$$(COT)PtI_2 \xrightarrow[\substack{Et_2O, C_6H_6, \\ \text{r.-t.}}]{RMgI} \begin{array}{l} (COT)PtRI \quad (664) \\ \text{or } (COT)PtR_2 \quad (665) \\ \text{and/or } (COT)Pt_2R_4 \quad (666) \end{array}$$

(226*b*)

Table 82

Yield (%)

R	(664)	(665)	(666)
Me	—	5	36
Et	4	—	—
Ph	—	—	70
o-Me.C_6H_4	—	64	—
p-Me.C_6H_4	—	29	—
1-Naphthyl	—	65	—

7. Homotropylium species

These are produced by protonation of COTs (see pp. 21–2, 99), or from 7,8-dichlorocyclo-octa-1,3,5-trienes by treatment with proton or Lewis acids (see p. 236).

For the ^{13}C n.m.r. spectrum of the homotropylium cation (35), see refs. 848, 849.

Ring-inversion of the *endo*-8-chloro-derivative (see p. 23) is catalysed by *cis*-7,8-dichlorocyclo-octa-1,3,5-triene.[850]

Reactions. U.v. irradiation of the homotropylium ion (35) leads to a species identified as the bicyclic cation (667).[851,852]

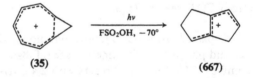

(35)   (667)

For the electrochemical reduction of (35), see ref. 267.

8. Cyclo-octa-1,3,5-triene, bicyclo[4.2.0]octa-2,4-diene and cyclo-octa-1,3,6-triene

Cyclo-octa-1,3,5-triene (62) and cyclo-octa-1,3,6-triene (63) are produced from COT by various reduction processes (see pp. 28–30); they may be purified *via* their silver nitrate complexes.[63,264]

At moderate temperatures cyclo-octa-1,3,5-triene equilibrates with bicyclo-[4.2.0]octa-2,4-diene (64) (see table 83), and pure bicyclo[4.2.0]octa-2,4-diene may be obtained by slow fractional distillation of the mixture;[853] if pure cyclo-octa-1,3,5-triene is required, the bicyclic valence tautomer may be removed by treatment with maleic anhydride (in the cold).[260]

Cyclo-octa-1,3,5-triene also equilibrates (slowly) with the 1,3,6-triene (*via* 1,5-hydrogen shifts[736,854]), but the equilibrium mixture contains only *ca.* 0.5–1 % of the 1,3,6-triene at 100–129°.[855] This almost complete conversion of the 1,3,6-triene into the 1,3,5-triene can be catalysed by potassium t-butoxide in t-butanol,[63] or potassium hydroxide in ethanol.[264] (For the conformations of (62) and (63), see ref. 857.)

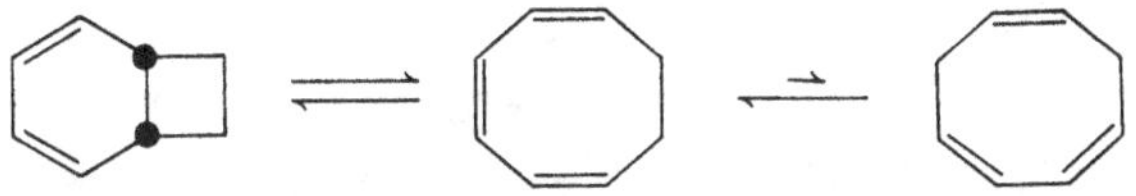

| (64) | (62) | (63) |

Table 83

Temp. (°C)	Proportion of (64) (%)	Ref.
20	9	856
60	10.8	136
80–100	15	260

Reactions. The chemistry of the cyclo-octatrienes (up to 1963) has been reviewed[265] (see also ref. 858); some of this early work would appear to need reinvestigation.

Thermal dimerisation. See p. 225.

Thermolysis. At temperatures above *ca.* 225°, cyclo-octa-1,3,5-triene gives a complex mixture of hydrocarbons from which the following constituents may be isolated:[853, 859] benzene, *E,E*-octa-1,3,5,7-tetraene (668), 1-vinylcyclohexa-1,3-diene (669), 5-vinylcyclohexa-1,3-diene (670), and tricyclo[3.2.1.0^{2,7}]oct-3-ene (671). The reaction sequence of scheme 73 is suggested.[853]

Scheme 73

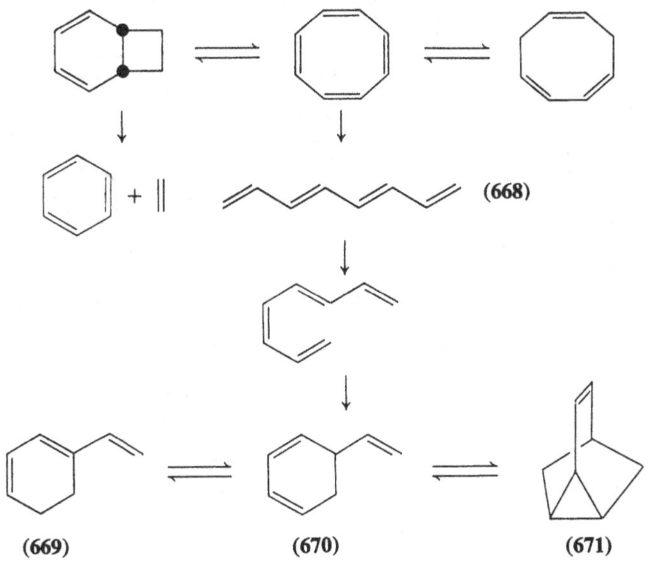

| (669) | (670) | (671) |

Photolysis. U.v. irradiation of cyclo-octa-1,3,5-triene yields the photo-isomers **(672)** and **(673)**.[853,860,861]

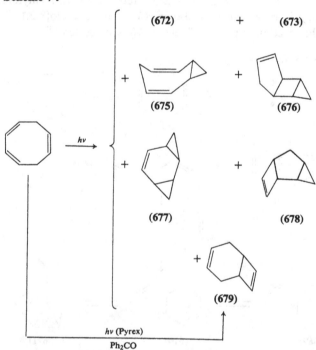

Z,Z-Octa-1,3,5,7-tetraene **(674)** has been detected as a transient intermediate in this reaction,[862,863] and under certain conditions the E,E-tetraene **(668)** may be isolated.[853]

(674)

Photolysis of bicyclo[4.2.0]octa-2,4-diene yields cyclo-octa-1,3,5-triene, together with benzene and ethylene.[853]

The photoproducts **(672)** and **(673)** are also obtained from cyclo-octa-1,3,6-triene, but in addition, five other isomers formulated as **(675)**–**(679)** may be isolated;[853,860] the benzophenone-sensitised photo-reaction, however, leads to isomer **(679)** exclusively[853] (scheme 74).

Scheme 74

Oxidation, and reactions with electrophiles. For oxidation reactions, see ref. 265. The cation (680) may be generated from cyclo-octa-1,3,5-triene in a 'super-acid' medium.[864]

(680)

Reaction of the 1,3,5- or the 1,3,6-triene with chlorine or bromine yields the dihalo-derivative (681);[265,865] further reaction then affords tetra- and hexa-halides.[264,265] The dibromide (681; X = Br)[3,865] with dimethylamine gives the bis(dimethylamino)-derivative (682) (47%);[3] with triethylamine in dimethyl sulphoxide, 2,3-homotropone (623) is formed.[866] (For reactions of (681; X = Cl), see ref. 265.)

For reactions with hydrogen chloride and hydrogen bromide, see ref. 265.

Reduction, and proton-abstraction. For catalytic hydrogenation of cyclo-octa-1,3,5-triene, see e.g. refs. 3, 264; of cyclo-octa-1,3,6-triene, see e.g. ref. 63.

Potassium amide in liquid ammonia abstracts a proton from cyclo-octa-1,3,5-triene to generate the cyclo-octatrienyl anion (683), which dispropor-tionates into COT^{2-} and cyclo-octa-1,3,5-triene.[867]

(683)

In the presence of $KOCPr_3^n-Pr_3^nCOD$, cyclo-octa-1,3,5-triene catalyses H–D exchange in COT (see p. 31).

Cycloadditions. Carbene-addition to cyclo-octa-1,3,5-triene, using di-iodo-methane and a zinc–copper couple,[868] or diazomethane in the presence of copper(I) chloride,[869] affords the isomeric products (684) and (685) (combined yield 42%; ratio 4.5:1[869]) (the initially formed adduct (685) is very susceptible to Cope rearrangement, and at 25° changes into (686) with a half-life of *ca.* 1 day[870]) (scheme 75).

Similar cyclopropanation[868,869] of cyclo-octa-1,3,6-triene yields a mixture of the products (687) (51%) and (688) (37%).[869] At 130°, the bicyclic diene (688) undergoes a 1,5-homodienyl hydrogen shift to form Z,Z,Z-cyclonona-1,3,6-triene (689).[869] Reversible 1,5-hydrogen shifts in the products (684) and (687) result in an equilibrium mixture at 160°, a third component being 'trishomobenzene' (690),[868,869] cyclopropanation of which gives 'hexahomo-benzene' (691).[871]

Scheme 75

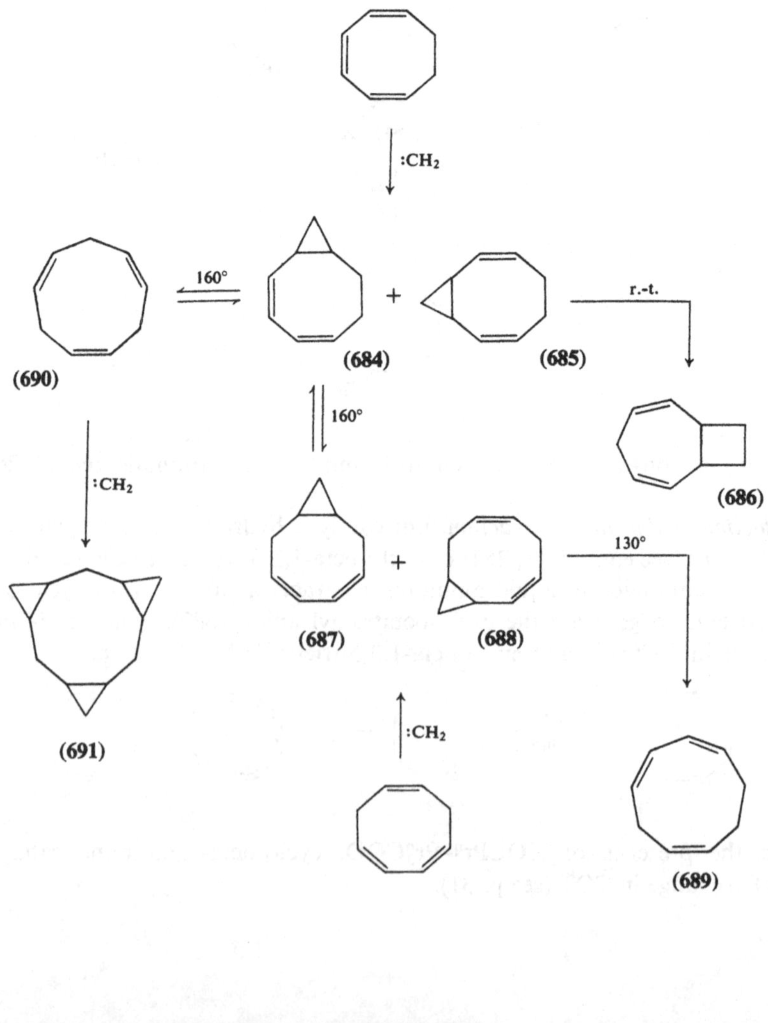

The addition of dibromocarbene to cyclo-octa-1,3,5-triene leads to the monoadducts **(692)** (36 %) and **(693)** (10.5 %), together with two bis-adducts.[870] Reduction of the product **(693)** using sodium in liquid ammonia gives the labile system **(685)** (see above), and reaction with methyl-lithium yields the tricyclic diene **(694)**.[870]

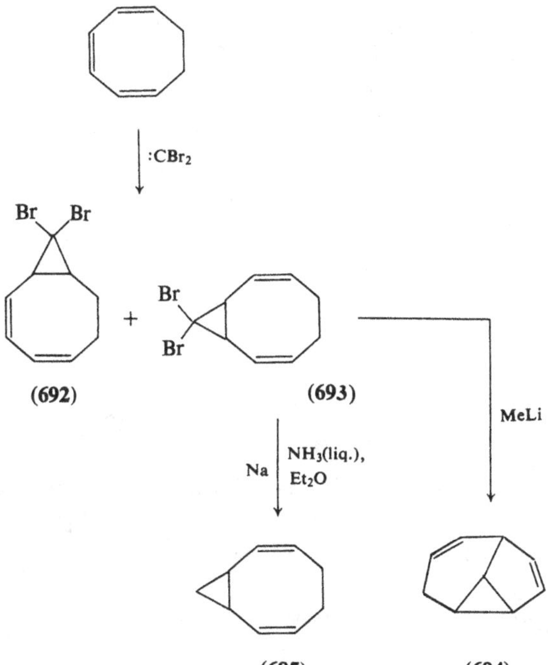

:CBr$_2$

Br Br

+

Br
  Br

(692)   **(693)**

MeLi

Na | NH$_3$(liq.), Et$_2$O

(685)   **(694)**

With cyclo-octa-1,3,6-triene, dichlorocarbene gives the adducts **(695)** and **(696)** (in a ratio of 1 : 1)[872] (an erroneous structure for the product was originally proposed[265, 858]).

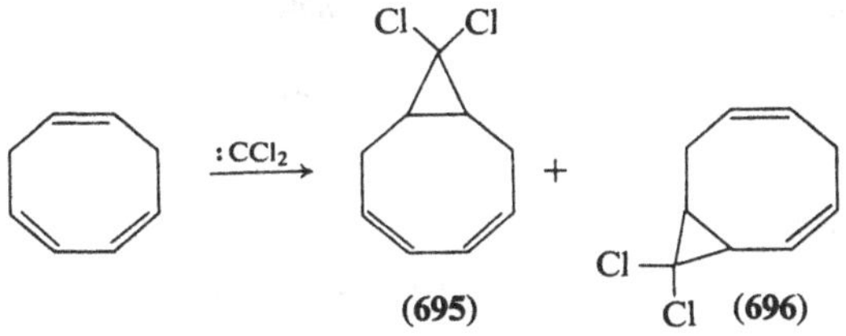

Cl Cl

:CCl$_2$

+

Cl
  Cl

(695)   **(696)**

[2 + 2] Cycloadditions with chlorosulphonyl isocyanate are mentioned later (see p. 222).

The addition of dienophiles such as maleic anhydride to the cyclo-octatrienes occurs *via* bicyclo[4.2.0]octa-2,4-diene **(64)**, so that both the 1,3,5- and the 1,3,6-triene yield the same adduct **(697)**;[264, 265, 858] this is identical with the product obtained by selective semihydrogenation of the cyclo-octatetraene–maleic anhydride adduct.[260] Similarly, structures (698) and (699) result from acetylenedicarboxylic and azodicarboxylic esters respectively. The known reactions of this type are listed in table 84.

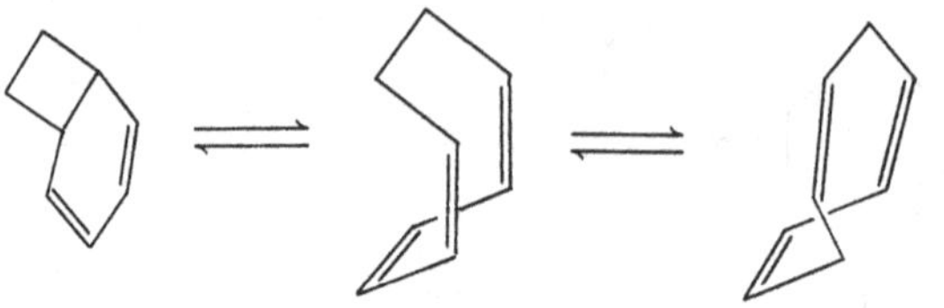

(64)

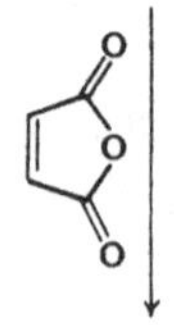

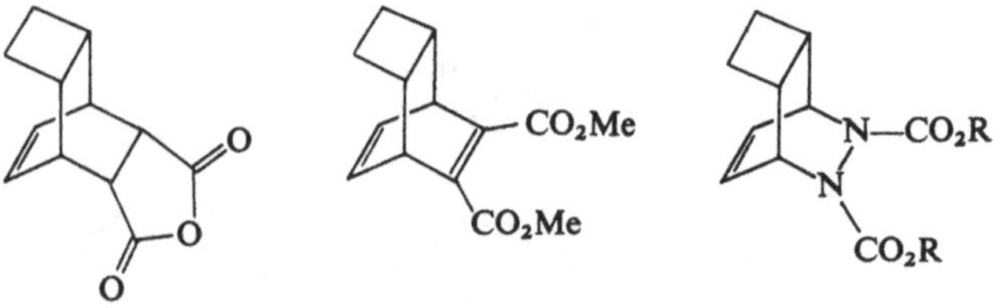

(697)   (698)   (699)

Table 84

Dienophile	Reaction conditions	Product	Yield (%)	Ref.
Maleic anhydride	C_6H_6, reflux	(697)	86[a], (64)	259, (873)
Dimethyl acetylene-dicarboxylate	C_6H_6, reflux	(698)	77	260
'Azodicarboxylic ester'	—	(699)	—	858
Dimethyl azodicarboxylate	—	(699; R = Me)	—	874

[a] Reaction temperature 100°.

Thermolysis of compound (698) provides a route to cyclobutene (see p. 352); for further reactions of (699; R = Me), see ref. 874.

Reaction of cyclo-octa-1,3,5-triene with benzyne (from the decomposition of benzenediazonium-2-carboxylate in dichloromethane at 50°) also occurs *via* the valence tautomer (64), affording the adduct (700) (35%) exclusively.[875]

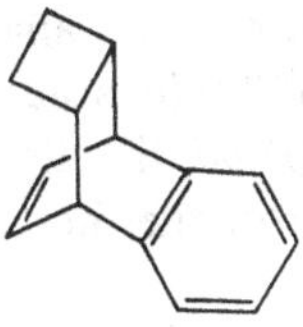

(700)

Nitrosobenzenes, however, also add to cyclo-octa-1,3,5-triene itself, and products of types (701)–(703) may be isolated[876] (see table 85). In refluxing ethanol, products of structure (701) isomerise to the conjugated system (702).[876]

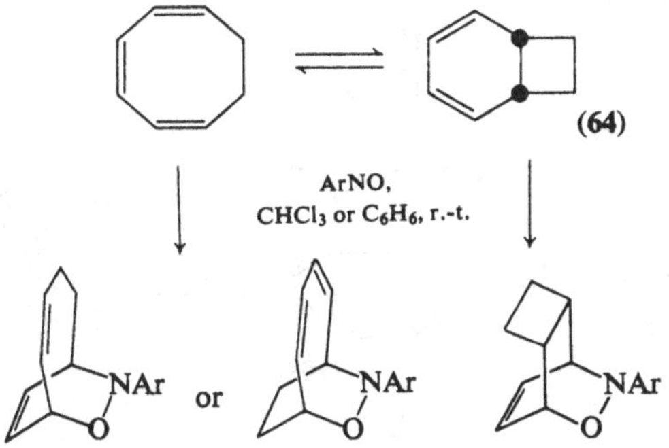

Table 85

	Ar	Yield (%)		
		(701)	(702)	(703)
a	Ph	—	26	15
b	*p*-Cl.C$_6$H$_4$	—	42	17
c	*p*-NO$_2$.C$_6$H$_4$		66[a]	
d	2,4-(NO$_2$)$_2$.C$_6$H$_3$	56	—	3

[a] Total yield; only (702c) was obtained pure. With purified cyclo-octa-1,3,5-triene, (701c) (25%) is isolable.

Cyclo-octa-1,3,6-triene reacts to form [4 + 2] adducts (704)[876] (table 86).

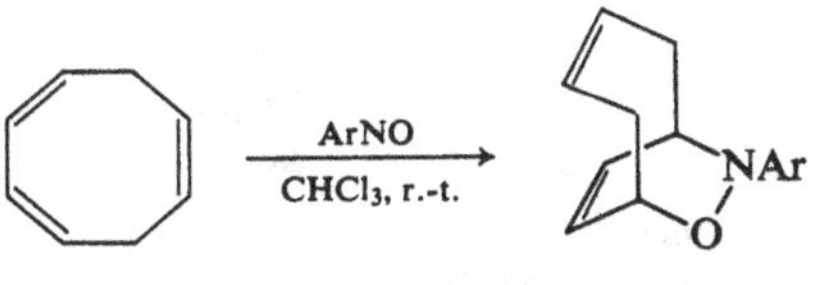

(704)

Table 86

Ar	Yield (%)
p-NO$_2$.C$_6$H$_4$	50
2,4-(NO$_2$)$_2$.C$_6$H$_3$	58

In polar solvents, chlorosulphonyl isocyanate reacts with cyclo-octa-1,3,5-triene to give the adducts (705)–(707) (characterised as the parent lactams) in varying proportions (see table 87) *via* the dipolar intermediate (708), adduct (707) being the ultimate product of thermodynamic control.[877] No reaction

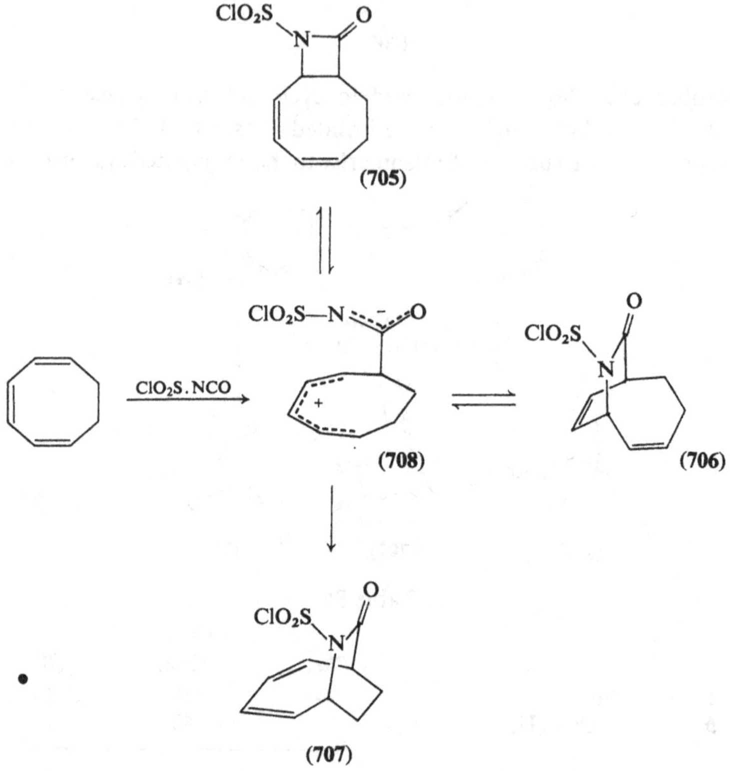

(705)

(708)

(706)

(707)

Table 87

Solvent	Temp. (°C)	Time (h)	Product ratio (706):(707)[a]	Total yield (%)
CH_2Cl_2	25	4	86:14	(Incomplete reaction)[b]
CH_2Cl_2	25	36	71:29	63
CH_2Cl_2	25	200	63:37	64
$MeNO_2$	85	4	50:50	61
$MeNO_2$	85	24	28:72	65

[a] Determined for parent lactams.
[b] Adduct (705) isolated as the lactam (2%).

occurs in cold non-polar solvents; on heating, adducts derived from bicyclo-[4.2.0]octa-2,4-diene are obtained (see below).

With bicyclo[4.2.0]octa-2,4-diene (64), the dipolar intermediate (709) cyclises not only through nitrogen, but also through oxygen; after 1 h in

carbon tetrachloride, the products are the adducts (710) (41 % (as lactam)) and
(711) (37%), but after 5 days these are completely transformed into the
isomers (712) (30% (as lactam)) and (713) (25%).[878]

With electron-deficient dienes, cyclo-octa-1,3,5-triene reacts *via* the valence
tautomer (64) acting as a dienophile. Thus hexachlorocyclopentadiene and its
derivatives give adducts of structure (714)[879] (table 88).

Table 88

X	Yield (%)
CCl₂	60
C(OMe)₂	60
⟨C with dioxolane⟩	58

Similar additions take place with tetracyclone and 2,5-dimethyl-3,4-diphenylcyclopentadienone, although the initial reaction is followed by Cope rearrangement to give products of structure (715); some direct addition to cyclo-octa-1,3,5-triene itself may also occur, since hexa-1,3-dienylbenzenes (716) are co-products[879] (table 89).

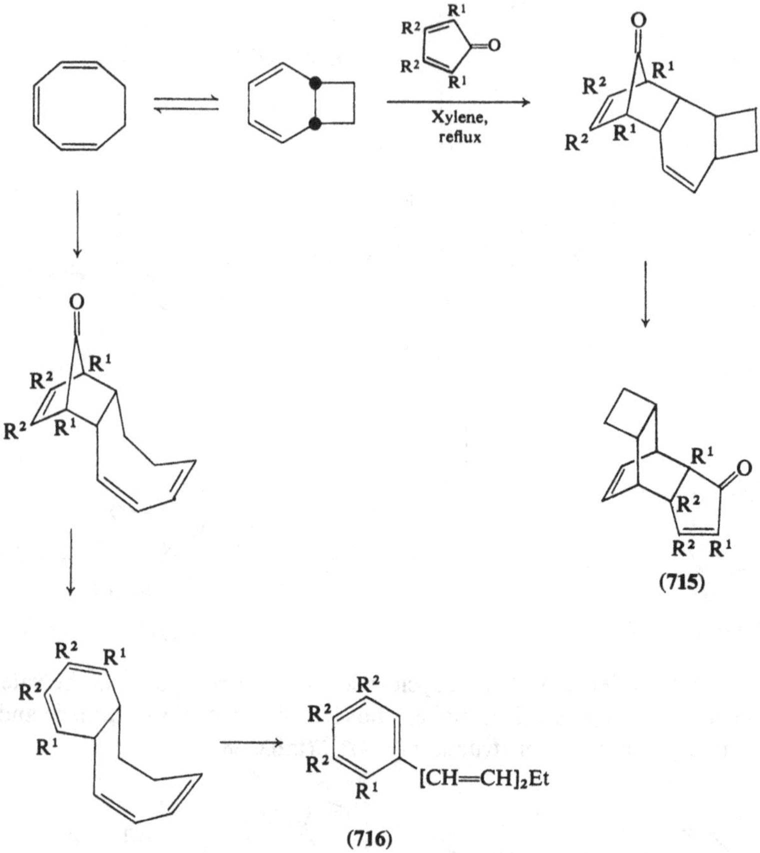

(715)

(716)

Table 89

R¹	R²	Yield (%)	
		(715)	(716)
Me	Ph	34	(Very low)
Ph	Ph	36	28

Electron-deficient dienes also add to cyclo-octa-1,3,6-triene. With e.g. hexachlorocyclopentadiene, the isolated product has the cage-like structure (717), evidently formed by initial addition to the 6,7-double bond of the 1,3,6-triene, followed by an intramolecular Diels–Alder reaction[880] (table 90).

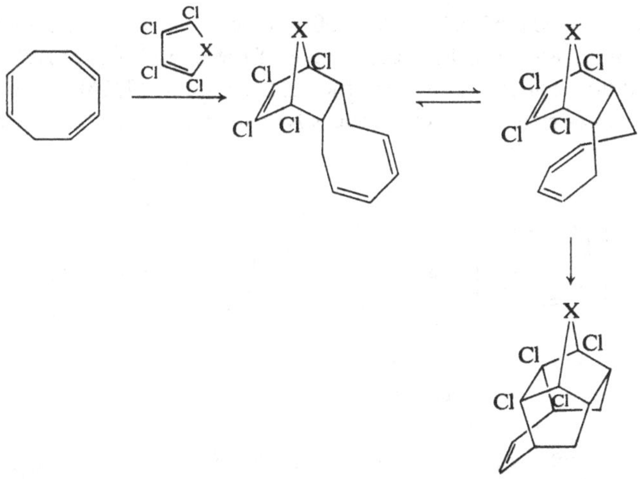

(717)

Table 90

X	Conditions	Yield (%)	Ref.
CCl$_2$	Toluene, *ca.* 100°	10	880
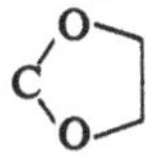 (dioxolane)	Toluene, *ca.* 100°	68	880
CO.CCl=CCl	Xylene, reflux	33	881

Cyclo-octa-1,3,5-triene undergoes dimerisation *via* Diels–Alder addition. If either of the cyclo-octatrienes is subjected to prolonged heating in the presence of e.g. potassium t-butoxide, dimer (718) is produced; however, in the absence of base a different product, allegedly of structure (719), is formed.[858]

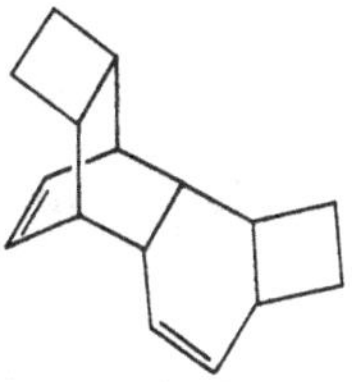

(718) 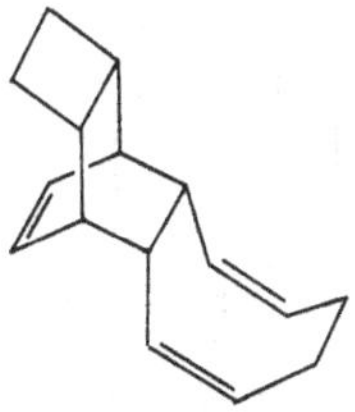   (719)

Tropone reacts with cyclo-octa-1,3,5-triene to give the [6 + 4] adduct (720) (21 %) of the valence tautomer (64).[882]

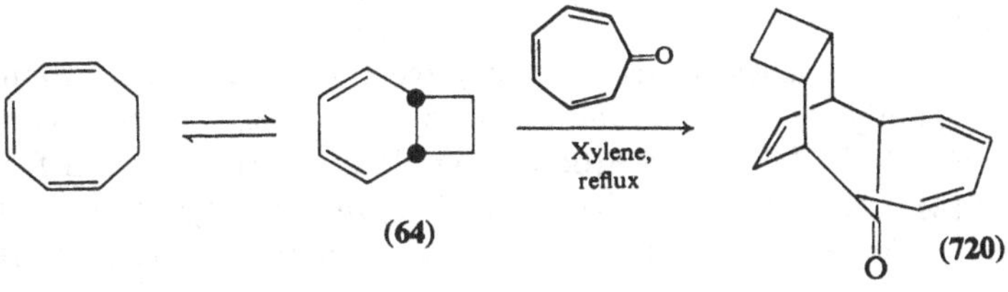

(64)   Xylene, reflux   (720)

Reactions with metal derivatives. With 50% aqueous silver nitrate, cyclo-octa-1,3,5-triene gives an immediate precipitate of the complex (721a);[63,264] with silver tetrafluoroborate, however, the 2:1 product (721b) is formed.[396] Cyclo-octa-1,3,6-triene with ethanolic silver nitrate gives the 1:3 complex (722).[264]

(721) a: X = NO$_3$, n = 1 b: X = BF$_4$, n = 2	(722)

Reaction of cyclo-octa-1,3,5-triene with (CPD)VCl$_3$ in the presence of isopropylmagnesium bromide affords a mixture of the complexes (723) and (724) (total yield 34%), which may be converted quantitatively into the product (724) by platinum-catalysed dehydrogenation.[883]

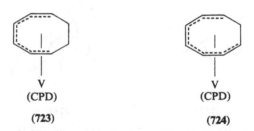

$$\xrightarrow[\text{Et}_2\text{O, } -50° \to \text{r.-t.}]{\substack{(CPD)VCl_3 \\ Pr^iMgBr,}} (C_8H_{10})V(CPD) + (C_8H_9)V(CPD)$$

$$(723) \qquad\qquad (724)$$

Pt
n-Hexane, 20°

The crude structures of (723) and (724) are presumably as shown.[883]

V (CPD)	V (CPD)
(723)	(724)

For ion-molecule reactions of (CPD)V(CO)$_4$ and cyclo-octa-1,3,5-triene, see ref. 422.

The cyclo-octatrienes (mixed isomers) react with the hexacarbonyls of chromium, molybdenum and tungsten to form complexes of types (725) and (726) (table 91).[884,885] The tricarbonyl complexes (725) also result from the action of (MeCN)Cr(CO)$_3$, (diglyme)Mo(CO)$_3$ and (NH$_3$)$_3$W(CO)$_3$ in dioxan or tetrahydrofuran.[886] The chromium complex (725; M = Cr) has the structure

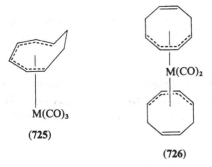

$M(CO)_6$ | Hydrocarbon solvent, 125–145°

(Cyclo-octatriene)$M(CO)_3$ + (Cyclo-octatriene)$_2M(CO)_2$
     (725)   (726)

Table 91

Product	M	Yield (%)
(725)	Cr	13
	Mo	69
(726)	Mo	17
	W	16

shown,[887] but it is suggested that the dicarbonyl complexes (726; M = Mo, W) contain cyclo-octa-1,3,6-triene ligands.[885] (For the mass spectrum of (725; M = W), see ref. 433.)

$M(CO)_3$

(725)

$M(CO)_2$

(726)

The molybdenum complex (725; M = Mo) absorbs carbon monoxide to form the product (727).[804]

$Mo(CO)_4$

(727)

With trityl tetrafluoroborate the complexes (725) undergo proton abstraction to give crystalline salts (728)[886] (table 92).

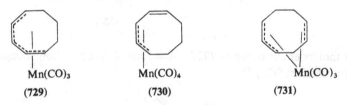

(725)   (728)

Table 92

M	Yield (%)
Cr	73
Mo	72
W	62

Cyclo-octa-1,3,5-triene, on treatment with the hydridomanganese carbonyl [HMn(CO)$_4$]$_3$, gives cyclo-octadienyl complexes (729) and (730) in minute yield.[439]

$$\xrightarrow[\substack{\text{Hexane,}\\ \text{reflux}}]{\text{[HMn(CO)}_4\text{]}_3}$$
(Cyclo-octadienyl)Mn(CO)$_3$ (729)
+ (Cyclo-octadienyl)Mn(CO)$_4$ (730)

The proposed structures of these products are shown below; a third complex, isomeric with (729), is thought to possess structure (731) (possibly derived from cyclo-octa-1,3,6-triene present in the starting material).

Mn(CO)$_3$	Mn(CO)$_4$	Mn(CO)$_3$
(729)	(730)	(731)

For ion-molecule reactions of (CPD)Mn(CO)$_3$ and cyclo-octa-1,3,5-triene, see ref. 422.

U.v. irradiation of cyclo-octa-1,3,5-triene in the presence of Fe(CO)$_5$[888,889] (or Fe$_2$(CO)$_9$[888]) yields the tricarbonyl complex (732) (up to 56%).

$$\xrightarrow[\substack{C_6H_6,\\ h\nu\ (\text{Pyrex})}]{\text{Fe(CO)}_5}$$
(Cyclo-octatriene)Fe(CO)$_3$ (732)

The structure of the complex (732) is formulated as shown.

Fe(CO)₃

(732)

The cyclo-octatriene complex (732) is also formed thermally from iron carbonyls (up to 20 % yield),[890, 891] but when vigorous conditions or long reaction times are employed, the main product is the bicyclo[4.2.0]octa-2,4-diene complex (733).[450, 885, 888, 891] The isomerisation (732) → (733) is essentially complete in 20–24 h at *ca.* 100°.[889]

$$\text{(732)} \quad \xrightarrow{\textit{ca.} 100°} \quad \text{(733)}$$

Fe(CO)₃   Fe(CO)₃

(732)   (733)

(For evidence bearing on the stereochemistry of (733), see ref. 892.) Under certain conditions, the binuclear complex (734) may be produced in low yield (*ca.* 3 %).[893, 894]

$$\xrightarrow[\substack{\text{Hydrocarbon solvent,}\\ \text{reflux}}]{\text{Fe}_2(\text{CO})_9 \text{ or Fe}_3(\text{CO})_{12}} \quad \begin{array}{l} \text{(732) + (733)} \\ + (\text{Cyclo-octatriene})\text{Fe}_2(\text{CO})_6 \end{array}$$

(734)

The structure of the (fluxional) binuclear complex (734) may be represented as shown[488, 895] (*cf.* '*cis*'-(COT)Ru₂(CO)₆ (186) (see p. 65)); for ¹³C n.m.r. spectral studies, see ref. 896.

(OC)₃Fe〜Fe(CO)₃

(734)

Cyclo-octa-1,3,5-triene reacts with Ru₃(CO)₁₂ to give the complexes (735) (38 %) and (736) (11 %), the ruthenium analogues of the iron complexes (733) and (734) respectively.[897]

$$\xrightarrow[\substack{\text{Heptane,}\\ \text{reflux}}]{\text{Ru}_3(\text{CO})_{12}} \quad \begin{array}{l} (\text{Bicyclo-octadiene})\text{Ru}(\text{CO})_3 \quad \text{(735)} \\ + (\text{Cyclo-octatriene})\text{Ru}_2(\text{CO})_6 \quad \text{(736)} \end{array}$$

With $Ru(SiMe_3)_2(CO)_4$ or $[Ru(SiMe_3)(CO)_4]_2$, an additional product is the tetrahydropentalenyl complex (737) (*cf.* (199), p. 69); this type of product is also formed from $Ru(GeMe_3)_2(CO)_4$.[897]

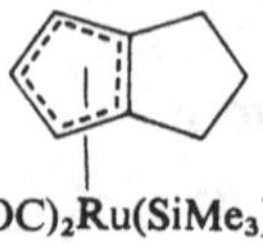

$(OC)_2Ru(SiMe_3)$ (737)

The tricarbonyl complex (735) is produced (in high yield) by the displacement of cyclo-octa-1,5-diene from $(COD)Ru(CO)_3$.[492]

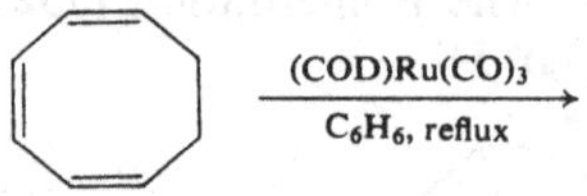

Reaction of iron(III) or ruthenium(III) chlorides with isopropylmagnesium bromide in the presence of cyclo-octa-1,3,5-triene, under the influence of u.v. light, affords complexes of the type (738)[898] (table 93). The ruthenium com-

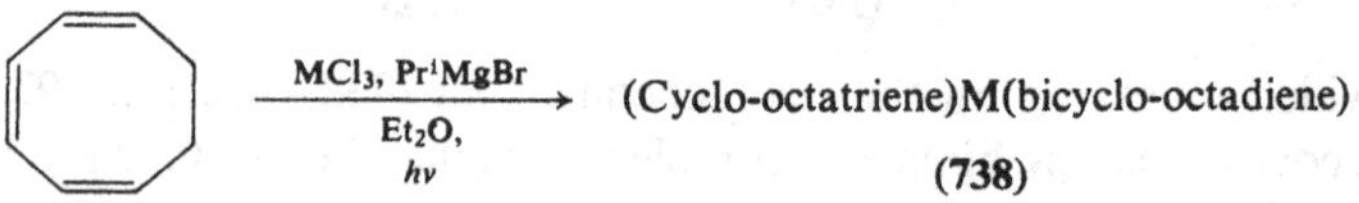

$\xrightarrow[\substack{Et_2O, \\ h\nu}]{MCl_3,\ Pr^iMgBr}$ (Cyclo-octatriene)M(bicyclo-octadiene)

(738)

Table 93

M	Yield (%)
Fe	29
Ru	2.2

pound (738; M = Ru) may be obtained in higher yield (up to 19%) by using (norbornadiene)$RuCl_2$[898] or $[(benzene)RuCl_2]_n$[899] in place of $RuCl_3$.

Structure (738; M = Fe) for the iron complex is revealed by X-ray crystallography.[900]

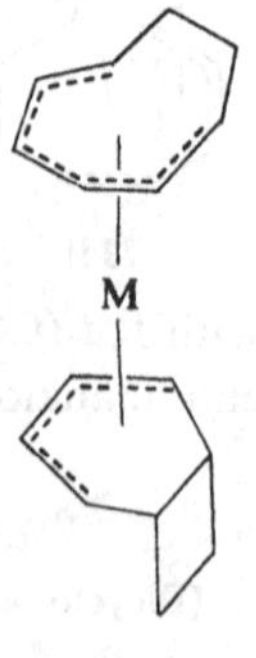

(738)

The 'mixed' complex (739) results from the use of (COD)RuCl$_2$ (yield 12%),[898] or RuCl$_3$ in the presence of cyclo-octa-1,5-diene (yield 6%).[901]

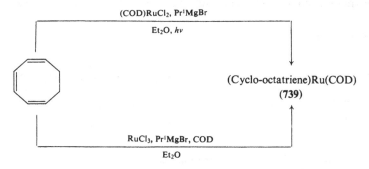

(Cyclo-octatriene)Ru(COD)
(739)

(COD)RuCl$_2$, Pr^{i}MgBr

Et$_2$O, *hv*

RuCl$_3$, Pr^{i}MgBr, COD

Et$_2$O

The cyclo-octatrienes (mixed isomers), on treatment with Co$_2$(CO)$_8$, yield the dimeric complex (740) (87%)[902] (the structure of the ligand is uncertain).

+

Co$_2$(CO)$_8$ | n-Heptane, *ca.* 130°

[(Cyclo-octatriene)Co(CO)$_2$]$_2$ (740)

Cyclo-octa-1,3,5-triene reacts with (CPD)Co(CO)$_2$ to give the 'mixed' products (741) (8%) and (742) (11.5%).[903]

(CPD)Co(CO)$_2$
⟶
Petroleum (b.p. 120–130°), reflux

(Bicyclo-octadiene)Co(CPD)
(741)

+ (Cyclo-octatriene)Co(CPD)
(742)

At higher temperatures only the latter (742) is isolated.[501,903,904] (*cf.* the iron complexes (732) and (733)). The proposed structures are as follows.

Co

Co

(741)   (742)

(For ion-molecule reactions of $(CPD)Co(CO)_2$ and cyclo-octa-1,3,5-triene, see ref. 422.)

The rhodium complexes (743)–(745) may be prepared as shown in scheme 76.

Scheme 76

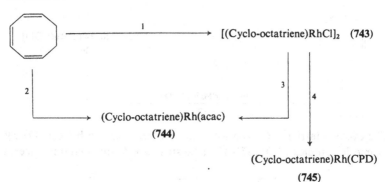

Reaction	Reagents and conditions	Yield (%)	Ref.
1	$[(CH_2{=}CH_2)_2RhCl]_2$, Et_2O, r.-t.	91	905
2	$(OC)_2Rh(acac)$, n-hexane, reflux	—	510
3	$Tl(acac)$, CH_2Cl_2, r.-t.	51	905
4	$(CPD)Tl$, C_6H_6, r.-t.	87	905

The species (743)–(745) are formulated as follows.[905]

(743)   (744)

(745)

The iridium complexes **(746)** and **(747)**, analogues of **(743)** and **(745)** respectively, result from the following reactions.[905]

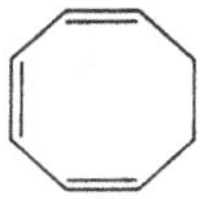

$$\text{[structure]} \xrightarrow{\text{Na}_2\text{IrCl}_6} \text{[(Cyclo-octatriene)IrCl]}_2$$

(746)

$$(\text{CPD})\text{Tl} \downarrow \begin{array}{l} \text{C}_6\text{H}_6, \\ \text{r.-t.} \end{array}$$

(Cyclo-octatriene)Ir(CPD)

(747)

9. Substituted cyclo-octa-1,3,5-trienes, cyclo-octa-1,3,6-trienes and bicyclo[4.2.0]octa-2,4-dienes

7-Bromocyclo-octa-1,3,5-triene, and 7,8-dichloro- or 7,8-dibromocyclo-octa-1,3,5-triene, result from the low-temperature addition of hydrogen bromide or halogens, respectively, to COT (see pp. 23–5); at higher temperatures equilibrium is established with the corresponding 7-substituted or 7,8-disubstituted bicyclo[4.2.0]octa-2,4-diene. The substituent halogen may of course be replaced by other groups, and the *trans*-7,8-diacetoxy-compound is available directly from COT and mercury(II) acetate (see p. 20).

5,8-Disubstituted cyclo-octa-1,3,6-trienes are formed by free radical attack on COT (see pp. 32–3), and by reaction of COT²⁻ with aldehydes or ketones (see pp. 178–80). Dialkyl- and disilyl-derivatives of both cyclo-octa-1,3,5-triene and cyclo-octa-1,3,6-triene are also available from COT²⁻ (see pp. 173–4).

The position of the cyclo-octa-1,3,5-triene ⇌ bicyclo[4.2.0]octa-2,4-diene equilibrium for various monosubstituted **(748)** and disubstituted **(749)** derivatives is shown in tables 94 and 95. (For energy-barriers to ring-inversion of **(749**; X = Y = D, Br), see refs. 231, 281, 564.)

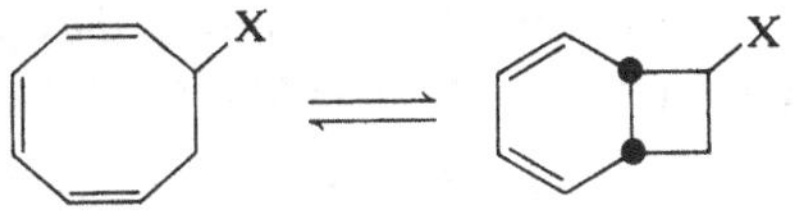

(748)

Table 94

X	Proportion of bicyclic form (%)	Temp. (°C)	Ref.
Br	30, (35)	30–35, (60)	233, 906, (136)
OAc	75, (53)	30–35, (60)	906, (136)
OH	90	30–35	906
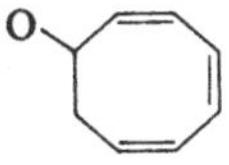	90	30–35	906

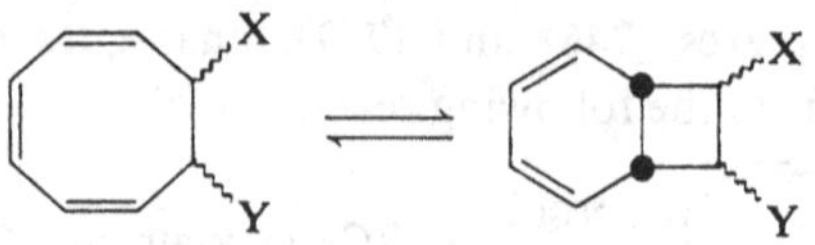

(749)

Table 95

X	Y	Relative stereochemistry of substituent groups	Proportion of bicyclic form (%)	Temp. (°C)	Ref.
Me	Me	*cis*	81	60	136
Me	Me	*trans*	94	60	136
OAc	OAc	*trans*	95	60	136
Br	Br	*trans*	100	20	231
Cl	Cl	*cis*	80	60	136
Cl	Cl	*trans*	99	−30	136

7-Substituted cyclo-octa-1,3,5-trienes

Some functional group transformations in this series are outlined in scheme 77.

Scheme 77

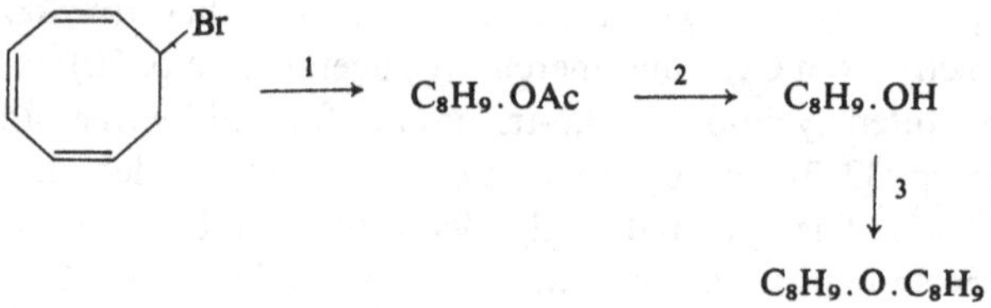

Reaction	Reagents and conditions	Yield (%)	Ref.
1	NaOAc, HOAc, 10–20°	47	906
2	LiAlH₄, Et₂O, −34° → 10°	54	906
3	50–70°/0.8 mm	—	906

Reaction of 7-bromocyclo-octa-1,3,5-triene with acetylenic Grignard reagents results in a valence tautomeric mixture of the open-chain, monocyclic and bicyclic derivatives, (750), (751) and (752)[907] (table 96).

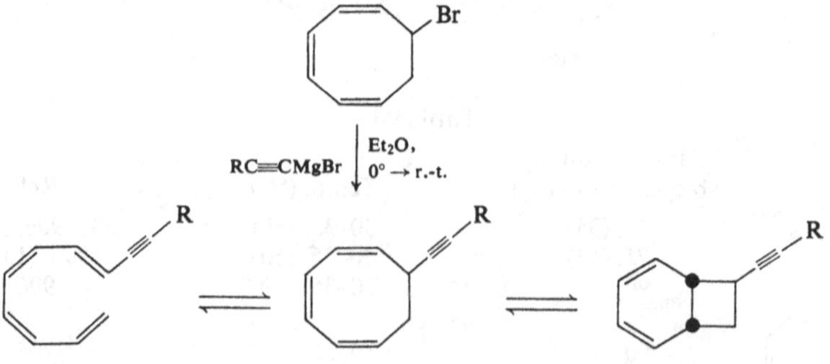

(750)   (751)   (752)

Table 96

R	Total yield (%)
Bu^t	45
Ph	28

7-Bromocyclo-octa-1,3,5-triene (see p. 25) is converted by the action of sodium azide into the azido-compound (753) (55%), thermolysis of which affords the pyrrolylbutadiene (754) (43%).[233]

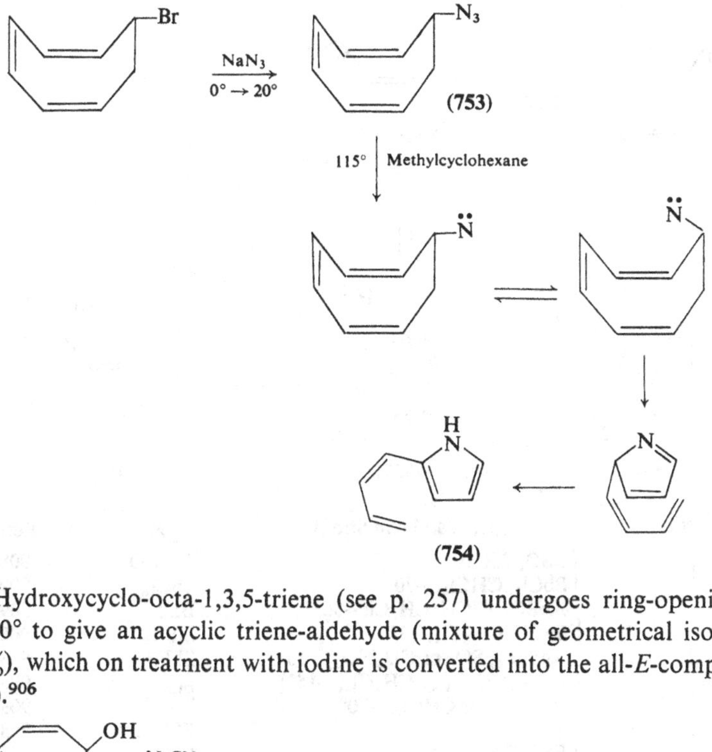

(753)

115° | Methylcyclohexane

(754)

7-Hydroxycyclo-octa-1,3,5-triene (see p. 257) undergoes ring-opening at *ca.* 80° to give an acyclic triene-aldehyde (mixture of geometrical isomers) (87%), which on treatment with iodine is converted into the all-*E*-compound (755).[906]

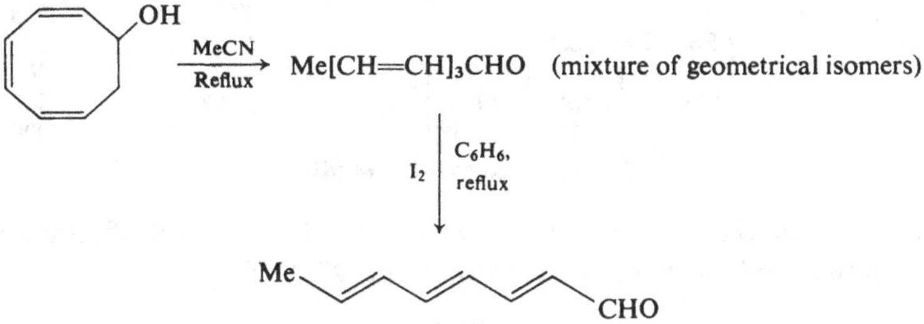

Me[CH=CH]$_3$CHO   (mixture of geometrical isomers)

I$_2$ | C$_6$H$_6$, reflux

(755)

7,8-Disubstituted cyclo-octa-1,3,5-trienes

Thermal dimerisation of the dimethyl stereoisomers (503) and (504) (see p. 173) occurs *via* their respective valence isomers (see p. 254).

The generation of *endo-* and *exo-*8-chlorohomotropylium salts, (756) and (757) respectively, from the two isomers (41) and (42) of 7,8-dichlorocyclo-octa-1,3,5-triene (see p. 23), by the action of proton or Lewis acids, is summarised in scheme 78. For the mechanisms of these reactions, see ref. 909. The salt (757; $X = SbCl_6^-$) may be isolated as a crystalline solid.[908]

Scheme 78

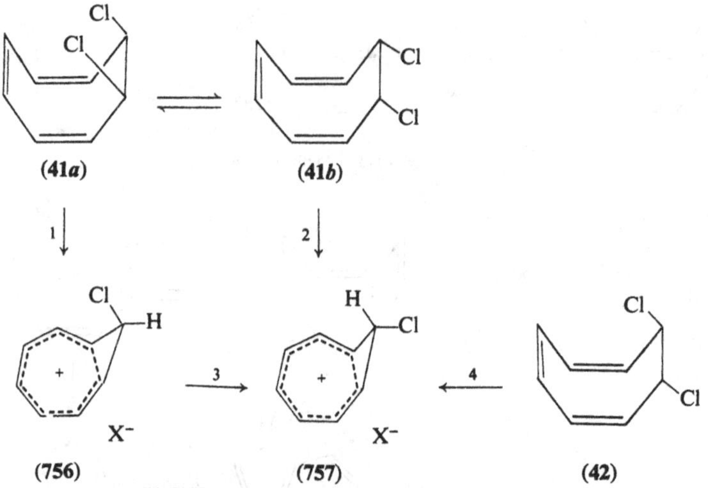

(41a)	(41b)		(756)	(757)	(42)

Reaction	Reagents and conditions	X	Ref.
1	$\begin{cases} FSO_2OH, 0° \\ SbCl_5, CHCl_3, -70° \\ AgBF_4, ClCH_2CH_2Cl, -30°^a \\ SnCl_4, CH_2Cl_2, -15° \end{cases}$	$\begin{array}{l} FSO_2O^- \\ SbCl_6^- \\ BF_4^- \\ SnCl_5^- \end{array}$	$\begin{array}{l} 908 \\ 909 \\ 909 \\ 908 \end{array}$
2	$\begin{cases} AgSbF_6, SO_2 \text{ or } CD_3NO_2, -15° \\ SbCl_5 \begin{cases} SO_2 \text{ or } CH_2Cl_2, -15° \\ \text{or } CHCl_3, -70° \end{cases} \end{cases}$	$\begin{array}{l} SbF_6^- \\ SbCl_6^- \end{array}$	$\begin{array}{l} 908 \\ 908 \\ 909 \end{array}$
3	30°	FSO_2O^-	908
4	$\begin{cases} FSO_2OH, -20° \\ SbCl_5 \begin{cases} CH_2Cl_2, -20° \\ \text{or } SO_2, -40° \\ \text{or } CHCl_3, -70° \end{cases} \end{cases}$	$\begin{array}{l} FSO_2O^- \\ SbCl_6^- \end{array}$	$\begin{array}{l} 908 \\ 908 \\ 908 \\ 909 \end{array}$

^{a} $AgBF_4$ added to (41a) $\rightleftharpoons$ (41b).

Treatment of the dibromo-derivative (758) with silver tetrafluoroborate results in the *exo-*8-bromohomotropylium salt (759).[230]

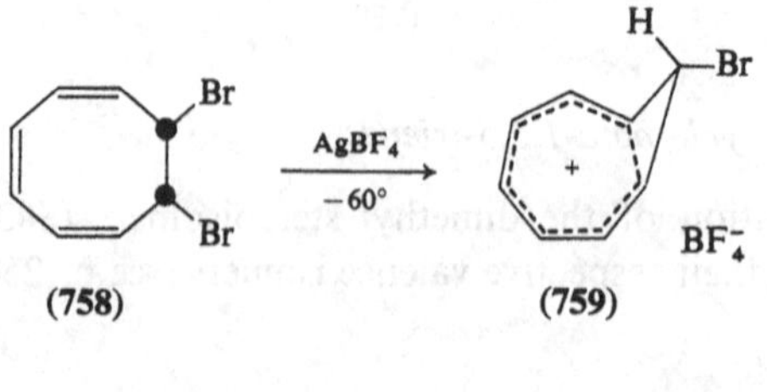

(758)   (759)

The low-temperature dehydrohalogenation of 7,8-dihalocyclo-octa-1,3,5-trienes with potassium t-butoxide in dichloromethane, which provides an excellent route to chloro- and bromocyclo-octatetraene, has already been described (see p. 83). With the dibromide (758) and sodium methoxide in methanol the reaction takes a different course, the main product being the cycloheptatriene derivative (29) (76%)[540] (see p. 20).

(758)   (29)

7,8-Dimethylenecyclo-octa-1,3,5-trienes are available by dehydrogenation of suitable 1,2-di(halomethyl)cyclo-octatetraenes (see p. 117).

The use of 7,8-dimethylenecyclo-octa-1,3,5-trienes for the synthesis of annulated COTs has already been mentioned (see pp. 124–7). U.v. irradiation of the chloro-derivative (321a) (see p. 117) gives rise to a mixture of the products (760) (ca. 10%) and (761) (ca. 3%).[910]

(321a)   (760)

(761)

5,8-Disubstituted cyclo-octa-1,3,6-trienes

At ca. 100° the dimethyl-derivative (505) (see p. 173) undergoes a [1,5]-hydrogen shift to form the 1,3,5-triene (762), which then equilibrates with the bicyclo-[4.2.0]octadiene (763) (this fragments at ca. 225° to give toluene).[379]

(505)   (762)   (763)

Similarly, compound (74) (see p. 33) rearranges in refluxing xylene to give the bicyclic product (764) (82%),[289] presumably *via* (765).

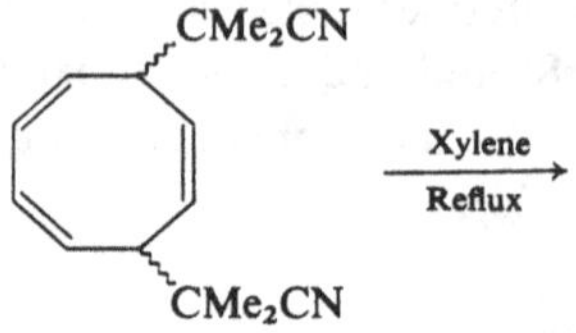

CMe₂CN

Xylene
Reflux →

CMe₂CN

(74)

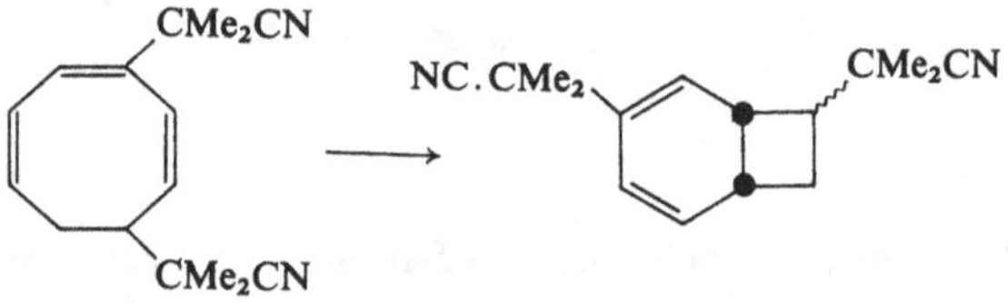

CMe₂CN

NC.CMe₂   CMe₂CN

CMe₂CN

(765)   (764)

Treatment of the diol (534a) (containing *ca.* 25% of the bicyclic isomer (535a)) (see p. 180) with acid results in the product (766) (48%); a *retro*-aldol type condensation is apparently involved.[725]

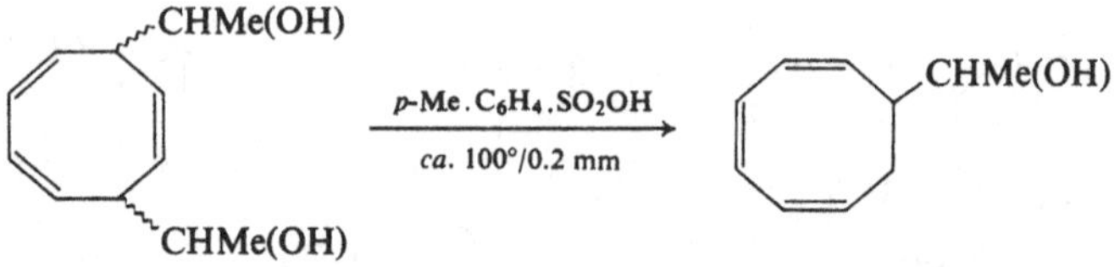

CHMe(OH)

p-Me.C₆H₄.SO₂OH
―――――――→
ca. 100°/0.2 mm

CHMe(OH)

CHMe(OH)

(534a)   (766)

Tertiary diols (534) (see p. 180), however, yield the tricyclic ethers (767)[725] (*c/.* refs. 739, 754) (table 97).

R¹R²
—OH

p-Me.C₆H₄.SO₂OH
―――――――→
Δ

—OH
R¹R²

R¹
R²
O—R²
R¹

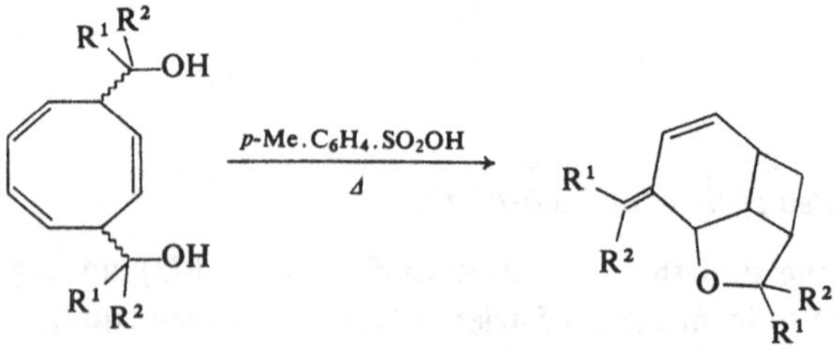

(534)   (767)

Table 97

R¹	R²	Reaction conditions	Yield (%)
Me[a]	Me	Δ/1 mm	28
Ph	Ph	C₆H₆, reflux	45
	Fluorenyl	C₆H₆, reflux	44

[a] Starting material contained 62% of (535c).

The 5,8-dinitro-compound (**75**) (see p. 33), on treatment with sodium methoxide followed by bromine, affords 1,4-dinitrocyclo-octatetraene.[615]

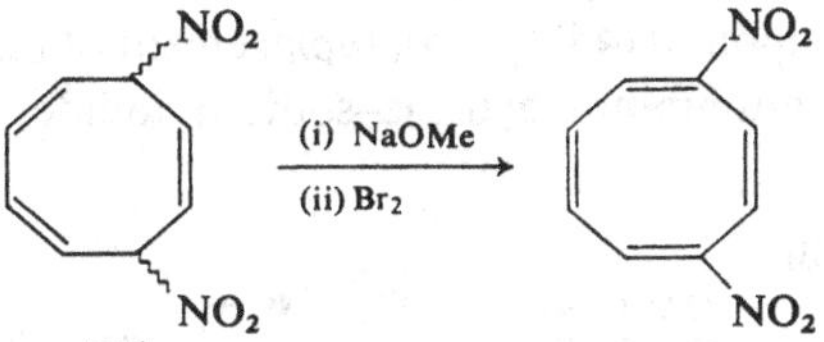

NO$_2$ NO$_2$

(i) NaOMe

(ii) Br$_2$

NO$_2$ NO$_2$

(**75**)

The 5,8-bis(trimethylsilyl)-compound (**506**) (see p. 174) reacts with Fe$_2$(CO)$_9$ (or Fe$_3$(CO)$_{12}$) to form the complexes (**768**) (8 %) and (**769**) (0.5 %).[911]

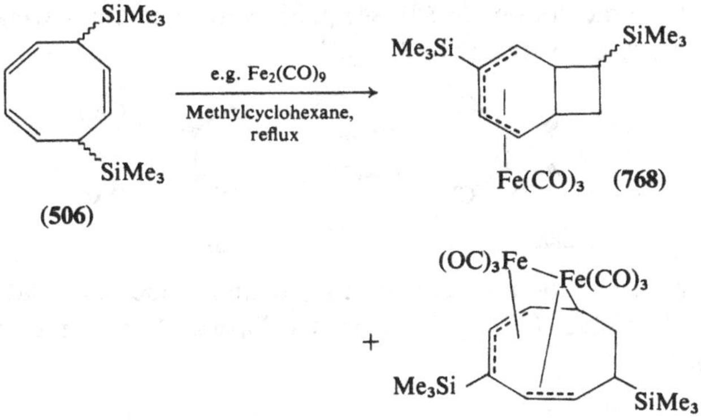

SiMe$_3$

e.g. Fe$_2$(CO)$_9$

Methylcyclohexane, reflux

Me$_3$Si   SiMe$_3$

SiMe$_3$

(**506**)

Fe(CO)$_3$ (**768**)

+

(OC)$_3$Fe—Fe(CO)$_3$

Me$_3$Si   SiMe$_3$

(**769**)

With Ru$_3$(CO)$_{12}$ or [Ru(SiMe$_3$)(CO)$_4$]$_2$, the complex (**770**) (6 %) is formed.[912]

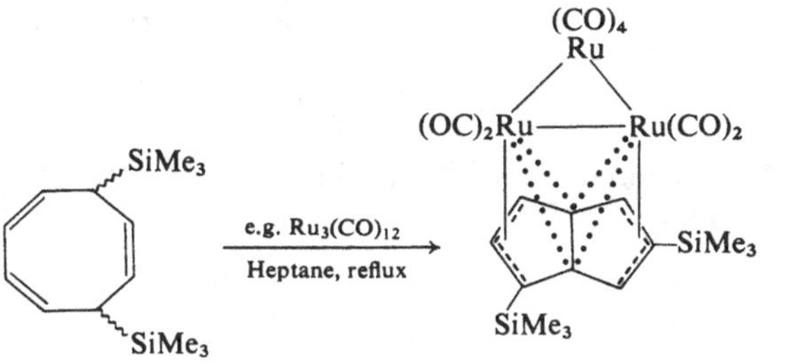

(CO)$_4$Ru

(OC)$_2$Ru——Ru(CO)$_2$

SiMe$_3$

e.g. Ru$_3$(CO)$_{12}$

Heptane, reflux

SiMe$_3$

SiMe$_3$

SiMe$_3$

(**506**)   (**770**)

A similar product is formed (15 %) from the trisubstituted starting material (**771**).[912]

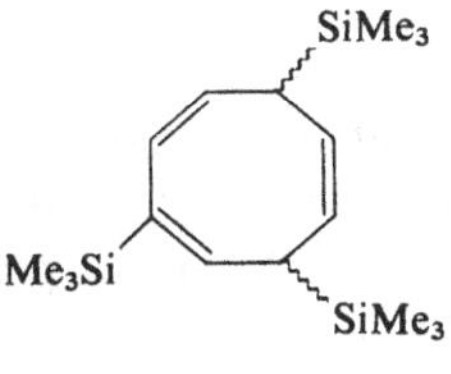

SiMe$_3$

Me$_3$Si

SiMe$_3$ (**771**)

7,8-Disubstituted bicyclo[4.2.0]octa-2,4-dienes

For thermal dimerisation, see p. 254.

In the dibromo-compound (44) (p. 24), replacement of halogen by hydrogen occurs with lithium aluminium hydride–sodium iodide, giving cyclo-octa-1,3,5-triene (32%).[260]

(44)

Reaction of the dichloride (38/39) (see p. 23) with potassium acetate gives the *trans*-diacetoxy-derivative (28) (39%).[11]

(38/39)   (28)

Sensitised photoaddition of oxygen to the dibromide (44) leads to the peroxy-bridged product (772) (78% from COT), together with a diepoxide (773) (3.5%).[913]

(44)   (772)   (773)

The peroxide (772) is a source of the bicyclic diendione (774)[913] and of cyclo-octa-2,5,7-trien-1,4-dione (775).[914]

(774)   (775)

Oxidative cleavage of the cyclohexadiene ring in the *cis*-dichloro-compound (38) affords the dichlorocyclobutanedicarboxylic acid (776),[553] of proved stereochemistry.[915] Analogous oxidation of the dibromide (44) results in the *trans*-dibromo-derivative (777),[915,916] which provides a source of 3,4-disubstituted cyclobutenes (scheme 79).

Scheme 79

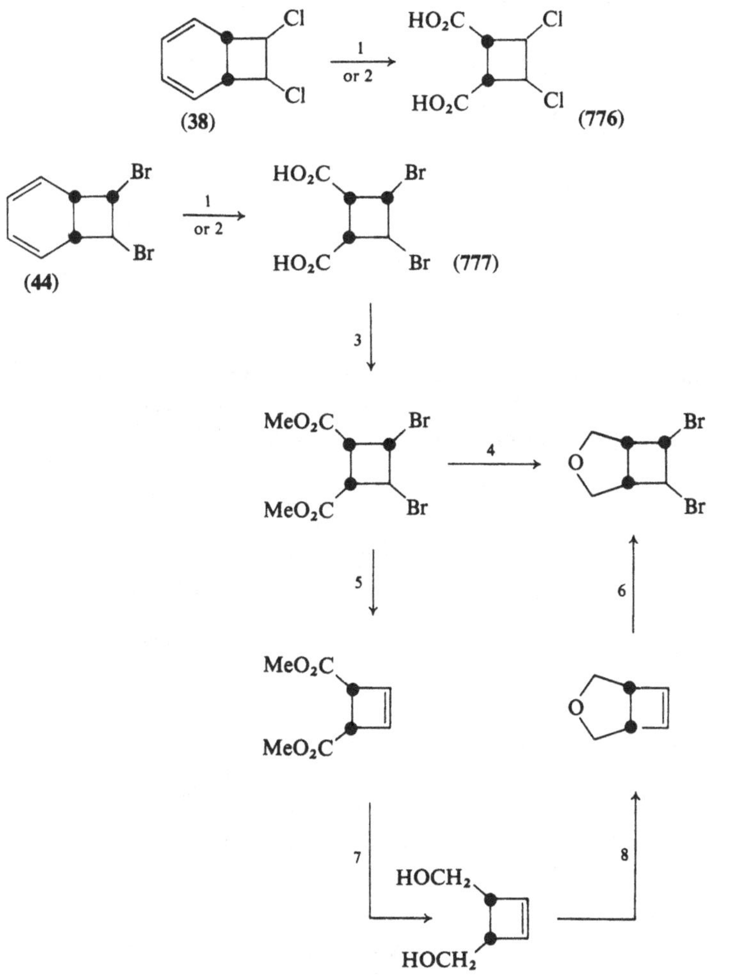

Reaction	Reagents and conditions	Yield (%)	Ref.
1	{KMnO₄, Me₂CO(aq.), 3–5° then r.-t.}	{(Dichloride) 35 (Dibromide) (27), 43}	553 (916), 917
2	O₃, EtOAc, −60°	{(Dichloride) 35 (Dibromide) 67}	553 916
3	CH₂N₂	97	916
4	{(i) LiAlH₄, Et₂O (ii) Acid hydrolysis}	58	918
5	Zn–Cu, MeOH, 75°	76	916
6	Br₂, CCl₄	70	918
7	{(i) LiAlH₄ (ii) Na₂SO₄(aq.)}	84	918
8	p-Me.C₆H₄.SO₂Cl, pyridine	—	918

Acid-catalysed dehydration of the diol (535c) containing 38 % of (534c) (see p. 180) yields the 1,2-divinylbenzene (778) (36%).[725]

(535c)   (778)

Iodine azide (from sodium azide and iodine monochloride in acetonitrile at −20°) reacts with the dibromide (44) to afford the product (779), characterised as the acetylenedicarboxylic ester-adduct (780) (50%).[919]

(44)   I   (779)

(780)

Schemes 80 and 81 show the products obtained on catalytic hydrogenation and subsequent dehalogenation of 7,8-dihalobicyclo[4.2.0]octa-2,4-dienes.

Catalytic hydrogenation of the diacetoxy-compound (28) affords the tetrahydro-derivative (781) (92%), which on acid-catalysed methanolysis gives the diol (782) (71%).[11, 206]

Scheme 80

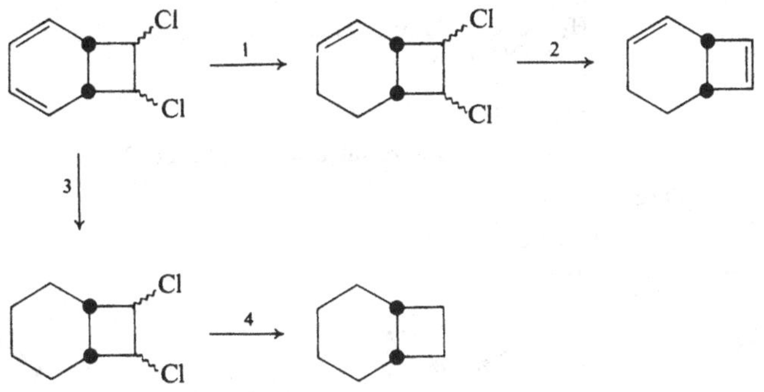

Reaction	Reagents and conditions	Yield (%)	Ref.
1	H₂, Pd–C, MeOH, EtOAc	73	(11), 920
2	Na, NH₃(liq.), Et₂O	83	920
3	{H₂, Pd–CaCO₃, MeOH, −5° to 0°	—	11
	{or H₂, Raney Ni, THF, 150 atm	—	11
4	H₂, Raney Ni, KOH(aq.), MeOH, 100°, 200 atm	75	11

Scheme 81

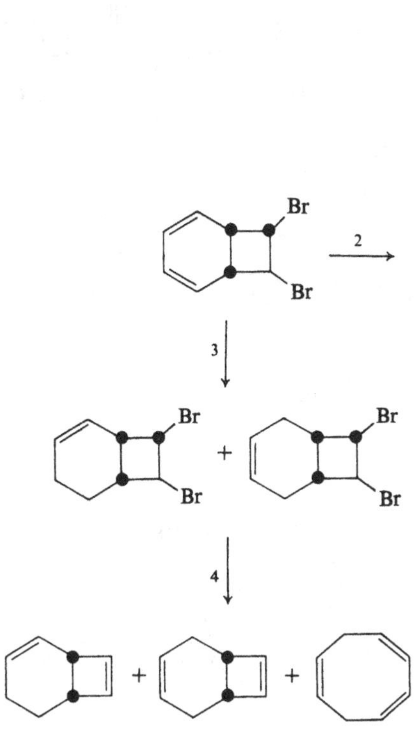

Reaction	Reagents and conditions	Yield (%)	Ref.
1	Mg	—	921
2	H_2, Pd–$CaCO_3$ or Pd–$BaSO_4$, MeOH, *ca.* 0°	—	11, 921
3	H_2, Pd	—	853
4[a]	Zn, EtOH	—	853

[a] The reaction products were subjected to g.l.c. at 100°.

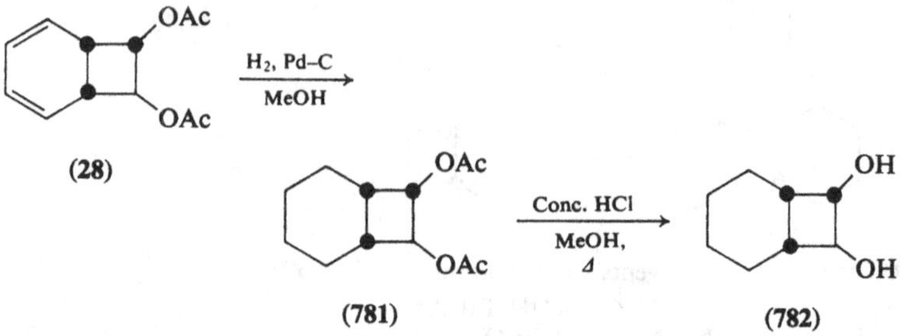
OAc
OAc
H₂, Pd–C
MeOH
(28)
OAc
OAc
Conc. HCl
MeOH,
Δ
OH
OH
(781)
(782)

The mixture of diacetoxy-derivatives, presumably (**28**) and (**32**), obtained by electrolysis of COT in acetic acid using a carbon anode (see p. 21), may be hydrogenated to give a product which on alkaline hydrolysis yields the diols

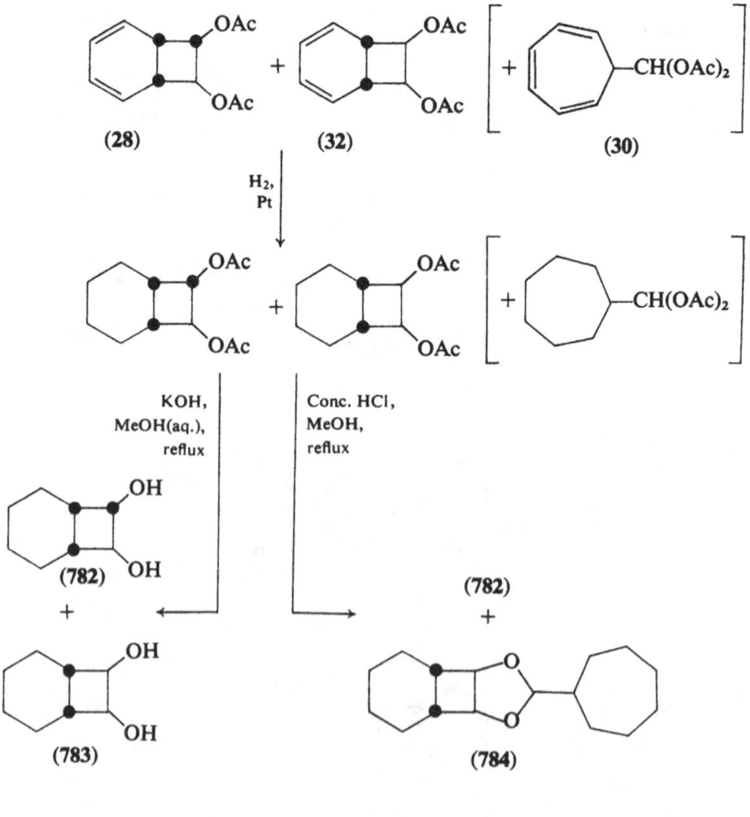
OAc
OAc
+
OAc
OAc
+
CH(OAc)₂
(28)
(32)
(30)
H₂,
Pt
OAc
OAc
+
OAc
OAc
+
CH(OAc)₂
KOH,
MeOH(aq.),
reflux
Conc. HCl,
MeOH,
reflux
OH
OH
(782)
+
OH
OH
(783)
(782)
+
O
O
(784)

(782) and (783) (29 % and 20 % respectively); in acid, however, a mixture of the *trans*-diol (782) (30 %) and the acetal (784) (62 %) is obtained.[922]

The low-temperature debromination of the dibromide (44) with disodium-phenanthrene generates bicyclo[4.2.0]octa-2,4,7-triene (4) (half-life 14 min at 0°);[32] with sodium iodide in refluxing acetone the product is COT.[39]

Dehydrobromination with potassium t-butoxide affords 1,2-didehydro-cyclo-octatetraene (236), which may be trapped with tetracyclone to give the adduct (785) (21 %)[624] (this process is not as efficient as with bromocyclo-octatetraene; see pp. 83–4, 164).

In general the 7,8-dihalobicyclo[4.2.0]octa-2,4-dienes are very susceptible to rearrangement. Hydrolysis of the dichloro-compound (38/39) (in water) furnishes phenylacetaldehyde,[923] methanolysis the dimethyl acetal (786),[923] and acetolysis the enol acetate (787);[923] in the presence of sodium methoxide, however, treatment with methanol gives the cycloheptatrienal derivative

(29).[206] (For possible rearrangement mechanisms, see ref. 923.) The diacetoxy-compound (28) also rearranges on treatment with aqueous acid, giving phenyl-acetaldehyde.[206]

In addition to these rearrangements, a number of ring-cleavage reactions, resulting in acyclic polyenes, are known.

U.v. irradiation of the diacetoxy-compound (28) yields the diacetoxy-tetraene (788),[924] and reaction with lithium aluminium hydride gives the triene-dialdehyde (789) (50–60%).[925]

With potassium cyanide, the dibromide **(44)** is converted into the tetraene-dinitrile **(790)** (27%),[926] and phenylmagnesium bromide affords the tetraene **(791)**;[927] extension of the polyene chain can be effected by the use of styryl-magnesium bromide, which gives the hexaene **(792)**[927] (see also below). Treatment with acetylenic Grignard reagents yields tetraene-diynes **(793)** in which the

two central double bonds generally have the *Z*-configuration (table 98); the proposed mechanism involves an initial coupling reaction of the mono-cyclic valence tautomer **(794)**, and subsequent conrotatory ring-opening of the

Table 98

R	Yield (%)
H[a]	—
Me	16
Bu^t	30[b]
CMe=CH$_2$	24
Ph	50
p-Me.C$_6$H$_4$	53
2,4,6-Me$_3$.C$_6$H$_2$	30
p-Br.C$_6$H$_4$	57

[a] Reaction in THF at r.-t.; product characterised by spectra only.
[b] The isolated product was the all-*E*-compound.

resulting *trans*-7,8-disubstituted cyclo-octa-1,3,5-triene (**795**).[928] The *E,Z,Z,E*-products (**793**) are converted into the all-*E*-isomers (**796**) by the action of heat or light in the presence of iodine[928] (table 99). Partial hydrogenation

(**793**)   (**796**)

Table 99

R	Reaction conditions	Yield (%)
Me	Light petroleum (b.p. 60–80°), C_6H_6, Me_2CO, hv	50
$CMe{=}CH_2$	C_6H_6, hv	10
p-Me.C_6H_4	C_6H_6, hv	38
Ph	C_6H_6, hv	75
2,4,6-Me_3.C_6H_2	C_6H_6, hv	20
	Xylene, reflux	10–20
p-Br.C_6H_4	C_6H_6, hv	85

(Lindlar catalyst) of these tetraene-diynes furnishes hexaenes;[927] complete hydrogenation (Pd–CaCO$_3$) gives long-chain alkanes.[928]

With e.g. maleic anhydride, stereoisomeric [4 + 2] cycloadducts (**797**) and (**798**) are produced from the *cis*- and *trans*-compounds (**799**) and (**800**) respectively; similar reactions with other dienophiles lead to structures (**801**)–(**808**)

(**799**)   (**797**)

(**800**)   (**798**)

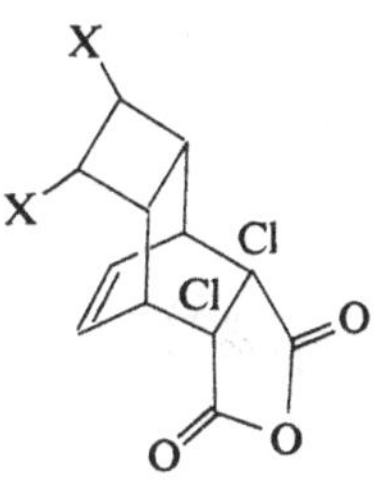

X
X
Cl
Cl
O
O O

(801)

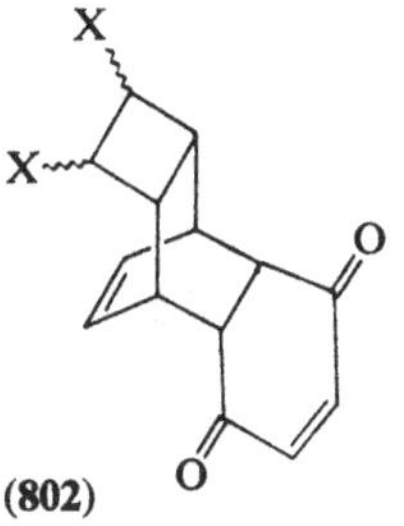

X
X
O
O

(802)

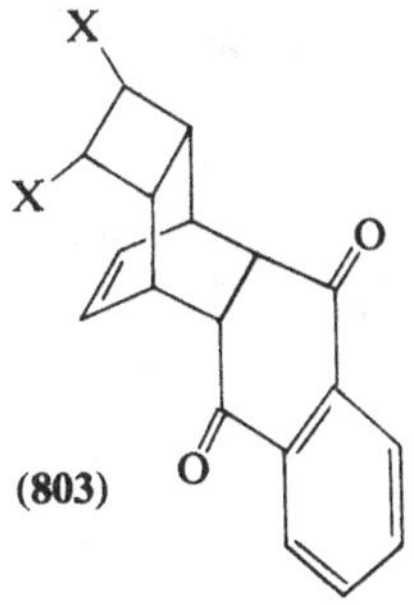

X
X
O
O

(803)

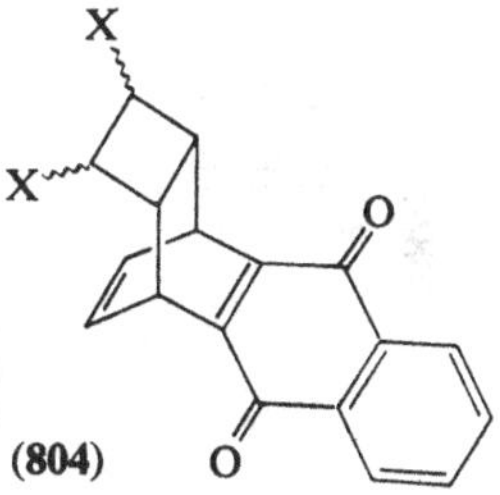

X
X
O
O

(804)

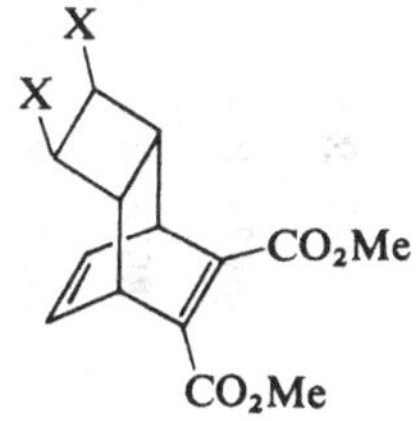

X
X
CO$_2$Me
CO$_2$Me

(805)

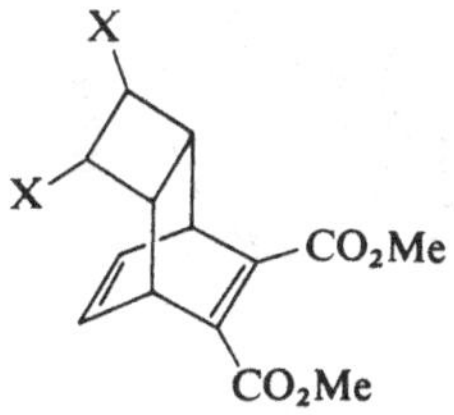

X
X
CO$_2$Me
CO$_2$Me

(806)

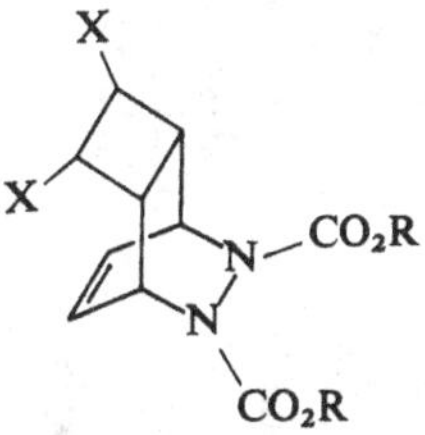

X
X
N—CO$_2$R
N
CO$_2$R

(807)

X
X
N
O
NR
N
O

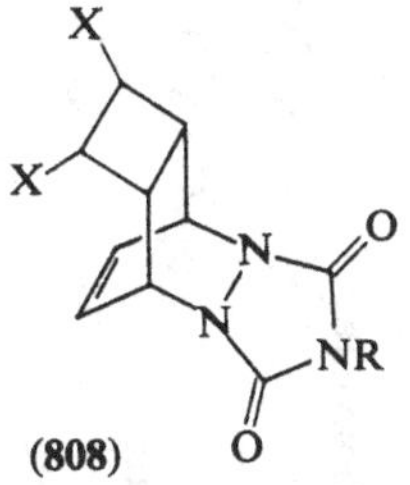

(808)

a: X = Cl
b: X = Br
c: X = OAc

(table 100). For the adduct of (**799**/**800**; X = Cl) and acrylonitrile, see ref. 39. For the adduct of (**800**; X = CO$_2$Me) (obtained from the dimethyl ester of (**532**) (see p. 179)) and *N*-phenylmaleimide, see ref. 756. For the adducts of the annulated derivatives (**111**) (see p. 40) with maleic anhydride and dimethyl acetylenedicarboxylate, see ref. 331.

Debromination of the adducts (**807b**) and (**808b**) leads to the products (**809**) and (**125**) (see p. 46) respectively (tables 101 and 102).

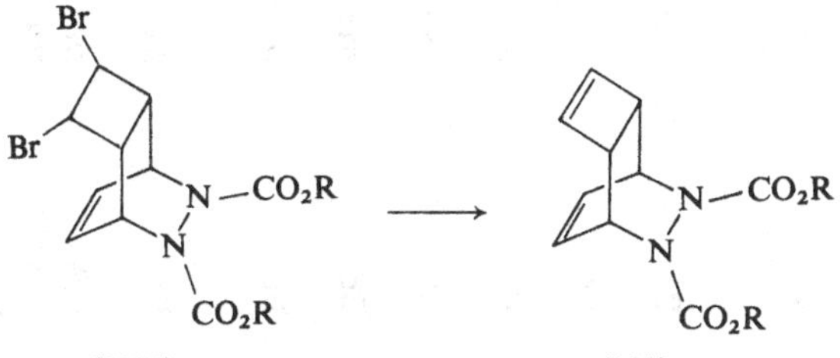

Br
Br
N—CO$_2$R
N
CO$_2$R

(807b)

$\longrightarrow$

N—CO$_2$R
N
CO$_2$R

(809)

Table 100

X	Stereochemistry	Dienophile	Conditions	Product	Yield (%)	Ref.
Cl	cis + trans	Maleic anhydride	C$_6$H$_6$, reflux	(797a) + (798a)[a]	37 + 40	11
Cl	cis + trans	p-Benzoquinone	C$_6$H$_6$, reflux	(802a)[b]	41	11
Cl	cis + trans	1,4-Naphthoquinone	C$_6$H$_6$, reflux	(803a)[c]	65	11
Cl	cis + trans	Dimethyl acetylenedicarboxylate	{ C$_6$H$_6$, reflux CCl$_4$, reflux	(806a) (805a) + (806a)[d]	51 75 (total)	11 228, (929)
Br	trans	Maleic anhydride	{ C$_6$H$_6$, reflux CH$_2$Cl$_2$, r.-t.	(798b)	{ 93 88	11 11
Br	trans	Dichloromaleic anhydride	Toluene, reflux	(801b)	—	372
Br	trans	Dimethyl acetylenedicarboxylate	C$_6$H$_6$, reflux	(806b)	55	11
Br	trans	Dimethyl azodicarboxylate	C$_6$H$_6$, 70°	(807b; R = Me)	86	40
Br	trans	Diethyl azodicarboxylate	C$_6$H$_6$, 70°	(807b; R = Et)	94	40
Br	trans	Dibenzyl azodicarboxylate	C$_6$H$_6$, 70°	(807b; R = CH$_2$Ph)	95	40
Br	trans	4-Methyl-1,2,4-triazoline-2,5-dione	Me$_2$CO, −70° to −50° → r.-t.	(808b; R = Me)	81	930
Br	trans	4-Phenyl-1,2,4-triazoline-3,5-dione	CH$_2$Cl$_2$, −78° → r.-t.	(808b; R = Ph)	87	583, (931)
OAc	trans	Maleic anhydride	C$_6$H$_6$, reflux	(798c)	100	11
OAc	trans	1,4-Naphthoquinone	C$_6$H$_6$, reflux	(803c)	84	11
OAc	trans	Dimethyl acetylenedicarboxylate	C$_6$H$_6$, reflux	(806c)	52	11, (207)

[a] Both isomers are isolable, but it is not certain which is which.
[b] The stereochemistry of the chlorine atoms is not certain.
[c] The isolated product is the dehydro-derivative (804) (stereochemistry of the chlorine atoms uncertain).
[d] The two isomers are separable.

Table 101 (see p. 249)

R	Reagents and conditions	Yield (%)	Ref.
Et	Zn–Cu, THF, reflux	68	40
	Li–Hg, THF, r.-t.	72	40
	Zn, DMF, reflux	92	932

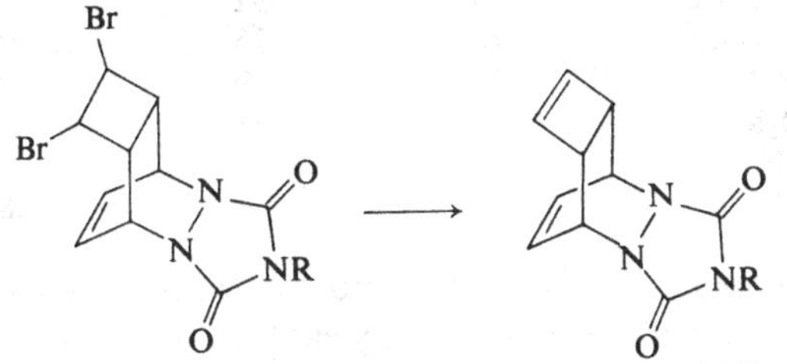

(808*b*)   →   (125)

Table 102

R	Reagents and conditions	Yield (%)	Ref.
Me	Zn–Cu, EtOH, reflux	77	930
Ph	Zn–Cu, DMF, reflux	99	569, (931)

The bicyclic dibromides (810) derived from monosubstituted COTs react with *N*-phenyltriazolinedione to give adducts which may be debrominated to afford the products (288)–(290)[569] (table 103). (For the photochemical trans-

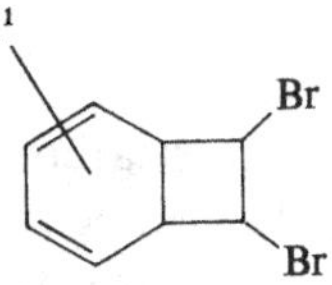

(810)   (Stereochemistry unspecified)

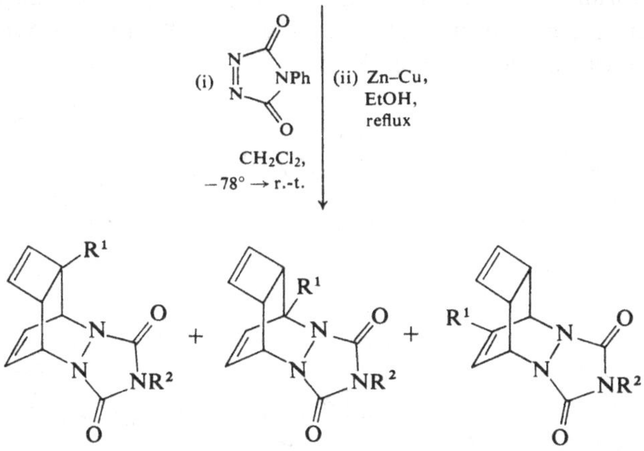

(i) [triazolinedione NPh structure] (ii) Zn–Cu, EtOH, reflux

CH₂Cl₂, −78° → r.-t.

(288)   +   (289)   +   (290)

R² = Ph

Table 103

R¹	(288)	(289)	(290)
		Yield (%)	
Me	30	—	—
Ph	6	—	9
CH₂OMe	23	32	—
CH₂OAc	10	35	10
CN	12	9	49
CO₂Me	—	31	—
F	90	—	—
Br	—	—	28

formation of compounds (288)–(290) into the 9,10-diazabasketanes (811), which by silver-catalysed rearrangement yield 9,10-diazasnoutanes (812), see pp. 355–6; subsequent conversion into monosubstituted semibullvalenes is described in ref. 716.)

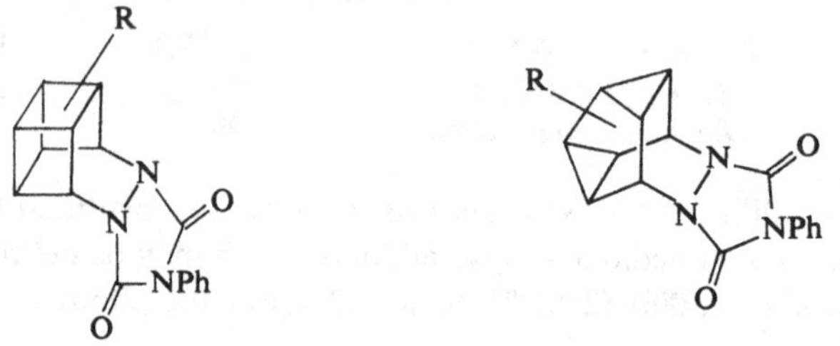

(811)   (812)

Cyclobutadiene (generated from its iron tricarbonyl complex by means of cerium(IV) ions) reacts with a mixture of the *cis*- and *trans*-dichloro-compounds (38) and (39) to give a mixture of the adducts (813) and (814) (52%).[933] (For further transformations, see ref. 933.) Similarly, benzyne (generated from anthranilic acid and isoamyl nitrite) gives the adducts (815) and (816) (40%)[934] (for further transformations, see ref. 934).

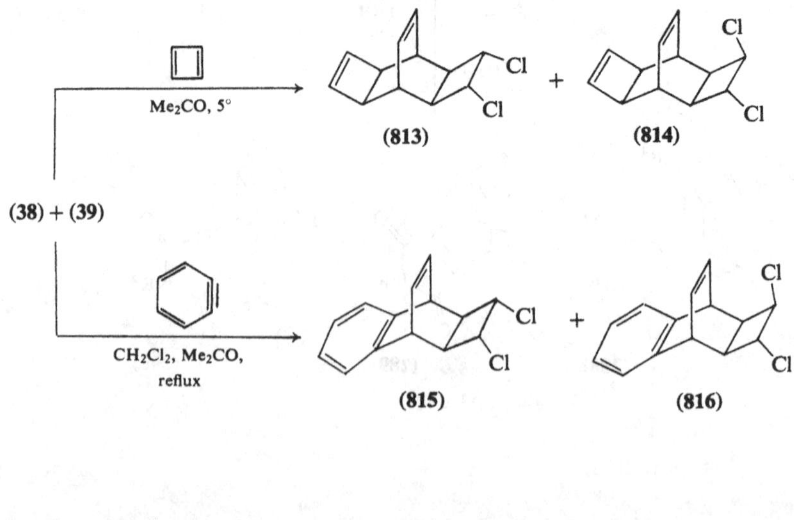

(813)   (814)

(38) + (39)

(815)   (816)

Cycloadditions in which the dibromo-compound (44) plays a dienophilic role result from the action of hexachlorocyclopentadiene or tetrachlorocyclopentadienone ketals, affording adducts of structure (817)[372] (table 104).

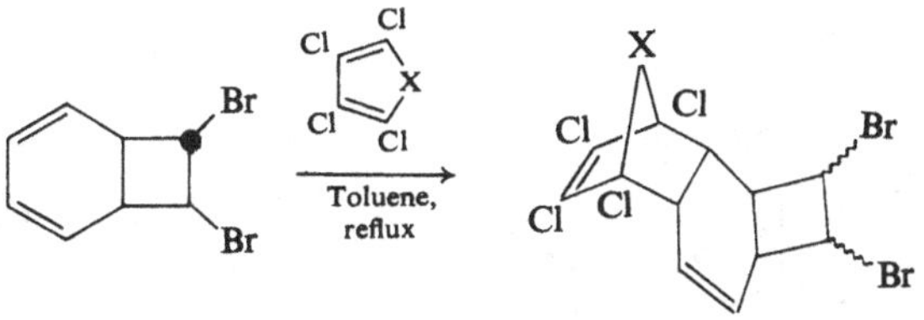

(44)   (817)

Table 104

	X	Yield (%)
a	CCl_2	77
b	$C(OMe)_2$	34
c	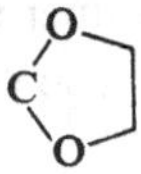	65

Hydrolysis of the ketal group in the adduct (817b) leads to the Cope-rearranged ketone (818) (90%),[372] which with sodium hydroxide in tetrahydrofuran gives the carboxylic acid (819) (63%).[881]

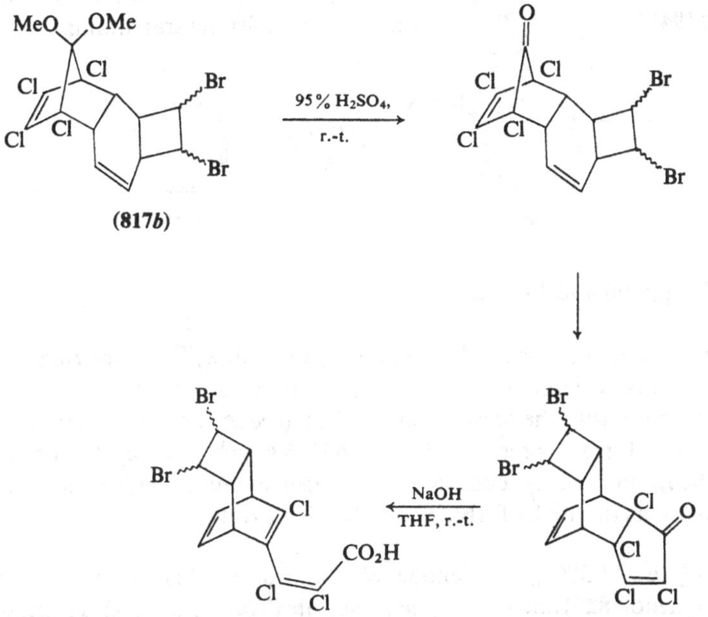

(817b)

(819)   (818)

Diels–Alder dimerisation of 7,8-disubstituted bicyclo[4.2.0]octa-2,4-dienes occurs on heating (or even on storage at room-temperature) to yield products

(820)   (111c)

of structure (820).[11,379] An analogous product is formed from the annulated derivative (111c), although fragmentation also occurs.[330,332]

10. Cyclo-octa-2,4,6-trienones

Cyclo-octa-2,4,6-trienone (821) is produced from the epoxide (27) of COT by base-catalysed isomerisation (see p. 314); it is also formed (63%) from compound (94) (see pp. 41–2) by treatment with moist neutral alumina.[317]

(94)   (821)

(See also pp. 96 and 143–4.)

The trienone (821) shows little tendency to enolise;[200,935] for ring-inversion studies, using variable-temperature n.m.r. spectroscopy, see ref. 935. It exists in equilibrium with the bicyclic isomer (822) (see scheme 82) (at 20°, *ca.* 5% of the bicyclic form is present;[200,856] at 60°, 6.6%[136]); the cyclobutenes (823) and (824), previously considered as valence tautomers,[936] may now be recognised as thermally forbidden cyclisation products.

Thermolysis. At 300°, the trienone (821) affords a 1:1:1 mixture of benzene, *o*-vinylphenol (825) and *o*-vinylphenyl acetate (826); scheme 82 is suggested.[936]

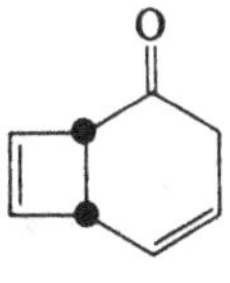

(823)

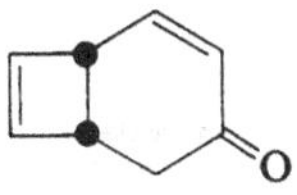

(824)

Scheme 82

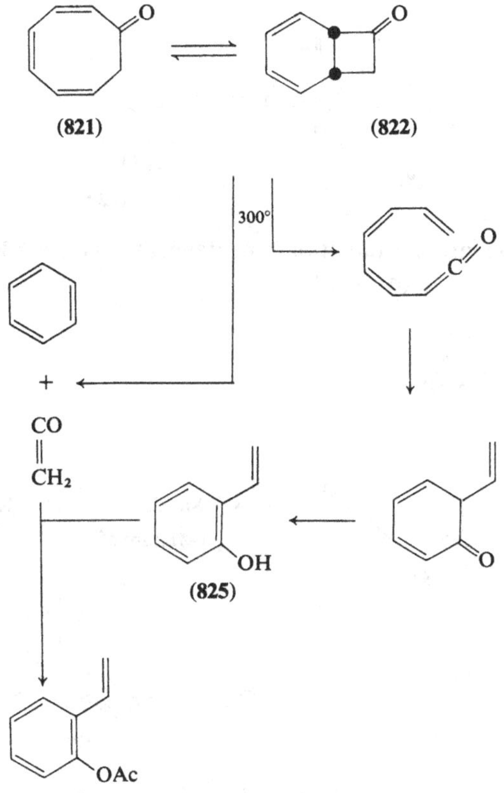

(821)   (822)

300°

+

CO
‖
CH₂

(825)

(826)

Photolysis. U.v. irradiation of the trienone **(821)** in pentane converts it into the thermally labile valence tautomer **(823)** (31 %), which reverts to the parent trienone on heating.[936] The photo-isomerisation occurs *via* the *E,Z,Z*-trienone **(827)**, which may be trapped with furan to give the adducts **(828)** and **(829)**.[937] Irradiation of the trienone **(821)** in methanol, however, yields a mixture of acyclic triene esters (geometrical isomers) **(830)** (31 %),[936] formed *via* the

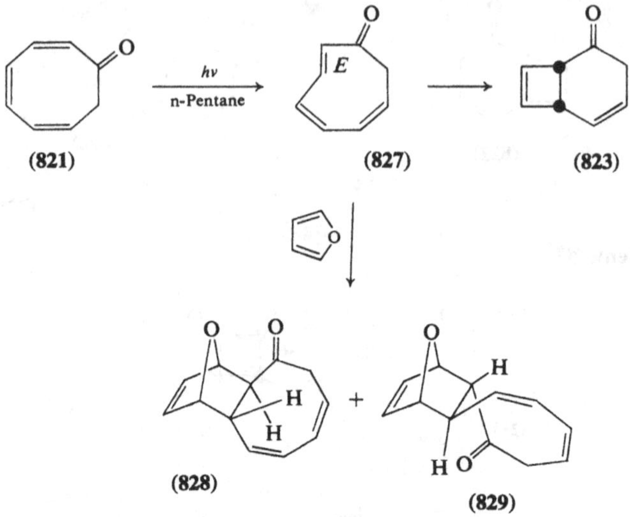

(821)   (827)   (823)

(828)   +   (829)

ketene (831).[936] On further irradiation in pentane, the mixture (830) is converted into the all-*E*-compound (832) (30%).[936]

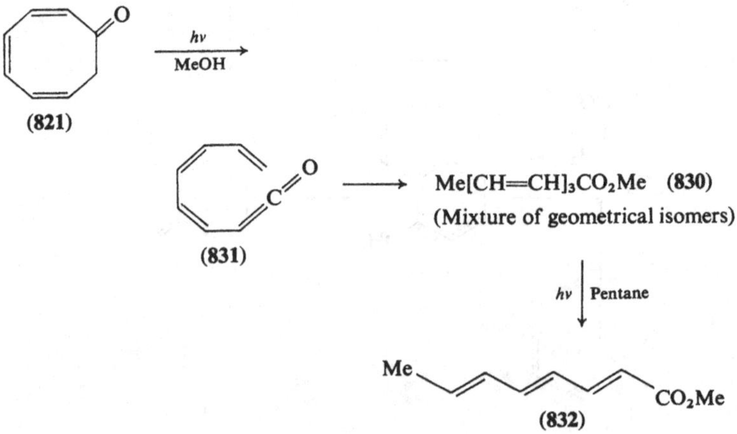

(821)

(831)   ⟶   Me[CH=CH]₃CO₂Me   (830)
(Mixture of geometrical isomers)

hv | Pentane

(832)

Protonation. Low-temperature protonation of cyclo-octatrienone in strongly acid media generates the hydroxyhomotropylium ion (833), which decomposes on warming to give the conjugate acid of acetophenone.[938]

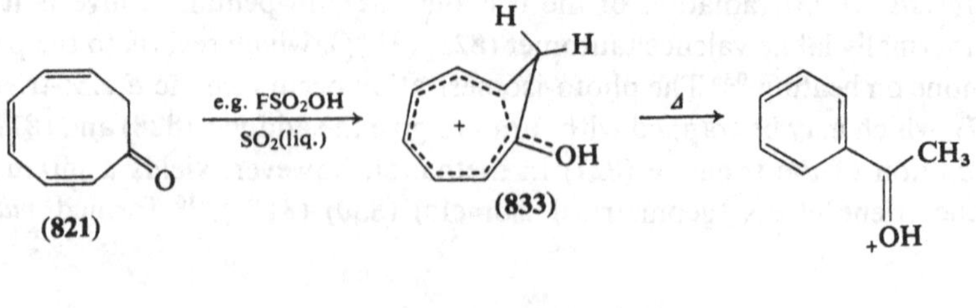

(821)   (833)

Reduction. Reduction of the trienone (**821**) with lithium aluminium hydride gives a low yield (**23**%) of impure trienol (**834**).[906]

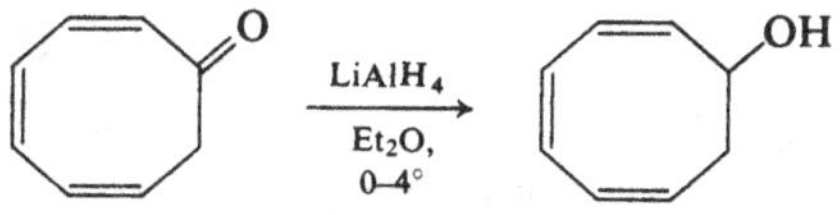

LiAlH$_4$
Et$_2$O,
0–4°

(**821**) (**834**)

Reactions with nucleophiles. Refluxing the trienone (**821**) with sodium hydroxide in water, or sodium alkoxide in the corresponding alcohol, gives low yields (up to 24%) of cyclohexa-1,4-diene derivatives (**835**).[939]

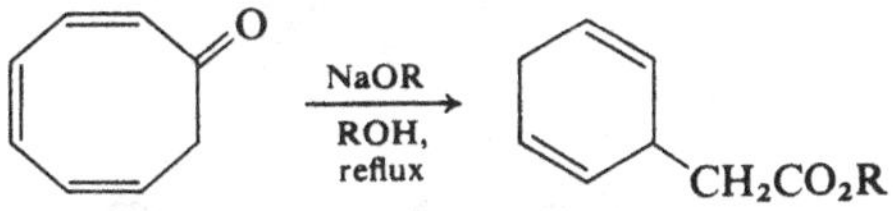

NaOR
ROH,
reflux

CH$_2$CO$_2$R

(**821**) (**835**)

R = H, Me, Et, CHMe$_2$

The trienone (**821**) reacts with orthoformic esters to yield the ketals (**836**), which exist predominantly in the bicyclic form (**837**) (see ref. 136) (table 105).

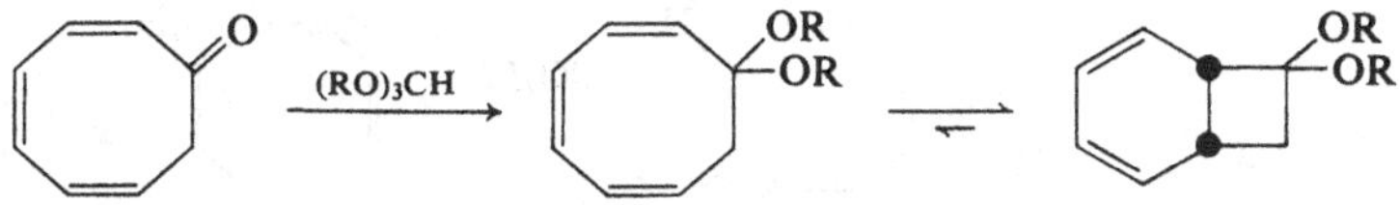

(RO)$_3$CH

OR
OR

OR
OR

(**821**) (**836**) (**837**)

Table 105

	R	Reaction conditions	Yield (%)	Ref.
a	Me	p-Me.C$_6$H$_4$.SO$_2$OH, MeOH, CH$_2$Cl$_2$	—	940
b	Et	FeCl$_3$, EtOH, r.-t.	69	941

Heating the diethyl ketal (**836b**) with aluminium t-butoxide gives ethoxycyclo-octatetraene (37%).[941]

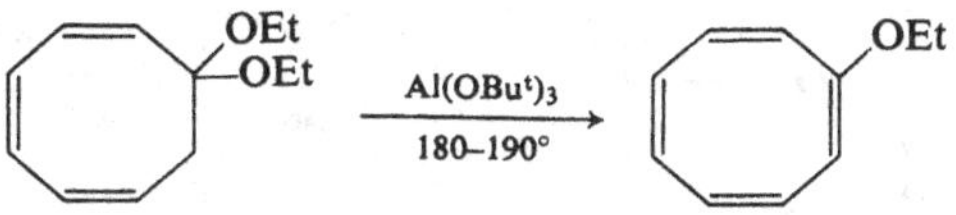

OEt
OEt

Al(OBu^t)$_3$
180–190°

OEt

(**836b**)

With phosphorus pentachloride or thionyl chloride in liquid sulphur dioxide, the dimethyl ketal (836a) affords the methoxyhomotropylium cation (271)[940] (see p. 99).

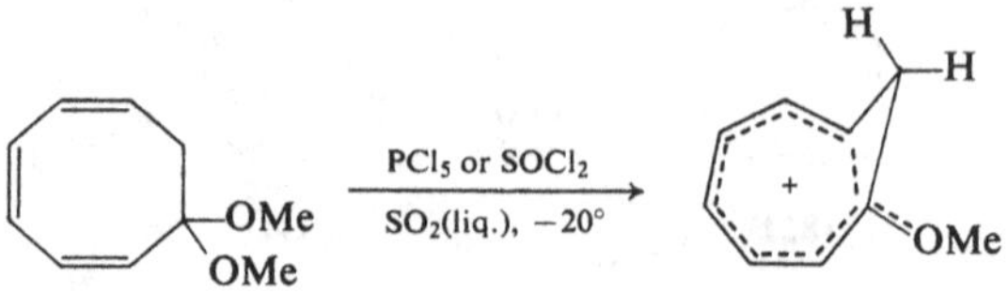

(836a)   (271)

The trienone (821) reacts with hydroxylamine to form a mixture of the stereoisomeric oximes (838) and (839);[200, 233] if these are converted into the benzenesulphonate derivatives (91%), one pure isomer (840) may be

Scheme 83

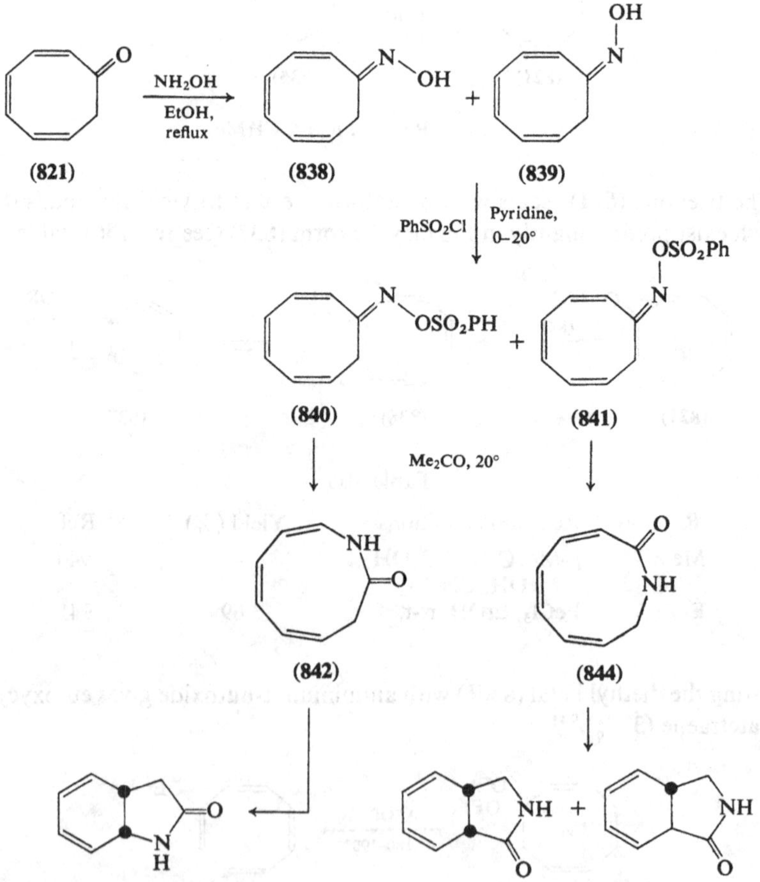

(821)   (838)   (839)

(840)   (841)

(842)   (844)

(843)   (845)   (846)

isolated.[233] Beckmann rearrangement of the benzenesulphonates (**840**) and (**841**) occurs in aqueous acetone; of the resulting lactams, (**842**) gives the *cis*-dihydro-oxindole (**843**), but (**844**) affords a mixture of the *cis*- and *trans*-dihydrophthalimidines (**845**) and (**846**)[233] (scheme 83).

With cyclic secondary amines in dimethyl sulphoxide, the trienone (**821**) gives the acyclic products (**847**) in almost quantitative yields.[942]

(**821**)   (**847**)

X = [CH$_2$]$_4$, [CH$_2$]$_5$, [CH$_2$]$_2$O[CH$_2$]$_2$

Reaction of the trienone (**821**) with cyanoacetic esters or malononitrile affords the condensation products (**848**) (table 106). (For the photochemical conversion of (**848**) into (**849**), see ref. 943.)

(**821**)   (**848**)

Table 106

R	Yield (%)	Ref.
CO$_2$Me	86[a]	943
CO$_2$Et	10[b]	941
CN	76	943

[a] A (separable) mixture of two geometrical isomers is formed.
[b] Only one geometrical isomer isolated from the mixture of products.

(**849**)

The Schmidt reaction of the trienone (**821**) with sodium azide in concentrated hydrochloric acid at 0° yields the *cis*-dihydrophthalimidine (**845**) (66%); the same reaction in trifluoroacetic acid, however, produces the tetrazoles (**850**) and (**851**) as the main products (19% and 6% respectively).[944]

Methylmagnesium iodide reacts with the trienone (821) to give the product (852) (23 %) but ring-opening readily occurs with the formation of the acyclic trienone (853; R = Me) (mixture of geometrical isomers); conversion to the all-*E*-compound (854) (82 %) is catalysed by iodine.[906] With ethyl- or phenyl-

magnesium bromide the initial product is not isolated, the open-chain ketone (853) being obtained directly (table 107). Similar results are obtained with

Table 107

R	Yield (%)	Ref.
Et	52	945
Ph	92	906

ethyl-lithium or triethylaluminium, but with lithium 'diethylcuprate' 1,4-addition occurs to give the cyclo-octadienone (855) (65.5%).[945]

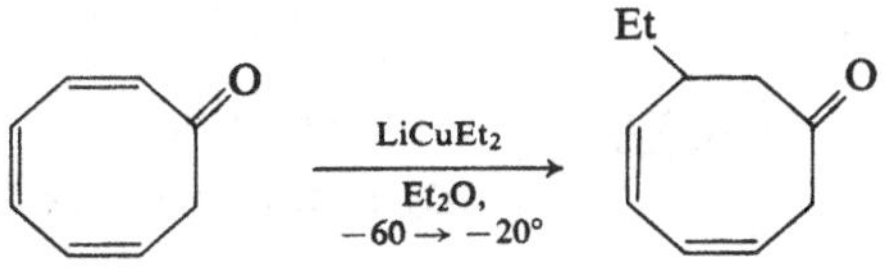

(821) (855)

Cycloadditions. Cyclo-octa-2,4,6-trienone readily forms Diels–Alder adducts with e.g. maleic anhydride, *via* the bicyclic valence tautomer (822); 1,4-

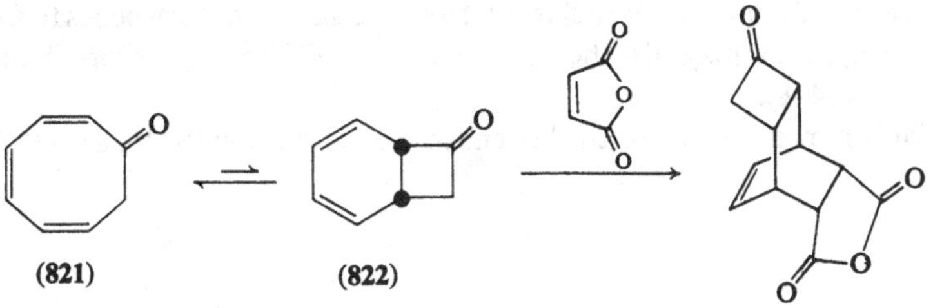

(821) (822)

(856)

naphthoquinone and dimethyl acetylenedicarboxylate afford the adducts (857) and (858) respectively[941] (table 108).

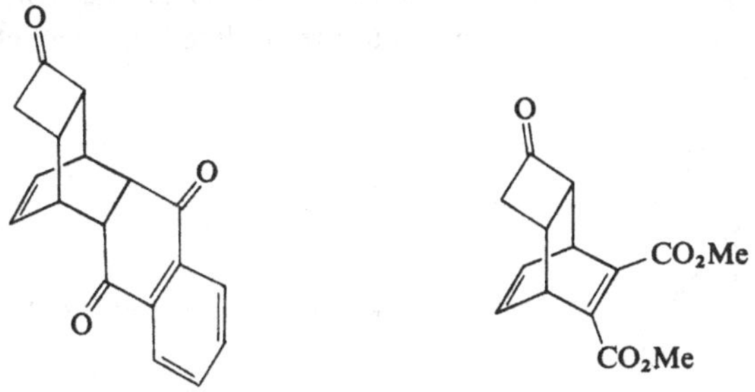

(857) (858)

Table 108

Dienophile	Conditions	Product	Yield (%)
Maleic anhydride	C_6H_6, 60°	(856)	84
1,4-Naphthoquinone	C_6H_6, reflux	(857)	80
Dimethyl acetylenedicarboxylate	50–90°	(858)	76

Reactions with metal derivatives. The trienone (821) reacts with $Fe_3(CO)_{12}$ (or $Fe_2(CO)_9$) to give the complexes (300) (13%) and (301) (2.4%)[893] (see p. 107).

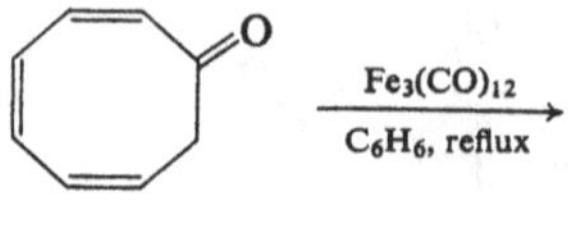

$$\xrightarrow[C_6H_6, \text{ reflux}]{Fe_3(CO)_{12}}$$

(Cyclo-octatrienone)Fe(CO)$_3$ (300)

+(Cyclo-octatrienone)Fe$_2$(CO)$_6$ (301)

(821)

11. Bicyclo[6.1.0]nona-2,4,6-trienes

Bicyclo[6.1.0]nonatrienes (82) result from the addition of carbenes to COT (see pp. 35–6); they may also be generated from COT^{2-} and *gem*-dihalides etc. (see pp. 174–8).

For n.m.r. evidence for the 'extended' conformation (82A), see ref. 314.

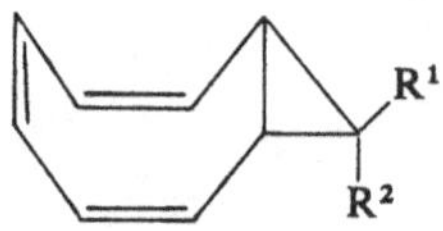

(82A) 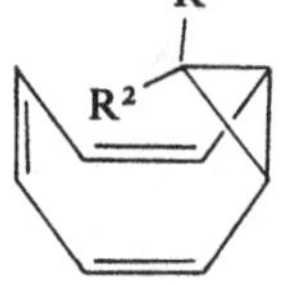   (82B)

Functional group transformations. Transformations of the *syn-* and *anti-9-*alkoxycarbonyl-derivatives (see pp. 35–6) are outlined in schemes 84 and 85.

Scheme 84*

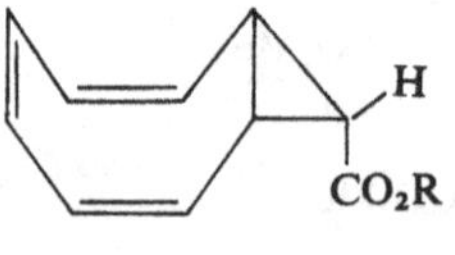

$\xrightarrow{R = Et}$ $C_9H_9 . CO_2H$ $\longrightarrow$ $C_9H_9 . COCl$

R = Me | LiAlH$_4$

$C_9H_9 . CH=N . NHTs$ $\longleftarrow$ $C_9H_9 . CHO$

$C_9H_9 . CH_2OH$

* No other experimental details are available, except spectral data for the alcohol,[314] and the m.ps. of the carboxylic acid[312] and the tosylhydrazone,[946] but see scheme 85 for analogous reactions.

Scheme 85

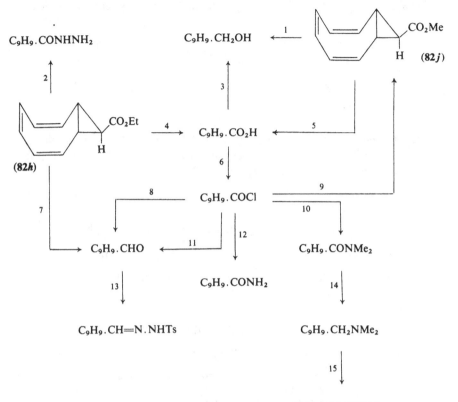

Reaction	Reagents and conditions	Yield (%)	Ref.
1	LiAlH$_4$	—	314
2	NH$_2$NH$_2$	—	310
3	LiAlH$_4$, Et$_2$O, r.-t.	100[a]	947
4	NaOH(aq.), 80–90°	45	309, (311)
5	NaOH(aq.), reflux	88	948
6	SOCl$_2$, reflux	88	(311), 948
7	{(i) LiAlH$_4$, Et$_2$O, reflux / (ii) CrO$_3$, pyridine, r.-t.}	30	948
8	LiAlH(OBu^t)$_3$, diglyme, −65°	39	948
9	MeOH	85[b]	311
10	Me$_2$NH, Et$_2$O	—	949
11	{(i) Aziridine, Et$_3$N, Et$_2$O, 0° / (ii) LiAlH$_4$, Et$_2$O, 0°}	24	948
12	NH$_3$	80[b]	311
13	H$_2$N.NHTs, EtOH, 60°	61	948
14	LiAlH$_4$	—	949
15	30% H$_2$O$_2$, MeOH	—	949

[a] Product not purified. [b] Overall yield from the carboxylic acid.

The bromo-derivative (82*l*) may be converted into the deuterio-derivative (82*m*) (*ca.* 90 %) by the action of butyl-lithium, followed by deuterium oxide.[749]

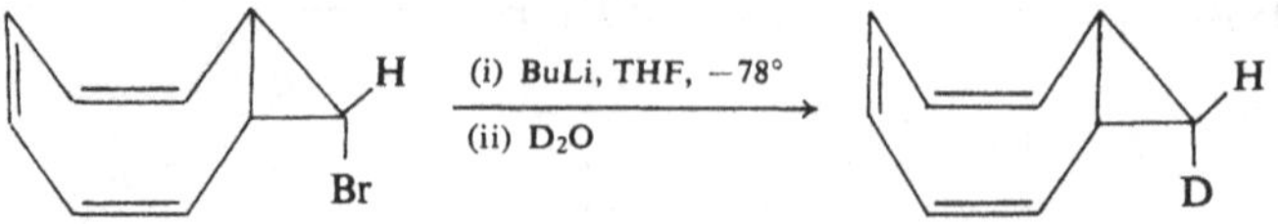

(i) BuLi, THF, −78°
(ii) D₂O

(82*l*)   (82*m*)

Thermal epimerisation. The *syn*-9-substituted derivatives (82; R^1 = H) undergo thermal equilibration with the corresponding *anti*-isomers (82; R^2 = H) (table 109) (for mechanistic discussions, see refs. 313, 747, 749).

(82) (R^1 = H)   (82) (R^2 = H)

Table 109

R	Temp. (°C)	Ref.
D	30	749
CO₂Me	100	313
CO₂Et	100	313
CN	139	748
OMe	r.-t.	747
F	0	747

Thermolysis. The thermal rearrangements of the system (82) have been studied intensively.

The products obtained from the hydrocarbons (82; R^1 = H or alkyl, R^2 = H or alkyl) and certain other derivatives are the *cis*- or *trans*-dihydro-indenes (859) or (860) (mixtures of epimers) (table 110). (For kinetic data on the

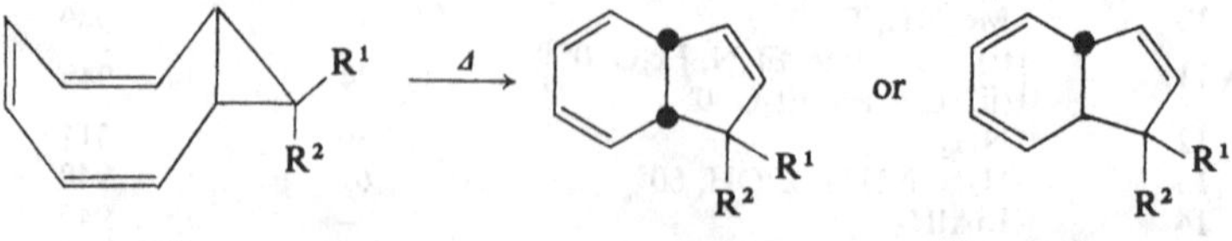

(82)   (859)   (860)

Table 110

R¹ (*anti*)	R² (*syn*)	Temp. (°C)	Predominant product[a]	Ref.
H (D)	H (D)	90	(859)	297, 299, (950), 951
Me	H	110–199	(859)	952, (953), 954
Buᵗ	H	179	(859)	746
CH₂OH	H	120, [150]	(859)	947, [955]
CO₂H	H	150	(859)	955
CO₂Me	H	130, [160]	(859)	[299], [311], 313
CO₂Et	H	130, [160–180]	(859)	[311], 313, (955)
CN	H	139	(859)	748
NMe₂	H	70	(859)[b]	747
Cl	H (D)	70, [75]	(859)	743, 956, [957]
H	Me	110–200	(859)[c]	952, (953), 954
H	Buᵗ	179	(860)	746
H	CO₂Me	130	(859)[d]	313
H	CO₂Et	130	(859)[d]	313
H	CN	139	(859)[d]	748
H	OMe	58	(859)[d]	747
H	F	35	—[d, e]	747
Me	Me	151	(860)	744
Me	Et	170	(860)	958
Et	Me	170	(860)	958

[a] Usually >90%, ignoring secondary products.
[b] The major product is the bicyclo[4.2.1]nonatriene (861; R = NMe₂) (see p. 269).
[c] Ratio (859):(860) = *ca.* 7:3. Epimerisation at C-9 may be involved.[313, 749]
[d] Epimerisation at C-9 precedes rearrangement.
[e] The products are indene and the bicyclo[4.2.1]nonatriene (861; R = F).

(861)

rearrangement of (82; R¹ = H, Ph, CO₂Me, CN, OMe, Cl; R² = H), see ref. 959.) Thermolysis of the deuterium-labelled species (862) and (863) affords the

(862)   (864)

products (864) and (865) respectively.[960] The diphenyl-derivative (866) gives the product (867).[961, 962]

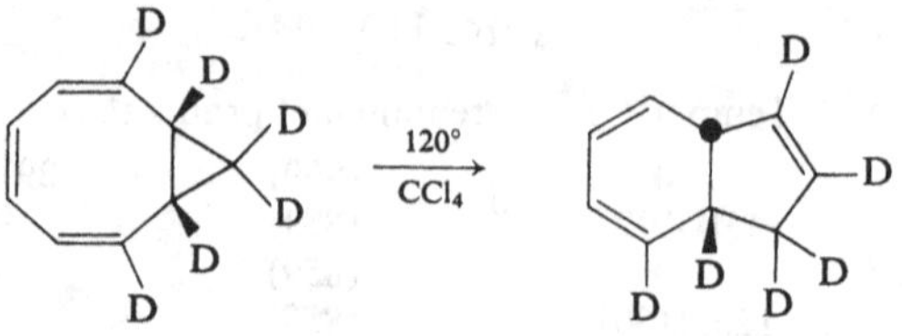

(863)   (865)

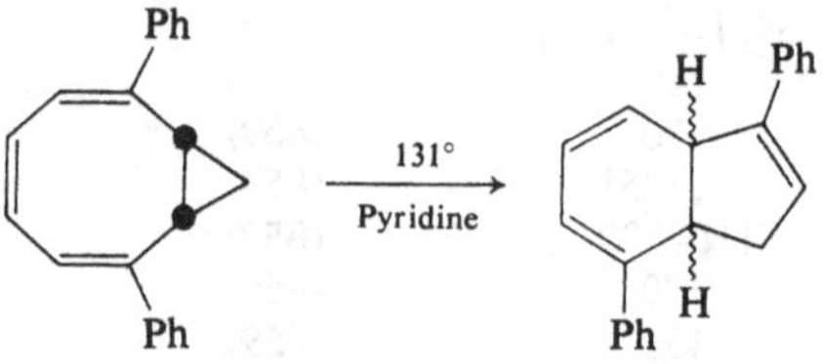

(866)   (867)

The rearrangements resulting in *trans*-products (860) may proceed by ring-opening to give the Z,Z,Z,E-cyclononatetraene (868), followed by a 6π electrocyclisation (scheme 86), both of the proposed steps being unexceptional from the viewpoint of orbital symmetry.

Scheme 86

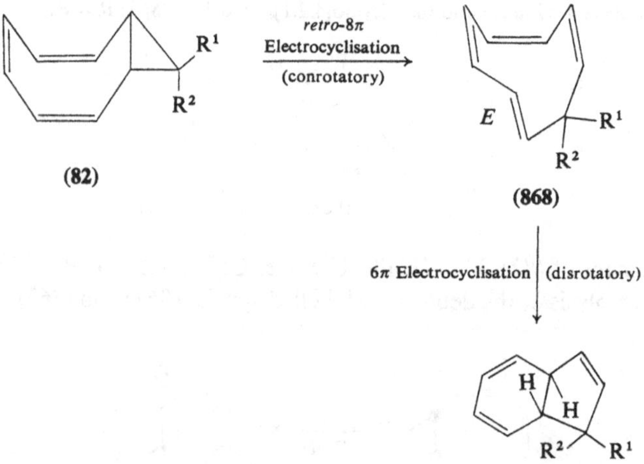

However, the formation of *cis*-products (859), which would seem to involve a thermally 'forbidden' step, is *kinetically preferred*, the 'allowed' pathway being followed only when the starting material (82) contains a bulky *syn*-9-substituent (R^2). This indicates that the adoption of the more congested conformation (82B) (scheme 87) is the initial step in the bond reorganisation leading

Scheme 87

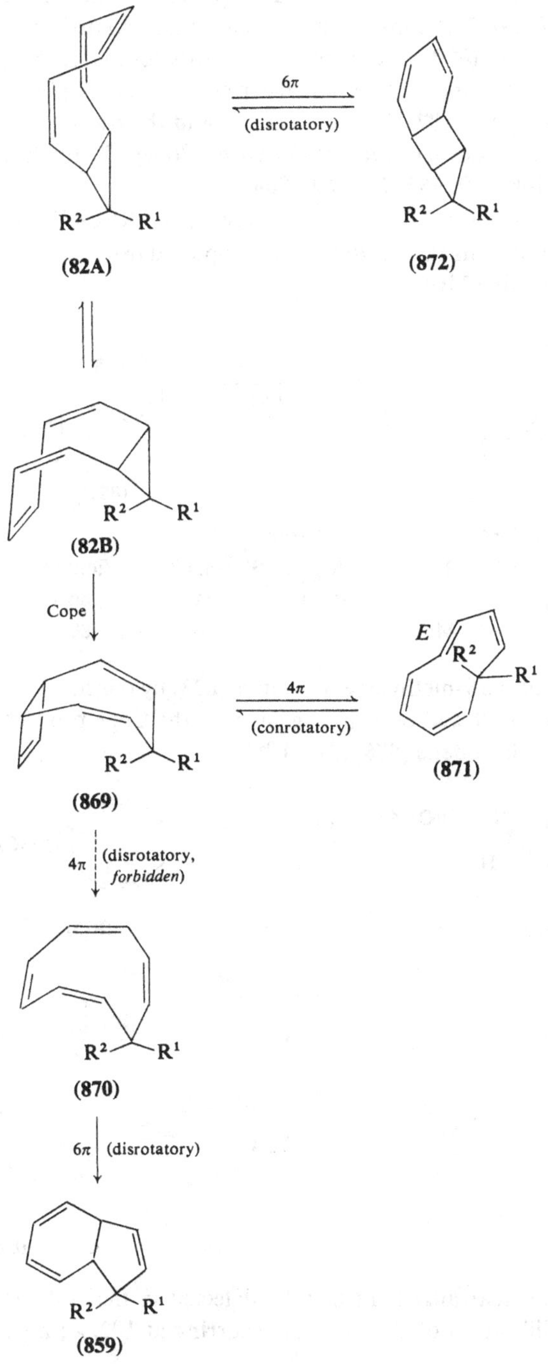

6π
(disrotatory)

(82A)

(872)

(82B)

Cope

E

4π
(conrotatory)

(871)

(869)

4π (disrotatory, *forbidden*)

(870)

6π (disrotatory)

(859)

to *cis*-products (**859**). The next step may be a Cope rearrangement, for which the conformer (**82B**) is an essential precursor, leading to the bicyclo[5.2.0]nonatriene system (**869**). The subsequent steps are more controversial. The all-*Z*-cyclononatetraenes (**870**) could be the immediate source of the observed products (**859**), but trapping experiments provide evidence only for the presence of the *Z,Z,E,Z*-system (**871**) (see pp. 286–7), and the tricyclic diene (**872**) (see p. 282). (The problem has been reviewed by Staley;[963] for later papers see e.g. refs. 313, 746, 749, 953, 954, 960, 964.)

The rearrangement (**82**) → (**859**) is catalysed by [RhCl(CO)₂]₂ (table 111); under these conditions the 9,9-dimethyl-compound (**82n**) also gives a *cis*-fused product[965] (*cf.* table 110).

(**82**)   (**859**)

Table 111

	R¹	R²	Temp. (°C)	Yield (%)
a	H	H	35	>99.5
n	Me	Me	140	80

Generation of the 9-methylene-derivative (**873**) by thermolysis of the amine-*N*-oxide (**874**) is followed by rearrangement to the Cope product (**875**) (21 %) and the *cis*-dihydroindene (**876**) (31 %).[949]

(**874**)   (**873**)

(**875**)   (**876**)

A degenerate rearrangement may be detected in the systems (**877**)–(**880**), complete equilibration of the isomers occurring at 102.5°; migration of C-9 (with inversion) around the C_8 ring takes place *via* a series of symmetry-controlled [1,7] sigmatropic shifts.[966]

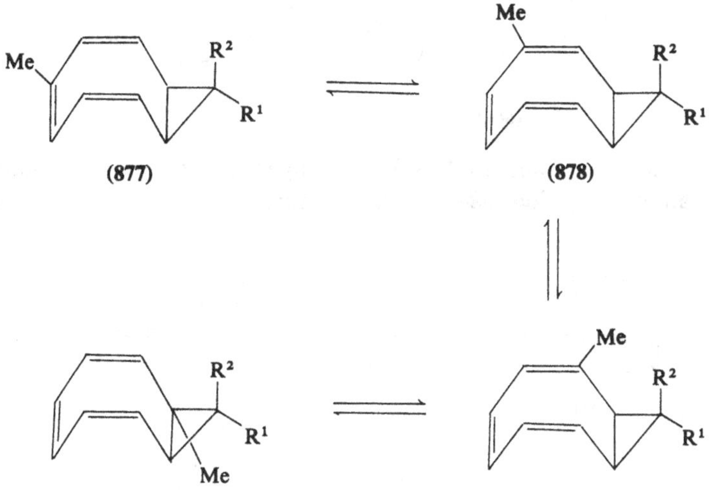

(877)　　　　　　(878)

(880)　　　　　(879)

R^1 = Me, R^2 = CN
or R^1 = CN, R^2 = Me

At 180–190°, however, the 9-cyano-compounds (82*d*), (82*o*) and (82*p*) undergo rearrangement to bicyclo[4.2.1]nonatrienes (881) (the reaction is not completely stereospecific) (table 112). The same type of rearrangement occurs

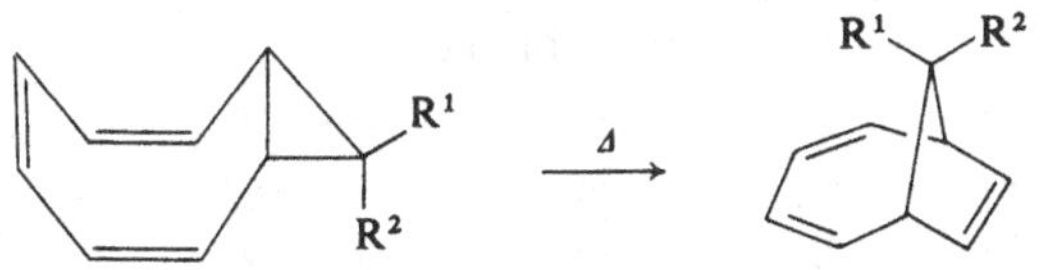

(82)　　　　　　(881)

Table 112

	R^1	R^2	Temp. (°C)	Yield (%)	Ref.
o	Me	CN	180	—	967
p	CN	Me	180	—	967
d	CN	CN	190	72	304, 305

with the *spiro*-cyclopentadiene (88*a*), which gives the system (882) (25–30 %);[315] the related *spiro*-fluorene behaves similarly.[968]

The *syn*-9-halo-derivatives (82*b*), (82*c*) and (82*e*) rearrange to the *cis*-dihydroindenes (883) (table 113), migration of halogen being involved (*cf.* the *anti*-9-chloro-compound (table 110)). It is suggested that ionisation of the

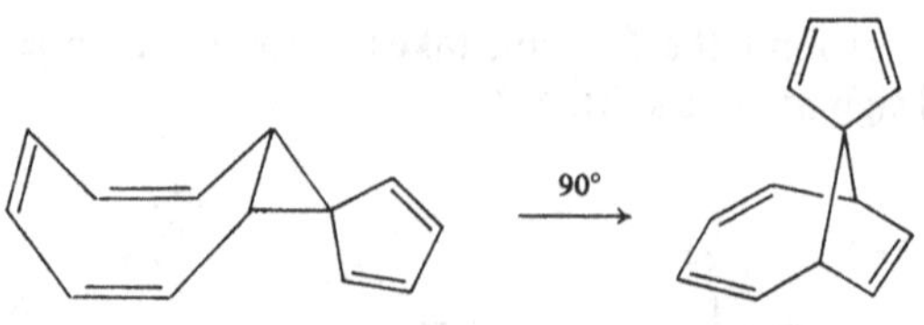

(88a)   (882)

tricyclic valence tautomer (872) is followed by rapid disrotatory ring-opening to give an allylic cation (884) which recaptures the halide ion on the same face.[957]

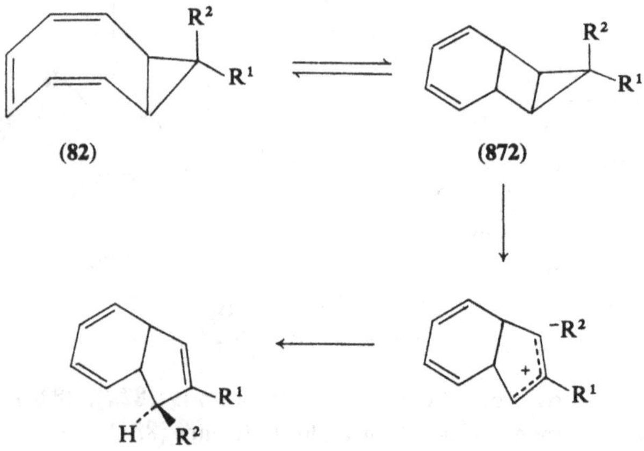

(82)   (872)

(883)   (884)

Table 113

	R^1	R^2	Temp. (°C)	Yield (%)	Ref.
b	Cl	Cl	80–90	ca. 100	(956), 957
c	Br	Br	80–90	ca. 100	297
e	H	Cl	(70), 75	(80)	(956), 957

Thermolysis of the 9-acetoxy-9-methyl-compound (522; R = Me) results in the elimination of acetic acid and the formation of 3-methylindene (79%).[377]

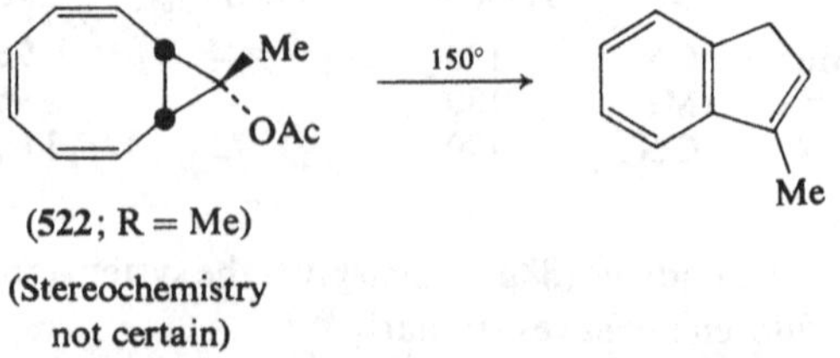

(522; R = Me)

(Stereochemistry
not certain)

Me

Generation of the carbene (885) by thermolysis of the sodium salt (886) of the *anti*-tosylhydrazone (887) leads to a variety of products, of which the bicyclic tetraene (**888**) is the most abundant (38%)[948,969] (*cf.* the *syn*-isomer below).

Similar thermal decomposition of the sodium salt (**889**) of the *syn*-tosyl-hydrazone (**890**) leads to the pyrazoline (**891**), which in turn thermolyses to give the bicyclic product (**888**) (53%) and bullvalene (**892**) (46%).[946]

Photolysis. U.v. irradiation of *cis*-bicyclo[6.1.0]nonatriene (**82a**) at −60° gives rise to a stationary state in which the starting material and a number of its isomers are in equilibrium.[970] The *trans*-fused isomer (**893**)[971] is obtainable in up to 11 % yield in methanol at 23°.[972]

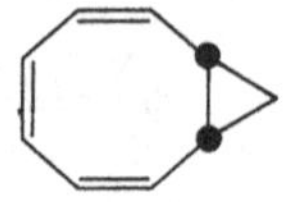

(**82a**)

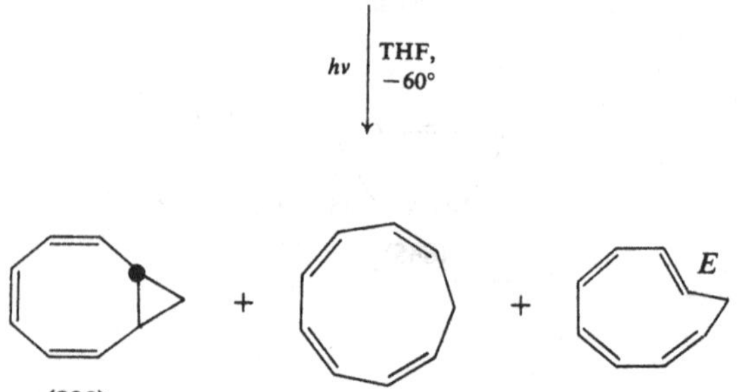

$h\nu$ THF, −60°

(**893**)   +   +   E

+   +

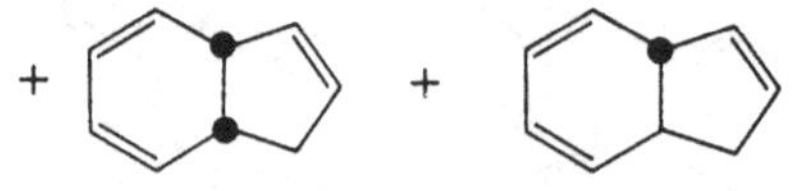

Irradiation of *syn*-9-substituted derivatives of the parent hydrocarbon (**82a**) effects their transformation into the *anti*-stereomers;[746,953,973] for example, photo-isomerisation of the *syn*-9-chloro-compound (**82e**) affords the *anti*-9-chloro-derivative (**82f**) (up to 85 %).[973]

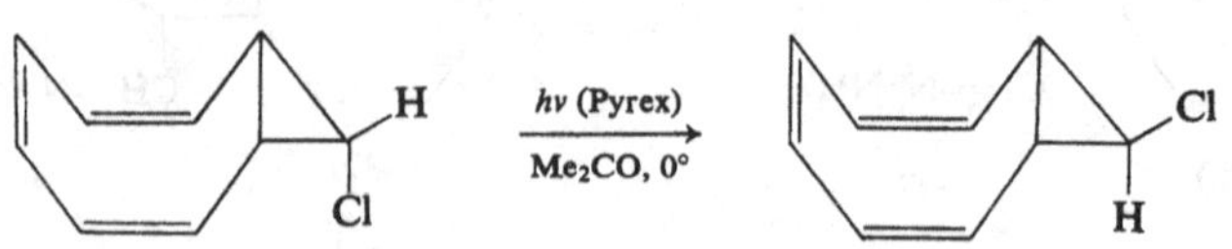

(**82e**) $h\nu$ (Pyrex)   (**82f**)
                 Me₂CO, 0°

At −15° in the presence of benzophenone as sensitiser, the chlorocyclonona-tetraene (**894**) is produced from either of the 9-chloro-stereomers (**82e**) or (**82f**); at 0°, thermal rearrangement of the cyclononatetraene (**894**) results in the *cis*-dihydroindene (**895**).[973]

U.v. irradiation of the ester (82*h*) yields a tricyclic product formulated as (896).[955]

Photolysis of the sodium salt of the *anti*-tosylhydrazone (887) results (after low-temperature work-up) in a mixture of the $C_{10}H_{10}$ hydrocarbons (888), (897) (major product) and (898)[312, 946] (*cf.* thermolysis, p. 271). (For photolysis of the pyrazoline (891), see ref. 946.)

Reactions with electrophiles. Protonation of bicyclo[6.1.0]nonatriene **(82a)** or its *anti*-9-methyl-derivative **(82q)** in a 'super-acid' medium occurs at C-3 ($\equiv$ C-6) to afford a 1,3-bis(homotropylium) ion **(899)**.[848, 864, 974]

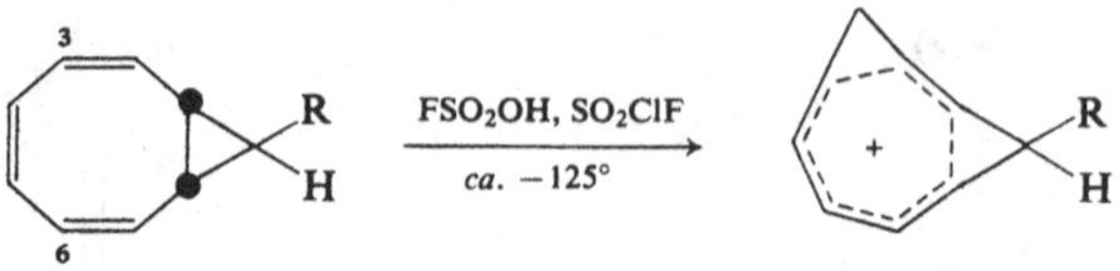

(82a)   (R = H)   (899)
(82q)   (R = Me)

The relative stereochemistry of the methylene bridges (*syn* **(899a)** or *anti* **(899b)**) in this species is uncertain. (For [13]C n.m.r. spectra, refs. 848, 849.)

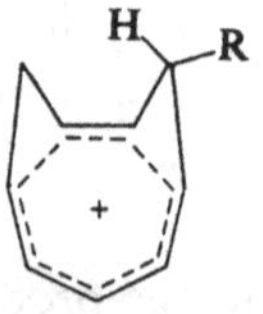

(899a)

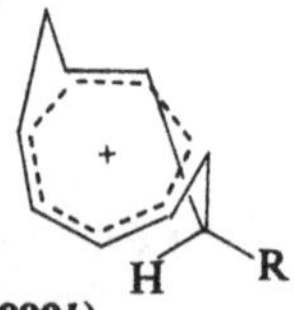

(899b)

The *syn*-9-methyl- and 9,9-dimethyl-derivatives fail to give bis(homo-tropylium) ions on protonation.[848]

Reaction of the bicyclic triene **(82a)** with iodine azide (from an excess of sodium azide and iodine monochloride) yields the monocyclic triazide **(900)** (80–90 %), characterised as the acetylenedicarboxylic ester-adduct **(901)** (40 %).[975,976]

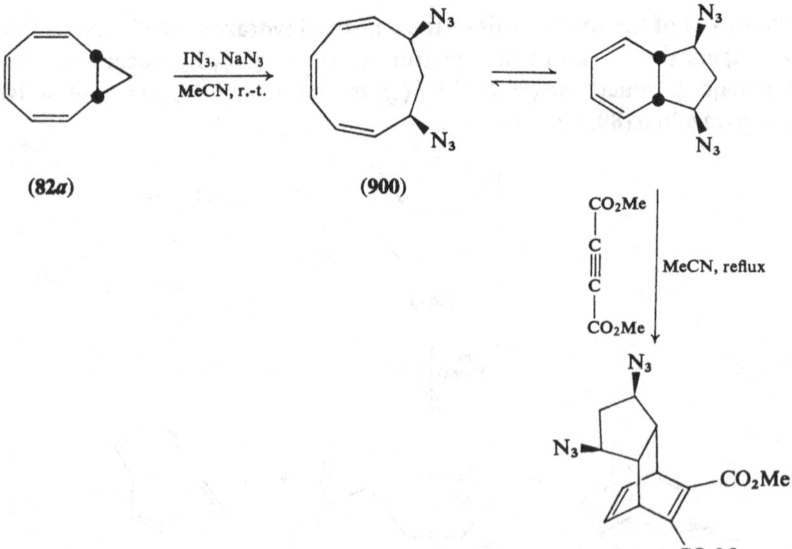

(82a)   (900)

(901)

Reduction, and reactions with nucleophiles. The addition of an electron to bicyclo[6.1.0]nonatriene (**82a**), using an alkali metal in e.g. tetrahydrofuran, or by means of electrolysis in liquid ammonia, generates the non-classical anion-radical (**902**), which accepts a further electron to form the dianion (**903**). With alkali metals in liquid ammonia, the anion-radical (**904**) of methylcyclo-octatetraene is produced. The preparations of these species are summarised in scheme 88. (For a polarographic study of the electrochemical reduction of (**82a**), see ref. 986. For a theoretical discussion of (**902**), see ref. 987.)

Scheme 88

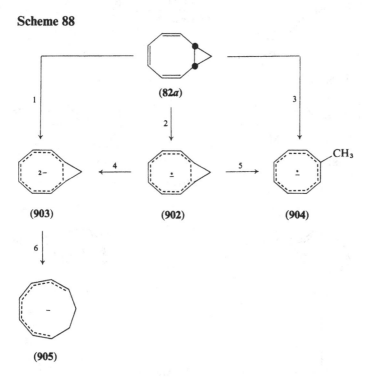

(903)   (902)   (904)

(905)

Reaction	Reagents and conditions	Ref.
1	2M, NH₃(liq.), −78° ⎫ ᵃ or 2Na, HMPA, THF, −30° ⎭	977
	or 2K, MeOCH₂.CH₂OMe	978, 979
2	1M, THF, −80°	972, 980, 981
	or electrolysis, NH₃(liq.), Me₄N⁺Cl⁻	982
3	1M, NH₃(liq.)	983
4	1K, THF, −80°	984
5	Electrolysis, NH₃(liq.), Me₄N⁺Cl⁻	982
6	NH₃(liq.), or HMPA	977, 985

ᵃ Protonation of (**903**) by the solvent results in (**905**) (see reaction 6).

Protonation of the dianion (**903**) (dipotassium salt in 1,2-dimethoxyethane) with methanol affords the bicyclic diene (**906**) (85 %),[978] whereas addition of solid ammonium chloride to the monoanion (**905**) (sodium salt in liquid ammonia) gives the monocyclic triene (**689**) (67%).[977]

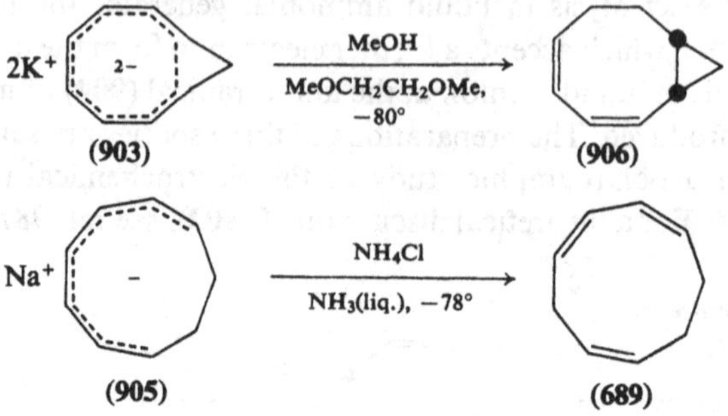
2K⁺
2–
MeOH
MeOCH₂CH₂OMe, −80°
(903)
(906)
Na⁺
–
NH₄Cl
NH₃(liq.), −78°
(905)
(689)

Scheme 89

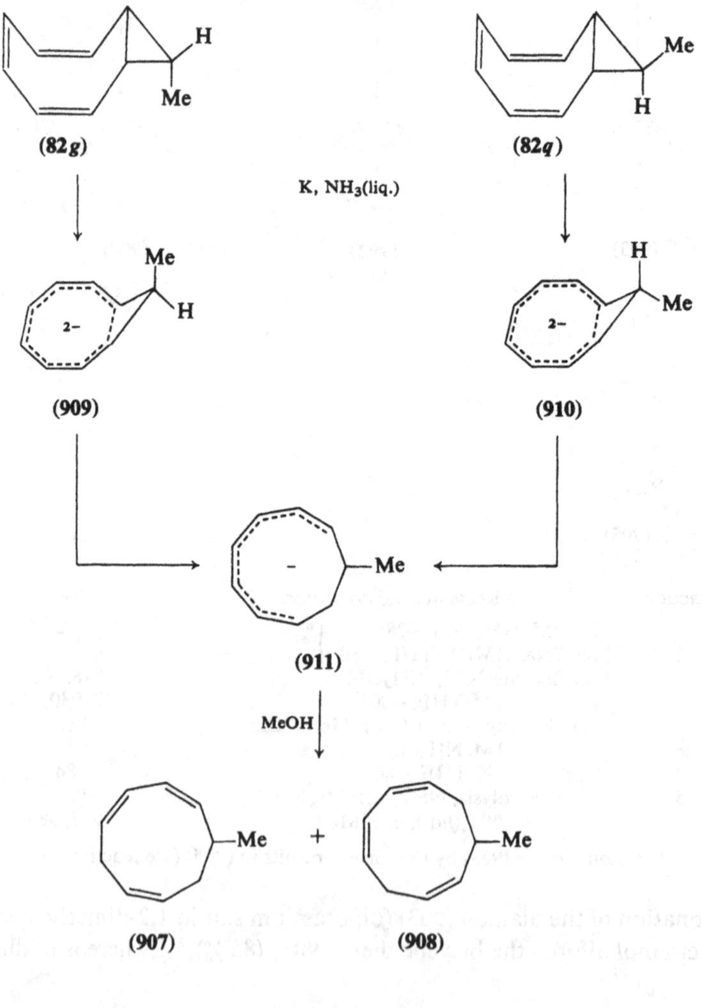
H
Me
(82g)
Me
H
(82q)
K, NH₃(liq.)
Me
H
2–
(909)
H
Me
2–
(910)
–
Me
(911)
MeOH
Me
+
Me
(907)
(908)

(For the reaction of (**903**) with carbon tetrachloride, see ref. 979. For the reactions of (**905**) with oxygen, potassium permanganate and methyl iodide, see refs. 985, 988.)

Reduction of the *syn-* or *anti*-9-methyl-derivative (**82g**) or (**82q**) with potassium in liquid ammonia, followed by quenching with methanol, yields a mixture of the isomeric cyclononatrienes (**907**) and (**908**), *via* stereochemically distinct dianions (**909**) and (**910**), but a common intermediate monoanion (**911**)[985] (scheme 89). (For the analogous reduction of the 1-, 2-, 3- and 4-methyl-compounds (**912**)–(**915**), see ref. 985; of the 9,9-dimethyl-derivative (**82n**), see refs. 977, 985.)

(**912**)

(**913**)

(**914**)

(**915**)

(**82n**)

For catalytic hydrogenation of the bicyclo[6.1.0]nonatriene system, see e.g. refs. 309, 310, 377.

Anions can also be generated from bicyclo[6.1.0]nonatriene by proton-abstraction. Thus reaction with e.g. lithium amide in liquid ammonia gives rise to the methylenecyclo-octatrienyl anion (**916**), which with water yields (mainly) a mixture of the methylenecyclo-octatrienes (**917**) and (**918**).[989] (For reaction of (**82a**) with potassium t-butoxide, see pp. 84–5.) Treatment with one equivalent of sodium in hexamethylphosphoramide-tetrahydrofuran,[977] or sodium hydride in dimethyl sulphoxide,[958] generates the cyclononatetraenyl anion (**919**); this gives all-Z-cyclononatetraene (**920**) (half-life 50 min. at 23°[990]) when it is quenched with water,[990] methanol,[991] or dilute acetic acid.[992]

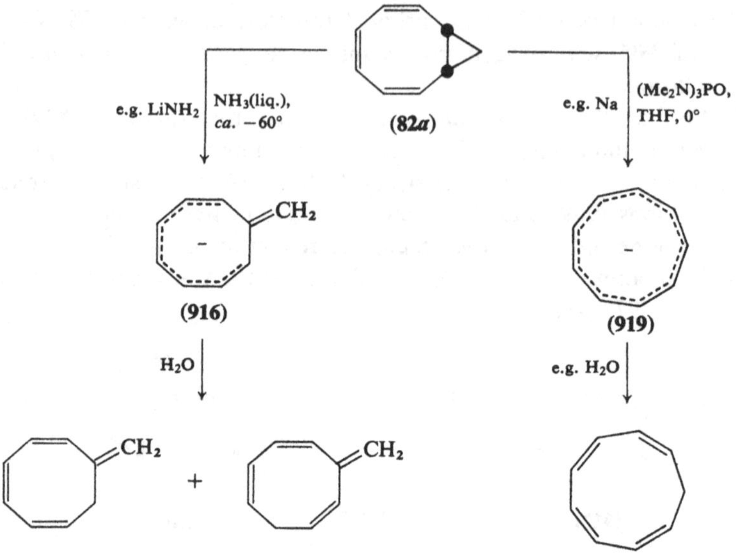

(82a)

(916)

(919)

(917)   (918)   (920)

The *anti*-9-substituted derivatives (82*j*) and (82*r*) are deprotonated by lithium di-isopropylamide to form anions represented as (921).[993]

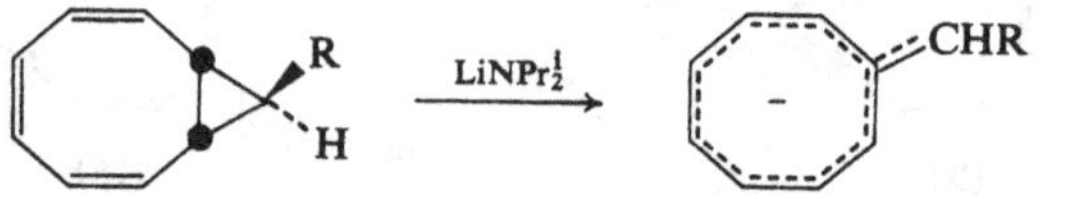

(82*j*)   (R = CO₂Me)   (921)
(82*r*)   (R = CN)

The *anti*-9-methoxy-compound (82*s*), on treatment with potassium at −40°, gives the anion (922) of Z,Z,Z,E-cyclononatetraene; at room-temperature this is completely converted into the all-Z-cyclononatetraenyl anion (923).[994]

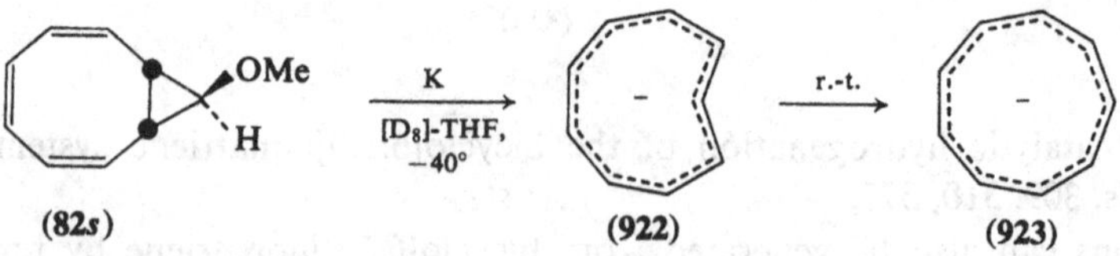

(82*s*)   (922)   (923)

The anion (923) is also produced from the *anti*-9-chloro-compound (82*f*),[743] or from a mixture of the *syn*- and *anti*-isomers,[956] by reaction with lithium in tetrahydrofuran; the yields are 60–65%.[956]

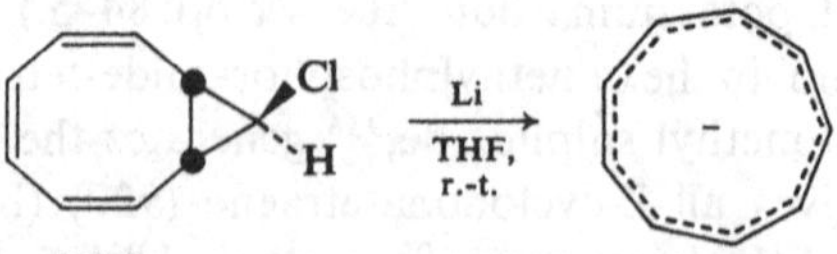

(82*f*)   (923)

Similarly, the chlorocyclononatetraenyl anion (**924**) may be prepared (*ca.* 50%) from 9,9-dichlorobicyclo[6.1.0]nonatriene (**82b**).[956]

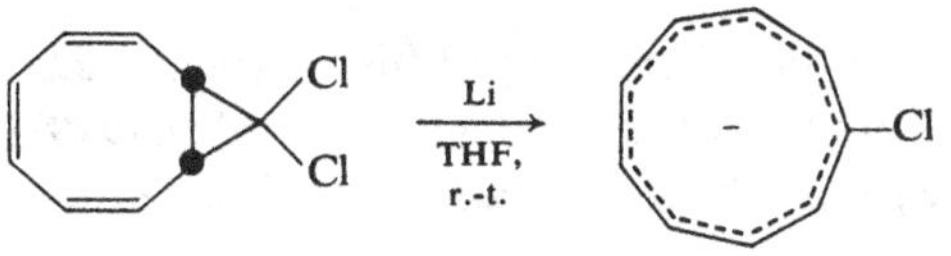

(**82b**)   Li, THF, r.-t.   (**924**)

In contrast with the non-classical radical anion (**902**) obtained by the addition of an electron to *cis*-bicyclo[6.1.0]nonatriene (**82a**), the radical anion similarly derived from the *trans*-fused isomer (**893**) has the classical structure (**925**).[995]

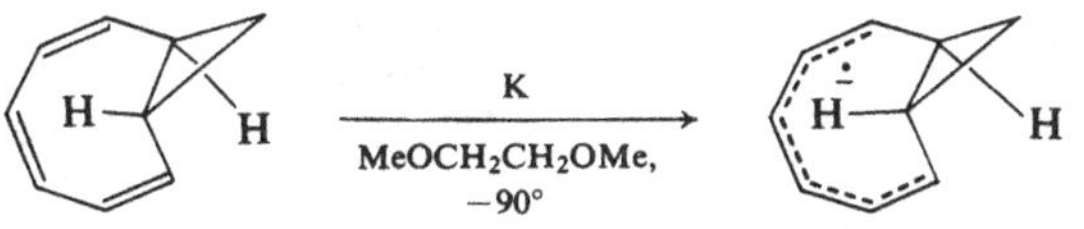

(**893**)   K, MeOCH₂CH₂OMe, −90°   (**925**)

In liquid ammonia the radical anion (**925**) apparently suffers protonation followed by further reduction to the monoanion (**926**); quenching with methanol leads to the diene (**927**)[984] (scheme 90). (For a polarographic study of the

Scheme 90

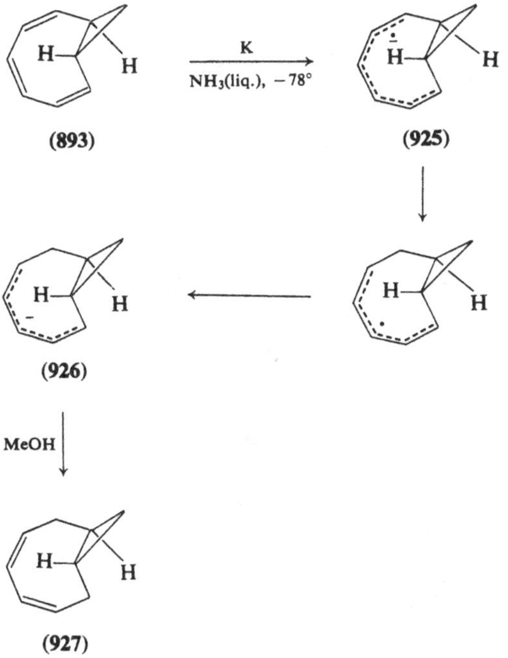

(**893**)   K, NH₃(liq.), −78°   (**925**)

(**926**)   ←

MeOH

(**927**)

electrochemical reduction of *trans*-bicyclo[6.1.0]nonatriene (**893**), see ref. 986.)

Cycloadditions. Reaction of bicyclo[6.1.0]nonatriene (**82a**) with dichlorocarbene results in a mixture of the adducts (**928**) (6%), (**929**) (15%), and (**930**) (41%); the relative geometry of the ring-fusions is uncertain.[301]

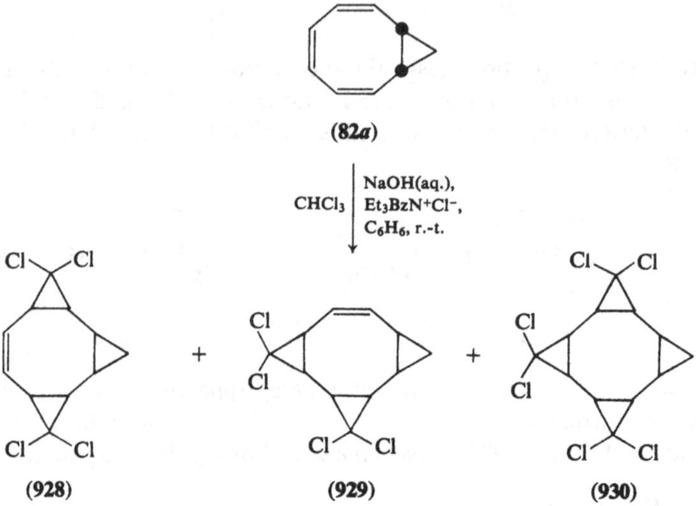

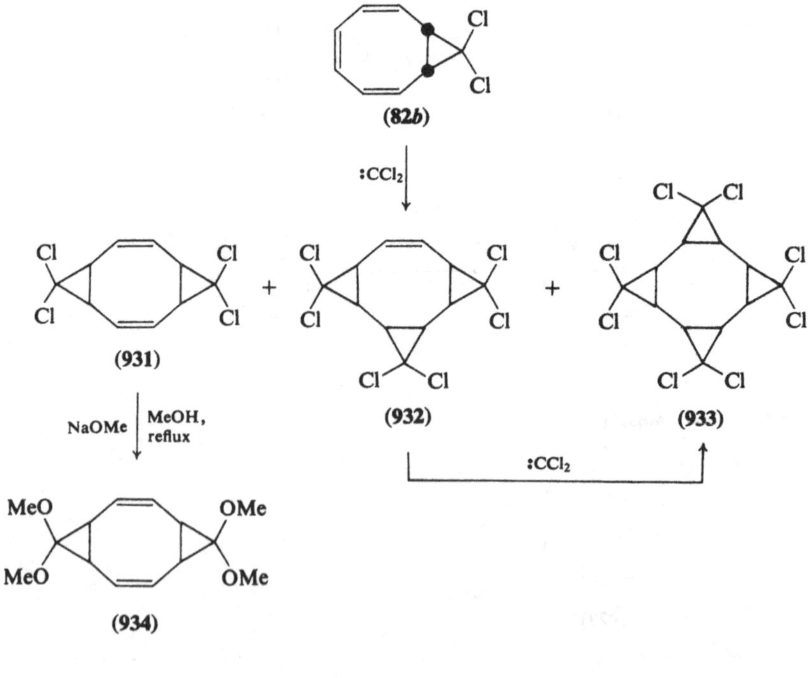

(**82a**)

CHCl$_3$ | NaOH(aq.), Et$_3$BzN$^+$Cl$^-$, C$_6$H$_6$, r.-t.

(**928**)   +   (**929**)   +   (**930**)

(**82b**)

:CCl$_2$

(**931**)   +   (**932**)   +   (**933**)

NaOMe | MeOH, reflux

:CCl$_2$

(**934**)

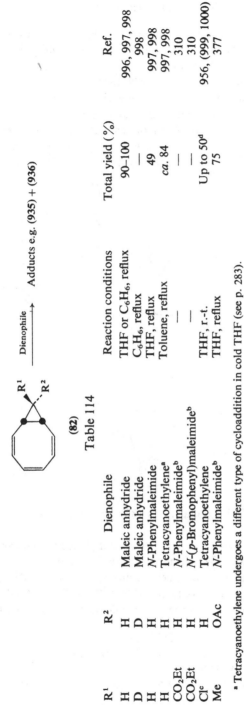

$$\text{(82)} \xrightarrow{\text{Dienophile}} \text{Adducts e.g. (935) + (936)}$$

Table 114

R¹	R²	Dienophile	Reaction conditions	Total yield (%)	Ref.
H	H	Maleic anhydride	THF or C_6H_6, reflux	90–100	996, 997, 998
D	D	Maleic anhydride	C_6H_6, reflux	—	998
H	H	N-Phenylmaleimide	THF, reflux	49	997, 998
H	H	Tetracyanoethylene[a]	Toluene, reflux	ca. 84	997, 998
CO₂Et	H	N-Phenylmaleimide[b]	—	—	310
CO₂Et	H	N-(p-Bromophenyl)maleimide[b]			310
Cl[c]	H	Tetracyanoethylene	THF, r.-t.	Up to 50[d]	956, (999, 1000)
Me	OAc	N-Phenylmaleimide[b]	THF, reflux	75	377

[a] Tetracyanoethylene undergoes a different type of cycloaddition in cold THF (see p. 283).
[b] Only one adduct isolated (structure uncertain).
[c] Starting material contaminated with the syn-isomer.
[d] Yield of major product (derived from the dihydroindene); for the minor product, see p. 284.

A similar reaction using the 9,9-dichloro-compound (82*b*) gives the products (931) (7%), (932) (26%), and (933) (1.7%); the pentacyclic adduct (933) may be obtained in 55% yield from (932).[302] Treatment of the adduct (931) with sodium methoxide affords the diketal (934) (8%).[302]

The reaction of bicyclo[6.1.0]nonatriene with dienophiles is complicated by its susceptibility to thermal rearrangement (see p. 264). Thus with maleic anhydride in refluxing tetrahydrofuran or benzene two adducts (935) and (936) are formed (ratio *ca.* 3:1), these products being derived from the tricyclic valence tautomer (937) and the *cis*-dihydroindene (938) respectively. The

| (82*a*) | (937) | (935) |

| (938) | (936) |

reactions of bicyclo[6.1.0]nonatrienes (82) with analogous dienophiles are listed in table 114.

Bicyclo[6.1.0]nonatriene reacts with dimethyl acetylenedicarboxylate to afford a single adduct (939). Dimethyl phthalate and the cyclopentadiene-adduct (940) are found as by-products, resulting from Alder–Rickert cleavage of the initially formed dihydroindene-adduct (941).[996, 1001] Analogues of the adduct (939) are formed from other acetylenic dienophiles (table 115).

Table 115

(82) R¹	R²	Dienophile	Reaction conditions	Yield (%)	Ref.
H	H	MeO_2C—C≡C—CO_2Me	90°	60	(996), 1001
H	H	F_3C—C≡C—CF_3	r.-t.	*ca.* 65	998
H	H	NC—C≡C—CN	THF, r.-t.	3	998
CO_2Et	H	MeO_2C—C≡C—CO_2Me	128°	—[a]	311

[a] Structure of adduct uncertain.

(82a) $MeO_2C\text{---}C\equiv C\text{---}CO_2Me$, THF, reflux (939) + (941)

(940) $MeO_2C\text{---}C\equiv C\text{---}CO_2Me$ + CO_2Me / CO_2Me

'Benzyne' forms an analogous adduct (942) (3%), but this is accompanied by a product of structure (943) (4%)[998] (*cf.* adducts of tetracyanoethylene (below) and chlorosulphonyl isocyanate (p. 285)).

(942) (943)

The *trans*-bicyclo[7.2.0]undecatriene system (944) results from the addition of tetracyanoethylene to bicyclo[6.1.0]nonatriene and *anti*-9-substituted-derivatives in tetrahydrofuran (preferably at room-temperature) (table 116). This type of product is apparently formed *via* the Z,Z,E,Z-cyclononatetraene (871; $R^2 = H$) (see scheme 87);[1002] the reaction of this with tetracyanoethylene may possibly proceed *via* the dipolar intermediate (945), which may possess bis(homotropylium) stabilisation.[999, 1000] *syn*-9-Substituted derivatives of bicyclo[6.1.0]nonatriene do not react,[1003] since they cannot attain the necessary conformation for the Cope rearrangement leading to system (871).

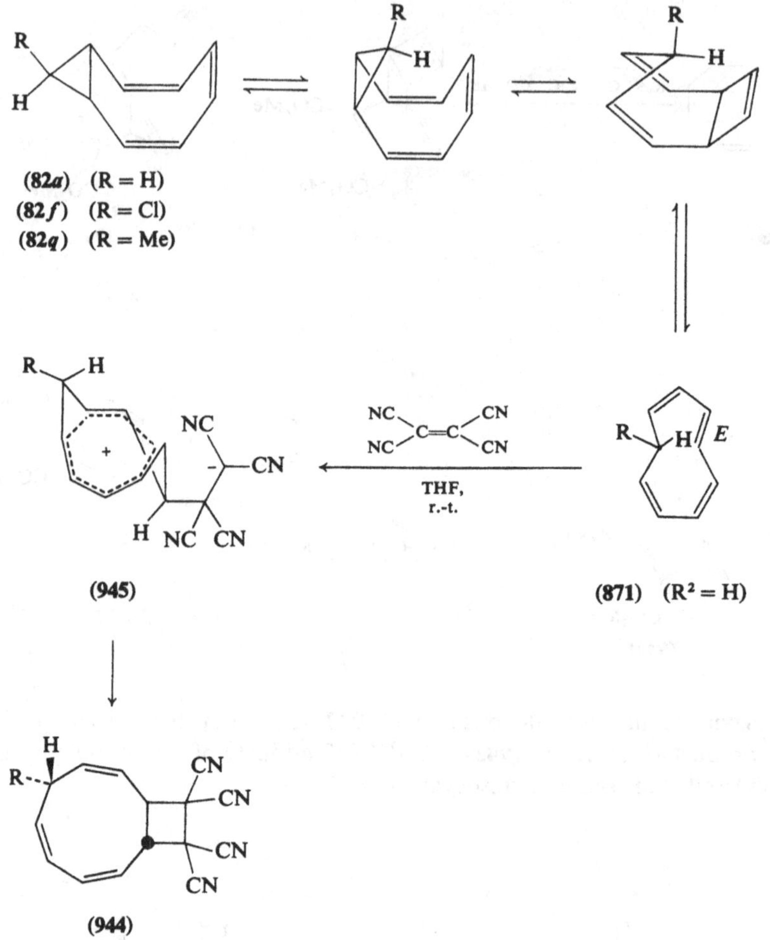

(82*a*) (R = H)
(82*f*) (R = Cl)
(82*q*) (R = Me)

(945)

(871) (R² = H)

(944)

Table 116

R	Yield (%)	Ref.
H	Up to 75	743, 996–1000
Me	69	999, 1000
Cl	—ᵃ	999, 1000

ᵃ The major product is a [4 + 2] adduct
of the dihydroindene (see table 114).

Similarly, *anti*-9-substituted bicyclo[6.1.0]nonatrienes on treatment with chlorosulphonyl isocyanate, followed by dechlorosulphonylation, afford the *trans*-fused β-lactams (946)[578,1004] (table 117). Kinetic studies appear to favour a mechanism similar to that suggested for the addition of tetracyano-ethylene[1005] (see above). (For analogous reactions with the 1-, 2-, 3- and 4-

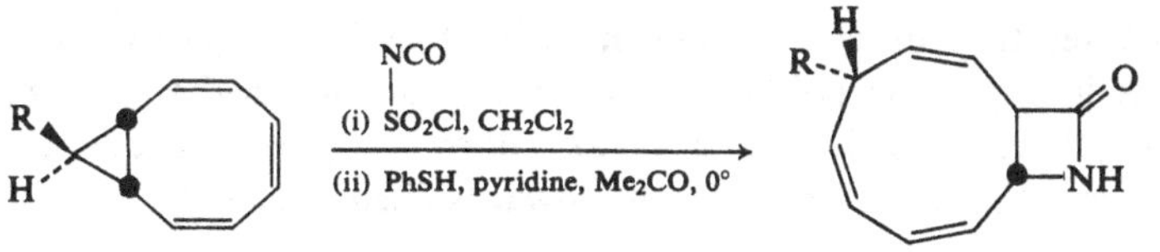

(82*a*) (R = H)   (946)
(82*f*) (R = Cl)
(82*q*) (R = Me)

Table 117

R	Reaction (i), temp.	Yield (%)
H	0° → 25°	60
Me	Reflux	56
Cl[a]	r.-t.	15

[a] Starting material contaminated with the *syn*-isomer.

methyl-derivatives (912)–(915), see refs. 578, 1003; the 2-methyl-compound gives a by-product (947) (5%).)

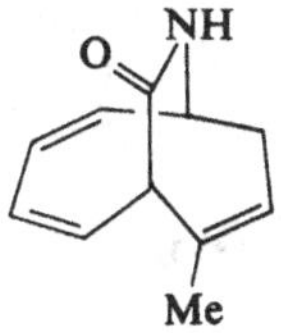

(947)

The 4-methoxy-compound (948) affords the product (949) (33.5%), which is readily converted into the ketone (950) (97%) in the presence of acid.[578]

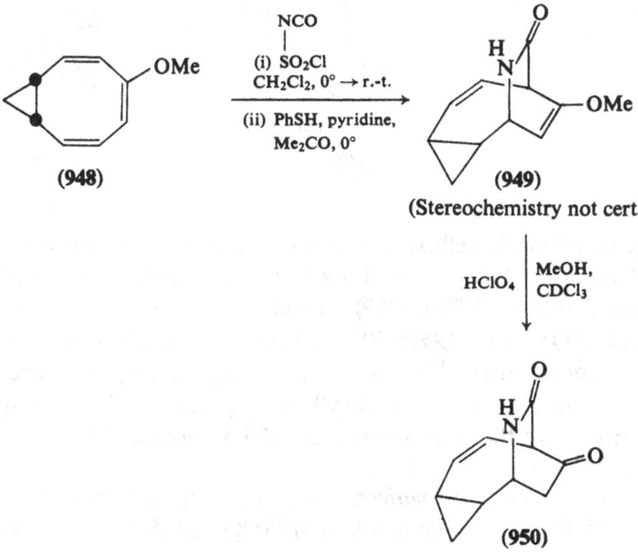

(948)   (949)
                              (Stereochemistry not certain)

(950)

The reaction of cyclopentadiene with bicyclo[6.1.0]nonatriene at 160° results in a product (14%) identical with that obtained from *cis*-dihydro-indene;[998] its original formulation as a norbornene derivative (951)[998] is not consistent with its lack of reactivity towards phenyl azide.[164]

(951)

It is likely that this reaction produces (952) and/or (953), partly if not entirely *via* Cope rearrangement of initially formed (954) and/or (955) (*cf.* hemicyclone below).

(82*a*)   (954)   (955)

(952)   (953)

With hemicyclone in refluxing benzene, bicyclo[6.1.0]nonatriene forms a product (956) or (957) (25%) resulting from Cope rearrangement of the *cis*-dihydroindene-adduct (958) or (959); in addition, adducts tentatively assigned structures (960) (45%) and (961) (30%) are produced, presumably arising from Z,Z,E,Z-cyclononatetraene[1006] (see scheme 87). Analogous products are apparently formed from 1,3-diphenylbenzo[c]furan.[1007] With α-pyrone at *ca.* 70°, a mixture of the adducts (962) and (963) is formed.[1008]

Reactions with metal derivatives. Bicyclo[6.1.0]nonatriene reacts with $Mo(CO)_3 . 3THF$ to give the complex (964) (90%); at 125° the ligand rearranges

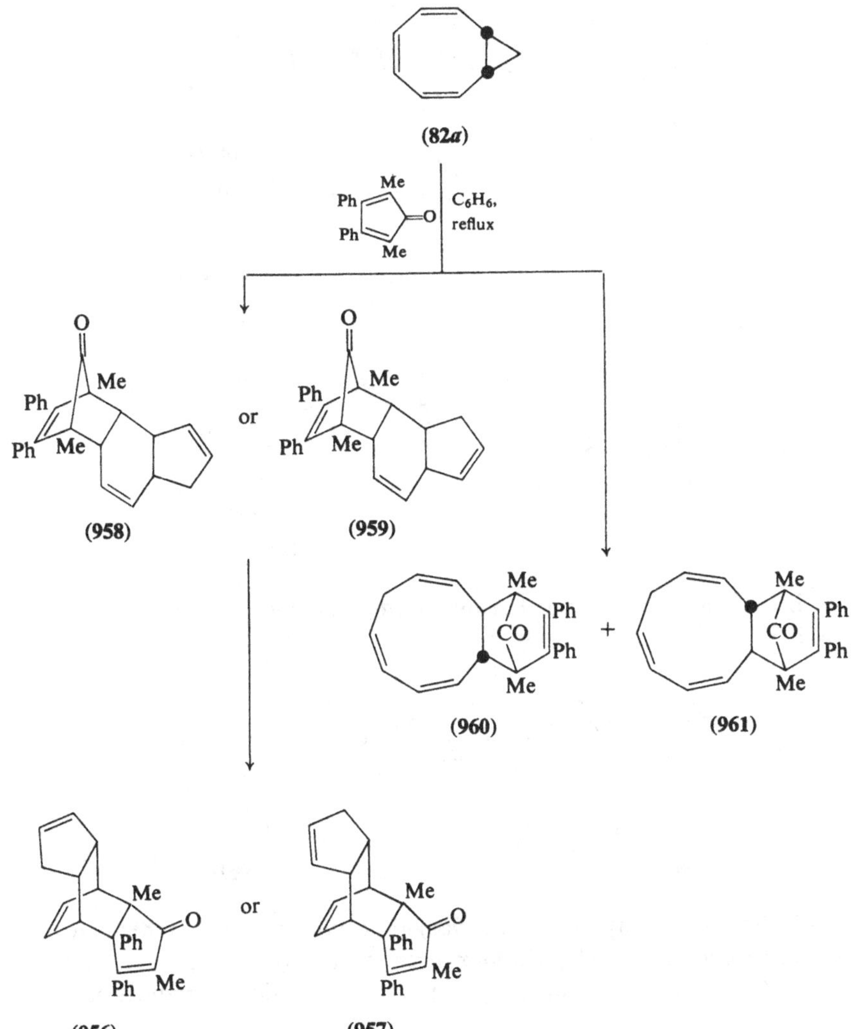

(82*a*)

Ph Me / Ph Me C₆H₆, reflux

(958)   or   (959)

(960)   +   (961)

(956)   or   (957)

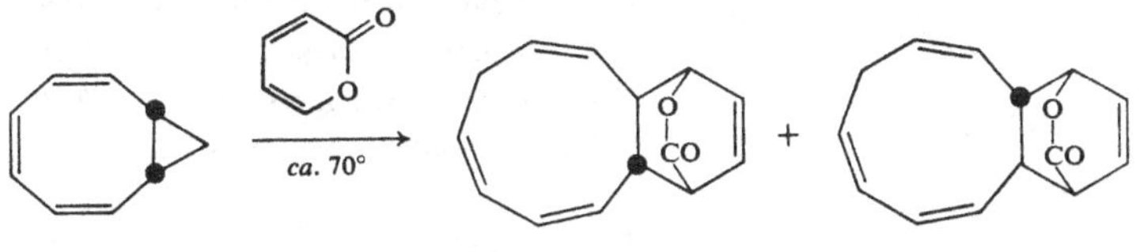

(82*a*)   *ca.* 70°   (962)   +   (963)

to form the bicyclo[4.2.1]nonatriene complex (965) (55%), which may be demetallated with diethylenetriamine to afford the hydrocarbon (966) (74%).⁹⁵⁰ Protonation of the molybdenum tricarbonyl complex (964) in a 'super-acid' medium occurs on the molybdenum atom, yielding a cation formulated as (**967**).[1009]

The iron tricarbonyl complexes (**968**)–(**971**) (yields in table 118) are formed by the reaction of bicyclo[6.1.0]nonatriene with $Fe(CO)_5$ under the influence of u.v. light.[1010] These complexes, together with additional products including

(Bicyclononatriene)Fe(CO)$_3$ (**968**)
+ (Cyclononatetraene)Fe(CO)$_3$ (**969**)
+ (Dihydroindene)Fe(CO)$_3$ (**970**)
+ (Bicyclononatriene)Fe$_2$(CO)$_6$ (**971**)

Table 118

Product	Yield (%)
(**968**)	15
(**969**)	35
(**970**)	6
(**971**)	1

(972), result from the thermal reaction of bicyclo[6.1.0]nonatriene with $Fe_2(CO)_9$ (table 119). The structures proposed[1010,1011] for these iron carbonyl

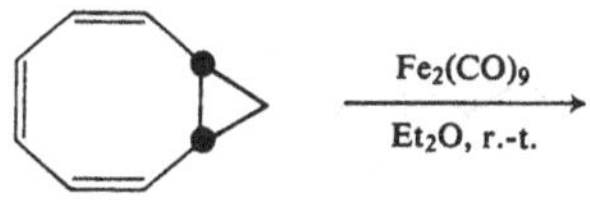

(82a)

$\xrightarrow[\text{Et}_2\text{O, r.-t.}]{\text{Fe}_2(\text{CO})_9}$ (968) + (969) + (970) + (971) + $(C_9H_{10})Fe(CO)_4$ (972)

+ two other products (not obtained pure)

Table 119

Product	Yield (%)	Ref.
(968)	15	1010
(969)	12 (7–9)	1010, (1011)
(970)	46 (30–32)	1010, (1011)
(971)	14 (20–25)	1010, (1011)
(972)	8–10	1011

complexes are shown below (for the ^{13}C n.m.r. spectrum of (971), see ref. 1012).

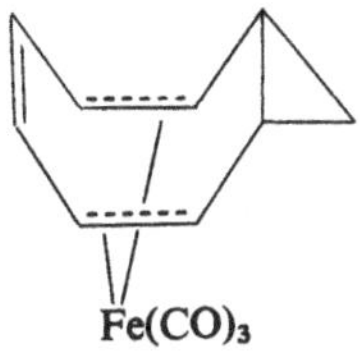

$Fe(CO)_3$ (968) $Fe(CO)_3$ (969) $Fe(CO)_3$ (970)

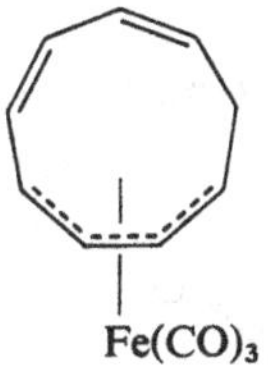

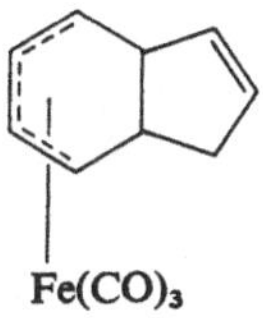

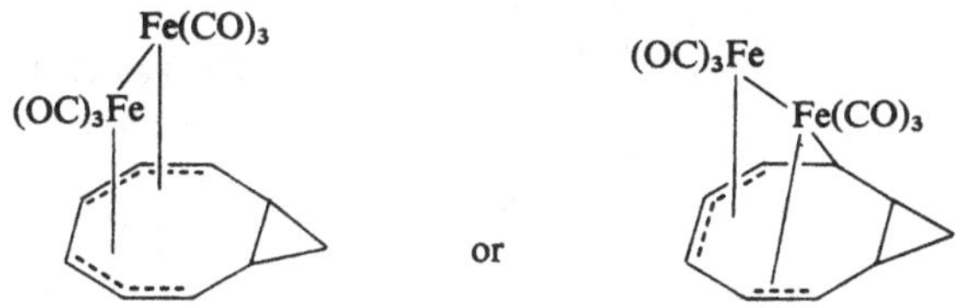

(971)

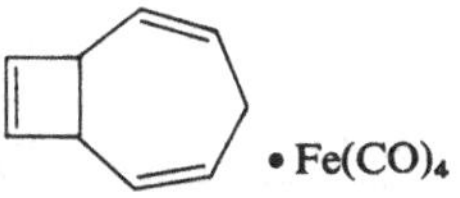

(972) · $Fe(CO)_4$

A thermal reaction with (benzylidene-acetone)$Fe(CO)_3$ leads to a mixture of the complexes (973), (970) and (969) (ratio *ca.* 10:5:1, total yield *ca.* 50%).[1013]

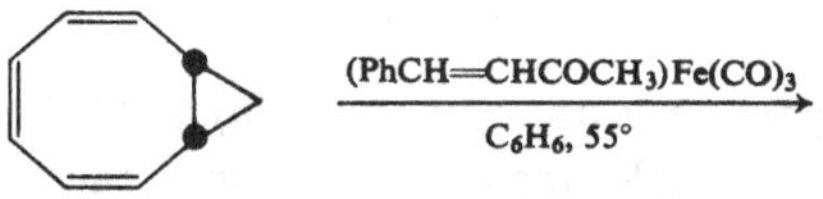

(82a)

$\xrightarrow[\text{C}_6\text{H}_6,\ 55°]{(\text{PhCH}=\text{CHCOCH}_3)\text{Fe}(\text{CO})_3}$ (Tricyclononadiene)$Fe(CO)_3$ (973)

+ (970) + (969)

The principal product (973) is a derivative of the valence tautomer (937) (p. 282) of bicyclo[6.1.0]nonatriene; the free hydrocarbon may be obtained by low-temperature demetallation.[1013]

(973)   (937)

Complex (969) rearranges at 101° to give the dihydroindene derivative (970).[1010] Complex (971) also undergoes thermal rearrangement to give (970) as the main product (61%); however, other products include (974) (1.5%), (975) (16%), and (976) (3%).[1014]

(971) $\xrightarrow[\text{Reflux}]{\text{Toluene}}$ (970) +

(OC)$_3$Fe —— Fe(CO)$_2$

(974)

(OC)$_2$Fe⸺⸺Fe(CO)$_2$

+   + [(C$_9$H$_9$)Fe(CO)$_2$]$_2$

Me

(975)   (976)

Complex (969) protonates to form the species (977).[1010]

$\xrightarrow[-120°]{\text{FSO}_2\text{OH, SO}_2\text{ClF}}$

Fe(CO)$_3$ Fe(CO)$_3$

(969)   (977)

9-Methoxybicyclo[6.1.0]nonatriene forms analogues of (970) and (971).[1015] Bicyclo[6.1.0]nonatriene and its 9,9-dimethyl-derivative react with Rh(CO)$_2$(acac) to yield complexes of structure (978)[510,965] (table 120).

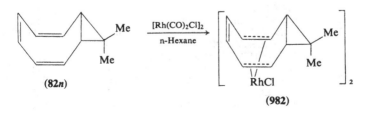

(82)   (R = H, Me)   Rh(acac)
                                                            (978)

Table 120

R	Yield (%)
H	77.5
Me	59

Reaction of bicyclo[6.1.0]nonatrienes with $Rh(CO)_2$(trifluoro-acac) gives the analogous complexes (979) (88.5–97%); these may be converted by catalytic hydrogenation into the corresponding bicyclo[6.1.0]nona-2,6-diene complexes

(82)   *a*: $R^1 = R^2 = H$
       g: $R^1 = H, R^2 = Me$
       q: $R^1 = Me, R^2 = H$
       n: $R^1 = R^2 = Me$

(979)   Rh
        (trifluoro-acac)

(981)   (980)   Rh
                      (trifluoro-acac)

(980) (87–90.5%), which on treatment with aqueous potassium cyanide afford the hydrocarbons (981) (68–85%).[1016] The 9,9-dimethyl-compound (82*n*), with $[Rh(CO)_2Cl]_2$, gives the dimeric complex (982) (82%).[965]

(82*n*)   RhCl   (982)

Scheme 91

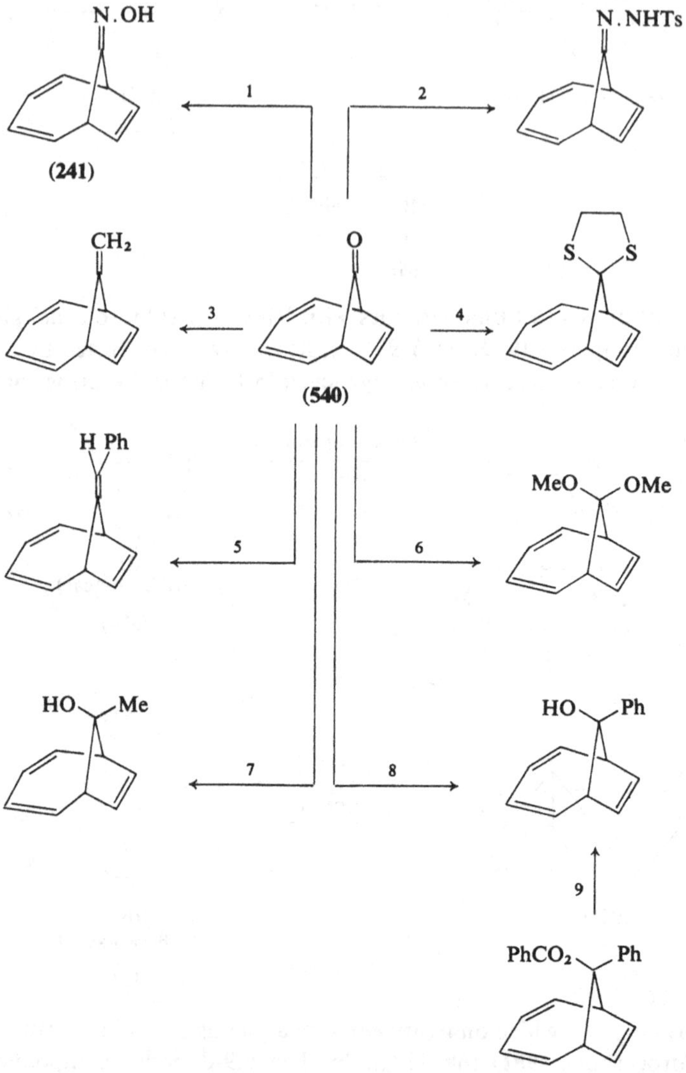

Reaction	Conditions	Yield (%)	Ref.
1	$NH_2OH(aq.)$, 0°	79	52
2	NH_2NHTs, MeOH, reflux	85	52
3	$Ph_3P=CH_2$, e.g. DMSO, 74°	38–77	758, 1017–1019
4	$HSCH_2CH_2SH$, $BF_3 . Et_2O$, 0°	79	52
5	$Ph_3P=CHPh$, THF, hexane, r.-t.	62	1020
6	$(MeO)_3CH$, NH_4NO_3	—	1021
7	MeMgBr (MeMgI), Et_2O, 0°	(77), 85	(758), 1017
8	PhLi, Et_2O, 5° → r.-t.	37	52
9	'Vigorous alkaline hydrolysis'	—	1022

12. Bicyclo[4.2.1]nona-2,4,7-trienes

Bicyclo[4.2.1]nona-2,4,7-triene (**966**) is not obtainable directly from COT, but may be prepared *via* bicyclo[6.1.0]nona-2,4,6-triene (see p. 288).

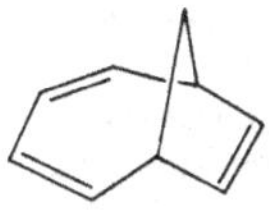

(**966**)

Bicyclo[4.2.1]nona-2,4,7-trien-9-one (**540**) is readily available from COT^{2-} and dimethylcarbamoyl chloride (see p. 181). For other derivatives, see pp. 176–80.

Functional group transformations. See schemes 91 and 92.

Scheme 92

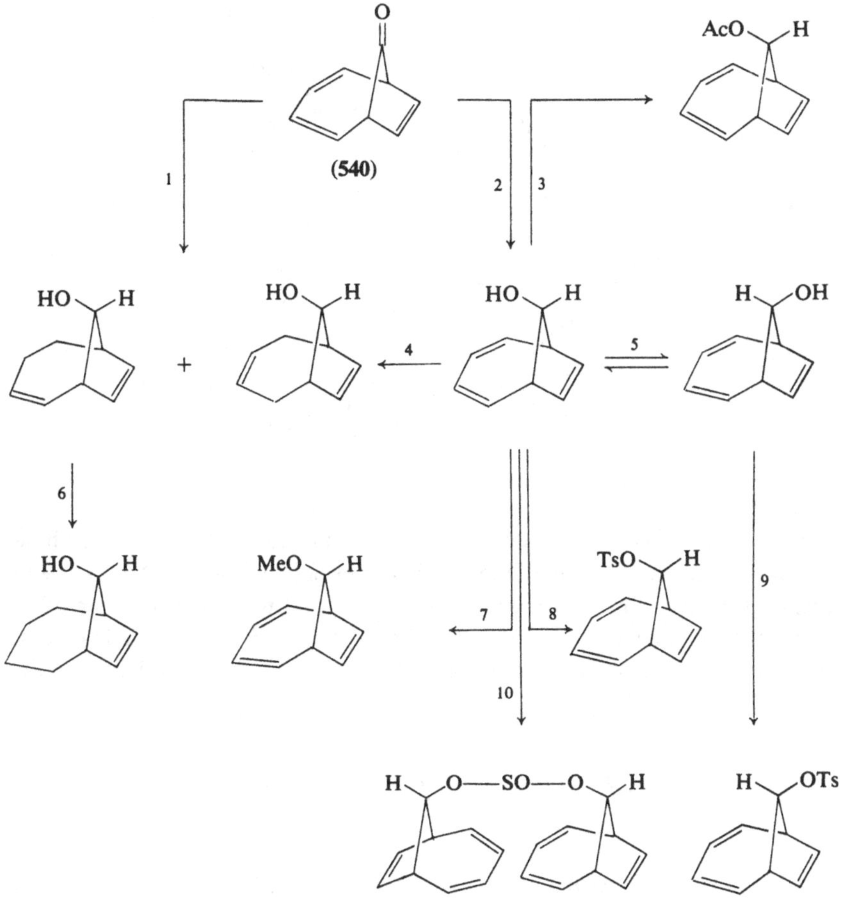

Reaction	Conditions	Yield (%)	Ref.
1	LiAlH$_4$, Et$_2$O, −78°	100[a]	757
2	NaBH$_4$, e.g. MeOH, 0° → r.-t.	74–95	52, 757, 758
3	Ac$_2$O, pyridine, reflux	70	52
4	LiAlH$_4$, Et$_2$O, e.g. reflux	—	757, 758, (1023)
5	Al(OPr^i)$_3$, Pr^{i}OH, Me$_2$CO, 150°	—	1024
6	LiAlH$_4$, THF, reflux	(95)	757, (1023)
7	MeI, NaH, MeOCH$_2$CH$_2$OMe, r.-t.	80	52
8	p-Me.C$_6$H$_4$.SO$_2$Cl, pyridine, e.g. −20°	86	52, (757, 1023)
9	—	—	1024
10	SOCl$_2$, pyridine, hexane, 0° → reflux	93	52

[a] Product ratio 4.4 : 1.

Thermal and other rearrangements. At temperatures above 290° bicyclo[4.2.1]-nonatriene (**966**) rearranges to the *cis*-dihydroindene (**938**), probably *via* an intramolecular Diels–Alder reaction and two successive [1,5] (homodienyl) hydrogen shifts.[1025]

The esters (**983**) rearrange thermally to give indenes (**984**) (table 121). These rearrangements may involve intermediate bicyclo[4.2.1]nona-2,4,7-trien-9-yl cations (**985**),[1017] and it is convenient at this point to collect together other rearrangements which may proceed *via* this species. (For mechanistic discussions, and the possible 'homoaromaticity' and 'bicycloaromaticity' of the

Table 121

R¹	R²	Conditions	Yield (%)	Ref.
MeCO	Me	Dimethyl sebacate, 200°	62	377
PhCO	Ph	*o*-Dichlorobenzene, reflux	68	377
p-Me.C₆H₄.SO₂	H	(G.l.c.)	—	52
p-Me.C₆H₄.SO₂	D	Dimethyl sulphoxide, 74°	>74	1017

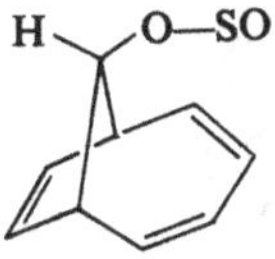	H	(G.l.c.)	—	52

cations (985), see refs. 1017, 1023, 1024, 1026.) Such rearrangements include those listed in table 122. A related process is the acid-catalysed rearrangement

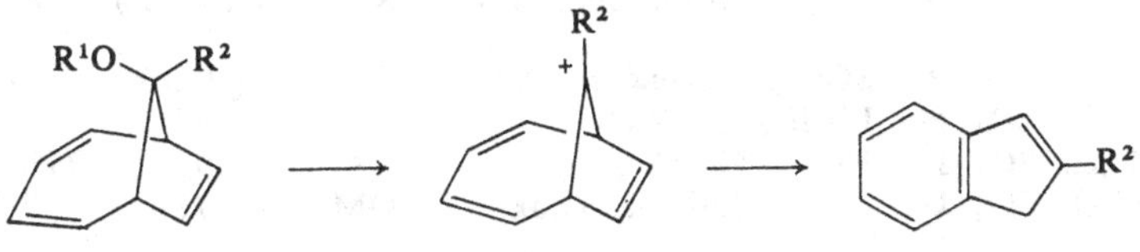

(983)   (985)   (984)

Table 122

R¹	R²	Reagents and conditions	Yield (%)	Ref.
H	Me	*p*-Me.C₆H₄.SO₂OH, C₆H₆, 74°	>80	1017, (1022)
p-Me.C₆H₄.SO₂	H	NaI, Me₂CO, reflux	59	52

of the 9-methylene-derivative (986) to 2-methylindene (>80%).[1017] (For the protonation of (986) in a 'superacid' medium at −135°, see ref. 1027.) The

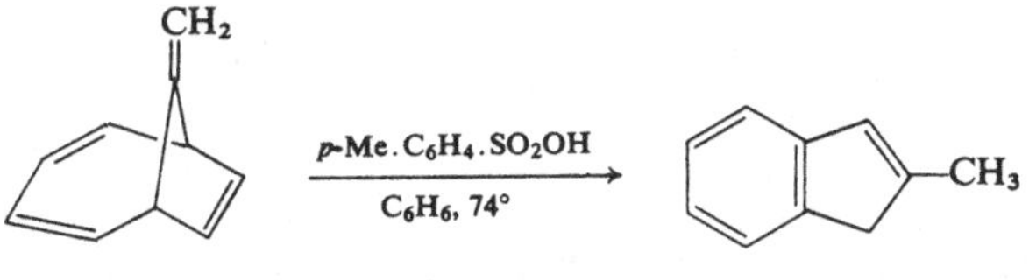

(986)

ketone (540), however, yields indan-1-one (39%), on treatment with boron trifluoride etherate[52] (or toluene-*p*-sulphonic acid[1017]).

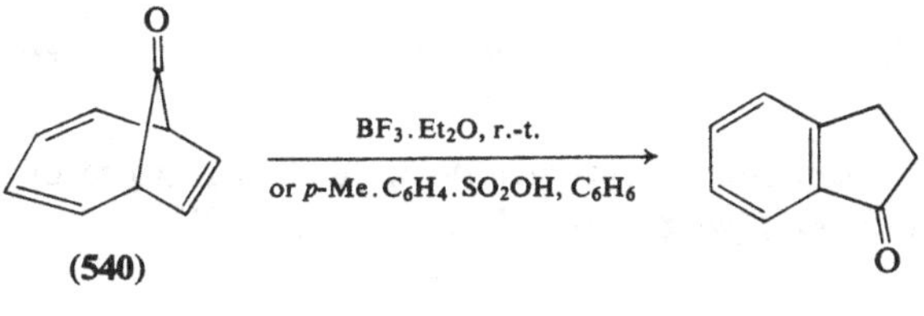

(540)

An alternative rearrangement mode, involving nucleophile participation, is observed with some systems, and may lead to products of structure (**987**) (table 123).

(**983**)   (**987**)

Table 123

R¹	R²	Reaction conditions	X	Yield (%)	Ref.
Hᵃ	Ph	SOCl₂, pyridine, Et₂O, r.-t.	Cl	(High)	1022
p-Me.C₆H₄.SO₂	H	LiAlH₄, Et₂O, 0°	H	85	52
p-Me.C₆H₄.SO₂	H	HOAc, NaOAc, 50°	OAc	99	757, 1023
p-Me.C₆H₄.SO₂	Hᵇ	MeOH, lutidine, reflux	OMe	77	1024

ᵃ The deuteriated compound (**988**) affords the product (**989**).[1028]

ᵇ The stereoisomer (**990**) gives a virtually identical result; the deuteriated derivative (**991**) affords (**992**).[1024]

(**988**)   (**989**)

(**990**)   (**991**)   (**992**)

Solvolysis of the *p*-nitrobenzoates (**993**) yields a mixture of the products (**994**), (**995**) and (**996**) (*ca.* 80, 10 and 10% severally).[757]

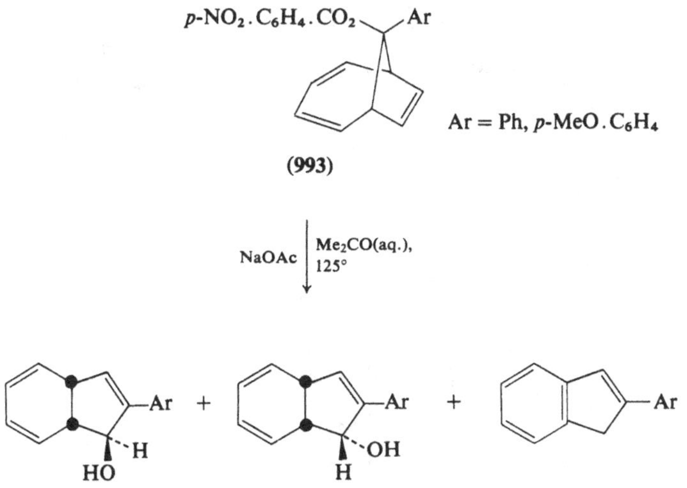

p-NO$_2$.C$_6$H$_4$.CO$_2$ Ar

Ar = Ph, p-MeO.C$_6$H$_4$

(993)

NaOAc | Me$_2$CO(aq.), 125°

Ar + Ar + Ar

HO ·H | OH H

(994) (995) (996)

Thermolysis of the sodium salt of the tosylhydrazone (997) affords indene (40%).[52]

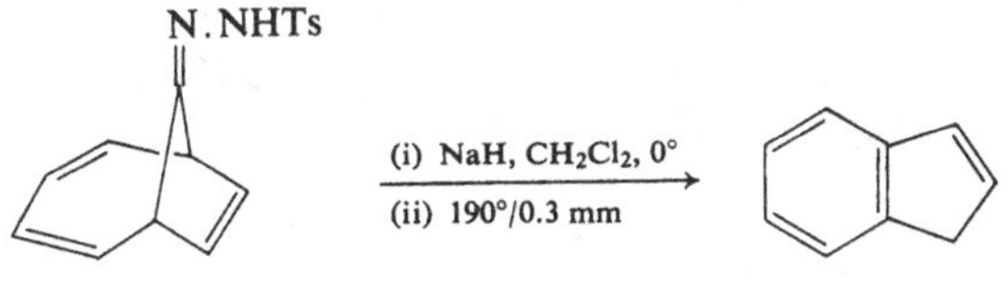

N.NHTs

(i) NaH, CH$_2$Cl$_2$, 0°

(ii) 190°/0.3 mm

(997)

An abnormal Beckmann rearrangement of the oxime (241) occurs on treatment with toluene-p-sulphonyl chloride, with the formation of cyano-cyclo-octatetraene (57%).[52]

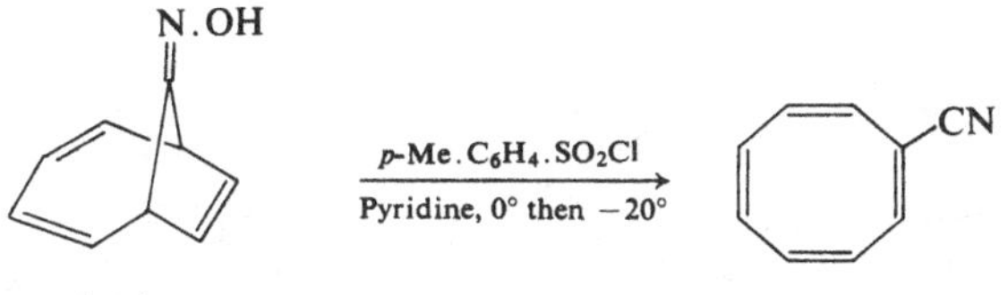

N.OH

p-Me.C$_6$H$_4$.SO$_2$Cl

Pyridine, 0° then −20°

CN

(241)

Photolysis. U.v. irradiation of bicyclo[4.2.1]nonatriene in acetone[1029] (or dioxan[1030]) affords the *endo-* and *exo-*tricyclic dienes (998) and (999) (20% and 13% respectively[1029]). In the presence of benzophenone, however, barbaralane (1000) is produced (up to 65%).[1031]

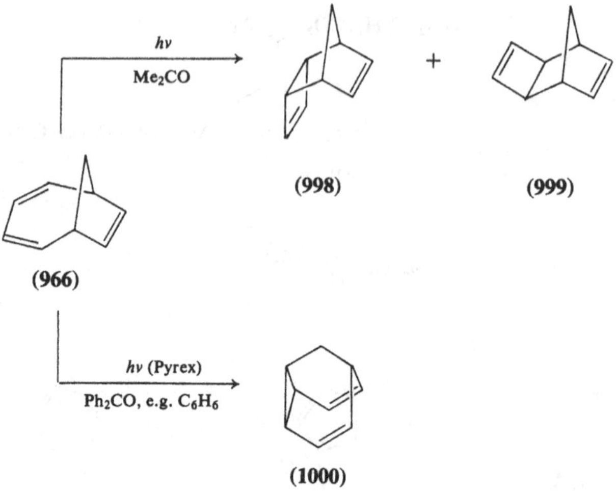
hν
Me2CO
+
(998)
(999)
(966)
hν (Pyrex)
Ph2CO, e.g. C6H6
(1000)

Non-sensitised irradiation of the *syn*-9-hydroxy-derivative (**1001**) gives the *endo*-tricyclic diene (**1002**) (50–60 %).[1032]

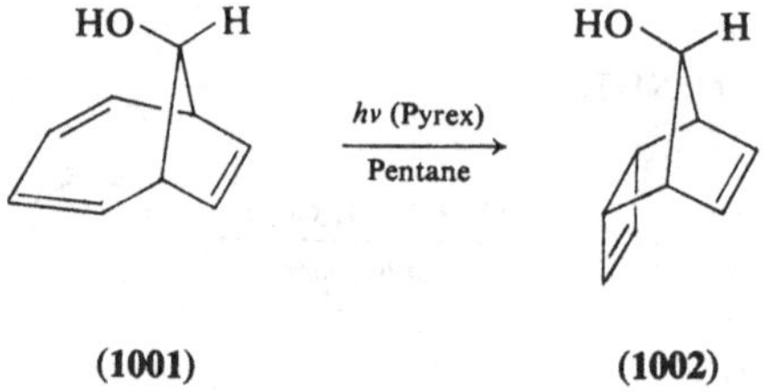
HO
H
hν (Pyrex)
Pentane
HO
H
(1001)
(1002)

In the absence of a sensitiser the 9-oxo-compound (**540**) affords COT as the main product (80–82 %),[51, 52] together with minor amounts of the norbornenone (**1003**) and barbaralone (**1004**).[52] In the presence of benzophenone, fluorenone or Michler's ketone, however, barbaralone is the principal product (up to 78.5 %).[51,52,604]

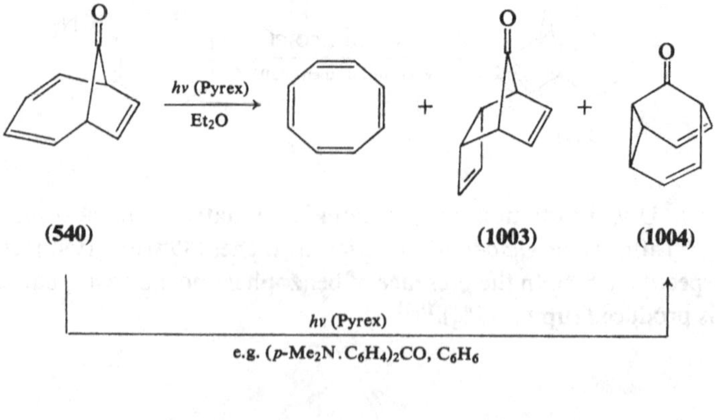
O
hν (Pyrex)
Et2O
+
O
+
O
(540)
(1003)
(1004)
hν (Pyrex)
e.g. (p-Me2N.C6H4)2CO, C6H6

Photolysis of the sodium salt of the tosylhydrazone (**997**) leads to the *syn*-9-methoxy-compound (**1005a**) (44%) and the dihydroindene (**1006**) (39%).[1024]

(**997**)   (**1005**)   (**1006**)

a: X = H
b: X = D

(In MeOD, the observed products were (**1005b**) and a 1:1 mixture of (**1007**) and (**1008**)[1024].)

(**1007**)   (**1008**)

Reduction. For examples of catalytic hydrogenation of the bicyclo[4.2.1]nonatriene system, see refs. 377, 756.

For partial reduction of the conjugated diene system in the 9-oxo-derivative (**540**) by lithium aluminium hydride, see scheme 92 (p. 293).

Reactions with diazoalkanes. Reaction of the ketone (**540**) with diazomethane affords the ring-expanded product (**1009**) (63%), together with the epoxide (**1010**) (27%).[52,1033] For the analogous reaction with diazoethane, see ref.

(**540**)   (**1009**)   (**1010**)

1034. The ketone (**1009**) is a useful source of bicyclo[4.2.2]decatetraene systems.[1033,1034]

Cycloadditions. Reaction of bicyclo[4.2.1]nonatriene with chlorosulphonyl isocyanate, followed by dechlorosulphonylation, gives the *exo*-β-lactam (**1011**) (80%).[1020]

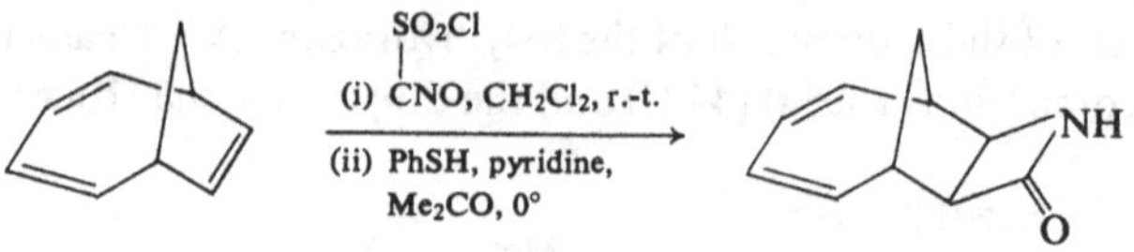

(966)   (1011)

The ketone (540) adds α-pyrone to afford the product (1012) (34%), which on thermal decarboxylation gives the tetraene (1013) (70%).[1019]

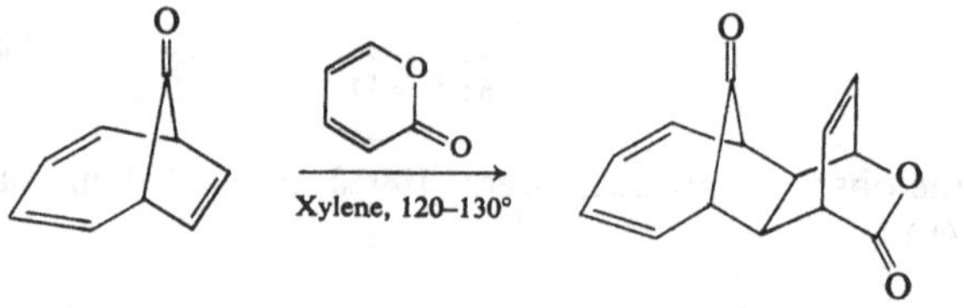

(540)   (1012)

(Stereochemistry uncertain)

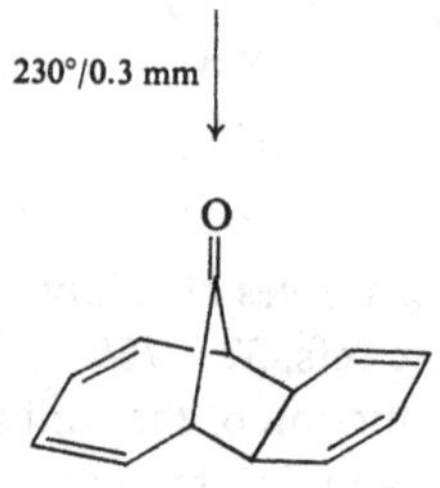

230°/0.3 mm

(1013)

13. Bicyclo[6.2.0]deca-2,4,6-trienes

The parent hydrocarbon (515a) may be prepared from COT^{2-} and 1,2-dibromoethane (see p. 175).

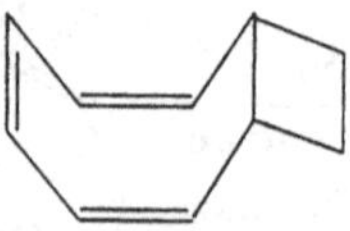

(515a)

9,9,10,10-Tetrahalo-derivatives of the bicyclo[6.2.0]deca-2,4,6-triene system are available from the [2 + 2] cycloaddition of 1,1-dichloro-2,2-difluoro- and chlorotrifluoroethylene to COT (see p. 39).

Functional group transformations. Transformations of 9,9,10,10-tetrahalo-derivatives are shown in schemes 93 and 94.

Scheme 93

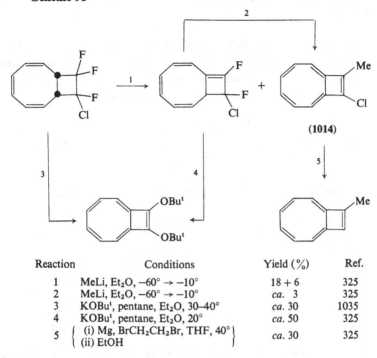

(1014)

Reaction	Conditions	Yield (%)	Ref.
1	MeLi, Et$_2$O, −60° → −10°	18 + 6	325
2	MeLi, Et$_2$O, −60° → −10°	ca. 3	325
3	KOBu^t, pentane, Et$_2$O, 30–40°	ca. 30	1035
4	KOBu^t, pentane, Et$_2$O, 20°	ca. 50	325
5	(i) Mg, BrCH$_2$CH$_2$Br, THF, 40° (ii) EtOH	ca. 30	325

Scheme 94

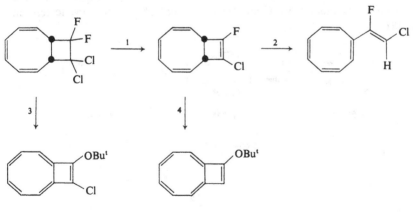

Reaction	Conditions	Yield (%)	Ref.
1[a]	MeLi, Et$_2$O, −20° to −10°	ca. 30	(324), 325
2	NaH, (Me$_2$N)$_3$PO, 5–10°	ca. 10	325
3	KOBu^t, pentane, Et$_2$O, 30–40°	50	1035
4	KOBu^t, pentane, Et$_2$O, 30–40°	ca. 10	1035

[a] Compound (**1014**) (see scheme 93) is obtained as a by-product (3%).

For further reactions of bicyclo[6.2.0]deca-1,3,5,7,9-pentaenes, see refs. 324, 325, 1035, 1036.

Thermolysis. cis-Bicyclo[6.2.0]deca-2,4,6-triene (**515a**) rearranges at 154° to give the *trans*-fused product (**1015**), which on further heating isomerises to the fully conjugated system (**1016**); the suggested mechanism involves the *Z,Z,Z,E*-cyclodecatetraene (**1017**)[1037] (*cf.* scheme 86, p. 266).

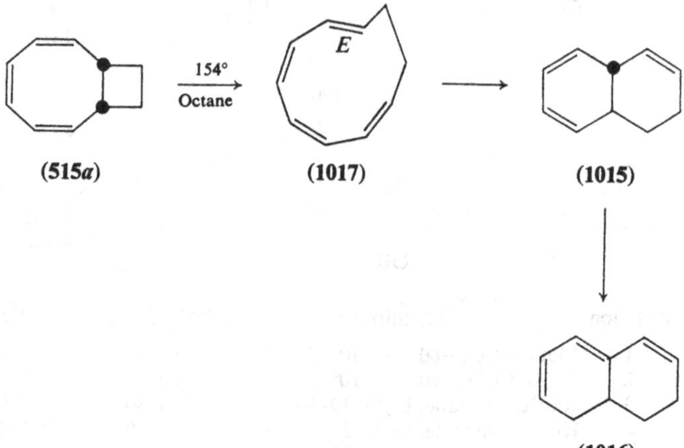

(515*a*)   (1017)   (1015)

(1016)

Photolysis. Photochemical rearrangement of *cis*-bicyclo[6.2.0]decatriene produces the *E,Z,Z,E*-cyclodecatetraene (**1018**), which equilibrates with the *trans*-fused bicyclic isomer (**1019**); at 98° *trans*-5,6-divinylcyclohexa-1,3-diene is formed, presumably by Cope rearrangement of the monocyclic tetraene (**1018**).[751]

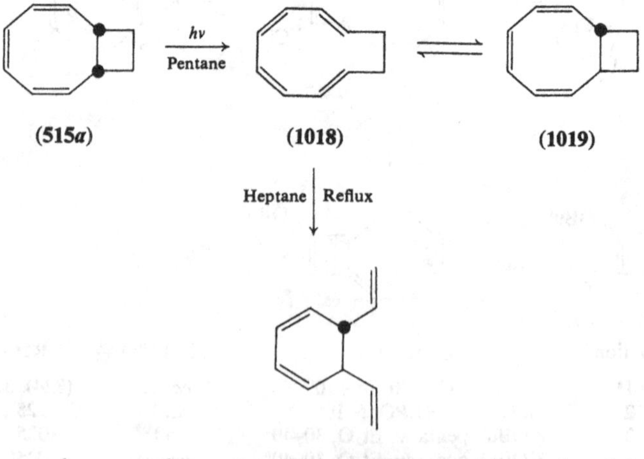

(515*a*)   (1018)   (1019)

Heptane | Reflux

Reduction and proton-abstraction. For a polarographic study of the electro-chemical reduction of the bicyclic triene (**515a**), see ref. 986.

For catalytic hydrogenation of the bicyclo[6.2.0]decatriene system, see e.g. refs. 1038, 1039.

Deprotonation of *cis*-bicyclo[6.2.0]decatriene with potassium amide in liquid ammonia gives the dianion (**1020**), which with iodine affords the cyclooctatetraene (**350**); reaction of the dianion (**1020**) with wet ether, followed by catalytic hydrogenation, yields the product (**1021**) (92%), identical with that obtained by direct hydrogenation of the starting material.[1038, 1040]

(**515a**)   (**1020**)   (**350**)

(**1021**)

Cycloadditions. Reaction of the chloromethyl-derivative (**515b**) with tetracyanoethylene yields an adduct (*ca.* 100%) assumed to possess structure (**1022**).[382]

(**515b**)

(**1022**)

The tetrasubstituted derivative (**103b**) reacts with acetylenedicarboxylic ester to give an adduct (presumably (**1023**)) (80%), which on thermolysis yields the bicyclo[2.2.0]hexene (**1024**) (65%); treatment of this with methyllithium generates the (explosive) Dewar-benzene (**1025**) (70%).[324] (The parent hydrocarbon (**515a**) does not react with chlorosulphonyl isocyanate[1003].)

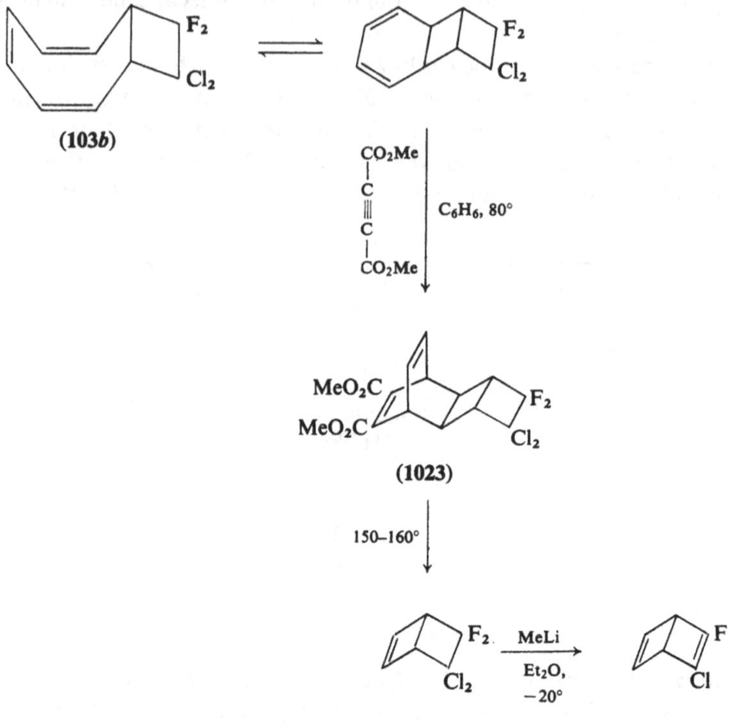

(103b)

C_6H_6, 80°

CO₂Me—C≡C—CO₂Me (drawn vertically with CO₂Me at top and bottom)

(1023)

150–160°

(1024)　　　　(1025)

$\xrightarrow[-20°]{\text{MeLi} \atop \text{Et}_2\text{O},}$

Reactions with metal derivatives. cis-Bicyclo[6.2.0]decatriene reacts with $Fe_2(CO)_9$ to form a plethora of complexes. At room-temperature the principal product is the binuclear compound (1026) (23%),[1041–1043] but at higher tem-

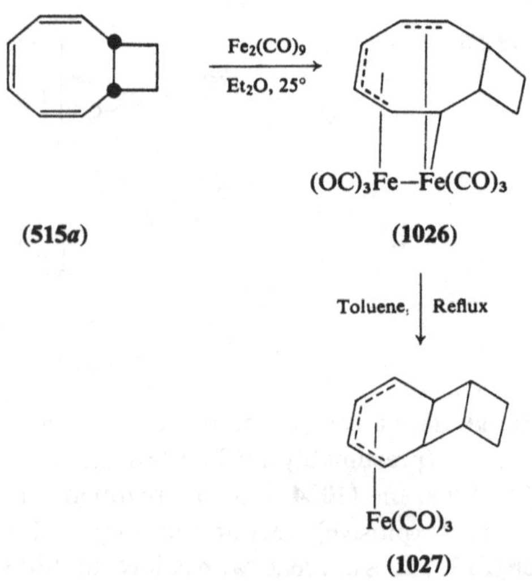

$\xrightarrow[\text{Et}_2\text{O, 25°}]{\text{Fe}_2(\text{CO})_9}$

(OC)₃Fe—Fe(CO)₃

(515a)　　　　(1026)

Toluene, Reflux

Fe(CO)₃

(1027)

peratures this is converted (mainly) into the tricyclic mononuclear derivative (**1027**) (64–66 %)[1041,1043] (this is also obtained (in very low yield) by u.v. irradiation in the presence of $Fe(CO)_5$[1013]). However, other complexes which have been isolated from this reaction (in minute yields) include (**1028**),[1041,1044] (**1029**),[1041,1043] (**1030**),[1045] (**624*a***)[1045] and (**1031**),[1045] although the last two may have resulted from contaminants in the bicyclo[6.2.0]decatriene.[1045]

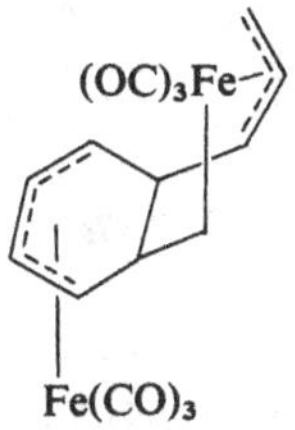
(OC)₃Fe
Fe(CO)₃

(**1028**)

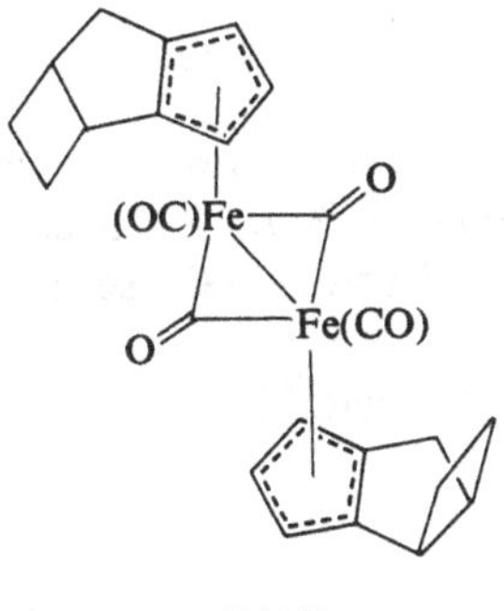
(OC)Fe
O
O
Fe(CO)

(**1029**)

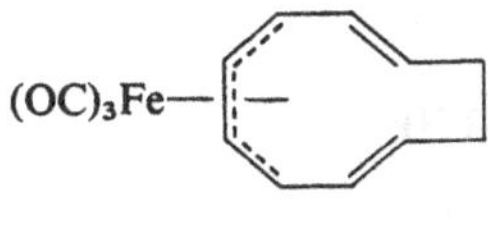
(OC)₃Fe

(**1030**)

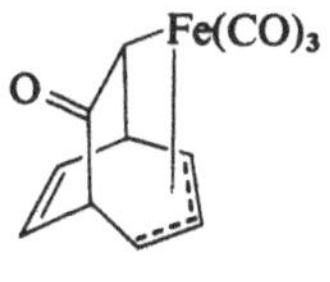
Fe(CO)₃
O

(**624*a***)

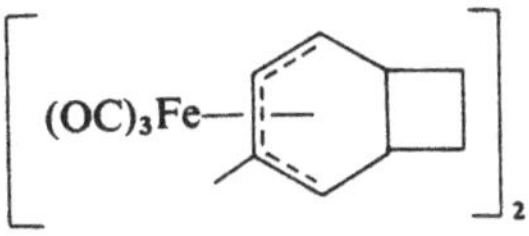
(OC)₃Fe
2

(**1031**)

With (benzylidene-acetone)$Fe(CO)_3$, the complex (**1027**) is formed in excellent yield (82 %); low-temperature demetallation then affords the tricyclic diene (**1032**) (84 %).[1013]

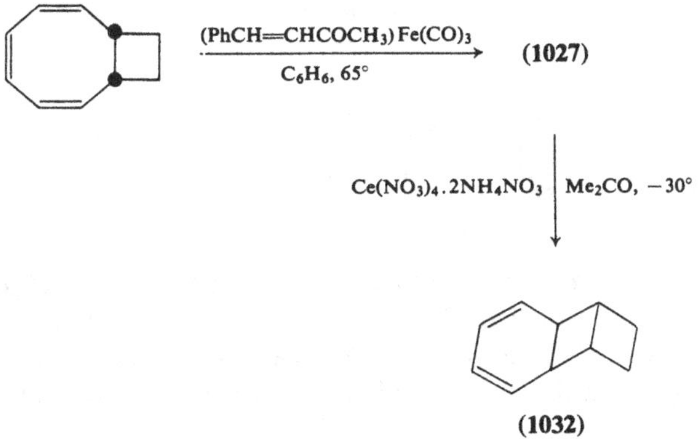
(PhCH═CHCOCH₃) Fe(CO)₃
C₆H₆, 65°
(1027)
Ce(NO₃)₄.2NH₄NO₃
Me₂CO, −30°

(**1032**)

14. Bicyclo[6.3.0]undeca-2,4,6-trienes

The bicyclo[6.3.0]undeca-2,4,6-triene system (**516a**) is formed (initially) from COT^{2-} and 1,3-dihalopropanes (see p. 176). Equilibration with the valence isomer (**517a**) is rapid at normal temperatures, the tricyclic diene (**517a**) predominating (*ca.* 97% at 58°).[752,753]

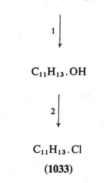

(516*a*)   (517*a*)

Functional group transformations. Transformations of the tetrahydropyranyl-oxy-derivative (**517b**) are given in scheme 95. (For dehydrochlorination of the chloro-derivative (**1033**), see ref. 382.)

Scheme 95

$$\text{(structure)} \sim\!\!\text{OTHP}$$

(517*b*)

THP = 2-tetrahydropyranyl

$$1\downarrow$$

$$C_{11}H_{13}.OH$$

$$2\downarrow$$

$$C_{11}H_{13}.Cl$$

(1033)

Reaction	Reagents and conditions	Yield (%)	Ref.
1	Conc. HCl, MeOH, reflux	51	382
2	$POCl_3$, pyridine, 0° → 100°	*ca.* 55	382

Reactions with metal derivatives. A mixture of the valence tautomers (**516a**) and (**517a**) (*ca.* 1:3) reacts with $(MeCN)_3Mo(CO)_3$ at room-temperature to give the complexes (**1034**) (33%) and (**1035**) (6.4%); the latter may be obtained in higher yield (44%) by using the pure ligand (**517a**) and a higher reaction temperature.[752,753]

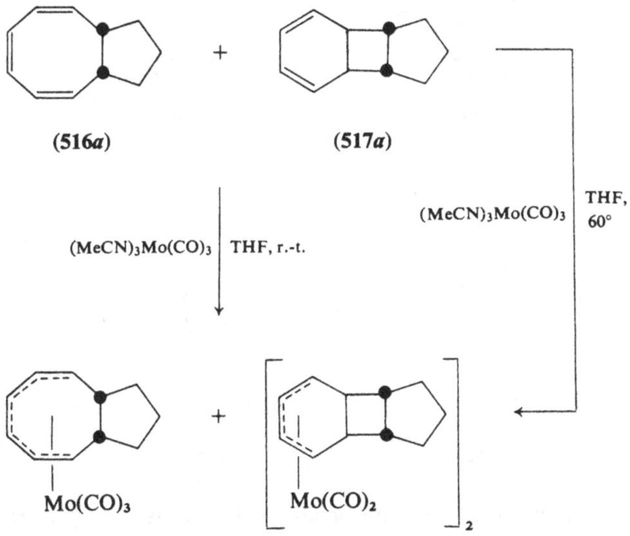

(516a)   (517a)

(MeCN)₃Mo(CO)₃ THF, r.-t.

(MeCN)₃Mo(CO)₃ | THF, 60°

Mo(CO)₃   Mo(CO)₂

(1034)   (1035)

With $Fe_2(CO)_9$, the only isolated product has structure (1036).[752,753,1046]

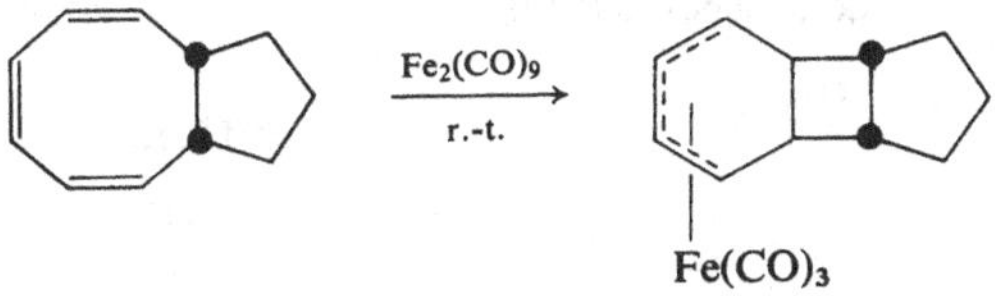

$Fe_2(CO)_9$
r.-t.

Fe(CO)₃

(516a)   (1036)

15. Bicyclo[6.4.0]dodeca-2,4,6-trienes

The parent hydrocarbon (518) is prepared from COT^{2-} and 1,4-dibromobutane (see p. 176). Valence tautomerism results in equilibration with the tricyclic diene (519) (*ca.* 50% at 114°).[752,753]

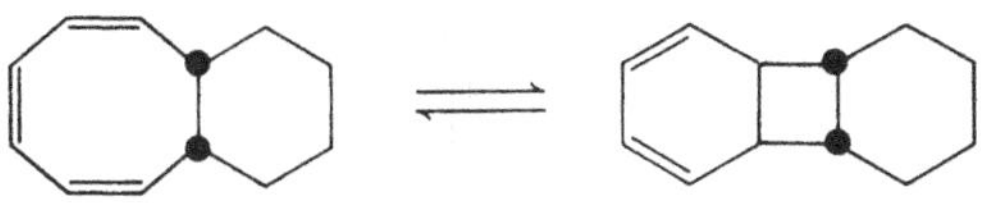

(518)   (519)

Reactions with metal derivatives. The mixture of valence tautomers (518) and (519) reacts with $(MeCN)_3Mo(CO)_3$ to afford the complexes (1037) (35%) and (1038) (9%).[752,753]

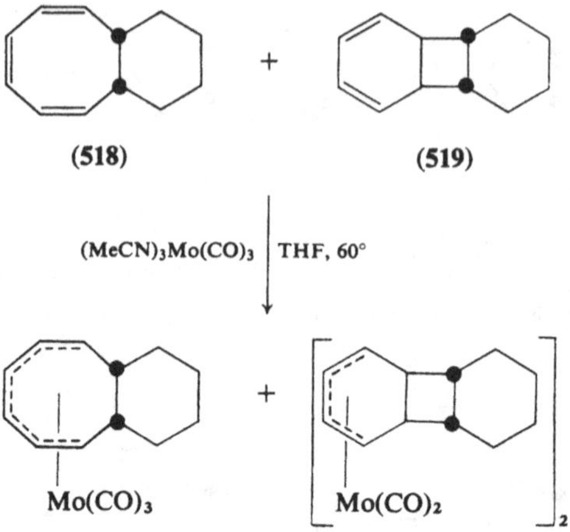

(518)   (519)

(MeCN)₃Mo(CO)₃ | THF, 60°

Mo(CO)₃   Mo(CO)₂

(1037)   (1038)

With $Fe_2(CO)_9$ (hexane, 50°) the products are (1039) (0.6%) (*cf.* (1026), p. 304) and (1040) (30%).[752, 753]

(Bicyclododecatriene)Fe₂(CO)₆

(1039)

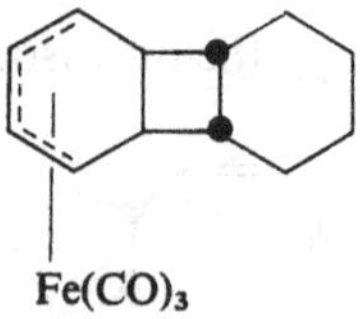

Fe(CO)₃

(1040)

16. 9-Azabicyclo[6.1.0]nona-2,4,6-trienes

The generation of nitrenes in the presence of COT leads to the 9-azabicyclo-[6.1.0]nona-2,4,6-triene system (92) (see p. 37).

Functional group transformations. See scheme 96.

Scheme 96

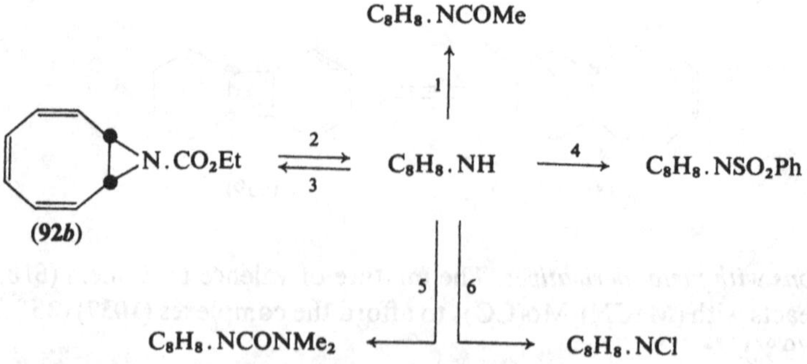

C_8H_8.NCOMe

1 ↑

N.CO₂Et $\underset{3}{\overset{2}{\rightleftharpoons}}$ C_8H_8.NH $\overset{4}{\longrightarrow}$ C_8H_8.NSO₂Ph

(92*b*)

5 | 6

C_8H_8.NCONMe₂ ←   → C_8H_8.NCl

Reaction	Reagents and conditions	Ref.
1	MeCOCl	1047
2	LiAlH$_4$, 0°	1047
3	ClCO$_2$Et	1047
4	PhSO$_2$Cl	1047
5	ClCONMe$_2$	1047
6	*N*-Chlorosuccinimide	1047

Thermolysis. The tendency for the *N*-substituted derivatives (92) to undergo thermal rearrangement to the valence isomers (1041) is very sensitive to the type of group attached to the nitrogen atom (table 124) (for a mechanistic discussion of this rearrangement, see ref. 317). This equilibrium may 'leak' (especially at

(92)   (1041)

Table 124

R	Proportion of (1041) (%)	Ref.
COMe	*ca.* 95	1048
CO$_2$Me	*ca.* 95[a]	1049
CO$_2$Et	*ca.* 95[a]	(319), 1048
CONMe$_2$	0	1047, (1050)
CN	—[b]	317
SO$_2$Ph	0	1047

[a] Starting with (1041).   [b] Polymerisation occurs.

higher temperatures) into other, irreversible rearrangement paths (see refs. 1048, 1049). Exceptionally, rearrangement of the compound (92c) occurs on hydrolysis, which leads to the 9-azabicyclo[4.2.1]nonatriene system (1042) (20%).[320]

(92c)   (1042)

Photolysis. U.v. irradiation of the ester (92b)[1051,1052] (or (1041; R = CO$_2$Et)[1051]) gives the thermally labile azonine (1043), which isomerises (even at room-temperature) to the *cis*-dihydroindole derivative (1044).[1051,1052] (For

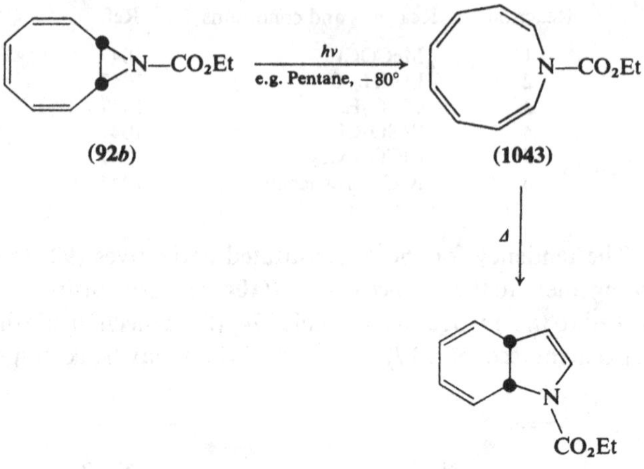

(92*b*)   (1043)

(1044)

conversion of (1043) into the parent azonine and other *N*-substituted derivatives, see refs. 1053, 1054.)

Reactions with metal derivatives. With $Rh(CO)_2$(trifluoro-acac), the ester (92*b*) affords the complex (1045) (41 %).[1055]

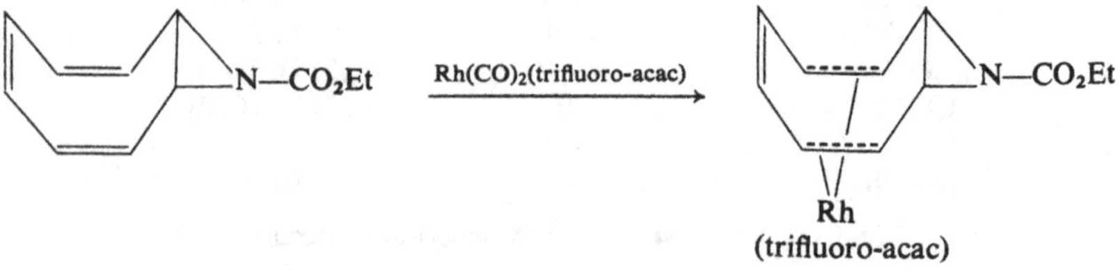

(92*b*)   (1045)

By catalytic hydrogenation and subsequent removal of the metal, complex (1045) may be converted into the diene (1046)[1055] (see pp. 291, 317).

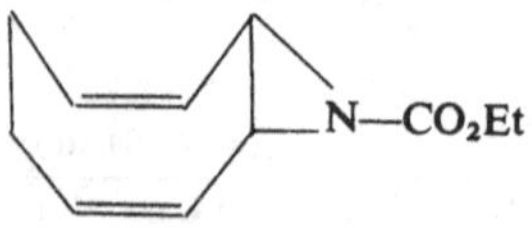

(1046)

17. 9-Oxabicyclo[6.1.0]nona-2,4,6-trienes

Treatment of COT with peracids gives the epoxide (27) (see p. 19).
For valence tautomerism, see p. 315.

Thermolysis. Thermolysis of the epoxide (27) at 260° yields 3-formylcyclohepta-1,3,5-triene (45%); at 330° this rearranges to the 1-formyl-compound (1047) (74%), and at 400° phenylacetaldehyde is formed.[936]

Photolysis. U.v. irradiation of the epoxide (27) at ordinary temperatures leads to the *cis*-bicyclic system (1048) (45%), its *trans*-fused stereomer (1049) (10%), and the cyclobutene derivative (1050) (6%), together with cycloheptatriene (12%).[1056-1059] At low temperatures all-Z-oxonin (1051) is formed,[1059,1060] together with a geometrical isomer (very likely the *Z,Z,Z,E*-compound (1052)),[1059] and these are converted thermally into the bicyclic products (1048) and (1049) respectively;[1059,1060] cycloheptatriene is known to be a photo-product of (1048).[1057,1058]

Oxidation. Further epoxidation of compound (27) gives the di- and tri-epoxides (1053), (1054), (1055) and (1056)[1061] (table 125). At *ca.* 200°, the product

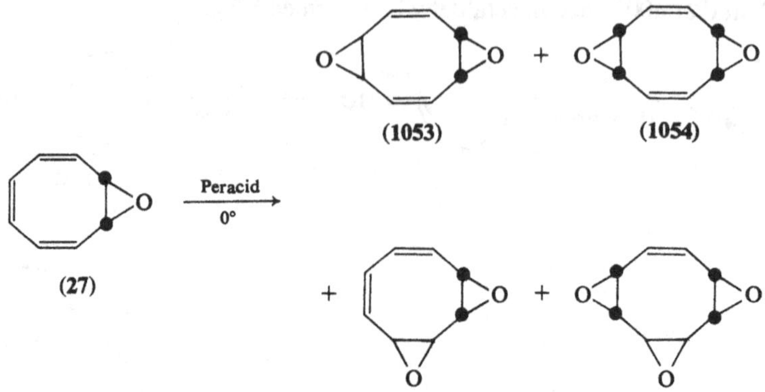

(1053)   (1054)

(27)

(1055)   (1056)

Table 125

	\multicolumn{4}{c}{Yield (%)}			
Peracid	(1053)	(1054)	(1055)	(1056)
m-Cl.C$_6$H$_4$.CO$_3$H	*ca.* 67	24	9	—
CF$_3$CO$_3$H	30	50	5	15

(1054) equilibrates with the tricyclic valence isomer (1057) (40%), whereas compound (1055) is slowly transformed into the isomer (1058).[1061]

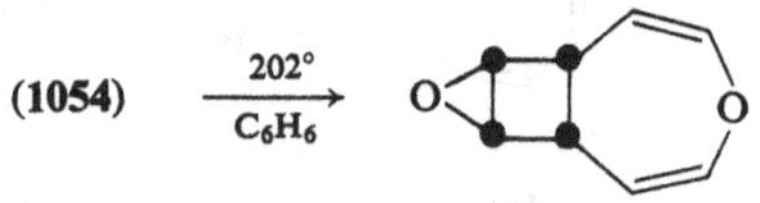

(1057)

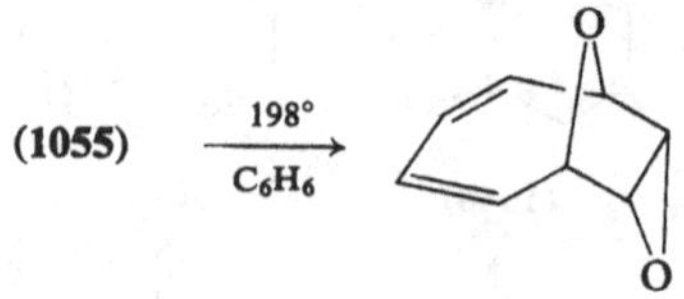

(1058)

Protonation. The epoxide (27) is very sensitive to acids, rapid rearrangement to phenylacetaldehyde occurring *via* 7-formylcyclohepta-1,3,5-triene.[1062,1063] The first, acid-catalysed stage may occur as follows,[909,1063] the hydroxyhomo-tropylium ion (1059) being detectable at −75° in fluorosulphonic acid. [909]

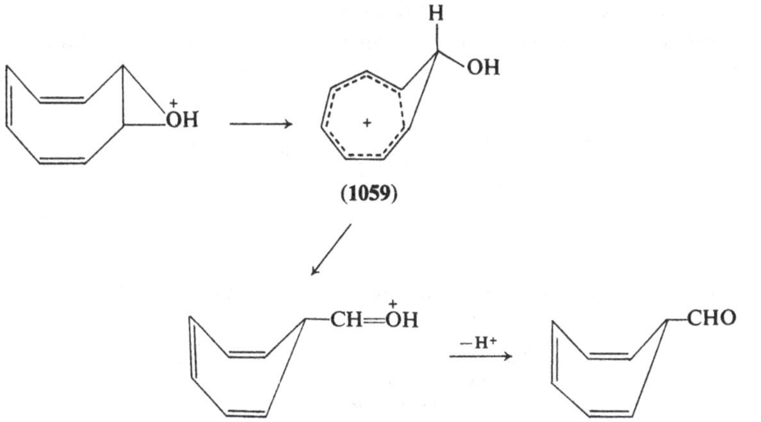

H
OH
+
OH
+
(1059)
+
CH=OH
−H⁺
CHO

The second stage of the rearrangement does not require a catalyst and proceeds slowly even at 0°.[965]

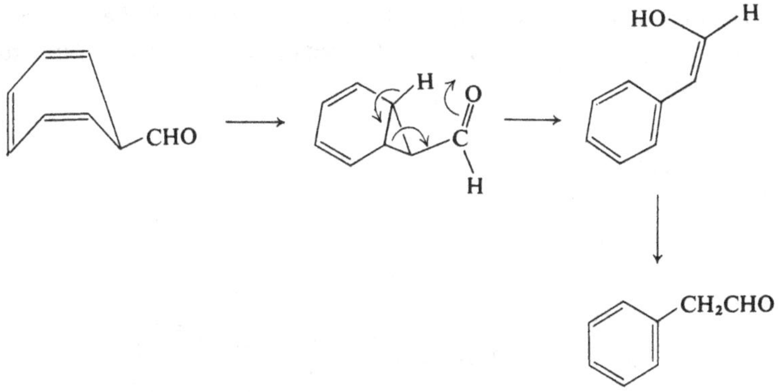

HO
H
CHO
H
O
C
H
CH₂CHO

(See also ref. 210.)

Oxidation of the epoxide (**27**) with acidic permanganate proceeds *via* the tropylium cation[1062, 1063] (*cf.* COT, p. 19).

Reduction. Reaction of the epoxide (**27**) with lithium aluminium hydride gives a mixture of the dienol (**1060**) and the acyclic trienal (**755**) (total yield 58 %).[1064]

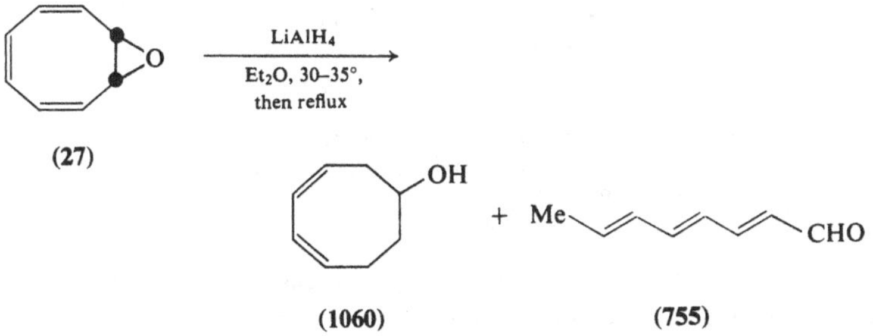

O
LiAlH₄
Et₂O, 30–35°,
then reflux
(27)
OH
+ Me
CHO
(1060)
(755)

Reactions with nucleophiles. In the presence of strong bases, preferably lithium diethylamide, the epoxide (27) isomerises to cyclo-octa-2,4,6-trienone (80–90%).[200,941]

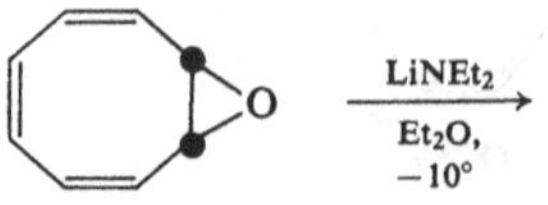

(27)

Reaction with ethyl-lithium gives the dienone (1061) (70%).[1064]

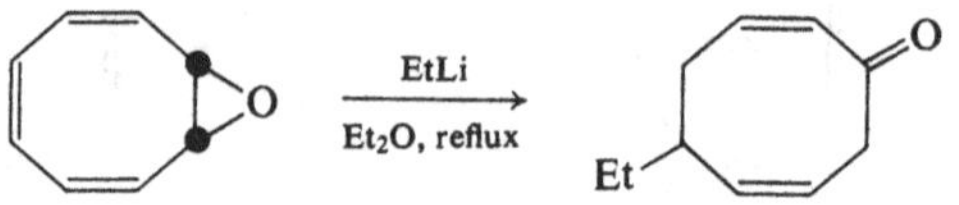

(27)   (1061)

The epoxide (27) reacts with Grignard reagents to afford the cycloheptatrienylcarbinols (1062), which on acid-catalysed dehydration give β-substituted styrenes (1063)[202] (table 126).

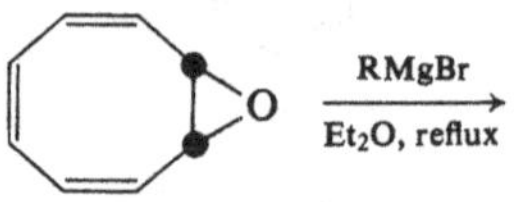

(27)

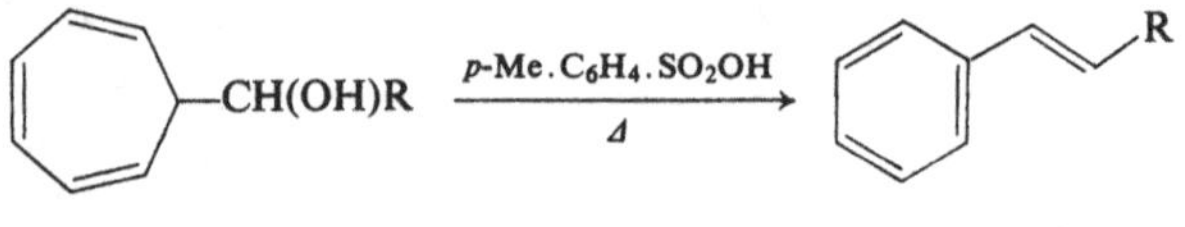

(1062)   (1063)

Table 126

Yield (%)

R	(1062)	(1063)
Et	83[a]	77
Bu[t]	23	—
Ph	73	37.5

[a] The same product is obtained (71%) from triethylaluminium in n-hexane.[1064]

A similar procedure using acetylenic Grignard reagents leads to the carbinols (1064), and thence to the enynes (1065)[1065] (table 127).

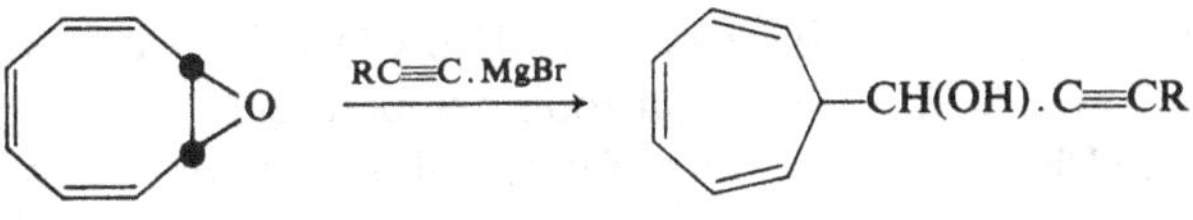

(27)   (1064)

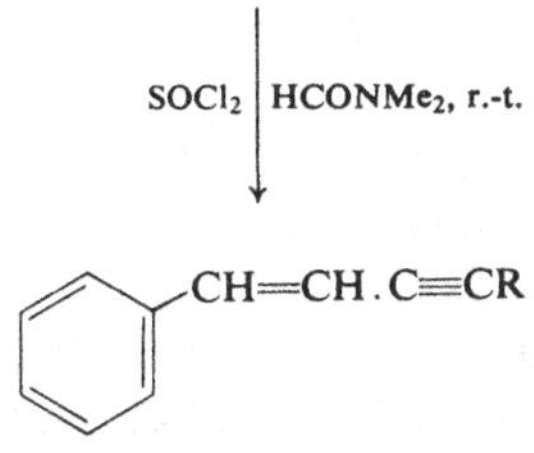

SOCl$_2$ | HCONMe$_2$, r.-t.

(1065)

Table 127

	Yield (%)	
R	(1064)	(1065)
Me	75	87
Me$_3$C	56	82
PhCH=CH	68	83
3,4-(MeO)$_2$.C$_6$H$_3$.CH=CH	52	78
Ph[CH=CH]$_2$	41	78
Ph	65	70
2,4,6-Me$_3$.C$_6$H$_2$	57	72
4-Br.C$_6$H$_4$	64	75

Cycloadditions. The epoxide (27) exhibits valence tautomerism, although the equilibrium concentration of the tricyclic isomer (1066) is low (0.007% at 20°;[856] 0.4% at 60°[136]). [4 + 2] Cycloaddition of the valence tautomer (1066) and maleic anhydride yields the adduct (1067)[11,199] identical with the product obtained by epoxidation of the cyclo-octatetraene–maleic anhydride adduct.[199]

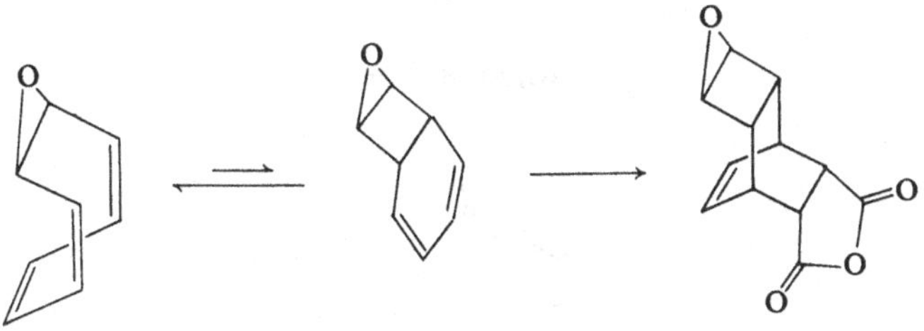

(27)   (1066)   (1067)

Reactions with metal derivatives. The rearrangement of the epoxide (27) to phenylacetaldehyde (see pp. 312–13) is promoted by e.g. lithium salts (in refluxing benzene),[1064] silver tetrafluoroborate (at 0°),[965] magnesium bromide (in refluxing ether[202] or benzene[1064]), $[Rh(CO)_2Cl]_2$ (at $-50°$),[965] and $(MeCN)_2PdCl_2$ or $(MeCN)_2PtCl_2$ (at room-temperature).[1066]

Complexes (1068) and (1069) are formed almost quantitatively at room-temperature, using copper(I) chloride and silver nitrate respectively.[1066]

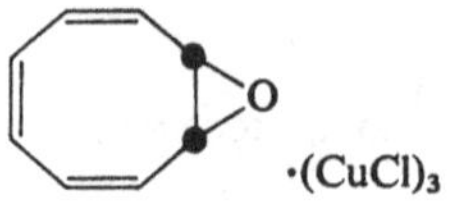

(1068) 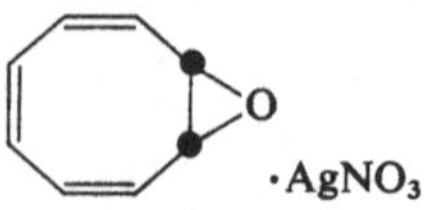   (1069)

The epoxide (27) reacts with $Fe(CO)_5$ in u.v. light to form the binuclear complex (1070) (70%), *via* the intermediate (1071) which may be isolated (*ca.* 5%) by working at low temperature, and shown to decarbonylate to give (1072) at 40°; demetallation of the complex (1070) affords 9-oxabicyclo[4.2.1]-nona-2,4,7-triene (1073).[1067]

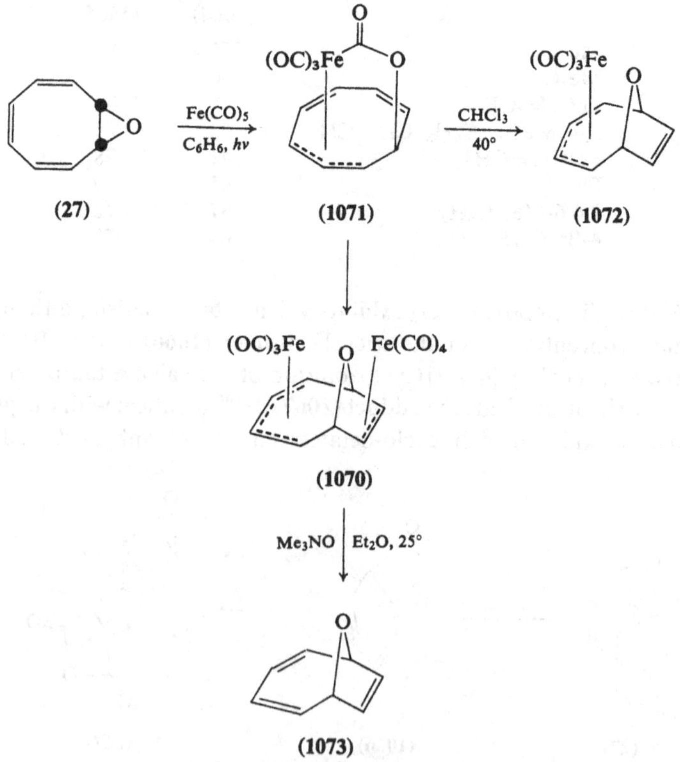

(27)   (1071)   (1072)

(1070)

(1073)

Reaction of the epoxide (27) with $Fe_2(CO)_9$ in ether yields the binuclear complex (1074) (up to 6%), together with smaller quantities of complexes containing formylcycloheptatriene, cyclo-octatrienone, and COT.[1068]

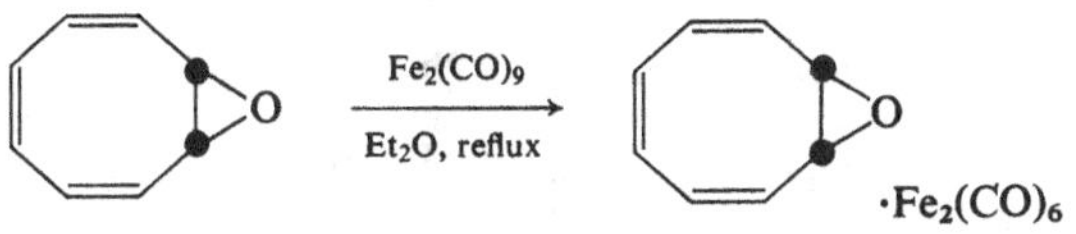

(27)   (1074)

The epoxide (27) reacts with $Rh(CO)_2(acac)$ and $Rh(CO)_2(trifluoro-acac)$ to form the complexes (1075) (table 128). Catalytic hydrogenation and

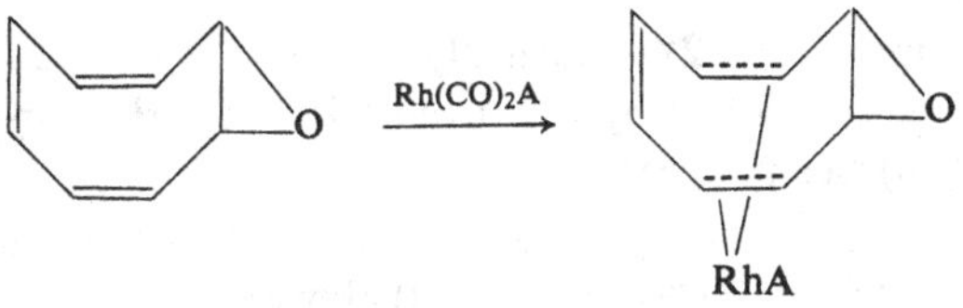

(27)   (1075)

Table 128

	A	Conditions	Yield (%)	Ref.
a	Acac	n-Hexane, reflux	65	510
b	Trifluoro-acac	—	85	1055

subsequent demetallation of (1075b) then affords the diene (1076)[1055] (see pp. 291, 310).

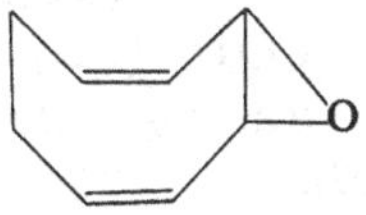

(1076)

With $(MeCN)_2MCl_2$ (M = Pd, Pt), the epoxide (27) gives the complexes (1077; M = Pd, Pt).[1066]

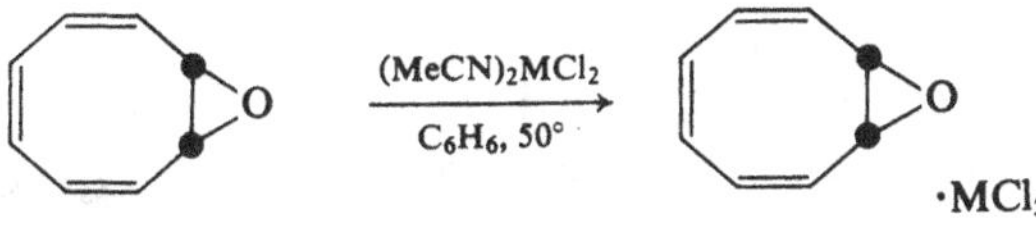

(27)   (1077)

18. 9-Phosphabicyclo[6.1.0]nona-2,4,6-trienes

The P-phenyl derivative (514) results from the reaction of COT^{2-} with dichloro-phenylphosphine (see p. 175).

Compound (514) readily rearranges (even at room-temperature, by a [1,5] sigmatropic shift) to the 9-phosphabicyclo[4.2.1]nonatriene system (1078) (54–69% from COT^{2-}).[380,750]

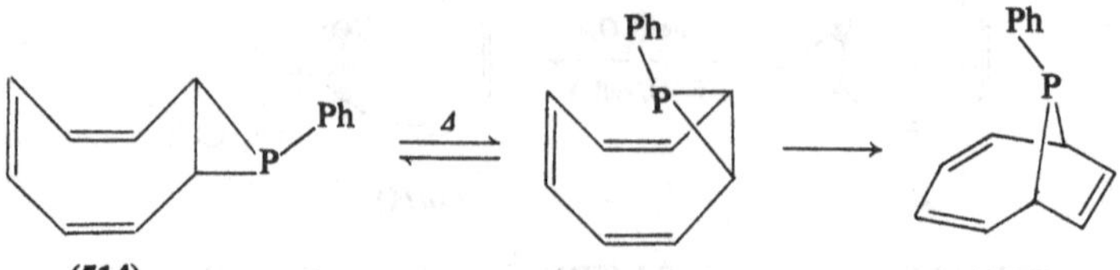
Ph
P
Δ
Ph
P
Ph
P

(514) **(1078)**

19. 9-Thiabicyclo[6.1.0]nona-2,4,6-trienes

The parent compound (**102**) is allegedly available from COT (see p. 38), or may be produced (*ca.* 80%) by u.v. irradiation of 9-thiabicyclo[4.2.1]nona-2,4,7-triene (**1079**) (see p. 325).

Rearrangement occurs at 56° (half-life *ca.* 1 h) to re-form 9-thiabicyclo-[4.2.1]nona-2,4,7-triene (**1079**)[323] (*cf.* (**514**) above).

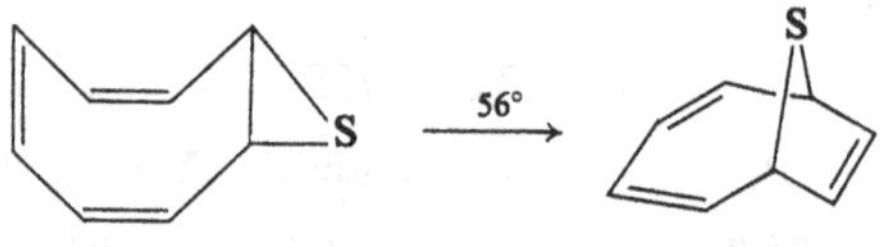
S
56°
S

(102) **(1079)**

20. 9-Azabicyclo[4.2.1]nona-2,4,7-trienes

The 9-azabicyclo[4.2.1]nona-2,4,7-triene system may be generated by the reaction of triplet cyanonitrene with COT (see p. 37), or more efficiently by the reaction of isoamyl nitrite with COT^{2-} (see p. 182); in one special case the formation of this system by the rearrangement of a 9-azabicyclo[6.1.0]nona-2,4,6-triene is known (see p. 309).

Functional group transformations. See schemes 97 and 98.

Scheme 97

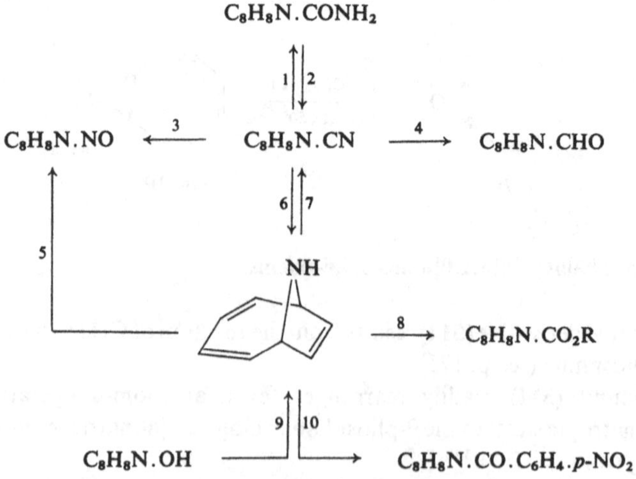
$C_8H_8N.CONH_2$
1
2
$C_8H_8N.NO$
3
$C_8H_8N.CN$
4
$C_8H_8N.CHO$
6
7
5
NH
8
$C_8H_8N.CO_2R$
9
10
$C_8H_8N.OH$
$C_8H_8N.CO.C_6H_4.p\text{-}NO_2$

Reaction	Reagents and conditions	Yield (%)	Ref.
1	30% H_2O_2(aq.), Me_2CO, Na_2CO_3(aq.), r.-t.	52	318
	or low-temp. alkaline hydrolysis	—	318
2	p-Me.C_6H_4.SO_2OH, pyridine, *ca.* 100°	100	318
3	(i) KCN, MeOH, reflux		
	(ii) AcOH(aq.), reflux	72	1069
	(iii) $NaNO_2$		
4	HCO_2H, BF_3.Et_2O, reflux	64	318
5	$NaNO_2$, AcOH	72	1070
6	NaOH(aq.), Me_2CO, reflux	71	318
7	BrCN, Et_3N, CH_2Cl_2, 0° → r.-t.	98, (47)	318, (320)
8	R = Me: $ClCO_2Me$	75	320
	R = Et: $ClCO_2Et$, Et_3N, Et_2O, 0° → r.-t.	81	318
9	Zn, AcOH	61	320
10	p-NO_2.C_6H_4.COCl, Et_3N, C_6H_6, reflux	82	318

Scheme 98

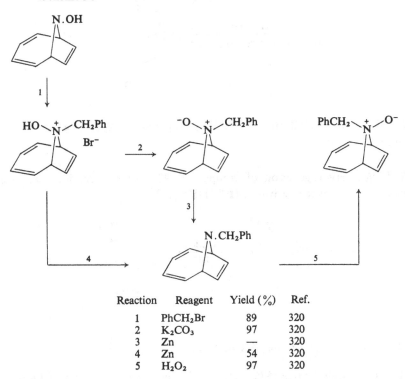

Reaction	Reagent	Yield (%)	Ref.
1	$PhCH_2Br$	89	320
2	K_2CO_3	97	320
3	Zn	—	320
4	Zn	54	320
5	H_2O_2	97	320

Photolysis. Light-induced rearrangement of the *N*-cyano-derivative (**93**) to the 9-azabicyclo[6.1.0]nonatriene system (**92a**) (*ca.* 51%) occurs in the absence of a photosensitizer.[1071] In the presence of benzophenone a [2 + 2] dimer is formed (70%), the orientation of which is believed to be as shown in structure (**1080**);[1072] in the presence of Michler's ketone, however, the product is the

9-azabarbaralane (1081) (*ca.* 70 %).[1073] (Compound (1080) may be transformed into (21)[1072] (see p. 16).)

Reduction. Hydrogenation of compound (93), using a rhodium catalyst, yields the hexahydro-derivative (1082) (80 %).[317]

21. 9-Silabicyclo[4.2.1]nona-2,4,7-trienes

The dimethyl-derivative (576) may be obtained from COT and dichlorodimethylsilane in the presence of magnesium (see p. 190).

Thermolysis. At 250° the dimethyl-derivative (576) is partly converted (46 % after 72 h) into the silabarbaralane (1083);[796] at 550°, the products are (95) (30 %) and (96) (15 %).[321]

Photolysis. U.v. irradiation of the compound (576) results in the tricyclic isomer (1085) (80%), together with the silabarbaralane (1083) (20%).[796]

Cycloadditions. Maleic anhydride and tetracyanoethylene fail to react with the system (576), but hexafluorobut-2-yne yields the *o*-disubstituted benzene (1086), possibly *via* the adduct (1087).[796]

22. 9-Phosphabicyclo[4.2.1]nona-2,4,7-trienes

9-Phenyl-9-phosphabicyclo[6.1.0]nona-2,4,6-triene (**514**), which is formed by the action of dichlorophenylphosphine on COT^{2-}, isomerises (even at room-temperature) to give the 9-phosphabicyclo[4.2.1]nona-2,4,7-triene system (**1078**) (see pp. 317–18).

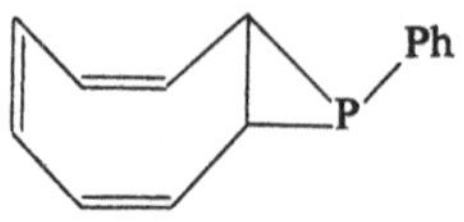

(514)

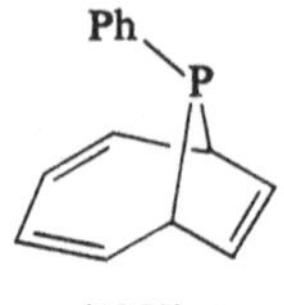

(1078)

Reactions at phosphorus. See scheme 99.

Scheme 99

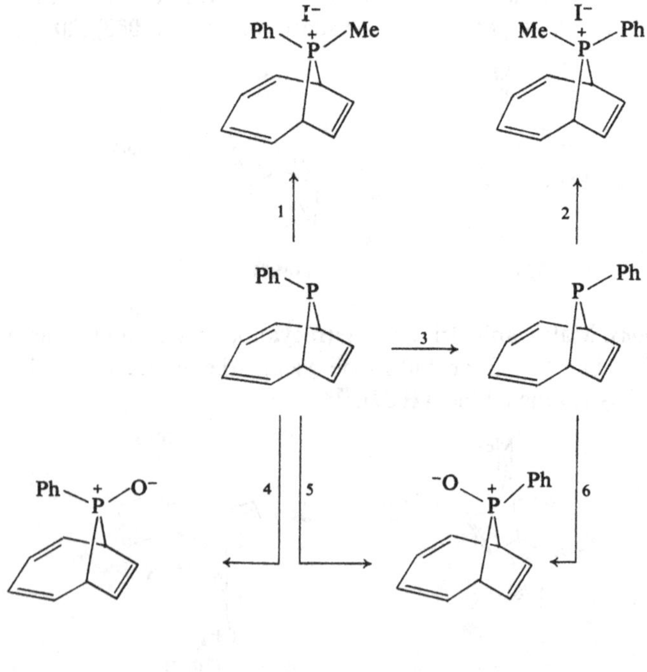

Reaction	Reagents and conditions	Yield (%)	Ref.
1	MeI, Me$_2$CO, cooling	81	380
2	MeI, Me$_2$CO, cooling	—	380
3	HCl, CHCl$_3$, 100°	—	380
4	30% H$_2$O$_2$(aq.), CHCl$_3$, 0°	(79), 90	(380), 750
5	Air, CHCl$_3$, r.-t.	42	380
6	⎰ 30% H$_2$O$_2$(aq.), CHCl$_3$, 0ᶜ	84.5	380
	⎱ or air, CHCl$_3$, r.-t.	—	380

Photolysis. U.v. irradiation of compound (**1078**) affords the 9-phosphabarbara-lane (**1088**) (25 %).[1074]

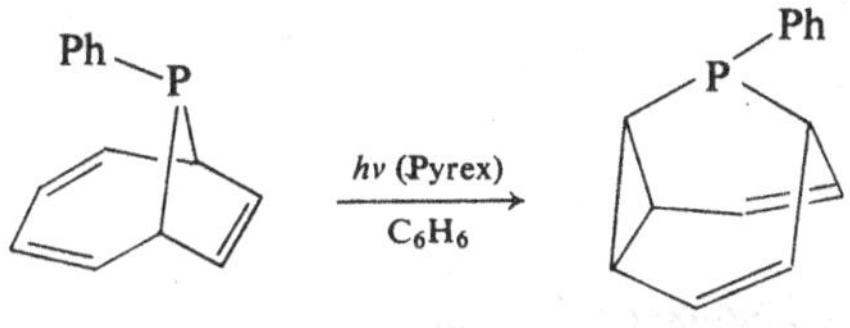

(**1078**)   (**1088**)

The phosphine oxide (**1089**), however, gives the tricyclic product (**1090**) (50 %), and on sensitised irradiation the phosphahomocubane (**1091**) (28–34 %).[750, 1074]

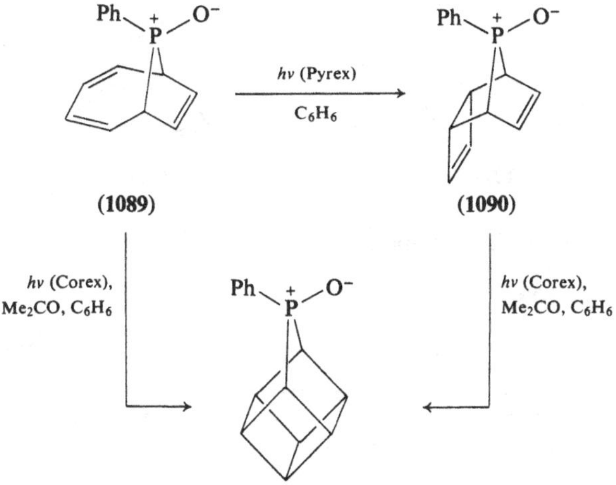

(**1089**)   (**1090**)

hv (Corex),
Me₂CO, C₆H₆

hv (Corex),
Me₂CO, C₆H₆

(**1091**)

Reduction. Catalytic hydrogenation (using Adam's catalyst) of the phosphine oxide (**1089**) gives the hexahydro-derivative (**1092**) (85.5 %).[750]

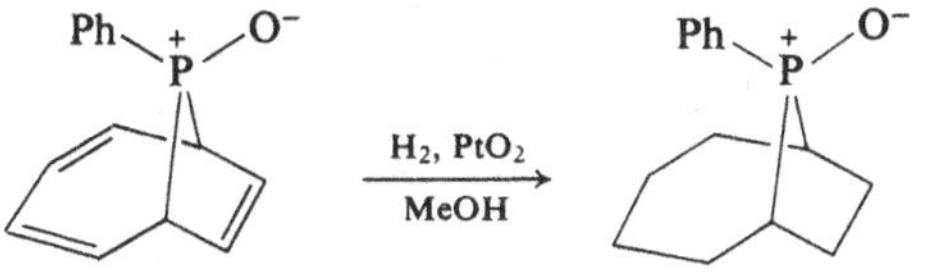

(**1089**)   (**1092**)

Reactions with metal derivatives. The epimeric phosphines (**1078**) and (**1093**) react with (PhCN)₂PdCl₂ to give complexes (**1094**) (82 %) or (**1095**) (42 %), and (**1096**) (33 %) respectively.[380]

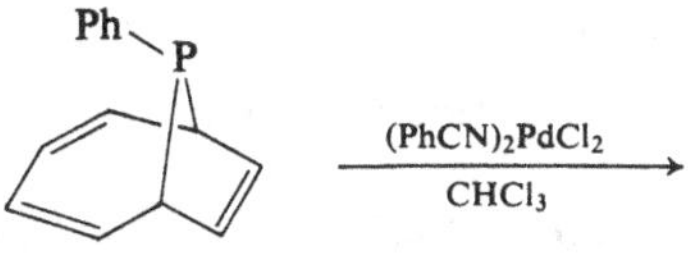

(**1078**)

$(C_{14}H_{13}P)_2PdCl_2$ (**1094**)

or $[(C_{14}H_{13}P)PdCl_2]_2$ (**1095**)

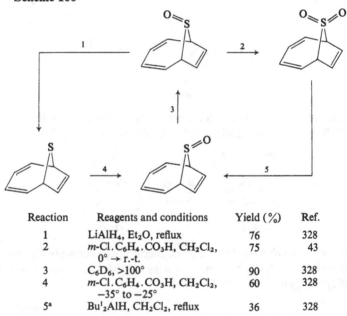

$$(1093) \xrightarrow[\text{CHCl}_3]{(PhCN)_2PdCl_2} (C_{14}H_{13}P)_2PdCl_2 \quad (1096)$$

23. 9-Thiabicyclo[4.2.1]nona-2,4,7-trienes

The sulphoxide (107) is formed by the addition of sulphur monoxide to COT (see pp. 39–40), and the sulphone (108) by the reaction of COT with sulphur dioxide in the presence of antimony(v) fluoride (see p. 40).

(107)   (108)

Reactions at sulphur. See scheme 100.

Scheme 100

Reaction	Reagents and conditions	Yield (%)	Ref.
1	LiAlH₄, Et₂O, reflux	76	328
2	m-Cl.C₆H₄.CO₃H, CH₂Cl₂, 0° → r.-t.	75	43
3	C₆D₆, >100°	90	328
4	m-Cl.C₆H₄.CO₃H, CH₂Cl₂, −35° to −25°	60	328
5[a]	Bu^{i}₂AlH, CH₂Cl₂, reflux	36	328

[a] Reaction carried out using the 1,6-dideuterio-derivative.

Thermolysis and photolysis. Thermolysis of the sulphone (108), or photolysis of either the sulphone (108) or the sulphoxide (107), gives rise to COT (see p. 5).

Photolysis of the parent compound (**1079**) gives 9-thiabicyclo[6.1.0]nona-2,4,6-triene (**102**) (*ca.* 80%).[323]

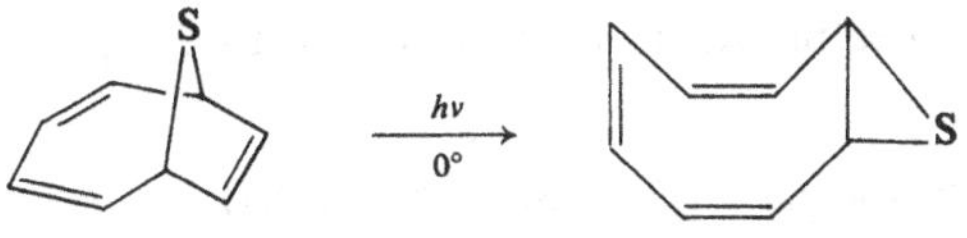
S
hν
0°
S

(**1079**) (**102**)

Reduction and deprotonation. Catalytic hydrogenation of the sulphoxide (**107**), using a rhodium catalyst, affords the dihydro-derivative (**1097**) (60%).[328] Reduction with di-imide gives a mixture of the products (**1098**) (*ca.* 50%) and (**1099**).[328]

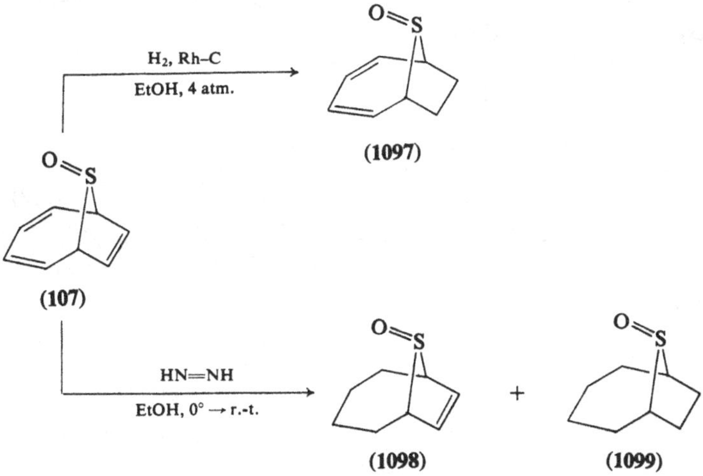
O=S
H₂, Rh–C
EtOH, 4 atm.
(1097)
O=S
(107)
HN=NH
EtOH, 0° → r.-t.
O=S
O=S
+
(1098)
(1099)

With n-butyl-lithium, the sulphone (**108**) forms a dianion (**1100**),[1075] which may be deuteriated (62%)[49, 328, 1075] or methylated (100%).[43, 1075] (It is also possible to prepare a mono-anion.[1076])

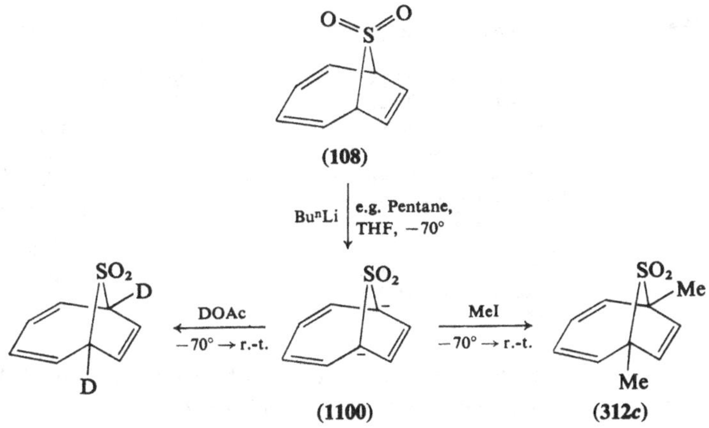
O=S=O
(108)
BuⁿLi
e.g. Pentane, THF, −70°
SO₂
D
D
DOAc
−70° → r.-t.
SO₂
MeI
−70° → r.-t.
SO₂
Me
Me
(1100)
(312c)

24. 9-Aza-10-oxobicyclo[4.2.2]deca-2,4,7-trienes

The 9-aza-10-oxobicyclo[4.2.2]deca-2,4,7-triene system (**128**) results from the addition of chlorosulphonyl isocyanate to COTs (see pp. 45–6).

Functional group transformations. See scheme 101.

Scheme 101

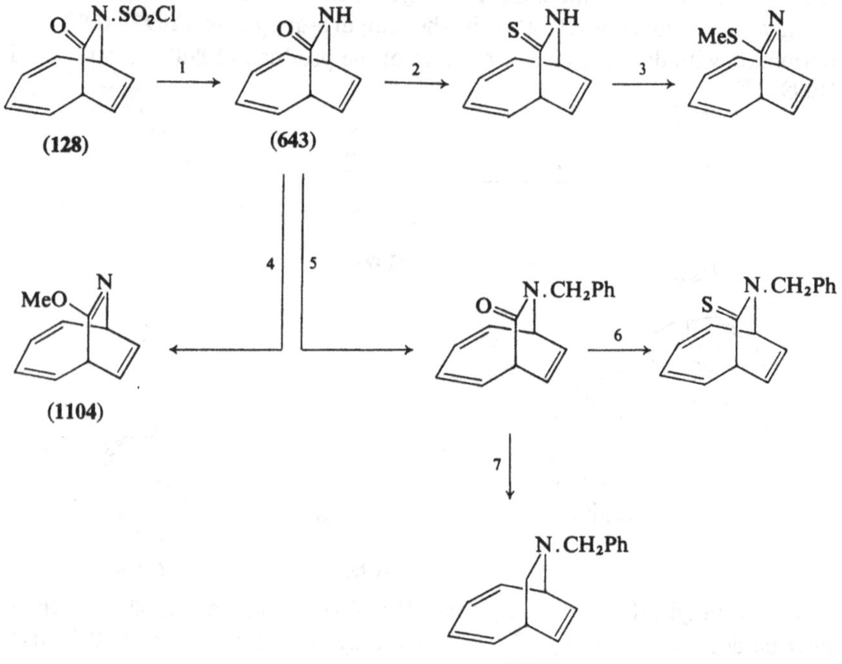

(**128**)   (**643**)

(**1104**)

(**1101**)

Reaction	Reagents and conditions	Yield (%)	Ref.
1	{ PhSH, pyridine, Me₂CO, 0°	98	(366), 367
	{ or NaOH, Me₂CO(aq.)	65	367
2	P₂S₅, pyridine, reflux	68	1077
3	Me₃O⁺ BF₄⁻, CH₂Cl₂, r.-t.	71	1077
4	Me₃O⁺ BF₄⁻, CH₂Cl₂, r.-t.	81	(366), 367
5	{ (i) NaH, DMF, 65°	93	(1078), 1079
	{ (ii) PhCH₂Cl, 45°		
6	P₂S₅	—	1078
7	{ (i) Me₃O⁺ BF₄⁻, CH₂Cl₂	67	1079
	{ (ii) NaBH₄, MeOH, 5° → r.-t.		

For reactions 1 and 4 with derivatives of monosubstituted COTs, see ref. 367.
For rearrangements of compound (**1101**), see ref. 1079.

Photolysis. Light-induced rearrangement of the system (643) leads to the lactam-bridged homotropilidene (1102) (54%), convertible into the azabullvalene derivative (1103) (92%);[366, 1077] an improved procedure[1077] involves the acetone-sensitised photorearrangement of the imino-ether (1104) (see scheme 101).

For similar transformations of derivatives of monosubstituted COTs, see ref. 1077.

Reduction. Catalytic hydrogenation of the lactam (643) affords the hexahydro-derivative (1105) (69%).[367]

25. Tricyclo[4.2.2.0²,⁵]deca-3,7-dienes

The tricyclo[4.2.2.0²,⁵]deca-3,7-diene system is formed by the reaction of olefinic dienophiles with COT, which undergoes [4 + 2] cycloadditions *via* its bicyclic valence tautomer (4) (see pp. 42–3).

Functional group transformations. See scheme 102.

Scheme 102

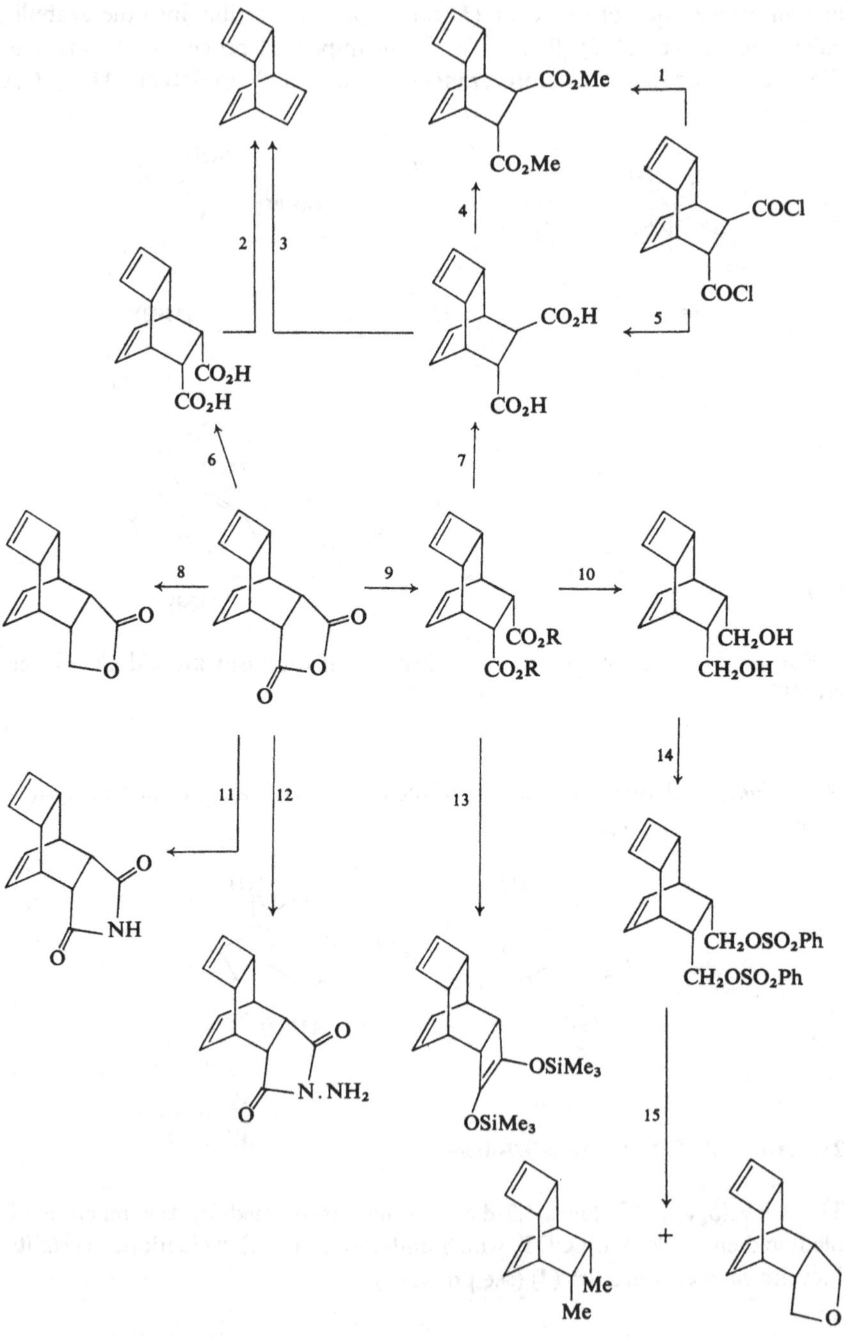

(1106)   (1107)

Reaction	Reagents and conditions	Yield (%)	Ref.
1	MeOH	—	340
2	{Pb(OAc)$_4$, pyridine, C$_6$H$_6$, 50° → 70°; or electrolysis, Pt electrodes, pyridine(aq.), Et$_3$N	10–15	341, 1080
3	Electrolysis, Pt electrodes, pyridine(aq.), Et$_3$N	37	1081
4	MeOH, H$_2$SO$_4$, reflux	53	1081
5	H$_2$O	80	11
6	NaOH(aq.), warm	—	340
7; R = Me	{(i) NaOMe, MeOH, reflux (ii) NaOH(aq.), reflux	85	11
8	LiAlH$_4$, THF, r.-t.	83	11
9; R = Me, Et, Bu^n, Bu^t, Me$_2$CH[CH]$_2$	ROH, H$_2$SO$_4$, reflux	—	931
10; R = Me	LiAlH$_4$, Et$_2$O, reflux	86–98	11
11	NH$_3$(aq.), 120°	98–99	341
12	NH$_2$NH$_2$, MeCN, HOAc	61	11
13; R = Me	Me$_3$SiCl, Na, xylene, 40° → 80°	74	1082
14	PhSO$_2$Cl, pyridine, −5° → 0°	—	1083
15	LiAlH$_4$, Et$_2$O, reflux	64[a] {47 (1106), ca. 50 (1107)}	341 341

[a] The cyclic ether (1107) is a by-product.

Photolysis. U.v. irradiation of the maleic anhydride adduct (**117**), or the di-ester (**1108**), in acetone yields the caged product (**1109**) $(40\%)^{355, 1084}$ or (**1110**) (35%);[1085] hydrolysis and subsequent oxidative bis-decarboxylation of these 1,1'-bishomocubane derivatives leads to basketene (**6**).[1084, 1085]

For the silver-catalysed rearrangement of the di-ester (**1110**) to the isomer (**1112**), see ref. 1086.

(**1112**)

A different type of cage-structure (**105**) results (90%) from irradiation of the *p*-benzoquinone adduct (**106**).[1087]

Oxidation. Epoxidation of the anhydride **(117)** or the dimethyl ester **(1108)** gives the *exo*-epoxide, **(1067)** (66 %)[199] or **(1113)** (61 %).[11]

PhCO₃H
CHCl₃, C₆H₆, 5°

(117)   **(1067)**

PhCO₃H
CHCl₃, r.-t.

CO₂Me
CO₂Me

(1108)   CO₂Me
CO₂Me

(1113)

Reactions with electrophiles. Electrophilic attack on the *endo*-9,10-disubstituted tricyclo[4.2.2.0²·⁵]deca-3,7-diene system usually occurs on the less hindered *exo*-face of the 3,4-double bond, and is frequently followed by transannular ring-closure involving the 7,8-double bond, a functional group or the solvent also participating. Exceptionally, the dimethyl ester **(1108)** reacts with sulphuric acid in methanol to afford the lactone-ester **(1114)** (90 %).[340]

H₂SO₄
MeOH,
reflux

CO₂Me
CO₂Me

(1108)   CO₂Me
O—
O

(1114)

Reaction of the anhydride **(117)** with chlorine in aqueous sodium hydroxide,[11] or t-butyl hypochlorite,[1088] gives the chloro-lactone-carboxylic acid **(1115; R = H)** (77.5 % or 97 % respectively); the dimethyl ester **(1108)** affords the analogous product **(1115; R = Me)** (93 %).[1088]

Reactions of the anhydride (117) etc. with bromine are summarised in scheme 103.

The *trans*-dicarboxylic acid (1117), in aqueous potassium hydroxide, reacts with bromine to give the bromo-lactone-carboxylic acid (1118),[340] (*cf.* ref. 1088).

With bromine in chloroform, the cyclic ether (1107) affords a mixture of the dibromides (1119) (9%), (1120) (1%), and (1121) (15%).[931]

Scheme 103

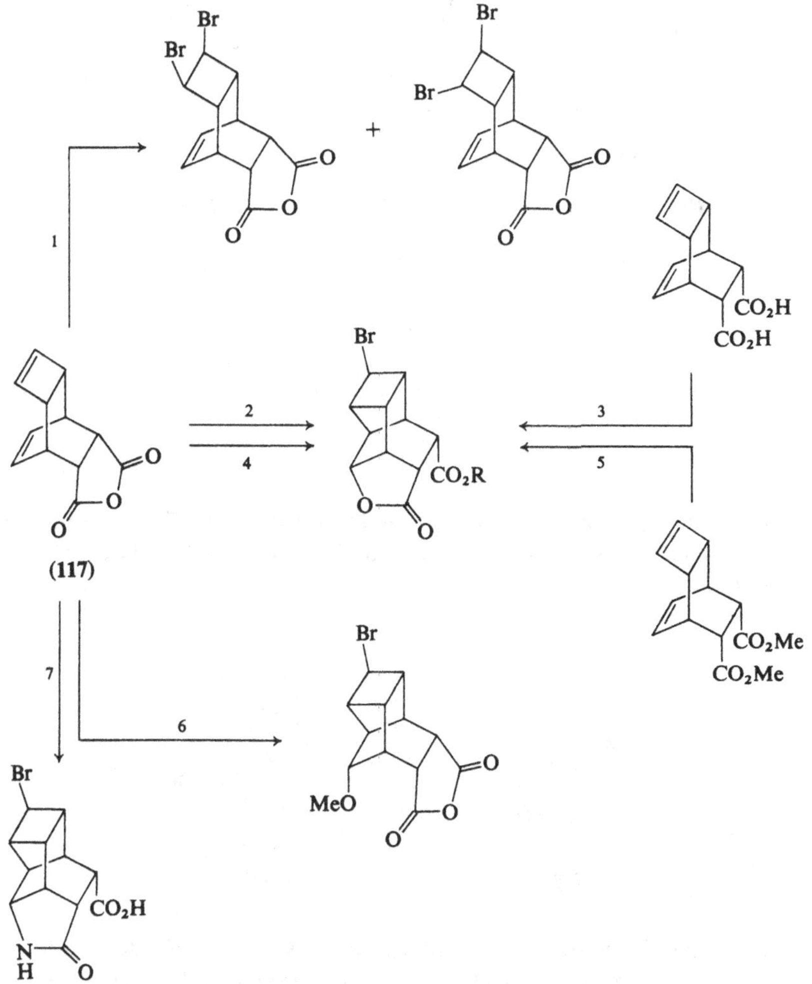

Reaction	Reagents and conditions	Yield (%)	Ref.
1	Br$_2$, CHCl$_3$, r.-t.	{ *ca.* 40 (*cis*) *ca.* 30 (*trans*) }	931
2	Br$_2$, MeOH KOH(aq.), r.-t.	86[a]	11
3	Br$_2$, MeOH, r.-t.	57[b]	11
4	Br$_2$, HOAc, r.-t.	95[a]	1088
5	{ Br$_2$, CHCl$_3$, r.-t. Br$_2$, MeOH, r.-t. }	86[b,c] 54[b]	(340), 1088 11
6	Br$_2$, MeOH, r.-t.	38	1088
7	{ Br$_2$, MeCN, r.-t. Br$_2$, PhCN, r.-t. }	98 } 80 }	1088

[a] R = H
[b] R = Me
[c] Formation of a dibromide, possibly of structure (**1116**) (*cf.* ref. 1088) may also occur.[11, 340]

(1107)   (1119)   (1120)

(1121)

With iodine monochloride in methanol, the anhydride (117) gives a mixture of the iodo-lactone-ester (1122; R = Me) (39%) and the carboxylic acid (1122; R = H) (20%).[1088]

(117)   (1122)

Iodine azide (from iodine monochloride and sodium azide) adds to the 3,4-double bond of the anhydride (117) to give the product (1123) (characterised as the acetylenedicarboxylic ester-adduct).[1088]

(117)   (1123)

The anhydride (**117**) and the di-ester (**1108**) react with mercury(II) acetate in methanol to form (after work-up as the chloro-mercurials) products originally formulated as (**1124**) and (**1125**) respectively[342] (but see appendix).

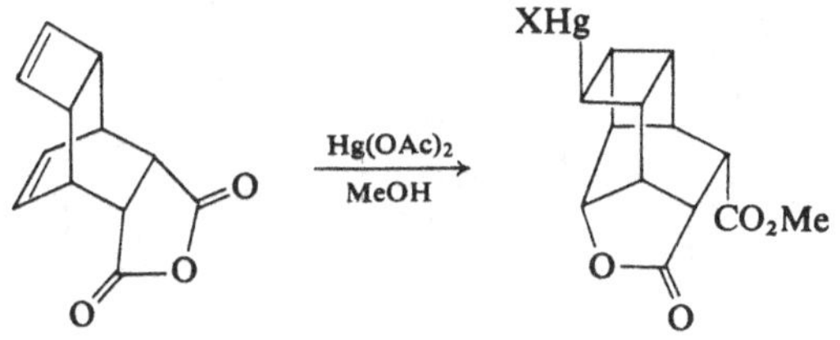

(**117**)   (**1124**)

(**1108**)   (**1125**)

X = OAc, Cl

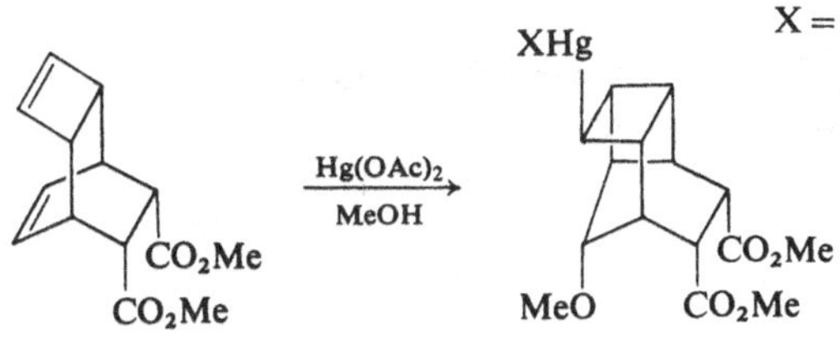

Reduction. Hydrogenation of the anhydride (**117**) in the presence of e.g. Adam's catalyst yields the tetrahydro-derivative (**1126**).[260] Selective hydrogenation of the 3,4-double bond to give the dihydro-derivative (**1127**) is readily achieved, however, by the use of e.g. palladium-on-carbon in aqueous potassium hydroxide (yield 89%).[11]

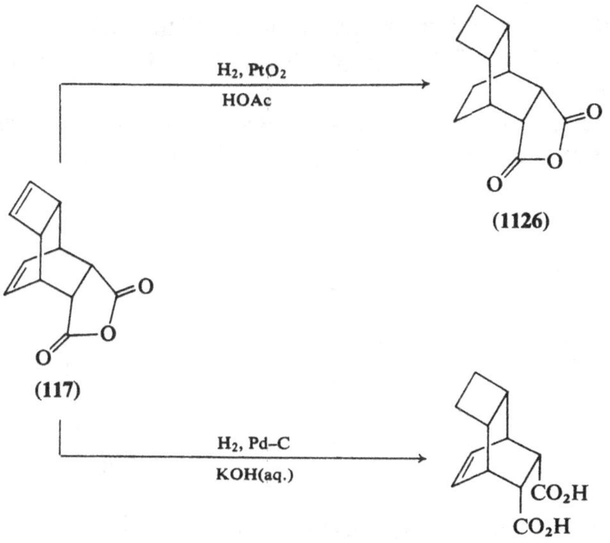

H₂, PtO₂
HOAc

(**1126**)

(**117**)

H₂, Pd–C
KOH(aq.)

(**1127**)

Reactions with free radicals. The radical addition of carbon tetrachloride and bromotrichloromethane to the anhydride (117) gives products (1128a) (70%)

and (1128b) (69%) respectively.[1089] Thiophenol apparently affords the analogous product (1129) (61%).[1089]

Cycloadditions. Intermolecular cycloadditions of the *endo*-9,10-disubstituted tricyclo[4.2.2.0^{2,5}]deca-3,7-diene system occur specifically on the *exo*-face of the 3,4-double bond.

The tetracyanoethylene adduct (1130) undergoes 1,3-dipolar addition with tetracyanoethylene oxide to give the product (1131) (66%).[337]

With the nitrone 3,4-dihydroisoquinoline *N*-oxide, the di-ester (1108) affords the adduct (1132) (82%).[335] Similarly, the anhydride (117) and the

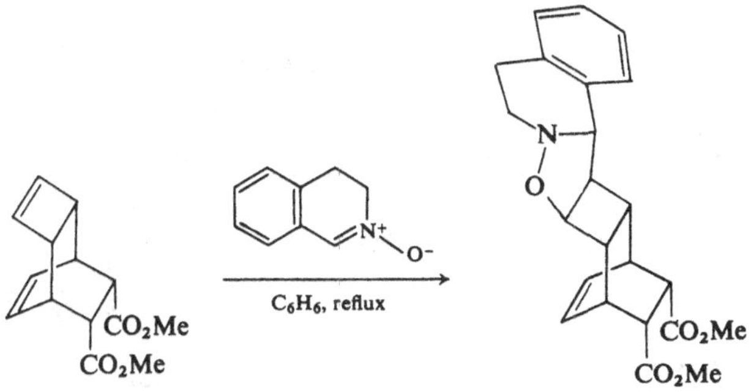

(1108)   (1132)

di-ester **(1108)** add benzonitrile oxide to give the products **(1133)** (80–85%) and **(1134)** (90%) respectively.[331,334]

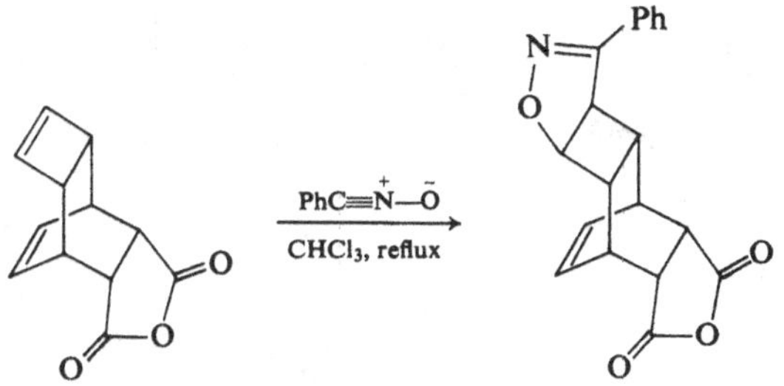

(117)   (1133)

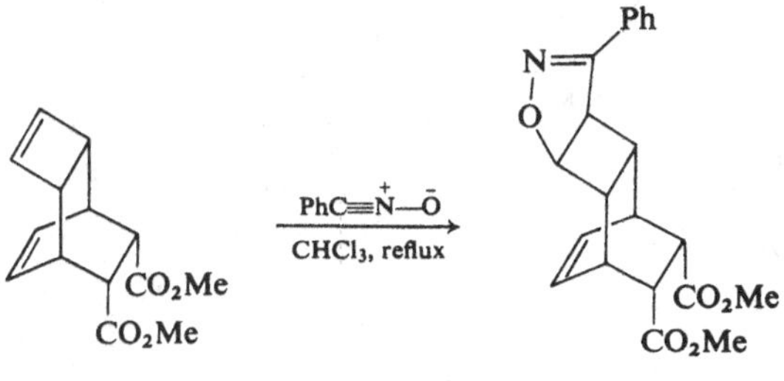

(1108)   (1134)

The 3,4-double bond of the tricyclo[4.2.2.0²,⁵]deca-3,7-diene system behaves as an excellent dienophile towards electron-deficient dienes.

1,2,3,4-Tetrachloro-5,5-dimethoxycyclopentadiene adds to the anhydride **(117)** exclusively in the *endo*-mode (*exo*-addition being inhibited by the bulky methoxy-groups) to form the product **(1135)** (79%); hydrolysis of the ketal group in **(1135)**, followed by thermal extrusion of carbon monoxide, results in the product **(1136)**.[1090]

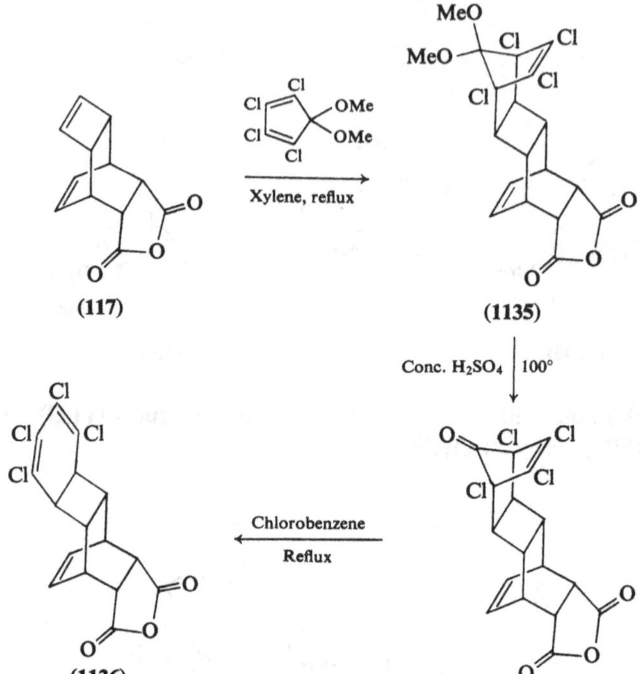
MeO
MeO
Cl
Cl
Cl
Cl
OMe
OMe
Xylene, reflux
Conc. H_2SO_4
100°
Chlorobenzene
Reflux
O

(117) (1135) (1136)

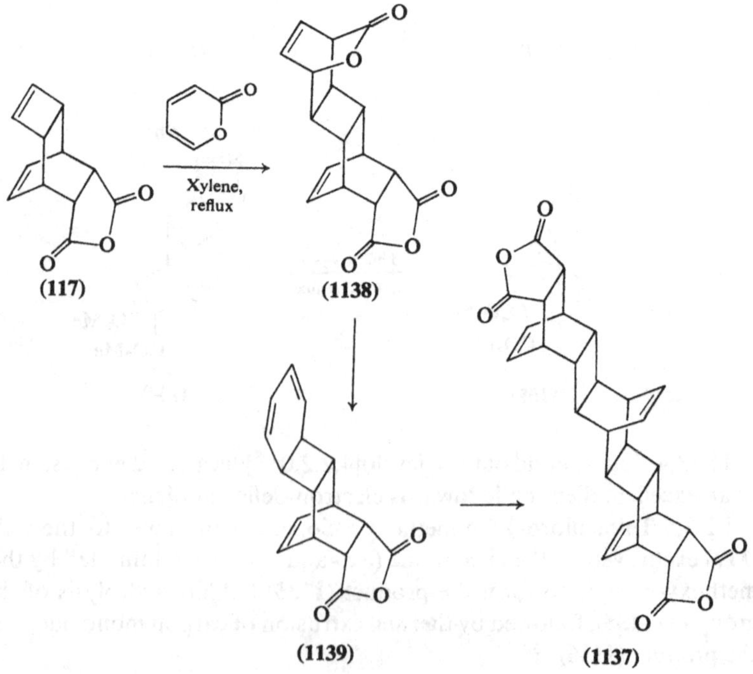
Xylene,
reflux
O

(117) (1138) (1139) (1137)

The reaction of α-pyrone with the anhydride (117) in refluxing xylene yields the product (1137) (86%), presumably *via* an initial lactone-bridged adduct (1138) (stereochemistry uncertain) and the decarboxylated derivative (1139), which then adds a second molecule of the dienophile[1091] (see pp. 340, 341).

The *trans*-dimethyl ester (1140) adds hemicyclone to give the adduct (1141) (stereochemistry uncertain) (76%); photolysis of this product leads to the bicyclic diene (1142) (62%).[1092]

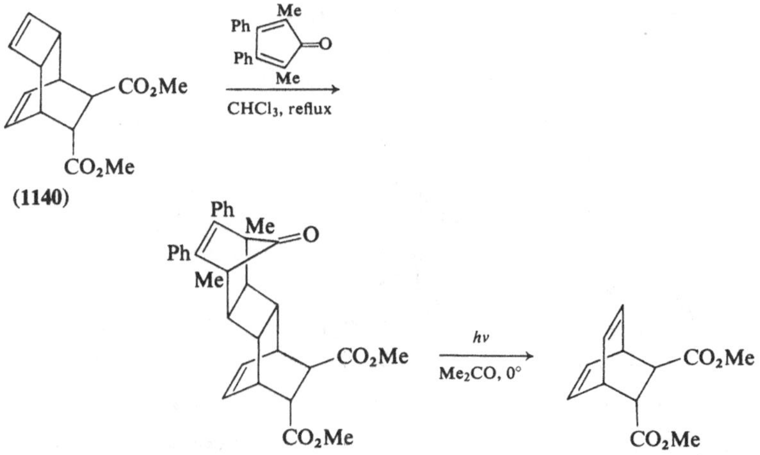

(1140)

(1141)

(1142)

Similarly, the addition of hemicyclone to the *p*-benzoquinone adduct (106) gives a product (1143) (80%) (stereochemistry uncertain); photolysis then affords the caged compound (1144) in excellent yield.[1093]

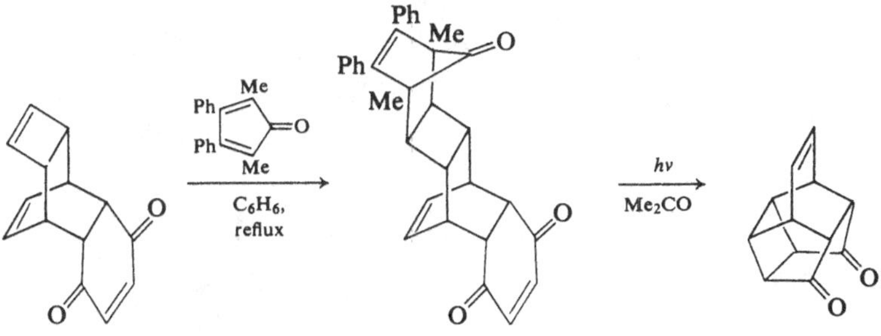

(106)

(1143)

(1144)

The di-ester (1108) reacts with tetrachloro-*o*-benzoquinone to form the adduct (1145), which on photolysis gives rise to the products (1146) (63%) and (1147) (12%); the formation of compound (1148) becomes significant with increase in the concentration of the starting material.[1094]

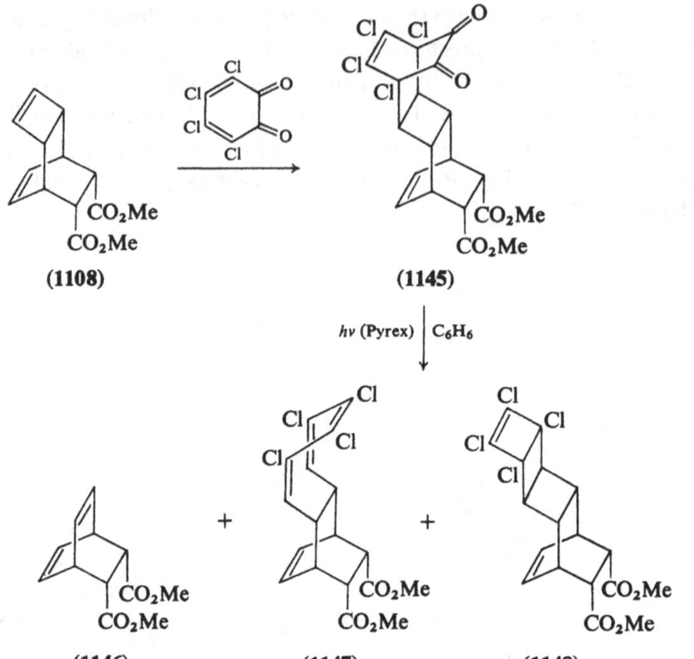

(1108)   (1145)

hv (Pyrex) | C_6H_6

(1146)   (1147)   (1148)

(Stereochemistry uncertain)

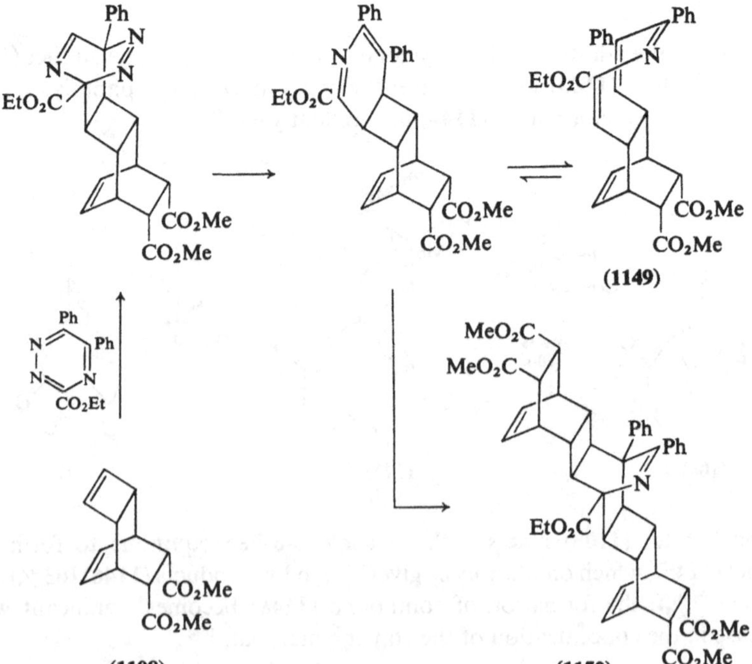

(1149)

(1108)   (1150)

With 3-ethoxycarbonyl-5,6-diphenyl-1,2,4-triazine, the di-ester (1108) yields a mixture of the products (1149) and (1150).[1095]

Reaction of the anhydride (117) with 3,6-di(2'-pyridyl)-*s*-tetrazine occurs at room-temperature to give the adduct (1151); an analogous product is obtained from the di-ester (1108).[1096] These 4,5-dihydropyridazine derivatives (which are prone to aerial oxidation) behave as dienes in [4 + 2] cycloadditions, and e.g. the adduct (1151) reacts with a second molecule of the anhydride (117) in refluxing toluene to afford a product formulated as (1152).[1096]

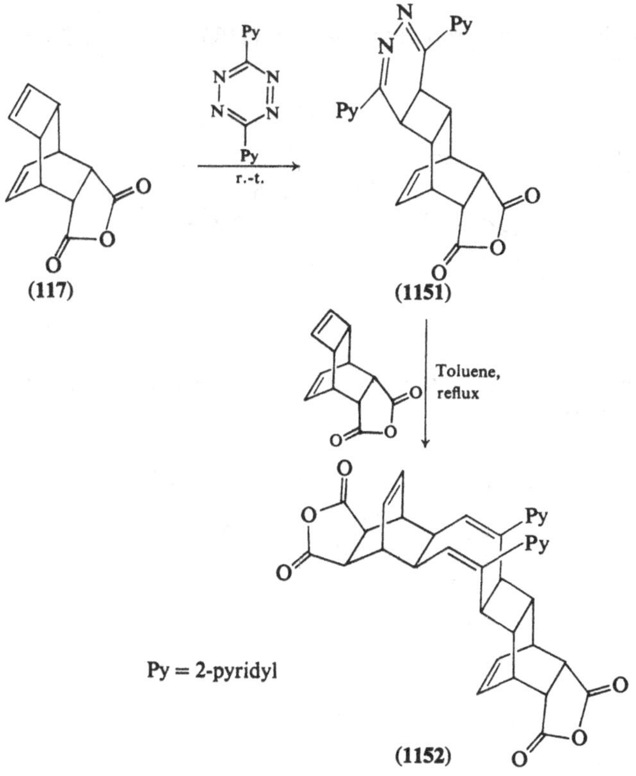

Py = 2-pyridyl

(Stereochemistry not certain)

Reactions with metal derivatives. The dimethyl ester (1108) forms a silver nitrate complex, presumably of structure (1153) (80.5%); the cyclic ether (1107) gives a similar product (70%).[341]

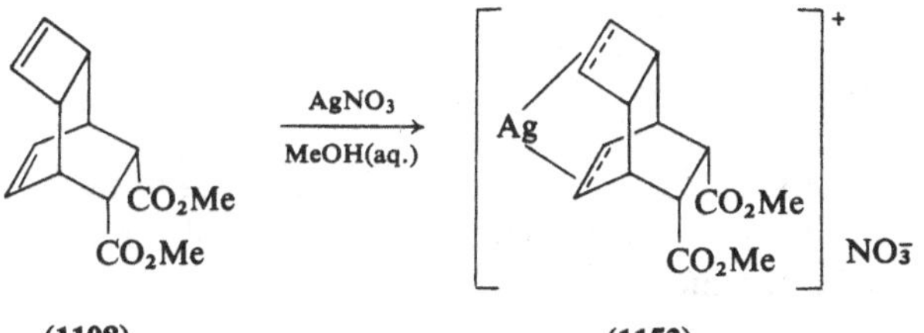

The dimethyl-derivative (1106) reacts with $Mo(CO)_6$ to give the tetra-carbonyl complex (1154) (68%).[341]

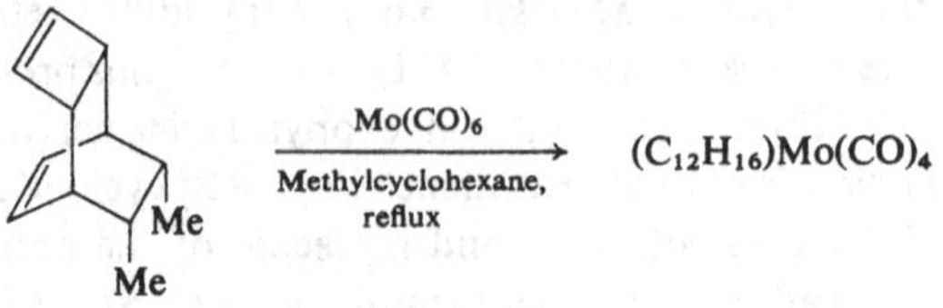

$$\xrightarrow[\text{Methylcyclohexane, reflux}]{Mo(CO)_6} (C_{12}H_{16})Mo(CO)_4$$

(1106)   (1154)

A (probably) dimeric rhodium complex (1155) (28%) is formed by the anhydride (117).[339]

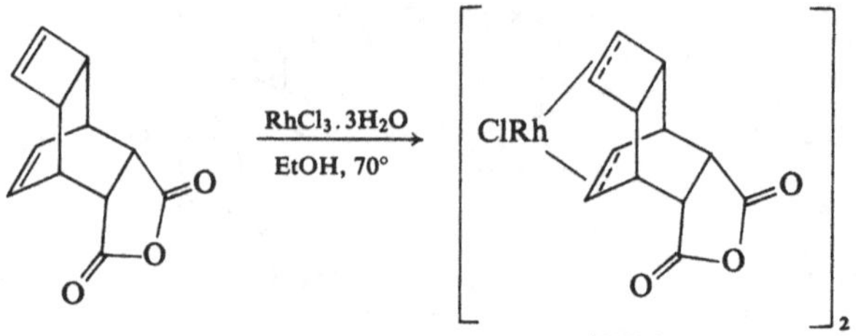

$$\xrightarrow[\text{EtOH, 70°}]{RhCl_3 . 3H_2O}$$

(1155)

Palladium complexes (1156) are readily formed from $(PhCN)_2PdCl_2$ and the system (1157)[341] (table 129).

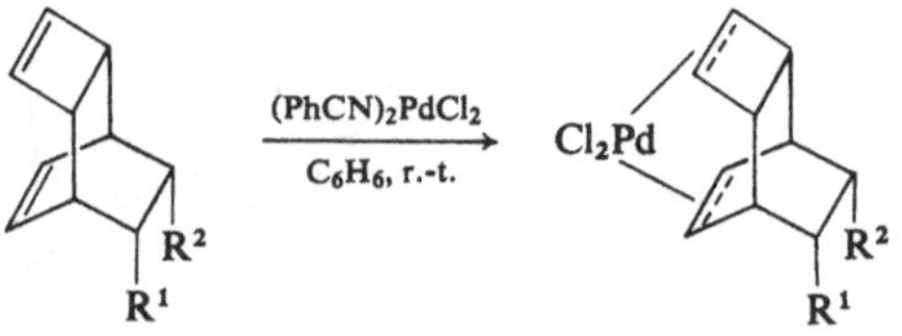

$$\xrightarrow[\text{C}_6\text{H}_6, \text{ r.-t.}]{(PhCN)_2PdCl_2}$$

(1157)   (1156)

Table 129

R^1	R^2	Yield (%)
Me	Me	92
CH_2OH	CH_2OH	63
$H_2C\diagdown_O\diagup CH_2$		63
CO_2Me	CO_2Me	72

26. Tricyclo[4.2.2.0^{2,5}]deca-3,7,9-trienes

The reaction of COT with acetylenic dienophiles leads to tricyclo[4.2.2.0^{2,5}]-decatrienes (see pp. 42–3); the parent compound (1158) ('Nenitzescu's hydro-carbon') may be obtained from the COT–maleic anhydride adduct (117) (see

p. 328). The triene **(1158)** is a member of a set of $C_{10}H_{10}$ isomers which lie on interconnected energy surfaces. Most of these hydrocarbons have COT as

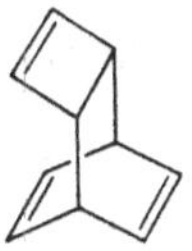

(1158)

their common progenitor; for a review of their numerous thermal, photochemical, and metal-catalysed interconversions, see ref. 1097.

Functional group transformations. See scheme 104.

Scheme 104

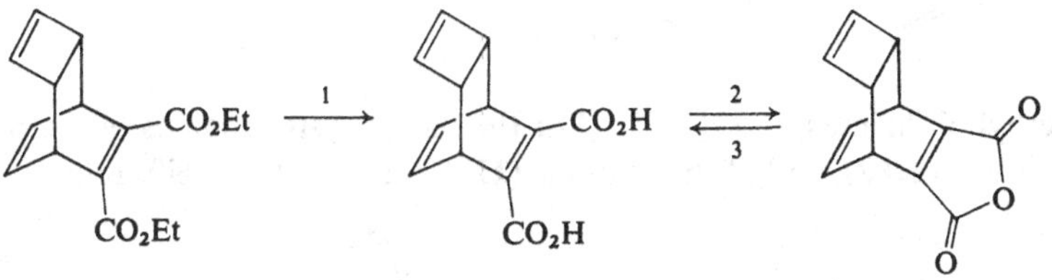

Reaction	Reagents and conditions	Yield (%)	Ref.
1	NaOH, MeOH(aq.), reflux	63	11
2	Ac$_2$O, C$_6$H$_6$, *ca.* 100°	—	11
3	NaOH(aq.)	—	11

Thermolysis. Thermolysis of the triene **(1158)** at *ca.* 300° affords *cis*-9,10-dihydronaphthalene, and further products resulting from hydrogen shift and transfer processes.[1080,1098,1099] Similar thermolysis of the diester **(7)** yields a

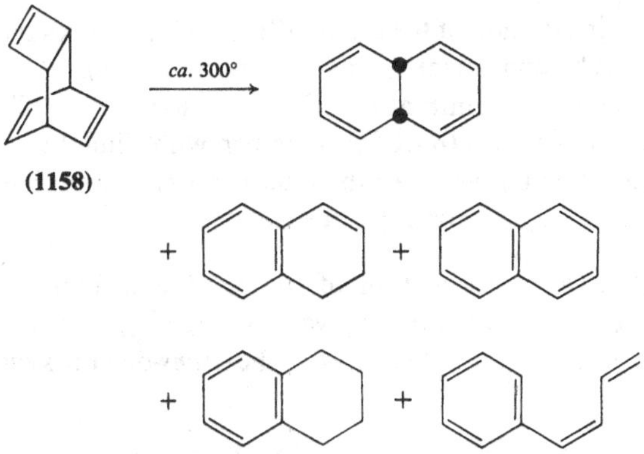

(1158)

mixture of the dihydronaphthalenes **(1159)** and **(1160)** (27%) and the naphthalene **(1161)** (17–25%),[342,352] together with buta-1,3-diene (15–17%) and

(7)   (1159)

(1160)   (1161)

dimethyl phthalate (30–32 %).[352] (For the thermolysis products of the benzo-derivatives (1162), and of the quinone (119), see refs. 352, 1099, 1100.)

(1162)   (119)

R = H, OMe, OAc

Photolysis. U.v. irradiation of the diester (7) at −50° produces the caged isomer (1163) (up to 61 % conversion); this reverts to starting material at higher temperatures (half-life 16 min at *ca.* 2°).[1101] Irradiation at 10° results in the cyclobutadiene dimer (5) (16–18 %), together with dimethyl phthalate (37–40 %), COT (5–7 %) and benzene; the intermediacy of cyclobutadiene in this photolysis is revealed by trapping experiments.[56]

Reduction. Selective hydrogenation of the 3,4-double bond in (7) may be effected to give the dihydro-derivative (698) (82 %),[353] which on further hydrogenation affords (after hydrolysis) the *trans*-dicarboxylic acid (1164) (83 %).[260]

Cycloadditions. Both the 3,4- and the 7,8-double bonds of the di-ester (7) take part in 1,3-dipolar cycloadditions; thus 3,4-dihydroisoquinoline *N*-oxide gives the adducts (1165) (43 %) and (1166) (37.5 %).[335]

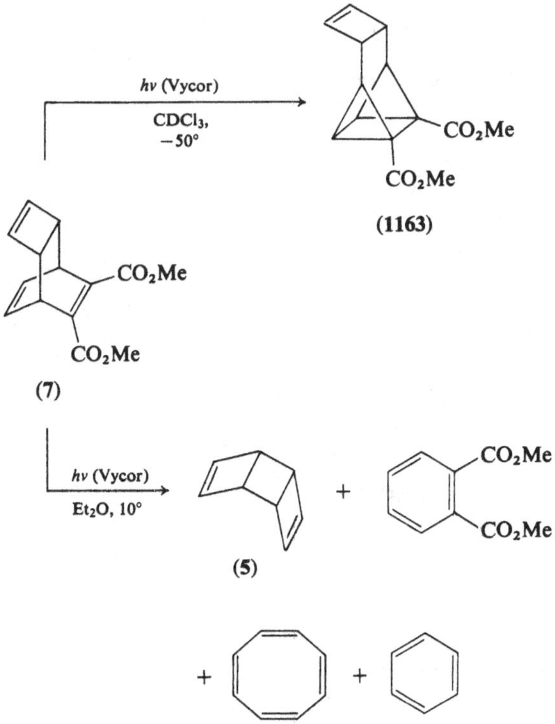
hν (Vycor)
CDCl3, −50°
CO2Me
CO2Me
(1163)
CO2Me
CO2Me
(7)
hν (Vycor)
Et2O, 10°
(5)
CO2Me
CO2Me

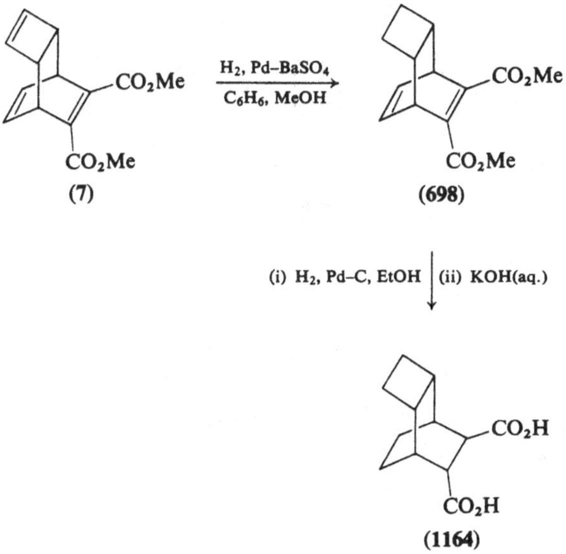
CO2Me
CO2Me
(7)
H2, Pd–BaSO4
C6H6, MeOH
CO2Me
CO2Me
(698)
(i) H2, Pd–C, EtOH
(ii) KOH(aq.)
CO2H
CO2H
(1164)

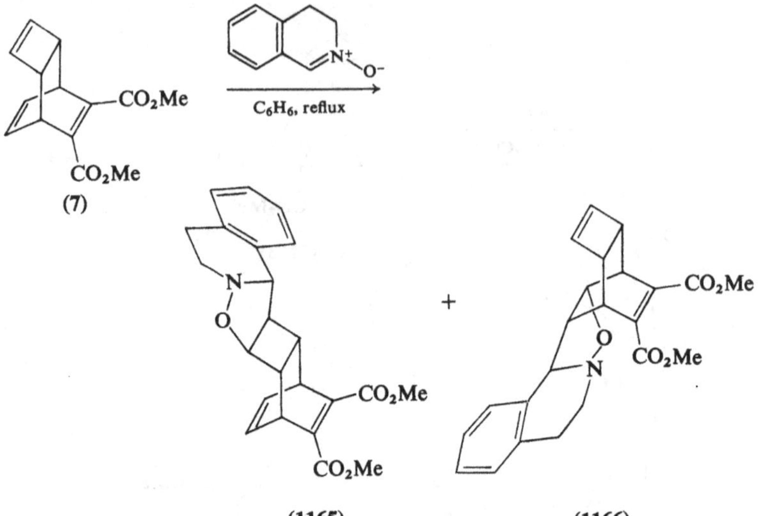

(7)

(1165)   (1166)

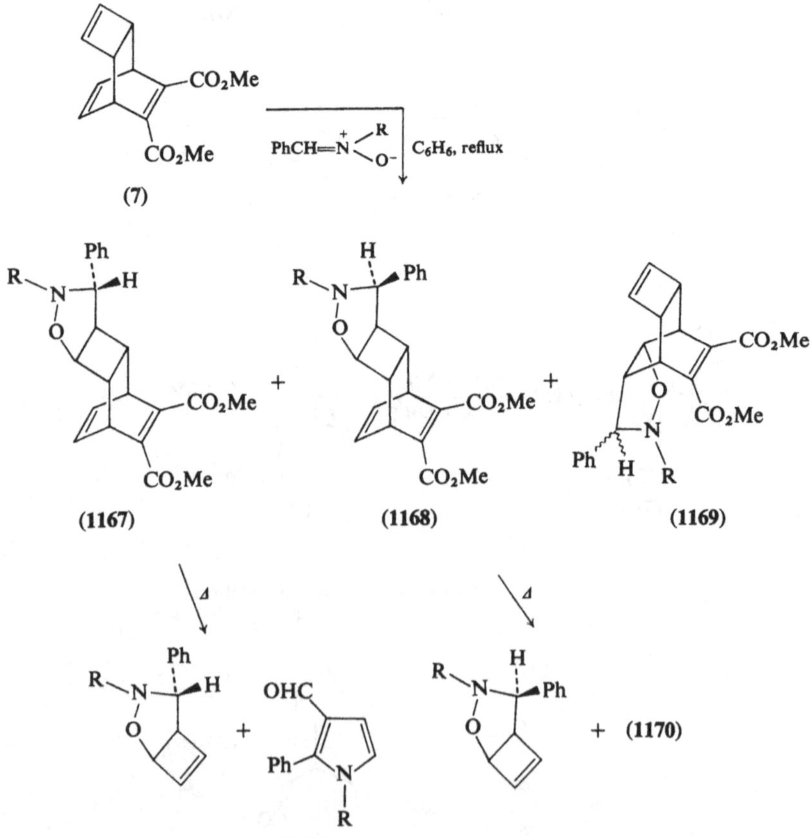

(7)

(1167)   (1168)   (1169)

(1170)

Table 130

Yield (%)

R	(1167)	(1168)	(1169)
Me	43	18	22
Ph	27	54	6
p-Cl.C_6H_4	25	57	(Very low)

Nitrones of structure PhCH=N(O)R afford a pair of stereoisomers (1167) and (1168), and a single stereoisomer of the other structural type (1169); when R is an aromatic group, Alder–Rickert fragmentation of the adducts (1167) and (1168) is accompanied by rearrangement to the 3-formylpyrrole (1170)[335] (table 130). The addition of the triphenylnitrone Ph_2C=N(O)Ph to the di-ester (7) is sluggish, and prolonged refluxing in toluene results in a mixture of the adduct (1171) (16%) and its thermolysis product (1172) (17%).[335]

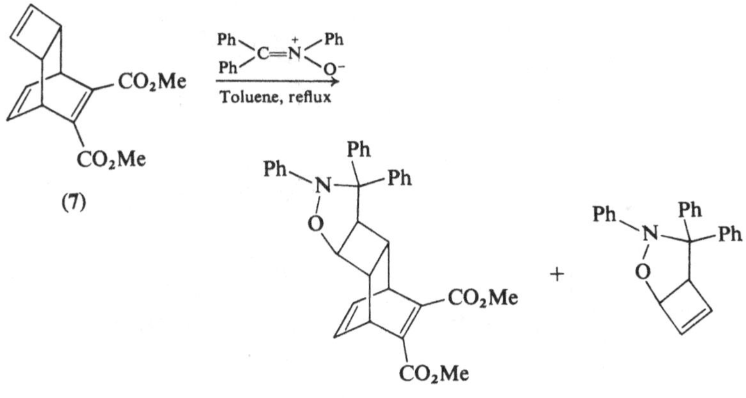

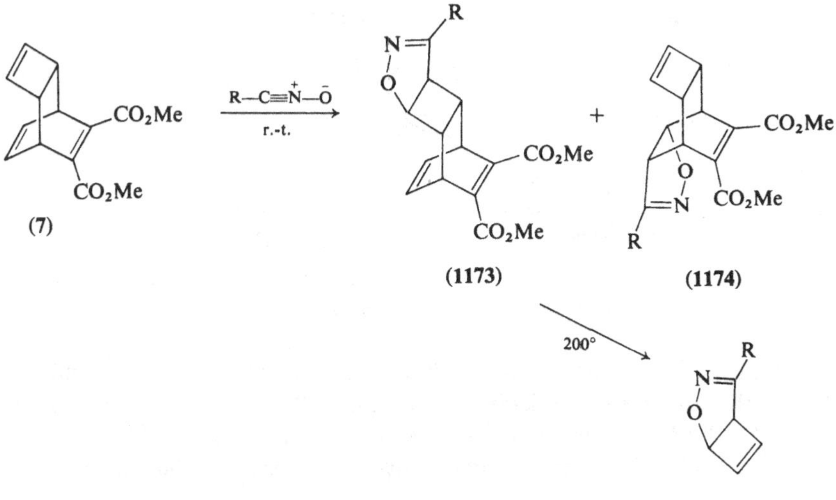

Table 131

Yield (%)

R	(1173)	(1174)
Ph	17	—
p-Br.C_6H_4	18.5	—

Nitrile oxides also afford two types of adduct, (1173) and (1174) (table 131); thermolysis of the former gives the expected products.[331]

The nitrile imine Ph—C≡N⁺—N⁻Ph cycloadds to each of the three double bonds in the di-ester (7); the isolated products are (1175) (18%), (1176) (51%), (1177) (18%) and the 2:1 adduct (1178) (2.3%).[336] (Further reactions of (1175)–(1177) are described in ref. 336.)

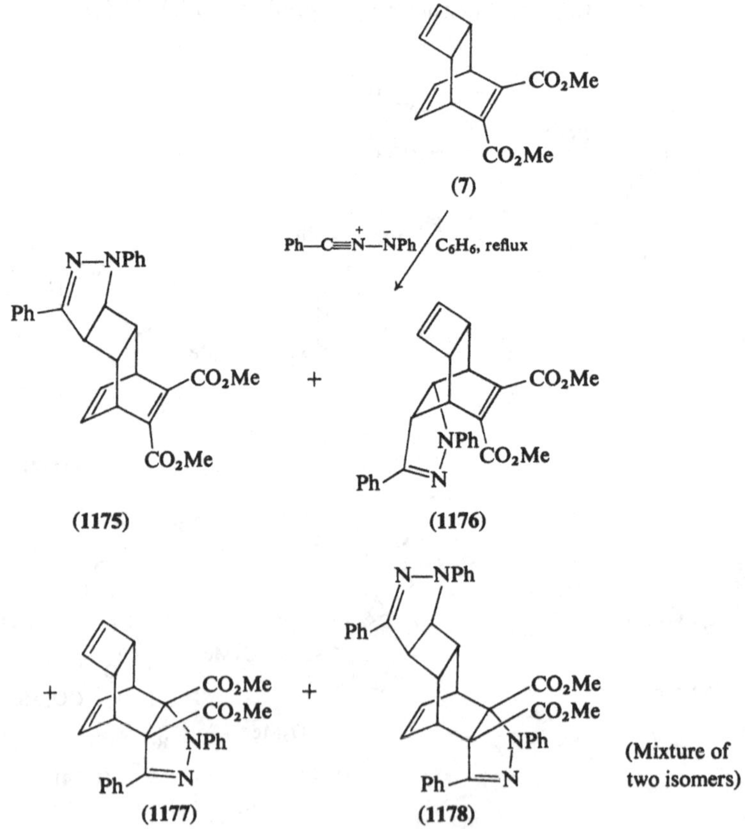

With 1,2,3,4-tetrachloro-5,5-dimethoxycyclopentadiene in refluxing chloroform, the di-ester (7) yields the *endo*-[4 + 2] adduct (1179) (up to 76%)[1090,1102,1103] Thermolysis of this product at *ca.* 200°,[1102,1103] or in refluxing xylene,[1090]) affords a quantitative yield of the cyclobutene (1180); if the initial Diels–Alder reaction is carried out in refluxing xylene, the product (**1181**) is formed.[1090] Photo-caging of the cyclobutene (**1180**) gives the homo-cubane derivative (**1182**) in nearly quantitative yield.[1090,1102,1103] (For further reactions of (**1182**), see refs. 1103, 1104.)

Hemicyclone reacts with the di-ester (**7**) to give a mixture of *endo*- and *exo*-adducts, (**1183**) and (**1184**) respectively, the *exo*-compound (**1184**) predominating (*ca.* 6:1).[1105] Alder–Rickert cleavage of (**1184**) affords the cyclobutene derivative (**1185**) in high yield.[1105]

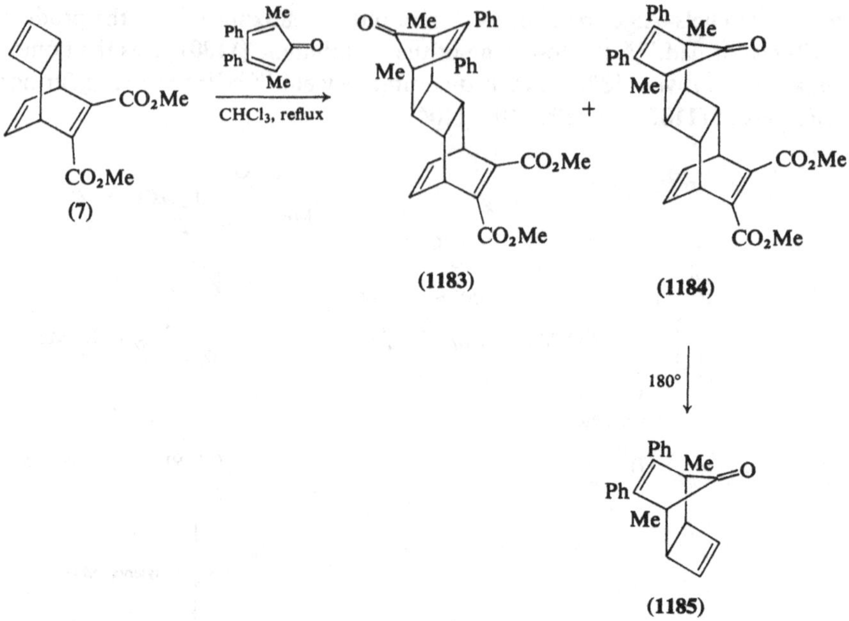
CO₂Me
CO₂Me
(7)
Me
Ph
Ph
O
Me
CHCl₃, reflux
O
Me
Ph
Me
Ph
CO₂Me
CO₂Me
(1183)
+
Ph
Me
O
Ph
Me
CO₂Me
CO₂Me
(1184)
180°
Ph
Me
O
Ph
Me
(1185)

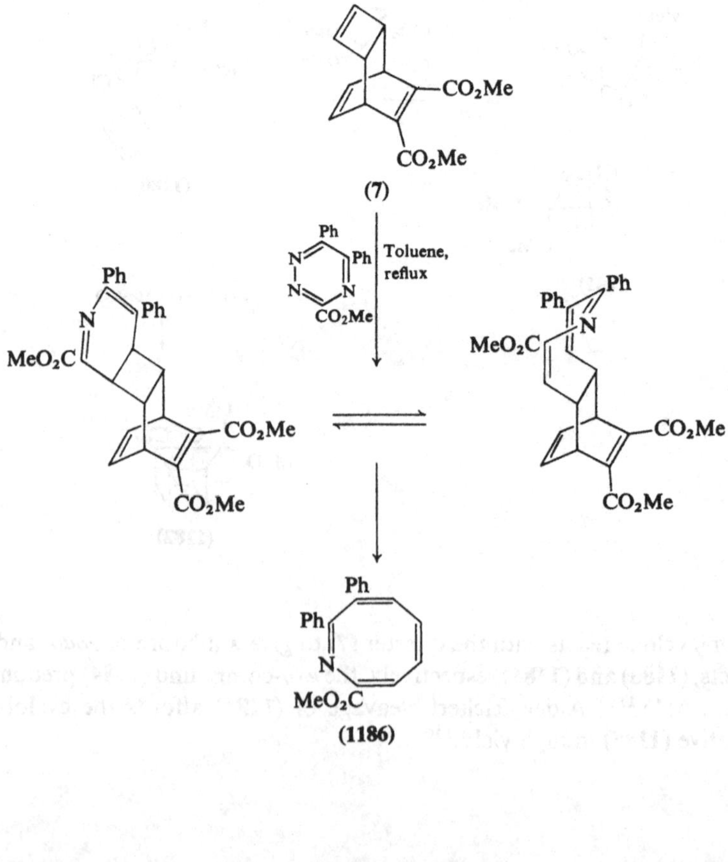
CO₂Me
CO₂Me
(7)
Ph
N
N
N
N
Ph
CO₂Me
Toluene,
reflux
Ph
N
Ph
MeO₂C
CO₂Me
CO₂Me
Ph
Ph
N
MeO₂C
CO₂Me
CO₂Me
Ph
Ph
N
MeO₂C
(1186)

Prolonged reaction of the di-ester (**7**) with 3-methoxycarbonyl-5,6-diphenyl-1,2,4-triazine in refluxing toluene gives the azocine (**1186**) (60%).[1095]

With 3,6-diaryl-*s*-tetrazines, the trienes (**7**) and (**1158**) form the products (**1187**), which have high thermal stability[1106] (table 132).

(**7**) (R = CO$_2$Me)   (**1187**)
(**1158**) (R = H)

Table 132

R	Ar	Yield (%)
CO$_2$Me	Ph	77
H	2-Pyridyl	—
CO$_2$Me	2-Pyridyl	80

Reactions with metal derivatives. The di-ester (**7**) forms a 1:2 complex (**1188**) with silver nitrate.[341] A dimeric rhodium complex (**1189**) is known.[339] With (PhCN)$_2$PdCl$_2$, a complex (**1190**) (70%) is produced.[341]

27. Tricyclo[4.2.2.0^{2,5}]deca-7,9-dienes

The tricyclo[4.2.2.0^{2,5}]deca-7,9-diene system results from the cycloaddition of bicyclo[4.2.0]octa-2,4-dienes to e.g. dimethyl acetylenedicarboxylate (see pp. 220, 248–9).

Thermolysis. Alder–Rickert fragmentation of the *cis*-disubstituted compounds (**805**) affords *cis*-3,4-disubstituted cyclobutenes (**1191**) (table 133). (*cis*-3,4-

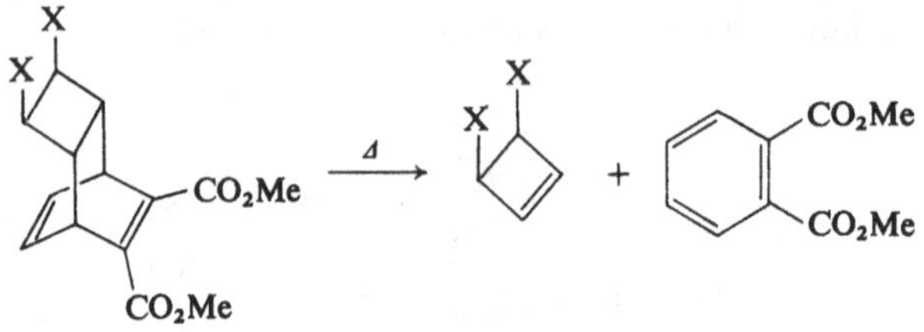

(805)   (1191)

Table 133

X	Conditions	Yield (%)	Ref.
H, (D)[a]	200°/100 mm	95[b]	260, (1107)
Cl[c]	190°/5 mm	—	228

[a] Hexadeuteriocyclobutene may be obtained from the analogous adduct prepared from decadeuteriocyclo-octa-1,3,5-triene.[262]

[b] Overall yield from COT is 34–39%.[353]

[c] A mixture of the *cis*- and *trans*-dichloro-compounds (**805a**) and (**806a**) may be used for the preparation of *cis*-dichlorocyclobutene (40–43% from COT).[228,929]

Dichlorocyclobutene is a convenient precursor of (cyclobutadiene)-Fe(CO)₃,[1108] and thence of cyclobutadiene itself.[1109])

This useful cyclobutene synthesis (see also p. 220) provides a route to cyclo-butenone from the adduct (**858**) of cyclo-octatrienone[1110] (see p. 261).

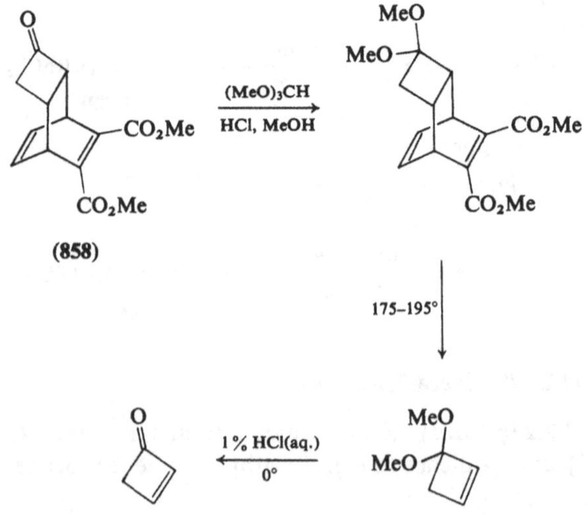

(858)

175–195°

1% HCl(aq.)
0°

In contrast, thermolysis of the *trans*-3,4-disubstituted compounds (**806**) leads to 1,4-disubstituted *E,E*-buta-1,3-dienes (**1192**), by conrotatory ring-opening of the intermediate cyclobutenes (**1193**) (table 134). (For synthetic uses of 1,4-diacetoxybutadiene, see ref. 207.)

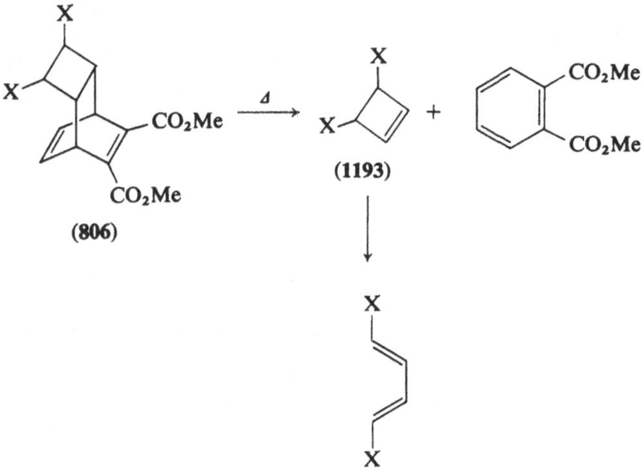

Table 134

X	Conditions	Yield (%)	Ref.
OAc	180–220°/18 mm	64	(11), 1111
	170–200°/18–20 mm	41–49[a]	207
Cl[b]	140–150°/17 mm	85	(11), 1111
Br	*ca.* 200°	80	11

[a] Overall yield from the diacetoxy-diene (**28**).
[b] The starting material was apparently a mixture of the *cis*- and *trans*-dichloro-compounds (**38/39**).

Reduction. For catalytic hydrogenation of various tricyclo[4.2.2.0²,⁵]deca-7,9-dienes, see ref. 11.

28. 9,10-Diazatricyclo[4.2.2.0²,⁵]deca-3,7-dienes

In general these diaza-compounds, of basic structure (**1194**), are best prepared *indirectly* from COTs, *viz.* by reaction with bromine, followed by [4 + 2] cycloaddition of azo-dienophiles, and finally debromination (see pp. 248–9).

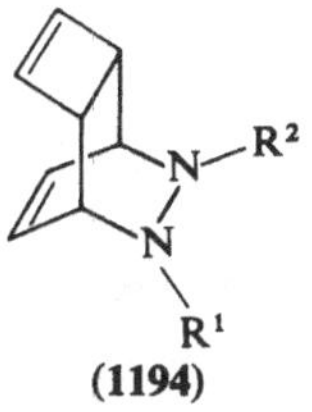

(1194)

Functional group transformations. See schemes 105 and 106.

Scheme 105

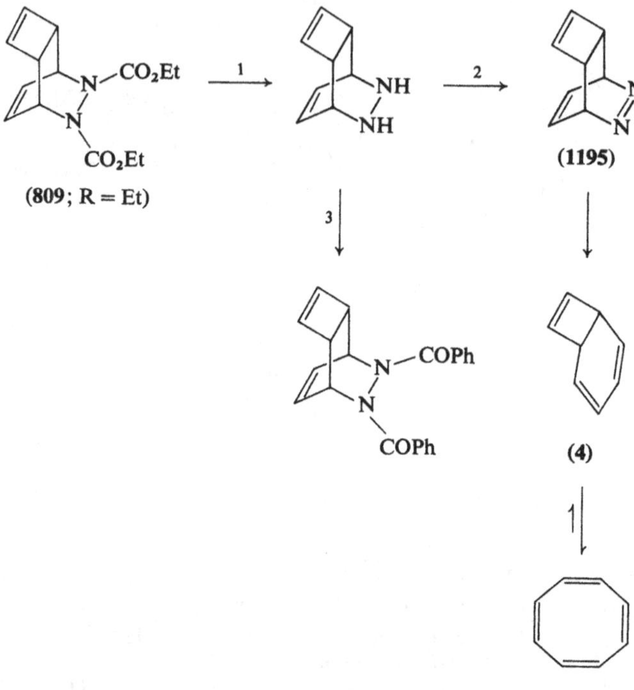

(809; R = Et)

(1195)

(4)

Reaction	Reagents and conditions	Yield (%)	Ref.
1	KOH, EtOH, 80°	78	40
2	MnO₂, n-hexane, −10° to −15°	—ᵃ	40
3	PhCOCl, NaOH(aq.), 0°	51	40

ᵃ The initial product (1195) extrudes nitrogen to give (4) (40–70%).

Scheme 106

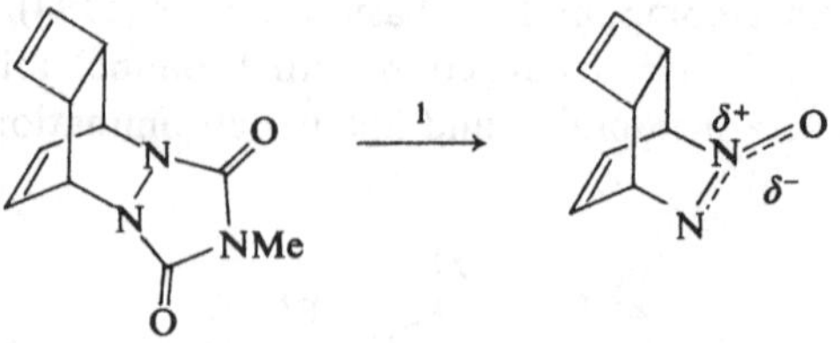

Reaction	Reagents and conditions	Yield (%)	Ref.
1	H₂O₂, KOH(aq.), 80–95°	65	1112

Thermolysis. At 350° the di-ester (**809**; R = Et) (see p. 249) rearranges to the bicyclic compound (**1196**) (73 %).[1113]

(**809**) (R = Et)   (**1196**)

Photolysis. U.v. irradiation of the 9,10-diazatricyclo[4.2.2.0²,⁵]deca-3,7-dienes (**1194**) in the presence of acetone furnishes the cage-compounds (**1197**) (table

(**1194**)   (**1197**)

Table 135

R¹	R²	Yield (%)	Ref.
CO₂Et	CO₂Et	80	40
	(acetyl-NMe-acetyl)	—	1102
	(acetyl-NPh-acetyl)	84–85	569, 1114

135). The *C*-substituted derivatives (**287**)–(**290**) (see pp. 104–5) likewise afford the products (**1198**)–(**1200**)[569] (table 136).

Silver-catalysed rearrangement of diazabasketanes, e.g. (**1197**), yields diazasnoutanes, e.g. (**1201**);[41, 569, 1114, 1115] for the transformation of diaza-snoutanes into semibullvalenes, see refs. 41, 716, 1114, 1115.

(287) $\xrightarrow{h\nu}$ (1198)

(288) $\xrightarrow{h\nu}$ (1199)

(289) $\xrightarrow{h\nu}$ (1200)

(290) $\xrightarrow{h\nu}$ (1199)

$R^2 = Ph$

Table 136

	R^1	Yield (%)		R^1	Yield (%)
(287)	Me	72	(288)	Me	93
	Ph[a]	30		Ph	42
	CH_2OAc	82		CH_2OAc[b]	72
	CH_2OMe	24		CH_2OMe	34
	CN	92		CN	80
(289)	CH_2OAc	70		F	56
	CH_2OMe	64			
	CO_2Me	35	(290)	Ph	21
	CN	50		CH_2OAc[b]	72

[a] *N*-Methyl-compound used.   [b] Mixture (1 : 1) of (288) and (290)

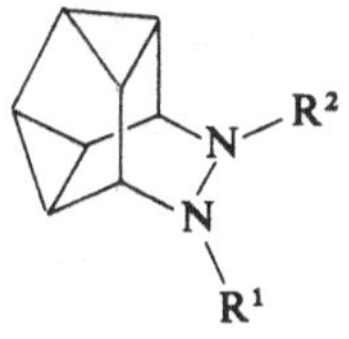

(1201)

Bromination. The addition of bromine to the derivative (125a) (p. 46) apparently leads to a mixture of the *cis*- and *trans*-dibromides, (1202a) (20%) and (1203) (40%).[931] (For the *cis*-dichloride (1202b), see ref. 362.)

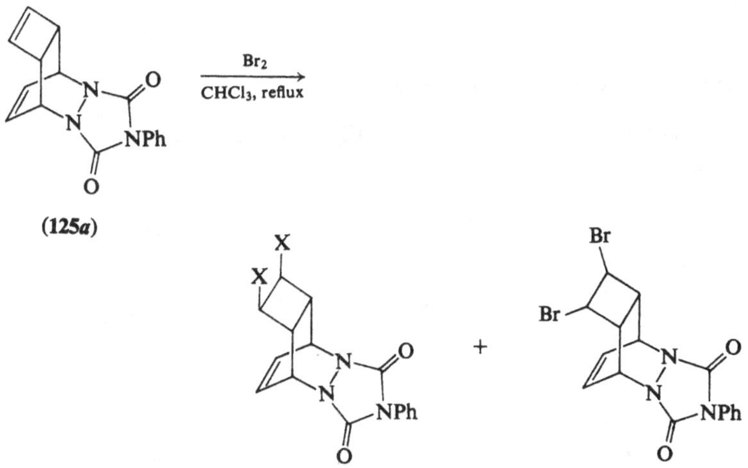

(125a)

(1202)   (1203)

a: X = Br
b: X = Cl

Reduction. Catalytic hydrogenation of the diaza-compounds (1194) in the presence of palladium-on-carbon affords the tetrahydro-derivatives (1204) (table 137). (For further reactions of system (1204), see refs. 874, 1117.)

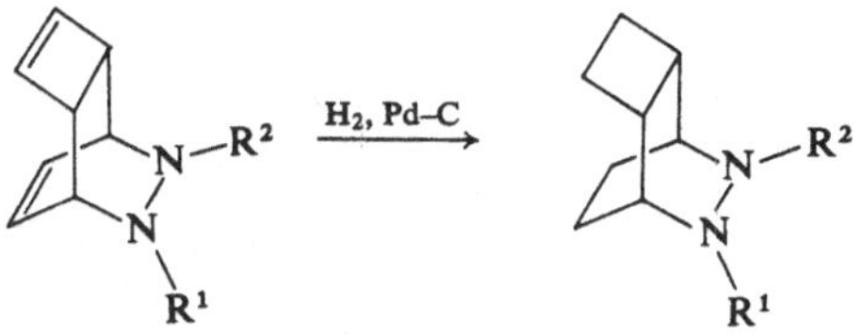

(1194)   (1204)

Table 137

R¹	R²	Solvent	Yield (%)	Ref.
COPh	COPh	EtOAc	82	40
CO₂Et	CO₂Et	MeOH	88	40
(see structure) NMe		EtOAc	100	930
(see structure) NPh		EtOAc	—	362

29. Oligomers of cyclo-octatetraenes

Four dimers, **(9)–(12)**, and a tetramer **(23)** result from thermal treatment of COT (see pp. 11–17).

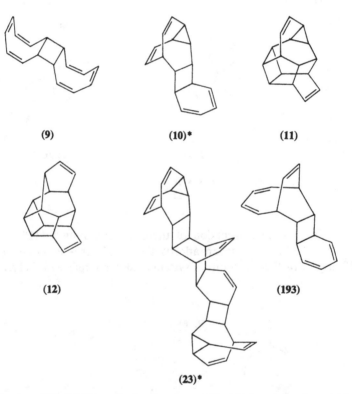

(9)   (10)*   (11)

(12)   (193)

(23)*

* Structures **(10)** and **(23)** are fluxional; only one of the interconverting forms is illustrated.

Additionally, a fifth dimer, (**193**), is produced by the action of $H_4Ru_4(CO)_{12}$ on COT (see pp. 67–8).

(a) Dimer (9)

Thermolysis. At 100°, dimer (**9**) slowly produces COT and dimer (**10**).[159, 169]

Photolysis. U.v. irradiation of dimer (**9**) provides a convenient route to [16]-annulene (**1205**) (60%);[1118, 1119] this polyene has a half-life of *ca.* 44 h at 20°, gradually transforming into the tricyclic isomer (**1206**) (90%).[1119]

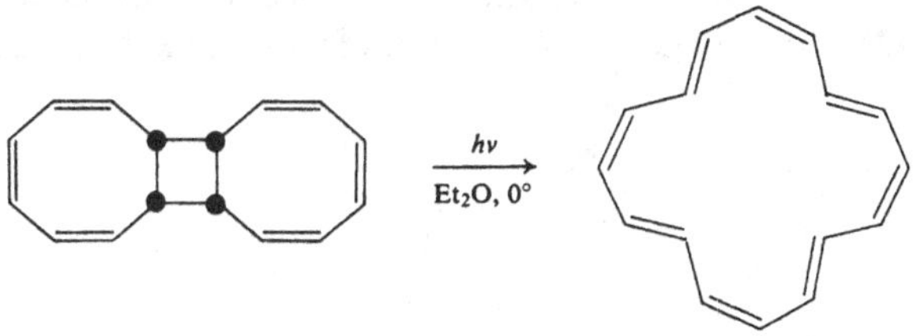

$$\xrightarrow[\text{Et}_2\text{O, 0}^\circ]{h\nu}$$

(9)   (1205)

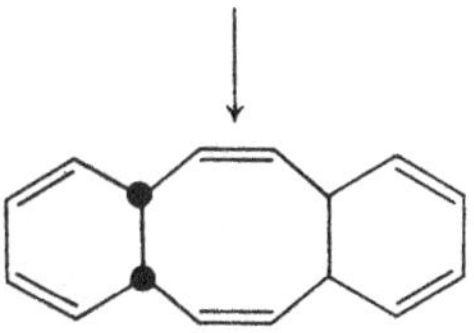

(1206)

Photolysis of the substituted dimers (**278**) (see p. 102) similarly generates the monosubstituted [16]annulenes (**1207**) in low yields.[160, 580]

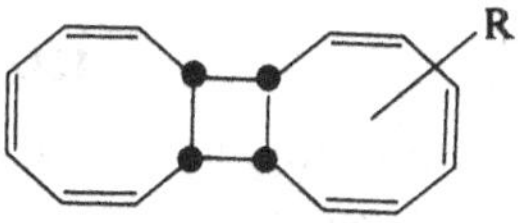

$R = Ph, CO_2Me, F, Cl$

(278)

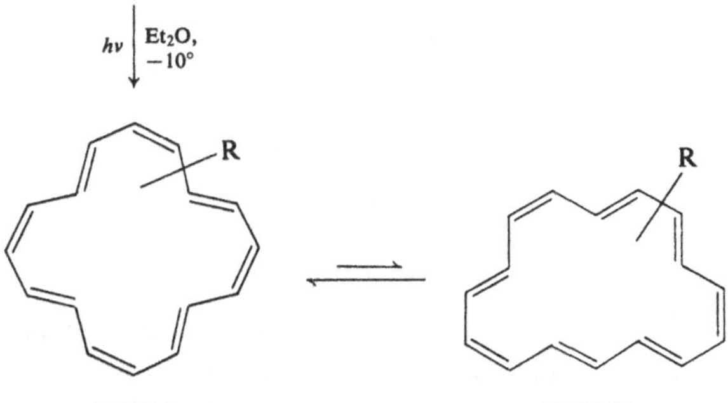

$$h\nu \left|\begin{array}{c}\text{Et}_2\text{O,}\\-10°\end{array}\right.$$

(1207a)   (1207b)

Photolysis of [16]annulene affords the stereoisomer **(1208)**.[1119] For the dication **(1209)**, see ref. 1120.

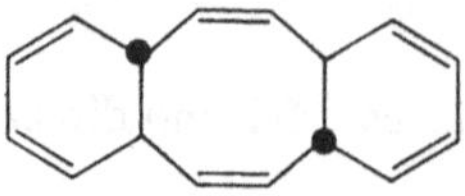

(1208) 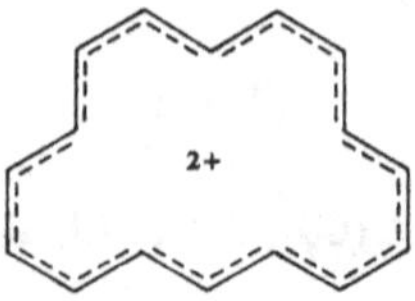   (1209)

Oxidation. Treatment of dimer **(9)** with peracetic acid gives the epoxide **(1210)** as the main product (*ca.* 45%),[1121,1122] together with small amounts of the isomers **(1211)** and **(1212)**.[1122] Photolysis of these epoxides leads to oxa[17]-

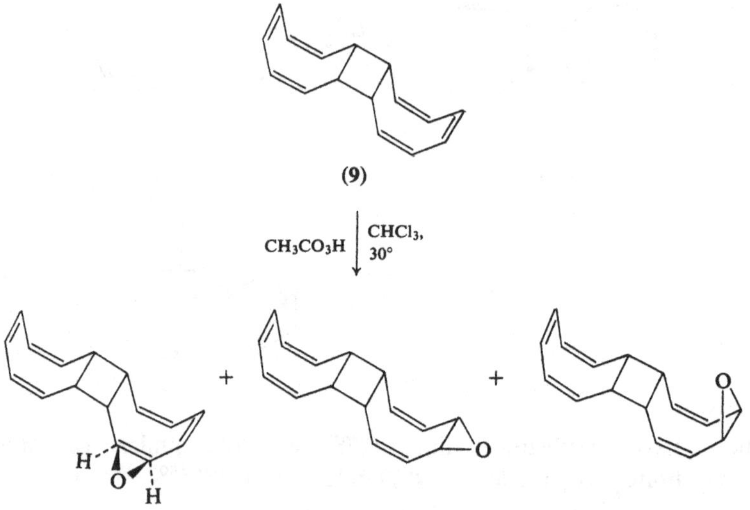

(9)

CH₃CO₃H | CHCl₃, 30°

(1210)   (1211)   (1212)

annulenes (differing in the sequence and individual numbers of Z and E double bonds), one of which may be assigned the provisional structure **(1213)**.[1122]

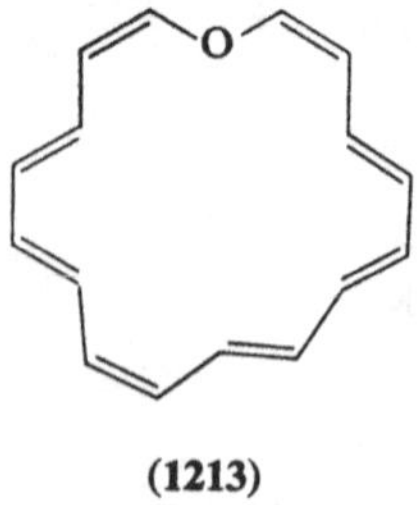

(1213)

Reduction and deprotonation. Dimer **(9)** reacts with alkali metals to give the radical anion of [16]annulene **(1214)**, which equilibrates with the dianion **(1215)** and the neutral hydrocarbon **(1205)**[1123] (*cf.* COT⁻, p. 28).

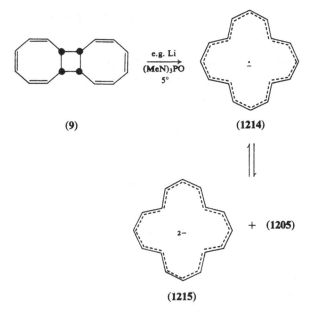

(9)   (1214)

+ (1205)

(1215)

Catalytic hydrogenation of dimer **(9)** gives a hexahydro-derivative,[156, 159] which may now be formulated as **(1216)**.

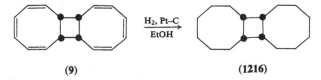

(9)   (1216)

Deprotonation of **(9)** with potassium amide in liquid ammonia affords the dianion **(1217)** (see p. 217), which with iodine gives the extremely labile product **(1218)**.[1038]

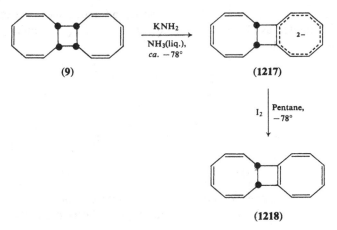

(9)   (1217)

(1218)

Dimer (**9**) catalyses H–D exchange in COT in the presence of KOCPr₃ⁿ–Pr₃ⁿCOD (see p. 31).

Cycloadditions. The reaction of dimer (**9**) with diazoacetic ester in the presence of copper powder results in the products (**1219**) (*ca.* 11 %) and (**1220**) (*ca.* 3 %), together with an inseparable mixture (*ca.* 6 %) of (**1221**) and (**1222**).[1124] For

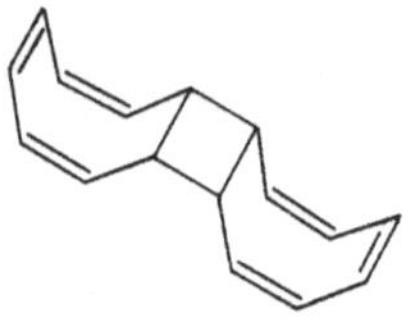

(9)

N₂CHCO₂Et $\bigg|$ Cu, 100°

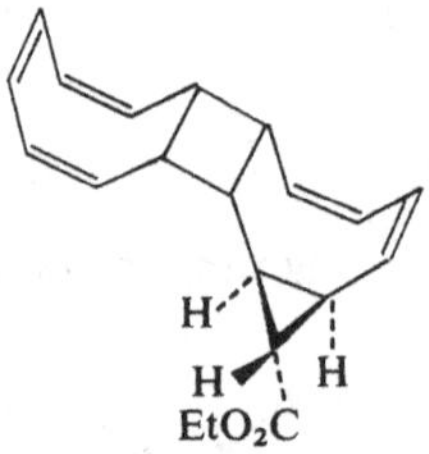

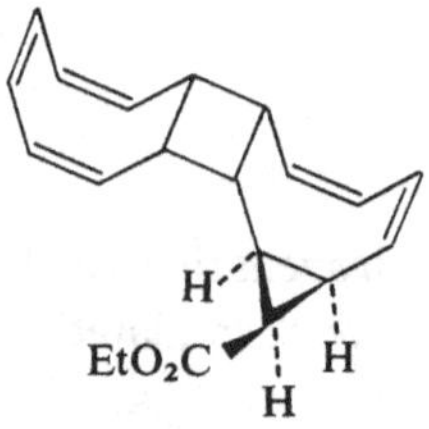

(1219)

(1220)

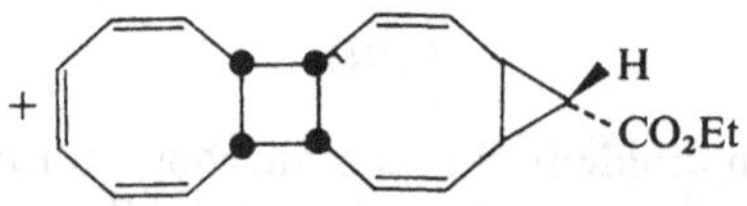

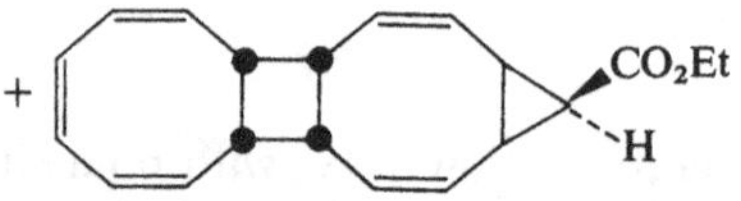

(1221)

(1222)

the conversion of these products into cycloheptadecaoctaenes, one of which may be formulated as (**1223**), and the generation of the [17]annulenyl anion (**1224**), see ref. 1124.

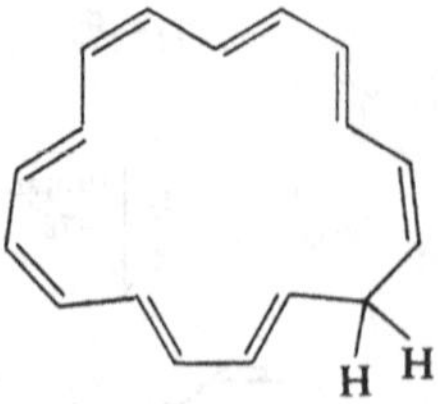

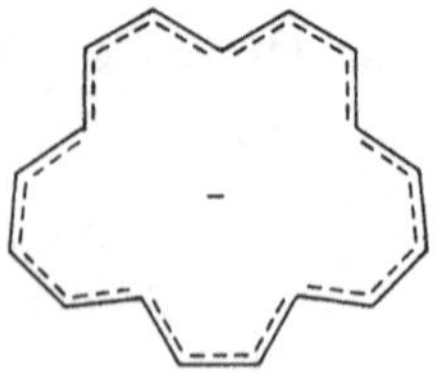

(1223)

(1224)

Addition of *N*-(ethoxycarbonyl)nitrene produces the adducts (**1225**) (*ca.* 20 %), (**1226**) (*ca.* 9 %), and (**1227**) (*ca.* 9 %), the latter probably resulting

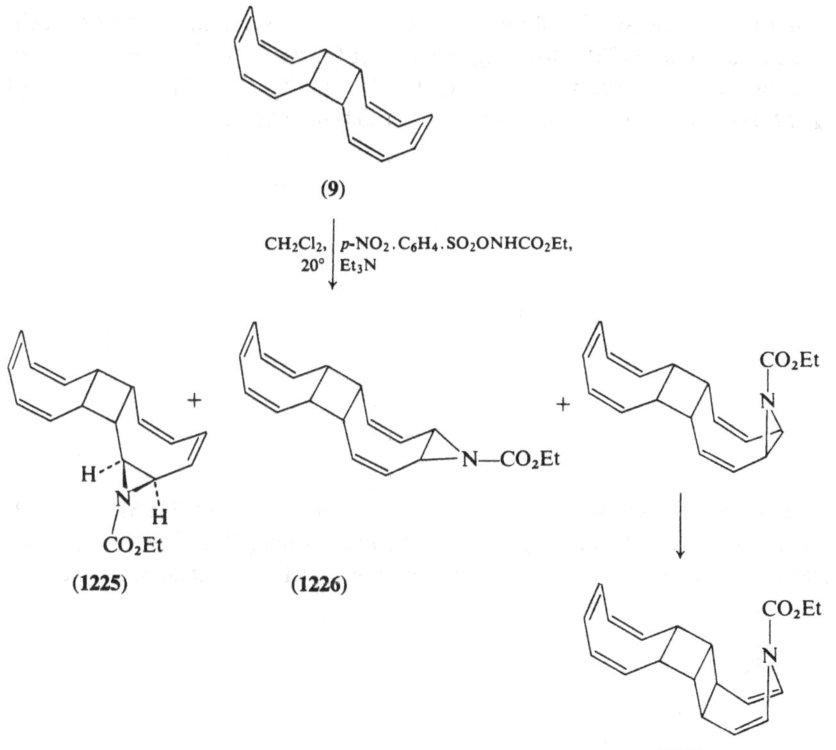

CH₂Cl₂, 20°
p-NO₂.C₆H₄.SO₂ONHCO₂Et, Et₃N
H
N
H
CO₂Et
N—CO₂Et
CO₂Et
N
CO₂Et
N

(9) **(1225)** **(1226)** **(1227)**

from a Cope rearrangement.[1125] Photolysis of the aziridines (**1225**) and (**1226**) affords a mixture of three aza[17]annulenes (configurations uncertain).[1125]

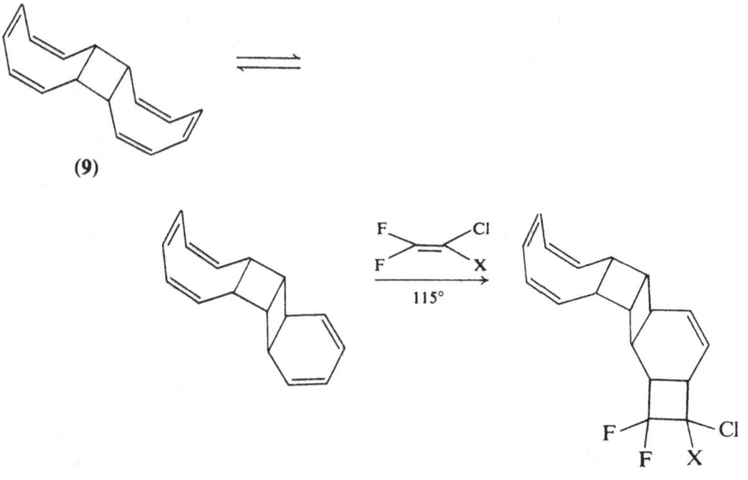

F
F
Cl
X
115°
F
F
Cl
X

(9)

(1228) **(1229)** *a*: X = Cl
b: X = F

With 1,1-dichloro-2,2-difluoroethylene, a [2 + 2] cycloaddion occurs *via* the valence tautomer (**1228**), forming the adduct (**1229a**) (21 %); chlorotrifluoroethylene reacts similarly to give (**1229b**) (*ca.* 0.8 %).[1126] The conversion of (**1229a**) into the 1,2-dihalo[18]annulene (**1230**) may be achieved.[1126]

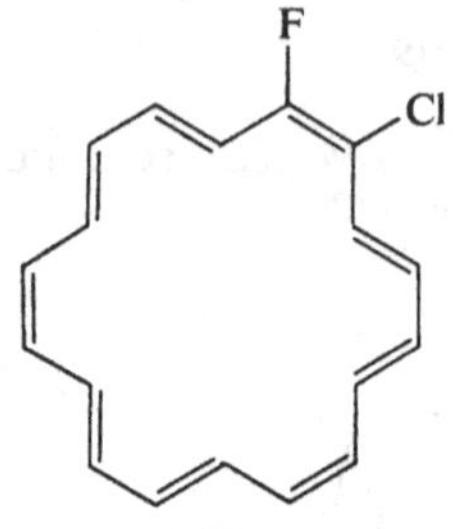

(**1230**)

Dimer (**9**) reacts with dienophiles *via* the valence tautomer (**1228**); thus with maleic anhydride, the initial product is the 1:1 adduct (**1231**) (stereochemistry presumed) (table 138). Further reaction with maleic anhydride produces the

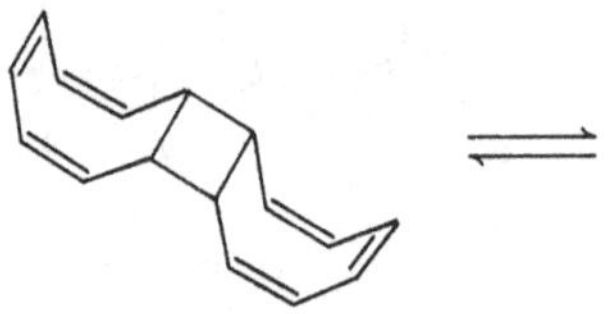

(**9**)

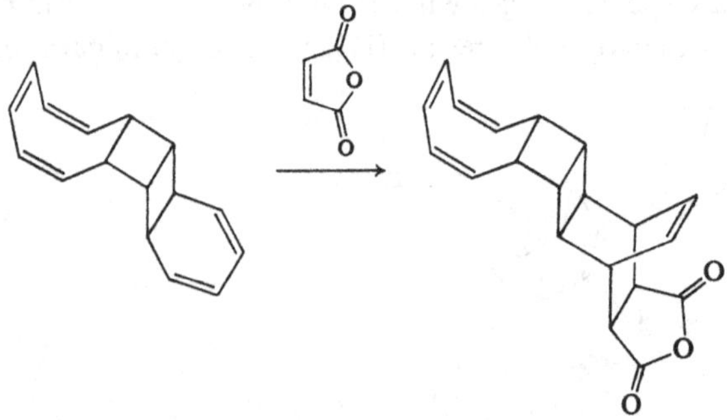

(**1228**)   (**1231**)

Table 138

Dienophile	Reaction conditions	Yield (%)	Ref.
Maleic anhydride	—	—	156
	THF, 100°	51	159
	C$_6$H$_6$, reflux	33	1127
p-Benzoquinone	—	—	156

2:1 adduct (**1232**) (84% from (**9**)), which at *ca.* 180° undergoes a symmetry-forbidden ring-opening to give the product (**1233**) (81%);[1127] an analogous product is formed by the reaction of dimer (**9**) with citraconic anhydride at 180°.[1127,1128]

(1231)

Toluene, reflux

o-Dichlorobenzene
Reflux

(1233)   (1232)

The reaction of dimer (**9**) with dimethyl acetylenedicarboxylate at 80° also gives 1:1 (**1234**) and 2:1 (**1235**) adducts, but with an excess of the dienophile at 100–120° compound (**1236**) is produced;[34] presumably, Alder–Rickert cleavage of the 2:1 adduct (**1235**) is followed by a 'forbidden' cyclobutene ring-opening and then further addition of the dienophile. At 150°, the 1:1 adduct (**1234**) fragments to give the hydrocarbon (**1237**); this adds acetylene-dicarboxylic ester to form the product (**1238**).[34] As a consequence of these transformations, the compounds (**9**), (**1234**), (**1235**), (**1237**) and (**1238**) all yield the product (**1236**) when treated with an excess of dimethyl acetylene-dicarboxylate at 100–120°.[34]

The thermal conversion of the hydrocarbon (**1237**) into the bridged isomer (**1239**) is described in refs. 34 and 933; photochemical rearrangement at −100° leads to the [12]annulene (**1240**).[1129]

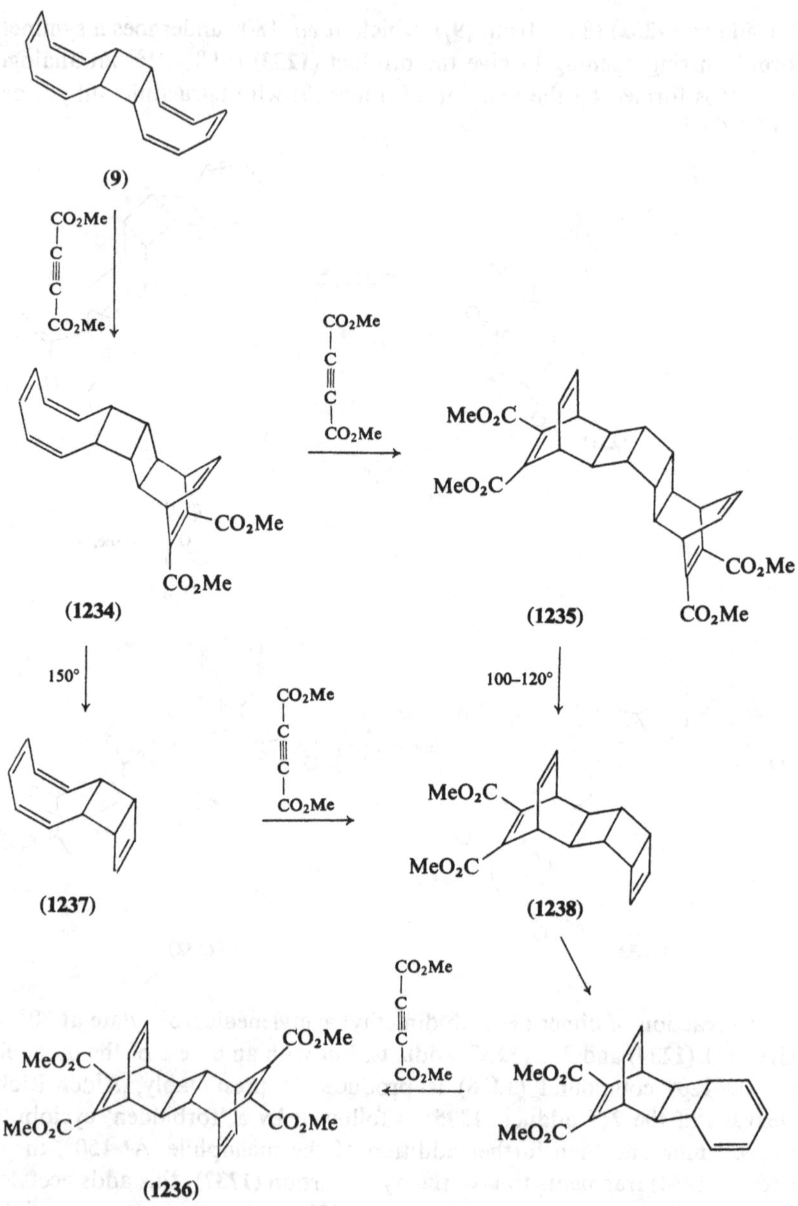
(9)
CO_2Me–C≡C–CO_2Me
CO_2Me–C≡C–CO_2Me
MeO_2C
MeO_2C
CO_2Me
CO_2Me
(1234)
(1235)
CO_2Me
CO_2Me
150°
100–120°
CO_2Me–C≡C–CO_2Me
MeO_2C
MeO_2C
(1237)
(1238)
CO_2Me–C≡C–CO_2Me
MeO_2C
MeO_2C
CO_2Me
CO_2Me
MeO_2C
MeO_2C
(1236)

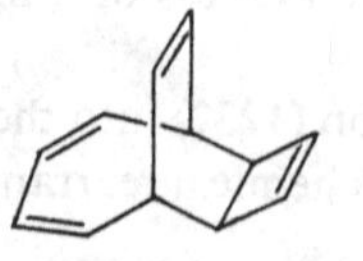

(1239)

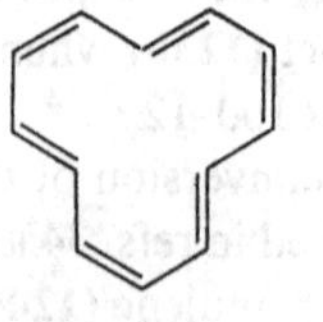

(1240)

Reactions with metal derivatives. Dimer **(9)** forms a silver nitrate complex **(1241).**[156]

(9)   (1241)

Iron carbonyl complexes of dimer (9) can be produced from COT and (COT)Fe(CO)$_3$ (see p. 64). Dimer (9) reacts with Fe(CO)$_5$ when heated and irradiated with u.v. light, giving the complexes (180) and (179)[470] (see p. 63); both (180) and (179) react further with Fe(CO)$_5$ to form the product (1242),[470, 1130] which on thermolysis gives (1243) and (1244).[470]

(b) *Dimer* (10)

Thermolysis. Thermal treatment of dimer (10) produces dimer (11), which rearranges to dimer (12) (see pp. 11–13); in addition, dimer (10) undergoes Diels–Alder dimerisation to afford the tetramer (23) (see p. 17).

Photolysis. The photolysis of dimer (10) provides an efficient synthesis of bullvalene (892) (75%).[1131]

(10)   (892)

Reduction. Catalytic hydrogenation of dimer (10) affords octahydro- and decahydro-derivatives of uncertain structure.[156, 159]

Reaction with base. With potassium t-butoxide, dimer (10) isomerises to give the product (1245) (70–80%) (which is convertible into phenylbullvalene).[1132]

(10)   (1245)

Cycloadditions. Dimer (10) is highly reactive as the diene component in Diels–Alder additions, e.g. it reacts with maleic anhydride to form the adduct (1246) (stereochemistry presumed). Similar adducts are obtained from other olefinic dienophiles (table 139). For conversion of the vinylene carbonate-adduct (1247) into the C$_{18}$-hydrocarbon (1248), and further reactions, see refs. 1134, 1135.

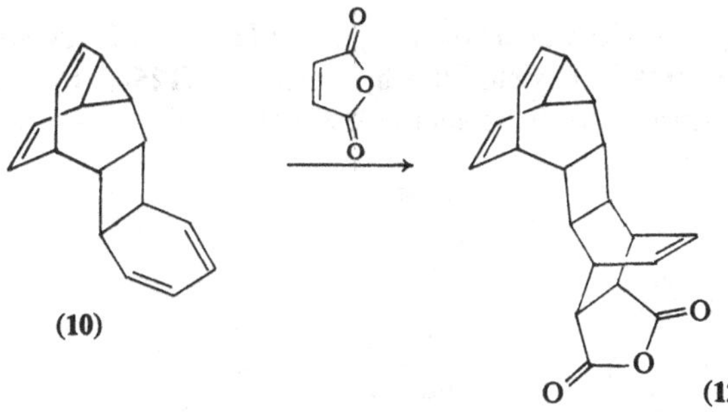

(10)

(1246)

Table 139

Dienophile	Reaction conditions	Yield (%)	Ref.
Maleic anhydride	Et$_2$O, 20°	85	159
Diethyl fumarate	Et$_2$O, 20°	73	159
Citraconic anhydride	o-Dichlorobenzene	25	1128
Tetracyanoethylene	C$_6$H$_6$, r.-t.	66	164
Cyclopent-2-enone	COT, reflux[a]	5	164
p-Benzoquinone	—	—	156
Vinylene carbonate	COT, 180°[a]	—	1133

[a] The dienophile reacted with the dimer (10) formed *in situ*.

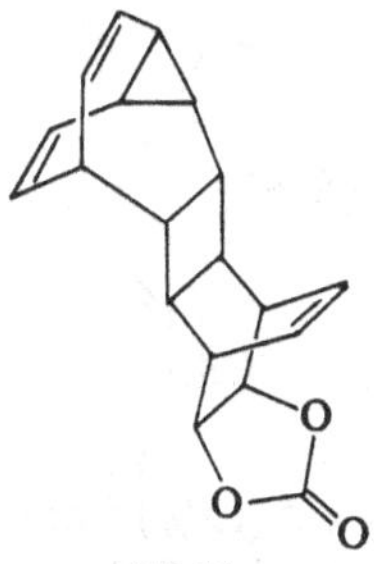

(1247)

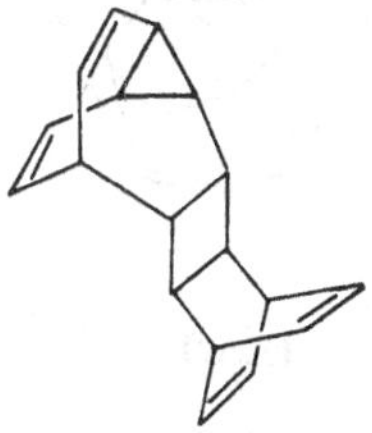

(1248)

The homotropilidene system of these Diels–Alder adducts is capable of undergoing a formal [2 + 2 + 2] cycloaddition with tetracyanoethylene; thus in refluxing toluene the vinylene carbonate-adduct (1247) reacts to give the product (1249) (53 %).[164]

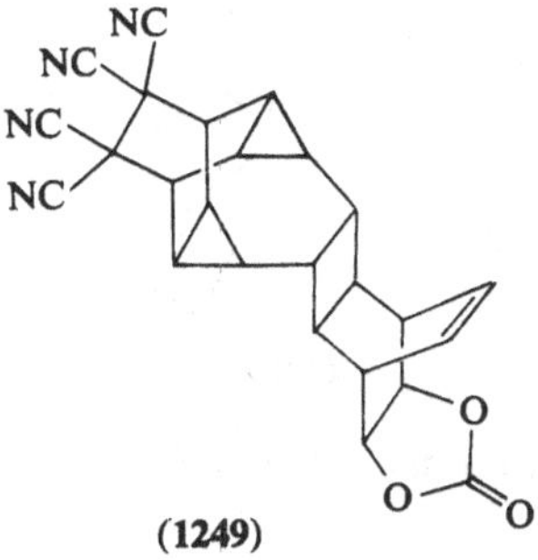

(1249)

With dimethyl acetylenedicarboxylate, dimer (**10**) affords the adduct (**1250**) (89%), which at 160–200° yields the hydrocarbon (**1251**) (67%);[159] at 480° this product rearranges to the isomer (**1252**) (31%).[1136]

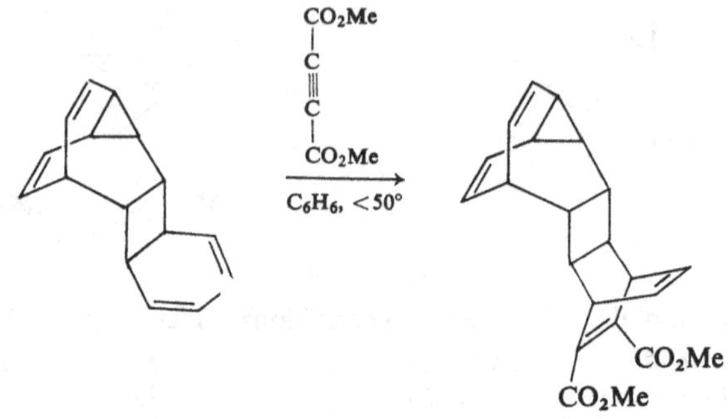
CO_2Me
C
C
CO_2Me
C_6H_6, <50°
CO_2Me
CO_2Me

(**10**) (**1250**)

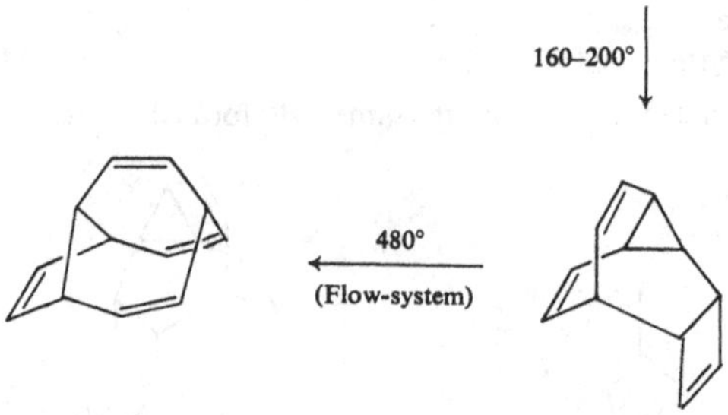
160–200°
480°
(Flow-system)

(**1252**) (**1251**)

An adduct of dimer (**10**) is formed in low yield (15%) from COT and cyclopentadiene at 190°.[998] An independent synthesis of the product indicates structure (**1253**), probably formed by Cope rearrangement following an initial Diels–Alder reaction in which dimer (**10**) adopts a dienophilic role.[164]

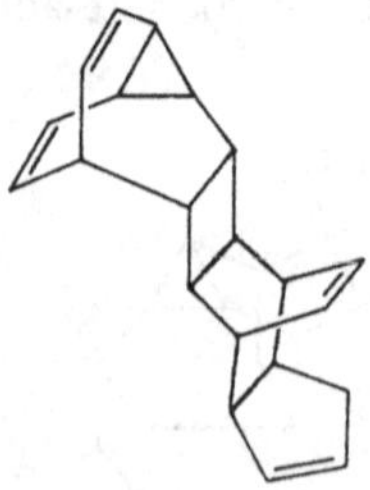

(**1253**)

The Diels–Alder dimerisation of dimer (**10**) has already been mentioned (see p. 14).

Reactions with metal derivatives. Dimer (10) reacts with silver nitrate to form a complex (1254).[156]

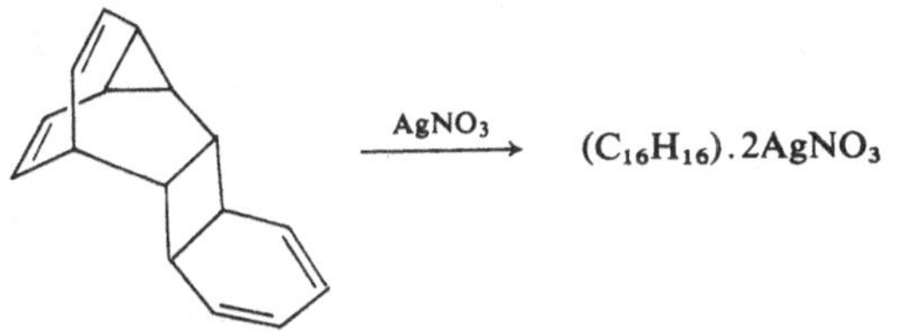

$AgNO_3$ $(C_{16}H_{16}) \cdot 2AgNO_3$

(10)   (1254)

Iron carbonyl complexes of dimer (10) are produced from $(COT)Fe(CO)_3$ and COT (see p. 64), and these may also be prepared directly from dimer (10). Thus under mild reaction conditions $Fe_2(CO)_9$ affords the product (1255) (15%), which readily loses carbon monoxide with the formation of the tricarbonyl complex (179); this is also produced from dimer (10) and $Fe_3(CO)_{12}$ in refluxing toluene.[470] Reaction of dimer (10) with $Fe(CO)_5$ under vigorous

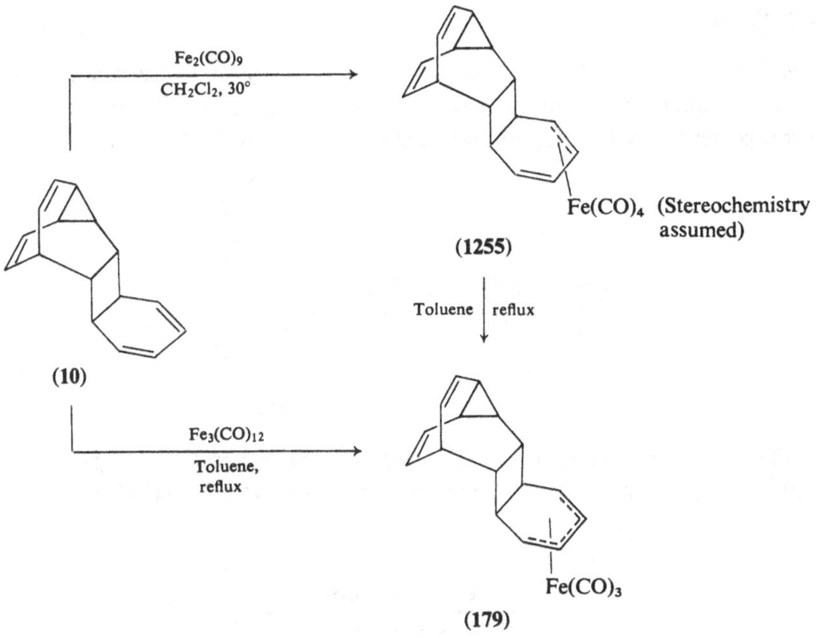

$Fe_2(CO)_9$
$CH_2Cl_2, 30°$

$Fe(CO)_4$ (Stereochemistry assumed)

(1255)

(10)

Toluene | reflux

$Fe_3(CO)_{12}$
Toluene, reflux

$Fe(CO)_3$

(179)

conditions produces the binuclear complex (1244) (3%), together with an apparent derivative (1243) of dimer (9) (4%), and a small amount of a complex (1256) (structure unknown).[470] Complexes (1244) and (1243) may also be obtained from (1242), which may be prepared from (179) and $Fe(CO)_5$[470] (see p. 367).

(10)   **(1244)**

(Stereochemistry assumed)

+   + $(C_{16}H_{16})Fe(CO)_3$

(1243)   **(1256)**

(c) Dimer (11)

Thermolysis and photolysis. Dimer (11) undergoes a thermal vinylcyclopropane → cyclopentene rearrangement to give dimer (12) (see p. 13); the same rearrangement may be effected photochemically (65–70%).[172]

(11)   **(12)**

Reduction. Reduction of dimer (11), using di-imide (generated from *p*-toluene-sulphonylhydrazine), affords the tetrahydro-derivative (1257) (93%).[164]

(11)   **(1257)**

Catalytic hydrogenation of dimer (11) gives tetrahydro- and hexahydro-derivatives of uncertain structure.[11,155]

Reactions with metal derivatives. Dimer (11) forms 1:1 (1258) and 1:2 (1259) complexes with silver nitrate.[155]

$$\xrightarrow[\text{EtOH, HOCH}_2\text{CH}_2\text{OH}]{\text{AgNO}_3} \quad (\text{C}_{16}\text{H}_{16}).\text{AgNO}_3$$

(1258)

$$\downarrow \text{AgNO}_3\text{(aq.)}$$

(11)

$$(\text{C}_{16}\text{H}_{16}).2\text{AgNO}_3$$

(1259)

(*d*) *Dimer* (12)

The two double bonds of structure (12) show a marked difference in reactivity, the cyclopentene double bond resembling that in a norbornene system.

Oxidation. Stereoselective epoxidation of the cyclopentene double bond may be effected to give the epoxide (1260) (84%).[164]

$$\xrightarrow[\text{CH}_2\text{Cl}_2, \text{ r.-t.}]{m\text{-Cl.C}_6\text{H}_4.\text{CO}_3\text{H}}$$

(12)   (1260)

Reduction. Catalytic hydrogenation of dimer (12) affords a tetrahydro-derivative,[155] presumably of structure (1261).

$$\xrightarrow[\substack{\text{EtOH,}\\ 50°}]{\text{H}_2, \text{PtO}_2}$$

(12)   (1261)

Cycloadditions. Reaction of the dimer (12) with dichlorocarbene results in the product (1262) (52%).[164] Cyclopropanation of both double bonds of the dimer **(12)**, followed by hydrogenolysis of the cyclopropane rings, affords a dimethyl-compound **(1263)**, which on treatment with aluminium bromide–t-butyl bromide rearranges to triamantane **(1264)** (2–5 %).[1137]

1,3-Dipolar addition of phenyl azide to dimer **(12)**, followed by thermal extrusion of nitrogen, leads to the *N*-phenylaziridine **(1265)** (69 %).[164]

[4 + 2] Cycloaddition of cyclohexa-1,3-diene to dimer (12) affords a 1:1 adduct, the catalytic hydrogenation of which results in the tetrahydro-derivative (1266); rearrangement in the presence of aluminium bromide–t-butyl bromide then gives, not the hoped-for tetramantane system, but the isomer (1267) ('bastardane') (5–8 %).[1138]

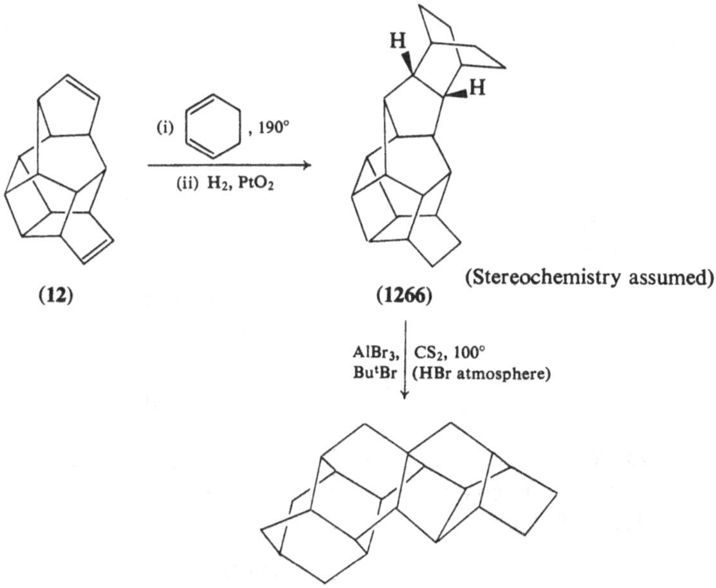

(i) [cyclohexa-1,3-diene], 190°

(ii) H₂, PtO₂

(12)   (1266)

(Stereochemistry assumed)

AlBr₃, | CS₂, 100°
BuᵗBr | (HBr atmosphere)

(1267)

Reaction with metal derivatives. Dimer (12) forms a complex (1268) with silver nitrate[155] (for the X-ray crystal structure of (1268), see refs. 161, 162).

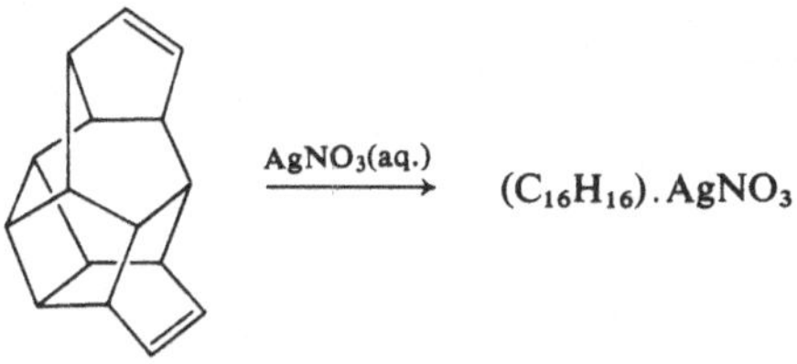

AgNO₃(aq.)
────────→ (C₁₆H₁₆).AgNO₃

(12)   (1268)

(e) *Tetramer* (23)

Thermolysis. The tetramer (23) fragments at *ca.* 300° to give dimer (12) in good yield (78 %), presumably *via* a *retro*-Diels–Alder reaction and subsequent reorganisation (see p. 13) of dimer (10).[164]

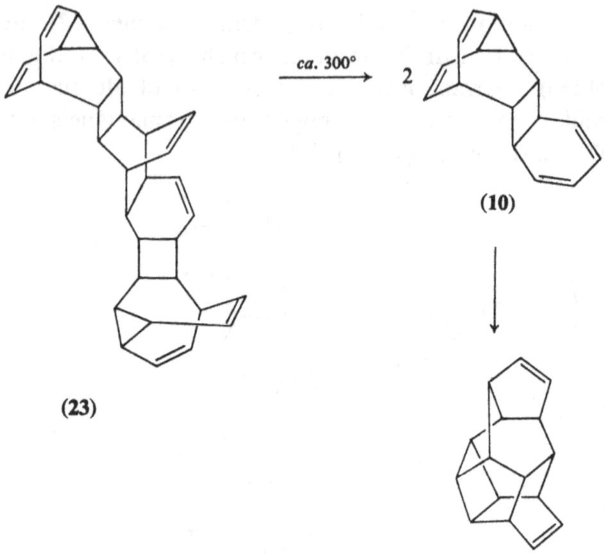

(23)

(10)

(12)

Cycloadditions. With tetracyanoethylene, both of the homotropilidene systems of the tetramer (23) undergo formal [2 + 2 + 2] cycloaddition, resulting in the 2:1 adduct (1269) (54%).[164]

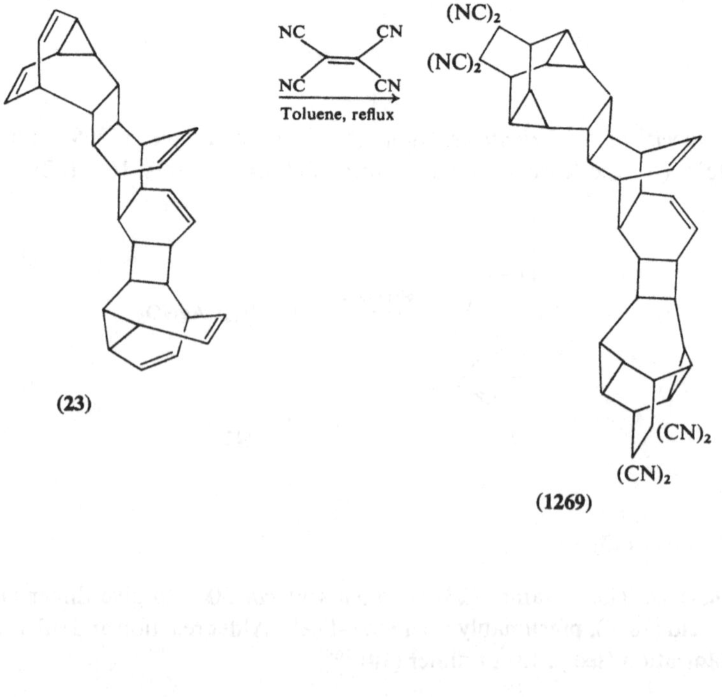

(23)

(1269)

Appendix

In order to make the work as complete and up-to-date as possible, the following additional material was collected after the completion of the main typescript; *Chemical Abstracts* was consulted to the end of vol. **86** and the literature has been covered, with a few inevitable omissions, up to the end of 1976 (only the *keyword* indexes for *Chem. Abs.* **85** and **86** were available).

The pages to which the new material relates are given in the margin.

page

1. Cyclo-octatetraene

2 The acetylene tetramerisation catalysis is now thought to involve a di-metal centre.[1116]

6 For electrochemical potentials (oxidation and reduction) of COT, using various electrodes and solvents, see ref. 1139.

11 Additions to table 5:

	Ref.
Molecular geometry	1140
Bond hybridisation	1141
Resonance	1142
Heat of formation	1140, 1143
Ionisation potential	1140, 1144
Magnetic properties	1145

13 The temperature-dependent 100 MHz p.m.r. spectrum of dimer (**10**) is shown in ref. 172.

22 For a theoretical calculation on the homotropylium ion (**35**), see ref. 1146.

28 For the effect of potassium iodide on the rate of electron-transfer between COT^{2-} (**60**) and $COT^{\cdot-}$ (**61**) in hexamethylphosphoramide, see ref. 1147.

E.s.r. spectroscopic studies of $COT^{\cdot-}$ produced by the action of X-rays on $2M^+ COT^{2-} . 2$ diglyme (M = K, Rb) are described in ref. 1148.

Further evidence for the planarity of $COT^{\cdot-}$ is provided by photo-ionisation experiments on alkali metal cyclo-octatetraenides.[1149]

For additional theoretical studies on $COT^{\cdot-}$ and COT^{2-}, see refs. 1150 and 1151 respectively.

29 A comparison of various platinum catalysts for the hydrogenation of COT is made in ref. 1152.

377

32 For a theoretical study of the reaction of COT with methyl and ethyl radicals, see ref. 1153.

35 The reaction of COT with CD_2N_2 (in the presence of copper(I) chloride) provides 9,9-dideuteriobicyclo[6.1.0]nonatriene (**82**; $R^1 = R^2 = D$).[950]

(**82**)

Products of structure (**82**; $R^1 = $ 1-naphthyl, 2-naphthyl, 9-anthryl; $R^2 = $ H) and (**82**; $R^1 = R^2 = $ Ph) are formed by u.v. irradiation of the appropriate diazo-compounds in the presence of COT.[1154] In one case the *syn*-isomer (**82**; $R^1 = $ H, $R^2 = $ 2-naphthyl) may also be isolated.[1154]

37 For the ^{13}C n.m.r. spectrum of compound (**93**), see ref. 1155.

38 Doubt has been cast on the authenticity of the photo-reaction claimed to result in the thi-iran (**102**).[1156]

40 For the reaction of COT with sulphur dioxide in the presence of antimony(v) fluoride, see also ref. 1157.

42 Additions to table 11:

Dienophile	Solvent	Temp. (°C)	Yield (%)	Ref.
Acrylonitrile	—	180	88[e]	1158
Acryloyl chloride	—	130–140	80[e]	1159
Maleic anhydride	—	170–180	95	1160
Hexafluorobut-2-yne	—	150	75	1161

[e] Mixture of *endo*- and *exo*-adducts, (**1270**) and (**1271**) respectively.

(**1270**)   (**1271**)

a: R = CN   *b*: R = COCl

46 For triazolinedione adducts (**126**; R = CH_2Ph, $C_6H_4 \cdot p$-OMe), see ref. 1162. Thiobenzophenone photo-adds to COT, giving the bridged product (**1272**) (58%).[1163]

(**1272**)

49 For the heats of formation of sodium and potassium cyclo-octatetraenides, see refs. 1164, 1165.

E.s.r. spectroscopic studies of ion-pairing in potassium cyclo-octatetraenide (in hexamethylphosphoramide) are described in ref. 1166.

For photo-ionisation of alkali metal cyclo-octatetraenides in a solvent 'glass' at 77 K, see ref. 1149.

50 COT reacts with copper(I) acetate to give complex (145c) (60%).[1167]

$$COT \xrightarrow[\text{Et}_2\text{O, r.-t.}]{\text{CuOAc}} (COT)(CuX)_2 \quad (145)$$

$$a: X = OAc$$

53 Lanthanide metal atoms (produced by high-temperature evaporation) react with frozen COT to form the complexes (1273).[1168]

$$COT \xrightarrow[-196°]{M(\text{vapour})} (COT)_3M_2 \quad (1273)$$

$$M = La, Nd, Er^{1169}$$

The neodymium complex (as the tetrahydrofuranate) possesses the ionic structure illustrated below.[1168]

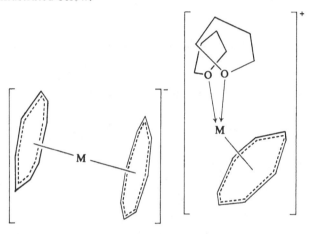

(1273; M = Nd) (THF-Solvated)

Reduction of cerium(IV) isopropoxide with triethylaluminium in the presence of an excess of COT gives the complex (551) (p. 184) (66%); the use of a limited amount of COT leads to a different product (1274) (67%), which may be converted into (551) by treatment with COT.[1170]

The complex (551) is isomorphous with (COT)₂M (155; M = Th, U) (see pp. 185–6).

$$COT \xrightarrow[\text{Pr}^i\text{OH, 140°}]{\text{Ce(OPr}^i)_4, \text{Et}_3\text{Al}} (COT)_2Ce \quad (551)$$

$$COT \mid Et_3Al, 130°$$

$$COT \xrightarrow[\text{e.g. Toluene, Pr}^i\text{OH, 110–115°}]{\text{Ce(OPr}^i)_4, \text{Et}_3\text{Al}} (COT)_3Ce_2 \quad (1274)$$

54   For the electronic absorption spectrum of $(COT)_2Ti$ (**159**), see ref. 1171.

55   For an X-ray structure determination of $(COT)ZrCl_2.THF$ (**161**), see ref. 1172.

57   Treatment of COT with chromium(III) chloride and isopropylmagnesium bromide gives the complex (**1275**) (72 %);[1173] the same product results from the condensation of COT with chromium vapour at $-196°$ (43 %)[1174] (see p. 379).

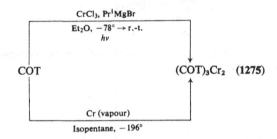

The structure of the product may be represented as follows.[1175]

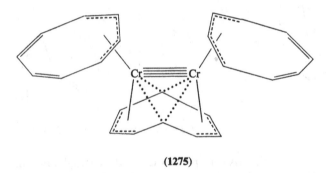

(**1275**)

59   For further ^{13}C n.m.r. studies of $(COT)M(CO)_3$ (**168**; M = Cr, Mo), see ref. 1176.

COT may be converted into the product (**1276**) (50 %) by means of the following reaction sequence.[1177]

$$COT \xrightarrow[\substack{(ii)\ EtOH,\ -100° \\ (iii)\ NH_4PF_6(aq.)}]{\substack{(i)\ [(Allyl)(benzene)MoCl]_2, \\ EtAlCl_2,\ C_6H_6,\ 0°}} [(Allyl)\ (benzene)\ Mo(COT)]^+\ PF_6^-$$

(**1276**)

The suggested structure of this salt is as shown.

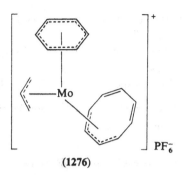

(1276)

COT reacts with [(CPD)Mo(CO)$_3$]$_2$ to give a purple complex **(1277)**.[1178]

$$COT \xrightarrow[\text{Octane, reflux}]{\text{[(CPD)Mo(CO)}_3]_2} (C_8H_8)Mo_2(CPD)_2(CO)_2 \quad \textbf{(1277)}$$

The structure of this product is as shown. It rearranges in polar solvents (seemingly by an intramolecular hydrogen-shift) to an orange isomer of structure **(1278)**.[1178]

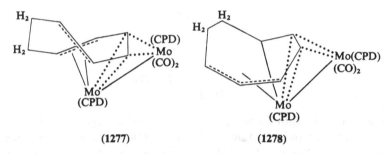

(1277)   **(1278)**

60 A light-induced reaction of (CPD)Mn(CO)$_3$ with COT leads to the thermally unstable product **(1279)**.[1179]

$$COT \xrightarrow[\text{n-Heptane, } h\nu]{\text{(CPD)Mn(CO)}_3} (COT)Mn(CPD) \quad \textbf{(1279)}$$

The product may be formulated as shown.

(1279)

62 The operation of 1,2-shifts of the iron atom in (COT)Fe(CO)$_3$ (**177**) is revealed by ^{13}C n.m.r. spectroscopy.[1180]
   The motion of the COT moiety in solid (COT)Fe(CO)$_3$ has been investigated using pulsed n.m.r. spectroscopy.[1181]
   For the He(I) photo-electron spectrum of (COT)Fe(CO)$_3$, see ref. 1182.

63 For the preparation of (COT)Fe(CO)$_3$ (86 %) using Fe$_2$(CO)$_9$, see ref. 1183.
   The same reagent has been employed for the preparation of (C$_8$D$_8$)Fe(CO)$_3$.[1184]

65 The proposed fluxional process in (COT)Ru(CO)$_3$ (**184**) is confirmed by ^{13}C n.m.r. studies.[1180]

66 For an X-ray structure determination of (C$_8$H$_9$)$_2$Ru$_3$(CO)$_6$ (**189**), see ref. 1185.

67 For pulsed n.m.r. studies of solid (COT)$_2$Ru$_3$(CO)$_4$ (**192**), see ref. 1181.

68 Details of the displacement of COD from (COD)Ru(CO)$_3$ to give (COT)Ru(CO)$_3$ (**184**) (*ca.* 100 %) have been published.[1186] The use of prolonged reaction times leads to a product (C$_{16}$H$_{16}$)Ru(CO)$_3$, for which structure (**1280**) is suggested [1186] (*cf.* (**180**), p. 63).

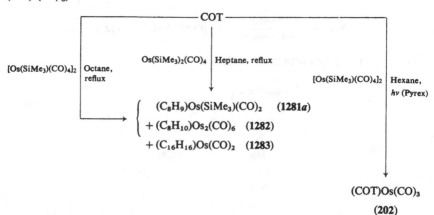

Ru(CO)$_3$

(**1280**)

70 Treatment of COT with Os(SiMe$_3$)$_2$(CO)$_4$ in reflexing heptane leads to the complexes (**1281a**) (46 %), (**1282**) (5 %), and (**1283**) (probably containing dimeric COT); (**1281a**) (62 %) and (**1282**) (4 %) are also obtained by a thermal reaction with [Os(SiMe$_3$)(CO)$_4$]$_2$, but a light-induced reaction gives compound (**202**) (23 %).[1187]

Similarly, with Os(GeMe$_3$)$_2$(CO)$_4$ in refluxing octane the main product is **(1281b)** (54%); another product is a tricarbonyl compound of formula (C$_{16}$H$_{16}$)Os(CO)$_3$.[1187]

Complexes **(1281a)** and **(1281b)** are apparently analogues of the ruthenium derivatives **(737)** (see p. 230), and **(1282)** of the iron and ruthenium compounds **(734)** and **(736)** (see p. 229).

(OC)$_2$Os(MMe$_3$)

(1281)

a: M = Si

b: M = Ge

(1282)

Reaction of COT with OsMe$_2$(CO)$_4$ gives a mixture of products, the main component being **(1284)** (21%).[1187]

$$\text{COT} \xrightarrow[\text{Octane, reflux}]{\text{OsMe}_2\text{(CO)}_4} \quad \text{(COT)Os}_2\text{(CO)}_6 \quad \textbf{(1285)}$$

$$+ \text{(C}_8\text{H}_6\text{)Os}_2\text{(CO)}_6 \quad \textbf{(1284)}$$

$$+ \text{(C}_8\text{H}_7\text{Me)Os(CO)}_3 \quad \textbf{(1287)}$$

$$+ \text{(C}_{16}\text{H}_{16}\text{)Os(CO)}_3 \quad \textbf{(1286)}$$

Compound **(1284)** contains an osma-indenyl system; **(1285)** is similar to the ruthenium complex **(186)** (p. 65); the complex of dimeric COT, **(1286)**, is probably an analogue of **(180)** (see p. 63); the formulation of **(1287)** is tentative.

Os(CO)$_3$

(OC)$_3$Os

(1284)

(OC)$_3$Os——Os(CO)$_3$

(1285)

Os(CO)$_3$

(1286)

71 The complex (cyclo-octatrienyl)Co(COT), originally formulated as (**203**), is now assigned structure (**1288**); interconversion of the enantiomeric forms (**1288a**) and (**1288b**) occurs in solution.[1188]

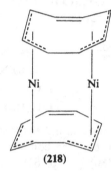

Co	Co
(COT)	(COT)
(**1288a**)	(**1288b**)

76 An X-ray structure determination of the complex (**218**) shows that it is dimeric, with the two nickel atoms sandwiched between COT ligands; the exact nature of the bonding is uncertain, but one possibility is shown below.[1189]

(**218**)

77 COT inhibits deuterium-exchange in the system cyclo-octene/Pd–Al$_2$O$_3$/Mg/D$_2$O/NaCl.[1190]

2. Substituted cyclo-octatetraenes

Monosubstituted derivatives

80 Addition to table 17:

R	Yield (%)	Ref.
CO$_2$Me	37	1191, 1192

Addition to table 18:

R	Yield (%)	Ref.
PhCH$_2$	13 (based on unrecovered COT)	1193

81 Addition to table 20:

Reagent	R	Yield (%)	Ref.
Bu^{t}Br(CuI)	Bu^t	13	1194

age

Bicyclo-octatetraenyl may be prepared (84%) from bromocyclo-octatet-raene *via* the lithio-derivative[1195] (*cf.* table 20).

82 Tritylation of (COT)Fe(CO)$_3$ (177), followed by demetallation of the product Ph$_3$C.C$_8$H$_7$.Fe(CO)$_3$ (1289) (see p. 406), gives tritylcyclo-octatetraene.[1196]

A few oxygenated COTs (1290) have been obtained by alkylation or acyla-tion of the enolate anion (1291), generated by deprotonation of the epoxide (27) with lithium di-isopropylamide (see table 143, p. 420).[1197]

(27)   (1291)   (1290)

R = Et, CO$_2$Et, Ac

90 Additions to scheme 29:

C$_8$H$_7$.CHO

$\downarrow$ 4

C$_8$H$_7$.CH(OH)CN $\xrightarrow{5}$ C$_8$H$_7$.CH(OH)CH$_2$NH$_3^+$ Cl$^-$

Reaction	Reagents and conditions	Yield (%)	Ref.
1	MnO$_2$, CCl$_4$, 25°	84	1191
4	NaCN, conc. HCl, Et$_2$O(aq.), 5°	96	1191
5	LiAlH$_4$, glyme, reflux	17	1191

Scheme 30, reaction 5: see also ref. 1191.

96 The hydrolysis of methoxycyclo-octatetraene provides a high-yield (94%) preparation of cyclo-octatrienone.[1198]

00 For the electrochemical and alkali metal reduction of the monosubstituted COTs (227; R = Bu^t, SiMe$_3$, GeMe$_3$, SnMe$_3$), see ref. 1194.

(227)

06 For e.s.r. spectral studies of ion-pairing in the potassium salts of monosubsti-tuted COTs (227; R = cyclo-octatetraenyl, OBu^t) (in hexamethylphosphora-mide), see ref. 1166.

In the radical anion and the dianion derived from bicyclo-octatetraenyl, delocalisation extends over both rings.[1166]

107 Addition to table 39:

R	Reagents and conditions	Yield (%)	Ref.
OMe	$Fe_3(CO)_{12}$, octane, reflux	79	1198

A binuclear complex (299; R = OMe) is also formed (10 %).

Reaction of bicyclo-octatetraenyl with iron carbonyls (*cf.* table 39) leads to the complexes (1292) (up to 16 %) and (1293), together with (COT)Fe(CO)₃ (177) and (COT)Fe₂(CO)₆ (178).[1195]

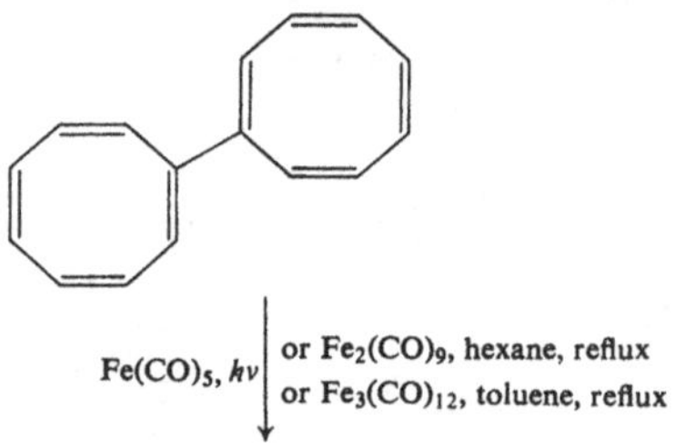

$Fe(CO)_5, h\nu$ | or $Fe_2(CO)_9$, hexane, reflux
or $Fe_3(CO)_{12}$, toluene, reflux

(Bicyclo-octatetraenyl)Fe(CO)₂ (1292)

+ (Bicyclo-octatetraenyl)Fe(CO)₃ (1293)

The structures of these products are as follows.

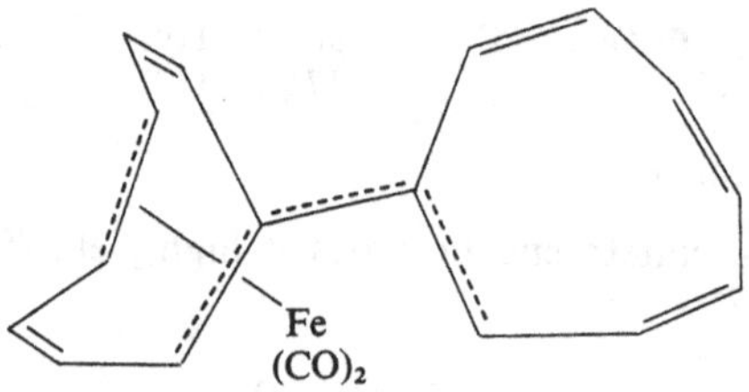

(1292)

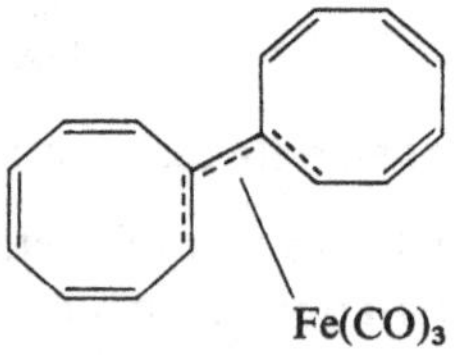

Fe(CO)₃

(1293)

(Bonding of metal to 1,3-diene system omitted.)

108 For full details of the preparation of (bicyclo-octatetraenyl)Ru₃(CO)₆ (305) (80 %), see ref. 1195; the unique structure of this product is shown below.

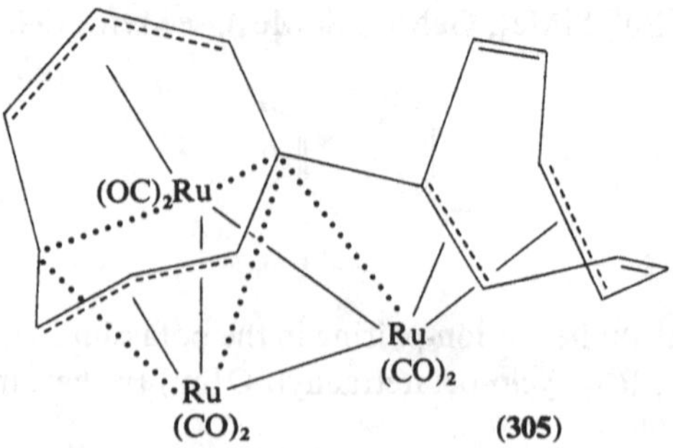

(OC)₂Ru

Ru
(CO)₂

Ru
(CO)₂

(305)

Bicyclo-octatetraenyl reacts with $[Ru(SiMe_3)(CO)_4]_2$ to give the complex (**1294**) (9 %).[1195]

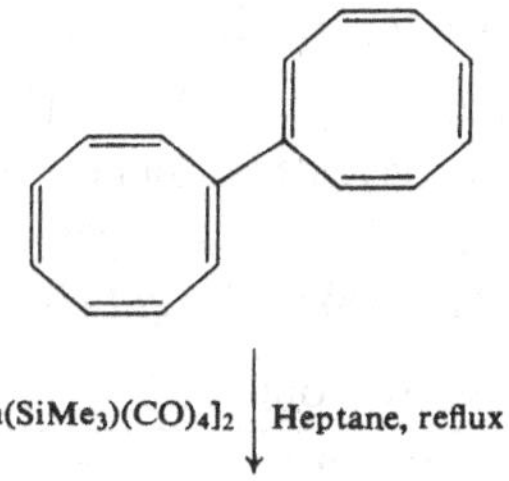

[Ru(SiMe₃)(CO)₄]₂ | Heptane, reflux

(Bicyclo-octatetraenyl)Ru₂(SiMe₃)₂(CO)₄   (**1294**)

This product appears to be an analogue of (**198b**) (p. 69). With $Os_3(CO)_{12}$ in refluxing xylene, bicyclo-octatetraenyl gives a minute yield of the complex (**1295**)[1195] (*cf.* (**1292**), p. 386).

(Bicyclo-octatetraenyl)Os(CO)₂   (**1295**)

Disubstituted derivatives

10 Addition to table 42:

R¹	R²	Yield (%)	Ref.
CO₂Me	CO₂Me	16–18	1199

11 Details of the conversion of monobromo- into 1,4-dibromocyclo-octatetraene (77 %) are given in ref. 1187; see, however, ref. 43.

The 1,4-disubstituted COTs (**1296**) have been mentioned as products resulting from thermolysis of the tricyclic dienes (**1297**).[1200]

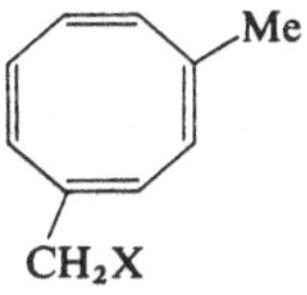

(**1296**)   X = CN, CO₂Me   (**1297**)

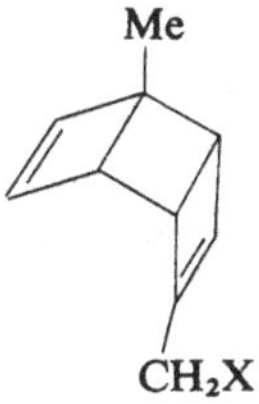

12 1,2-Dicyanocyclo-octatetraene is formed (quantitatively) by u.v. irradiation of the dicyanobarrelene (**1298**)[1201]

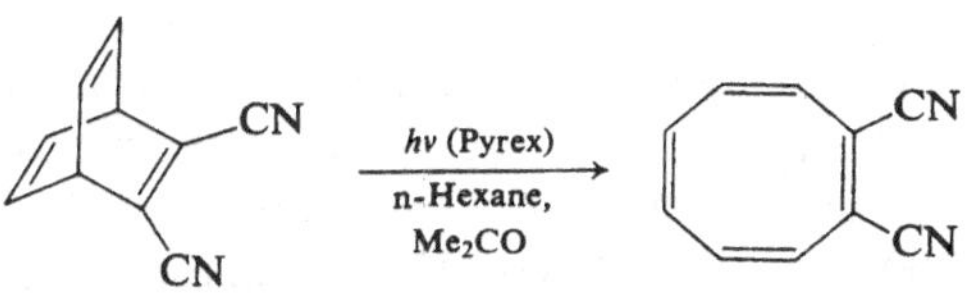

hv (Pyrex)
n-Hexane, Me₂CO

(**1298**)

388 *Appendix*
page
115 For the p.m.r. spectrum of 1,2-dimethylcyclo-octatetraene, see ref. 43.

120 For the radical anions of 1,4- and 1,5-dimethylcyclo-octatetraene, generated by low-temperature electrolysis, see ref. 1202.

122 Reaction of 1,4-dibromocyclo-octatetraene with $Na_2[Fe_2(CO)_9]$ or $Fe_2(CO)_9$ at room-temperature gives the benzocyclobutadiene complex (**1299**) (64% or 22% respectively); with $Fe_2(CO)_9$ in refluxing heptane, however, the main product is the ferra-indenyl system (**1300a**)[1187] (*cf.*(**1284**), p. 383).

Na₂[Fe₂(CO)₉]

THF, r.-t.

Br

Fe₂(CO)₉

THF, r.-t.

Fe(CO)₃

Br

(**1299**)

Fe₂(CO)₉ | Heptane, reflux

M(CO)₃

(OC)₃M   + (**1299**)

(**1300**)

a: M = Fe
b: M = Ru

Similarly, $Ru_3(CO)_{12}$ in refluxing hexane yields the ruthenium analogue (**1300b**) (18%).[1187]

Annulated and bridged derivatives

128 Thermolysis of the propellatriene (**1301**) leads to a mixture (*ca.* 1:1) of the bond-shift isomers (**1302a**) and (**1302b**) (74%).[1203]

130 Treatment of the diyne (**1303**) with $Ni(CO)_4$ affords the bisannulated COT (**1304**) (80%).[1204]

A similar product is obtained (38%) from the naphthalene analogue.[1204]

Treatment of the di-ester (**7**) with acecyclone etc. in refluxing xylene leads to 1,2-annulated 3,8-disubstituted COTs (**1305**), presumably *via* initial adducts of structure (**1306**) which undergo decarbonylation and Alder–Rickert cleavage to give (**1307**), valence tautomers of the observed products.[1205]

(1301)

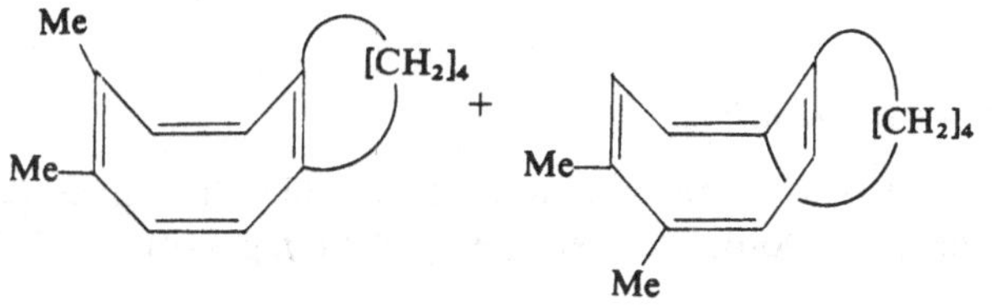

(1302a) + (1302b)

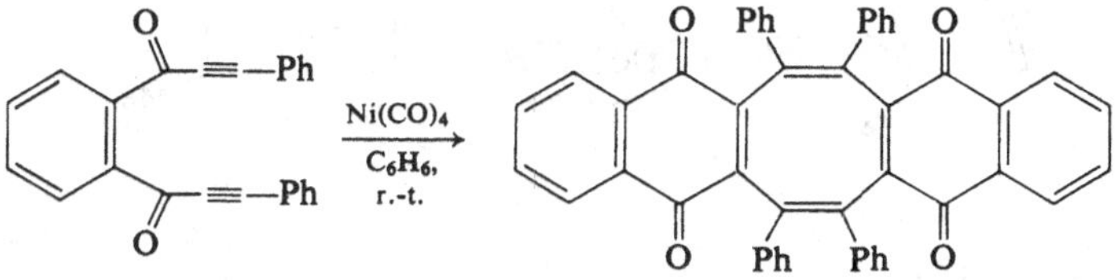

(1303) $\xrightarrow[\text{r.-t.}]{\text{Ni(CO)}_4 \atop \text{C}_6\text{H}_6,}$ (1304)

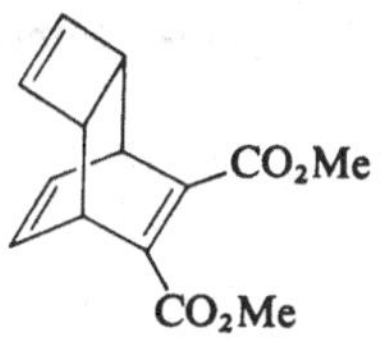

(7) $\xrightarrow{\text{Xylene, reflux}}$ (1306)

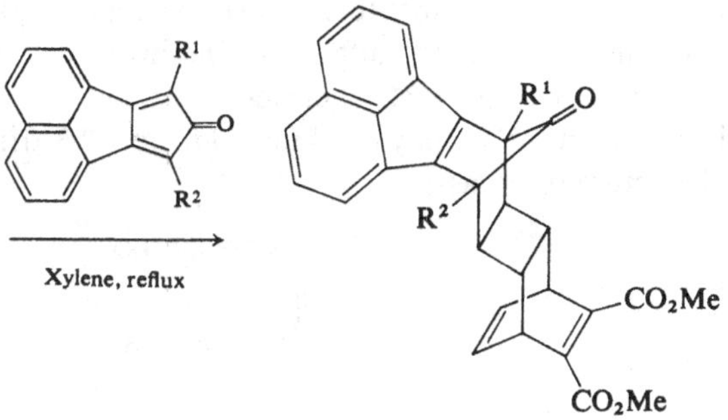

(Stereochemistry uncertain.)

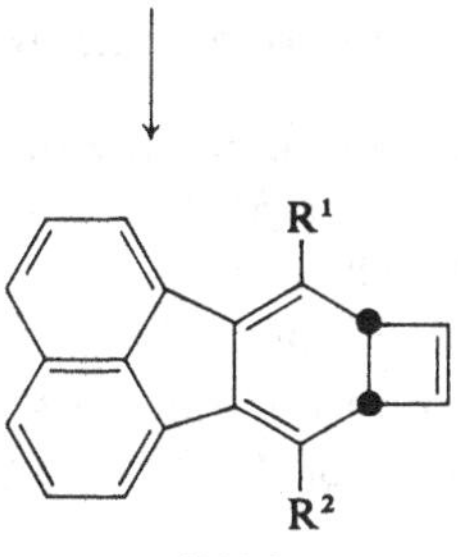

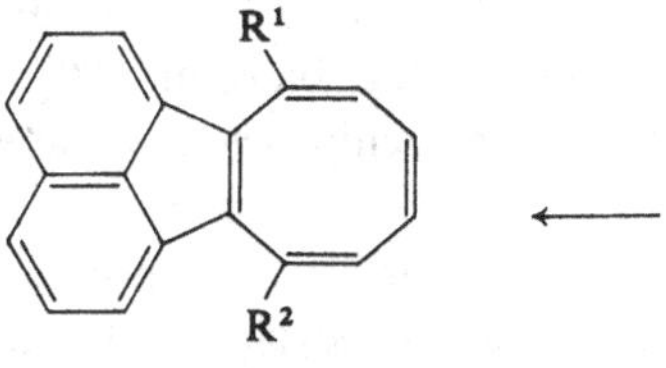

(1305) ← (1307)

$R^1 = R^2 = Me, Pr^i, Ph, CO_2Et$ $R^1 = Me, R^2 = Ph$
$R^1 = Me, R^2 = Pr^i$ $R^1 = Et, R^2 = CO_2Et$

page

132 For e.s.r. spectral data on the radical anion of compound (**350**), see ref. 1206.

(**350**)

Trisubstituted derivatives

134 Bromination–dehydrobromination of the bicyclic diene (**1308**) leads to 1,2,3-trimethylcyclo-octatetraene $(69\%)^{1207}$ (*cf.* p. 140)

Me

Me

Me

(**1308**)

(i) PyHBr₃
(ii) LiCl, Li₂CO₃, (Me₂N)₃PO

Me

Me

Me

(+)-1,2,3-Trimethylcyclo-octatetraene may be obtained by using (+)-(**1308**).[1207]

The ability of this substituted COT to support optical activity results from the inhibition of bond-shift and ring-inversion. Racemisation occurs at 160° in diglyme $(\Delta G^{\ddagger} = 31.7\ \mathrm{kcal\,mol^{-1}})$.[1207]

The dianion (**1309**) may be obtained from 1,2,3-trimethylcyclo-octatetraene by the action of potassium in ND₃.[1207]

2–

Me

Me

Me

(**1309**)

Tetrasubstituted derivatives

137 Photolysis of a mixture of the *endo-* and *exo-*dicarbonyl-bridged compounds (**1310**) and (**1311**) (see p. 426) affords the bicyclo[4.2.0]octa-2,4,7-triene (**1312**), which isomerises to 1,2,3,8-tetrachlorocyclo-octatetraene at 60° (half-life 15 min.).[1208]

In contrast, thermolysis of the carbonyl-bridged compound (**1313**) leads to a mixture of 1,2,3,8- and 1,2,3,4-tetrachlorocyclo-octatetraene[1209] (see p. 392).

139 For details of the preparation of 1,2,4,7- and 1,3,5,7-tetraphenylcyclo-octatetraene from the quaternary ammonium salt (**403**), see ref. 1210.

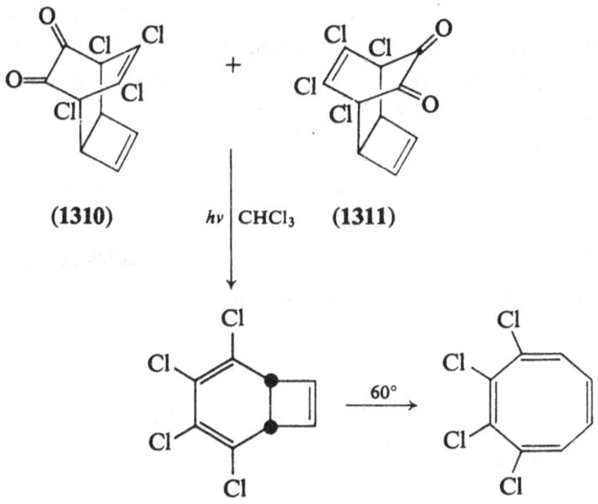

(1310)   *hv* | CHCl₃   **(1311)**

(1312)

(1313)   Xylene reflux   **(1312)**

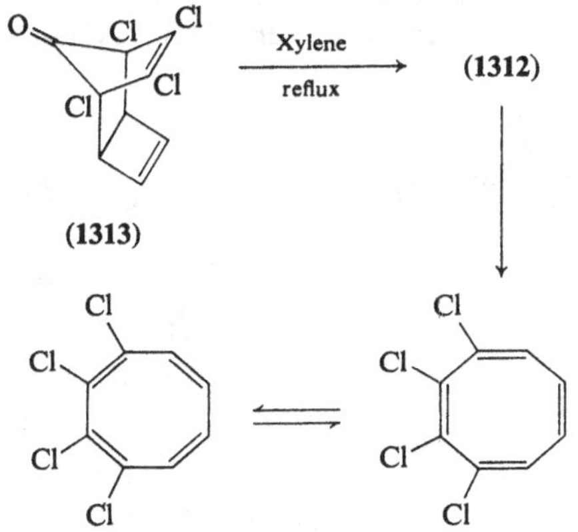

(1314)   (i) PyHBr₃, CCl₄, HOAc   (ii) LiCl, Li₂CO₃, (Me₂N)₃PO

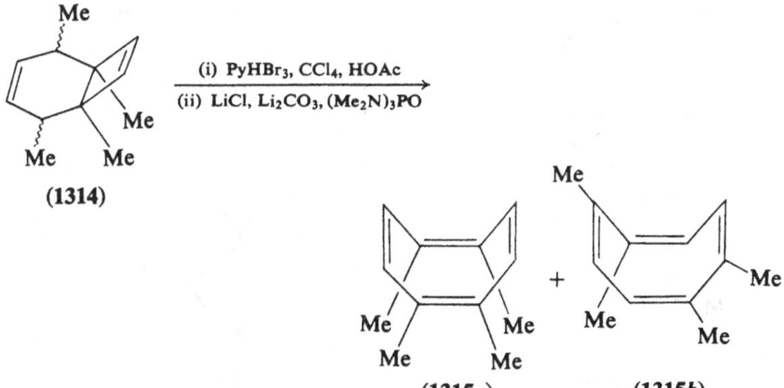

(1315a)   **(1315b)**

For full details of the preparation of 1,2,3,4- and 1,2,3,8-tetramethylcyclo-octatetraenes from compounds (409) and (408) respectively, see ref. 1203.

Bromination–dehydrobromination of the bicyclic diene (1314) leads to a mixture of the bond-shift isomers (1315a) and (1315b) (ratio *ca.* 1:10; 92%)[1203] (*cf.* (416a) ⇌ (416b), p. 142).

142 For full details concerning the isolable bond-shift isomers (415a) and (415b), see ref. 1203.

Like the tetramethyl-analogue (415a), 1,2,3,8-tetrachlorocyclo-octatetraene is stable, although equilibration with its bond-shift isomer may be induced by heating or irradiation.[1208]

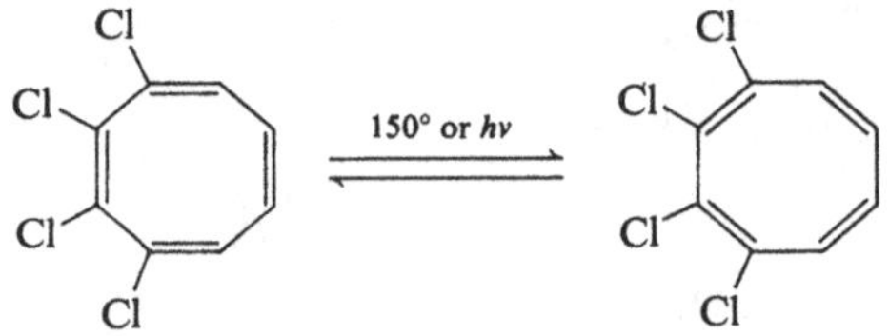

$$ \text{(structure)} \quad \xrightarrow{150° \text{ or } h\nu} \quad \text{(structure)} $$

15 The photo-product of 1,3,6,8-tetraphenylcyclo-octatetraene, originally formulated as (423), is now known to be an equimolar mixture of 1,2,4,7-tetra-phenylcyclo-octatetraene and its valence tautomer (1316) (yields up to 58%).[1210]

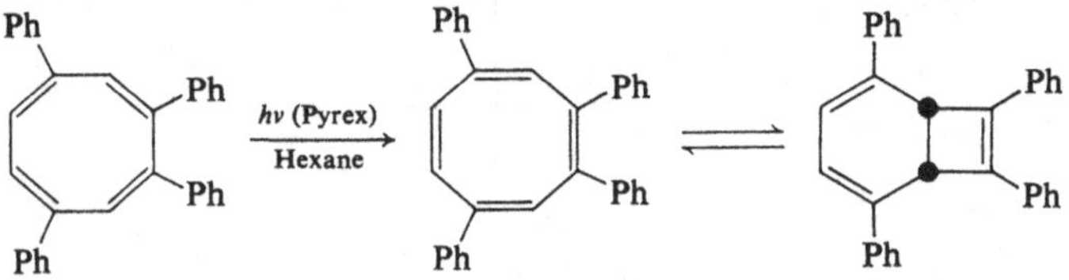

(1316)

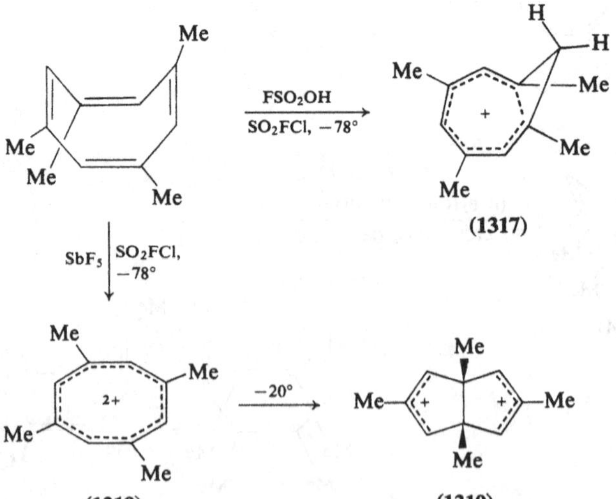

(1317)

(1318)

(1319)

For epoxidation of 1,2,3,4-tetramethylcyclo-octatetraene, see ref. 1203.

Protonation of 1,3,5,7-tetramethylcyclo-octatetraene occurs in FSO_2OH–SO_2FCl at $-78°$ to give the homotropylium ion (1317). In SbF_5–SO_2FCl, however, the dication (1318) is produced; at $-20°$ this rearranges to (1319)[1211] (*cf.* (486), p. 162).

(For a theoretical study of COT^{2+}, see ref. 1151.)

For e.s.r. spectral studies on the radical anion of 1,3,5,7-tetramethylcyclo-octatetraene, produced by the action of X-rays on $2K^+ TMCOT^{2-}$. 2 diglyme, see ref. 1148.

For details of the preparation of the dianion (424), see ref. 1203.

46 The catalytic hydrogenation of an equimolar mixture of 1,2,4,7-tetraphenyl-cyclo-octatetraene and its valence tautomer (see p. 392) yields the product (1320) quantitatively.[1210]

(1316)   (1320)

47 For details of the *N*-phenyltriazolinedione adducts (433) and (434), and their use in the separation of 1,2,3,4- and 1,2,3,8-tetramethylcyclo-octatetraene, see ref. 1203.

The mixture of 1,2,4,7-tetraphenylcyclo-octatetraene and its valence tautomer (see p. 392) reacts with maleic anhydride to give the adduct (1321) (90%).[1210]

(1316)   (1321)

A similar adduct is obtained (78%, based on unrecovered tetraphenylcyclo-octatetraene) with dimethyl acetylenedicarboxylate.[1210]

1,2,3,8-Tetrachlorocyclo-octatetraene gives the *N*-phenyltriazolinedione adduct (1322).[1208]

A similar reaction with 1,2,3,4-tetrachlorocyclo-octatetraene (in admixture with its bond-shift isomer) affords the adduct (1323).[1208]

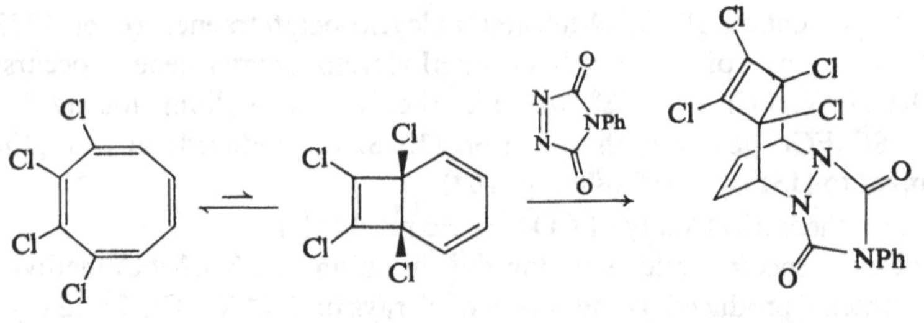

(1322)

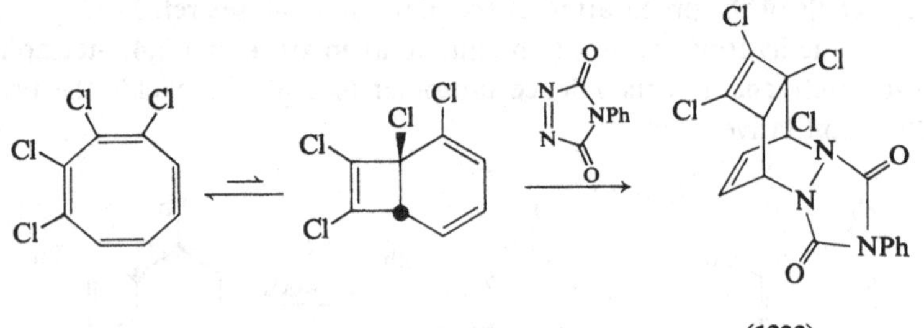

(1323)

148 For full details of the preparation of the iron tricarbonyl complexes (437) and (438), see ref. 1203.

Pentasubstituted derivatives

151 For full details of the preparation of 1,2,3,4,6- and 1,2,3,5,8-pentamethyl-cyclo-octatetraene, see ref. 1203.

Octasubstituted derivatives

155 For further details of the preparation of octakis(trifluoromethyl)cyclo-octatetraene, see ref. 1212.

Octamethylcyclo-octatetraene results (85%) from the bicyclic triene (1324) at room-temperature.[1213]

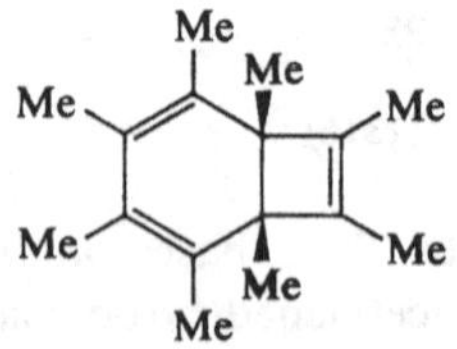

(1324)

159 U.v. irradiation of the tetrachloro-anhydride (1325) leads to octachloro-cyclo-octatetraene[1214] (*cf.* (477)).

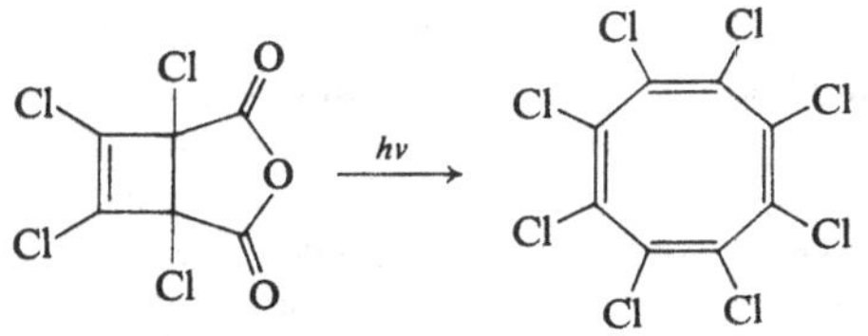

(1325)

Similarly, photolysis of the tetrafluoro-anhydride **(1326)** in the gas-phase gives octafluorocyclo-octatetraene; in the presence of nitrogen, the cyclo-butadiene dimer **(1327)** may be isolated (41 %), and subsequently subjected to thermal rearrangement.[1215]

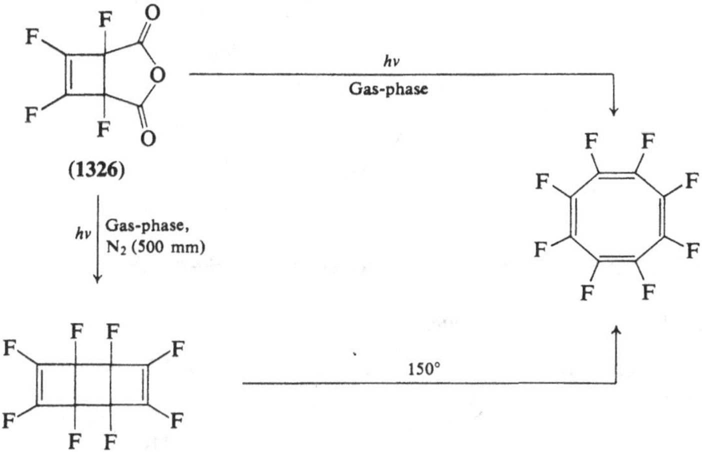

(1326)

hv Gas-phase

hv Gas-phase, N₂ (500 mm)

150°

(1327)

463 Octamethylcyclo-octatetraene reacts with maleic anhydride to give the adduct **(1328)**; analogous products are formed from other dienophiles (table 140) (for reactions with the isolated valence tautomer **(1324)**, see ref. 1213).

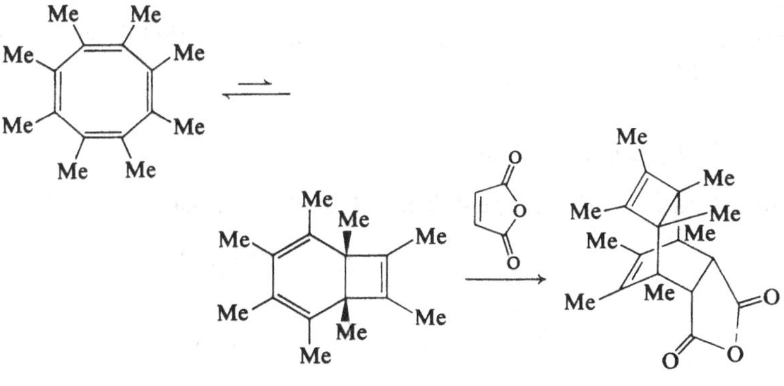

(1324)

(1328)

Table 140

Dienophile	Conditions	Yield (%)	Ref.
Maleic anhydride	Toluene, reflux	96	1213
Tetracyanoethylene	Benzene, reflux	74	1213
Dimethyl acetylenedicarboxylate	Toluene, reflux	50	1213
Dicyanoacetylene	Benzene, r.-t.	68	1213
Dimethyl azodicarboxylate	Toluene, reflux	88[a]	1213
N-phenyltriazolinedione	Benzene, reflux	83	1213

[a] A by-product of structure (1329) is isolable.

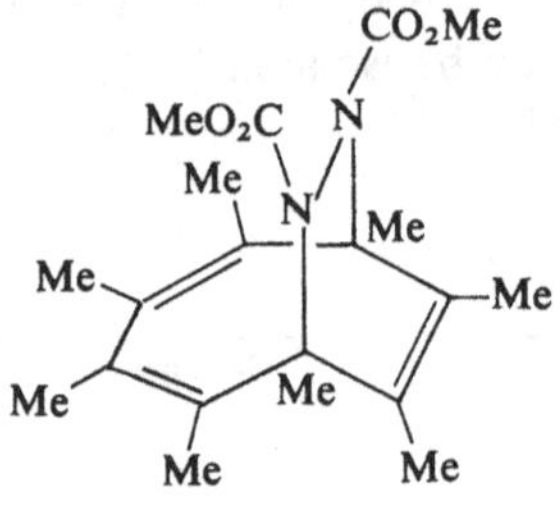

(1329)

3. Further reactions of compounds derived from cyclo-octatetraenes

Bicyclo[4.2.0]octa-2,4,7-trienes

165 For the preparation of the tetrachloro-derivative (1312), which is stable at room-temperature, see p. 391.

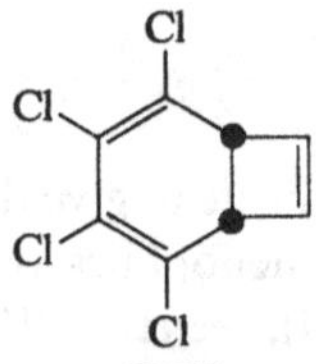

(1312)

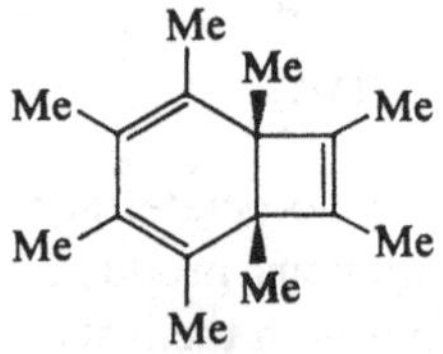

(1324)

For the (low-temperature) preparation of the octamethyl-derivative (1324), see ref. 1213.

Bicyclo[3.3.0]octa-1,4,6-triene

168 Protonation of the dihydropentalene (24) leads to the cation (667).[852]

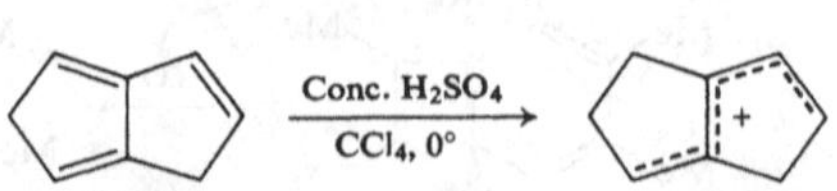

(24)

Conc. H_2SO_4 / CCl_4, 0°

(667)

For reaction of compound (24) with maleic anhydride, see also ref. 1216.

For the manganese complex (**1330**), see ref. 1216.

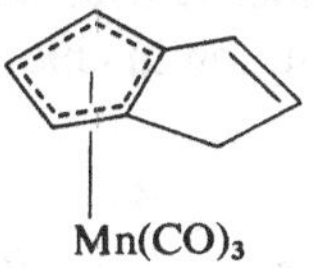

Mn(CO)₃

(**1330**)

Semibullvalenes

169 For the ¹³C n.m.r. spectrum of semibullvalene, see ref. 1217.
For estimation of the fluxional barrier, see ref. 1218.

170 The reaction of octamethylsemibullvalene with tetracyanoethylene is now
known to result in structure (**1331**) (88 %).[705, 1213]

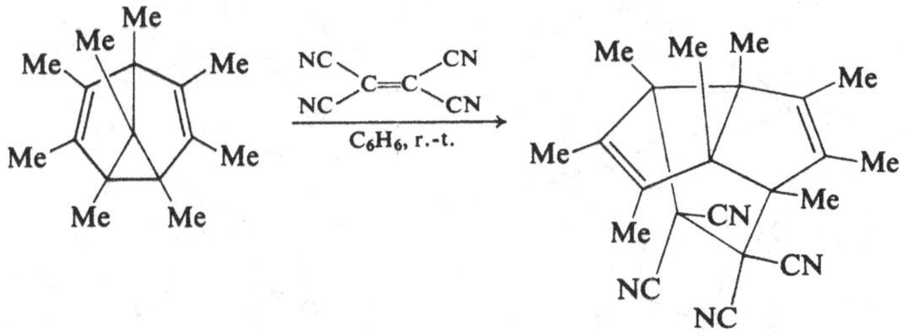

(**1331**)

Similar products (**1332**) are formed from azodicarboxylic esters (table 141).

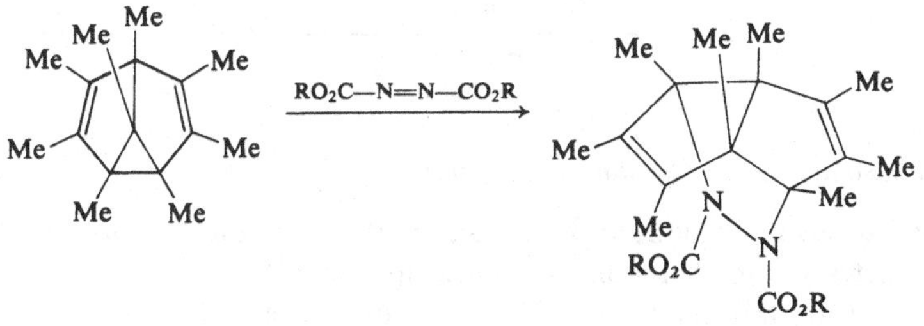

(**1332**)

Table 141

R	Conditions	Yield (%)	Ref.
Me	C₆H₆, 80°	81	1213
Et	80°	62	705, 1213
	Cyclohexane, *hv*(Pyrex)	—	1213
CH₂CCl₃	—	—	1219
CH₂Ph	C₆H₆, 30° → r.-t.	74	1213

With *N*-phenyltriazolinedione, however, a mixture of the adducts (1333) and (1334) is obtained; the former rearranges quantitatively to the latter in methanol at 60°, or when its solution in chloroform is treated with trifluoro-acetic acid or silica. [1219]

(1333)   (1334)

e.g. CF₃CO₂H

CHCl₃, r.-t.

Lithium, sodium and potassium derivatives

172 Anion radicals are produced by the action of COT^{2-} on *o*- or *p*-nitrophenyl thiocyanates and methyl *o*-nitrobenzenesulphenate.[1220]

For electron-transfer from COT^{2-} to azobenzene and azoxybenzene, see ref. 1221.

For the formation of benzylcyclo-octatetraene from the dianion by treatment with iodine, see ref. 1193.

174 For the reactions of $2Li^+$ COT^{2-} with benzal chlorides, leading to products of structure (82; R^2 = H) (*cf.* table 63), see the superscribed references.

178 Re-examination of the reaction of COT^{2-} with carbon dioxide has shown that the initial products are, after esterification, (1335) (6%) and (1336) (29%); these equilibrate in solution at room-temperature (ratio *ca.* 1:4).[1222]

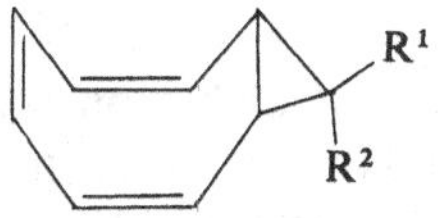

ge

(82)

$R^1 = Ph,^{1154,\ 1193}$ $p\text{-Me.C}_6H_4,^{1154}$ $p\text{-CF}_3.C_6H_4,^{1154}$ $p\text{-MeO.C}_6H_4,^{1154}$
$p\text{-Cl.C}_6H_4,^{1154}$ $p\text{-Br.C}_6H_4^{1154}$

$R^2 = H$

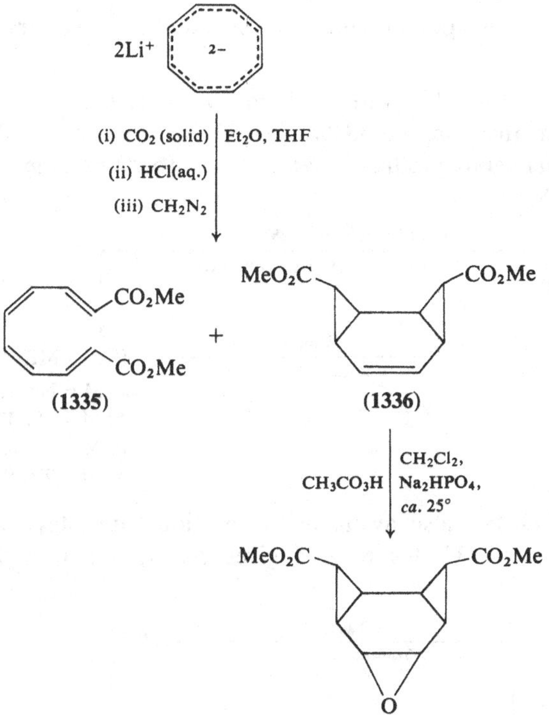

$2Li^+$ $2-$

(i) CO_2 (solid) | Et_2O, THF
(ii) HCl(aq.)
(iii) CH_2N_2

.CO$_2$Me

.CO$_2$Me

(1335)

+

MeO$_2$C

CO$_2$Me

(1336)

CH_3CO_3H | CH_2Cl_2, Na$_2$HPO$_4$, *ca.* 25°

MeO$_2$C

CO$_2$Me

O

(1337)

The tricyclic di-ester (1336) may be converted into the epoxide (1337) (84 %).[1222]

33 Reaction of 2,5-dithiocyanatothiophene with COT^{2-} in the presence of 18-crown-6 results in a linear polymer.[1223]

The complex (1338) results (almost quantitatively) from 2K$^+$ COT^{2-} and ScCl$_3$.3THF; the product reacts with more 2K$^+$ COT^{2-} to give the ionic complex (1339)[1224] (*cf.* (550), p. 184).

$$2K^+\ COT^{2-} \xrightarrow{\ ScCl_3.3THF\ } (COT)ScCl.THF\ \textbf{(1338)}$$

$2K^+\ COT^{2-} \downarrow$

$$K^+\ [(COT)_2Sc]^-\ \textbf{(1339)}$$

184 For MO calculations on [(COT)$_2$Ce]$^-$, see ref. 1225.

186 Magnetic susceptibility data are available on the following uranium complexes: (COT)$_2$U (**155**; M = U),[1226] (phenylcyclo-octatetraene)$_2$U (**554**; R = Ph, M = U),[1227] (TMCOT)$_2$U (**555a**),[1227] and (1,3,5,7-tetraphenylcyclooctatetraene)$_2$U (**555c**).[1227]

The crystal structure of the uranium complex (**555c**) is described in ref. 1228.

187 Full details of the preparation of [(COT)TiCl.THF]$_2$ (**560**) and [(COT)TiCl]$_4$ (**561**) are given in ref. 1229.

For magnetic susceptibility and electronic and e.p.r. spectra of (COT)$_2$V, see ref. 1171.

188 Reaction of 2K$^+$ COT^{2-} with niobium(v) chloride followed by phenyl-lithium affords the complex (**1340a**) (*ca.* 40%), together with the salt (**563a**) (*ca.* 20%); analogous products (**1340b–d**) are (best) obtained from RMCl$_4$ (M = Nb, Ta).[1230]

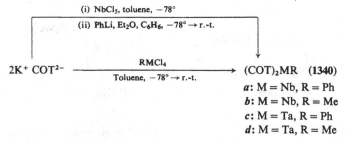

Complex (**1340d**) is also available by reaction with Me$_3$TaCl$_2$ at room-temperature; at −78° this reagent gives rise to the trimethyl-derivative (**1341**).[1230]

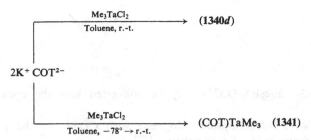

It is suggested that the general structure of (COT)$_2$MR (**1340**) is likely to be similar to that of (COT)$_2$Zr.THF (**160a**) (see p. 55), and that the observed fluxional behaviour involves the illustrated process.[1230]

The trimethyl-derivative (**1341**) is formulated as shown.

For an examination of the fluxional behaviour of (COT)Ru(NBD) (**566**), using ^{1}H and ^{13}C n.m.r. spectroscopy, see ref. 1231.

Reaction of COT^{2-} with [(arene)RuCl$_2$]$_2$ or (arene)RuCl$_2$.Py affords complexes of structure (**1342**) (30–50%).[1232]

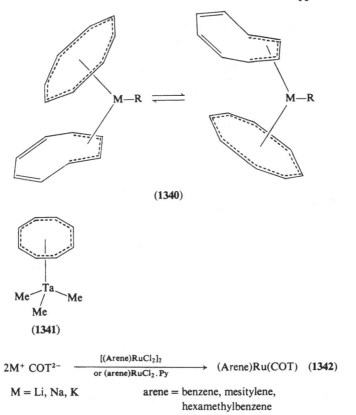

(1340)

(1341)

$$2M^+ \ COT^{2-} \xrightarrow[\text{or (arene)RuCl}_2\cdot\text{Py}]{[\text{(Arene)RuCl}_2]_2} \text{(Arene)Ru(COT)} \quad (1342)$$

M = Li, Na, K   arene = benzene, mesitylene,
                                       hexamethylbenzene

The structure of the (fluxional) hexamethylbenzene complex is as shown.[1232]

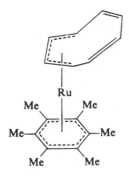

(1342; arene = hexamethylbenzene)

89 The rhodium and iridium complexes [(pentamethylcyclopentadienyl)MCl$_2$]$_2$
react with COT^{2-} at low temperatures to give low yields of the fluxional
products (1343); these isomerise at 20° in solution to afford complexes of
structure (1344).[1233]

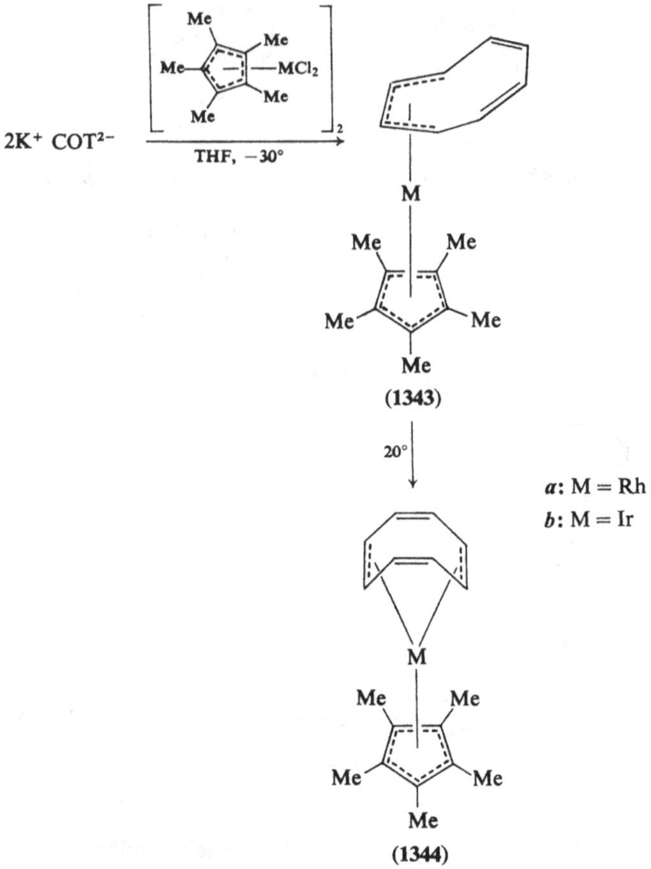

$$2K^+ \text{ COT}^{2-} \xrightarrow[\text{THF, } -30°]{}$$

(1343)

20°

a: M = Rh
b: M = Ir

(1344)

Reaction of COT^{2-} with the rhodium complexes $[(\text{COD})\text{RhCl}]_2$ and $[(\text{NBD})\text{RhCl}]_2$, followed by treatment with moist alumina, gives the cyclo-octatrienyl complexes (1345) (96 % and 49 % respectively).[1234]

$$2\text{Na}^+ \text{ COT}^{2-} \xrightarrow[\text{(ii) } H_2O-Al_2O_3]{\substack{\text{(i) } [(\text{Diene})\text{RhCl}]_2 \\ \text{THF, r.-t.}}} \text{(Cyclo-octatrienyl)Rh(diene)}$$

(1345)

a: diene = cyclo-octa-1,5-diene
b: diene = norbornadiene

Using a modified work-up procedure, the product $(\text{COT})\text{Rh}_2(\text{COD})_2$ (1346a) may be obtained (12 %), and in pentane solution the complex (1345a) is slowly converted into (cyclo-octatrienyl)$_2$Rh$_2$(COD) (1347) (76 %); complex (1345b) in pentane containing COT slowly gives (1346b) (43 %).[1234]

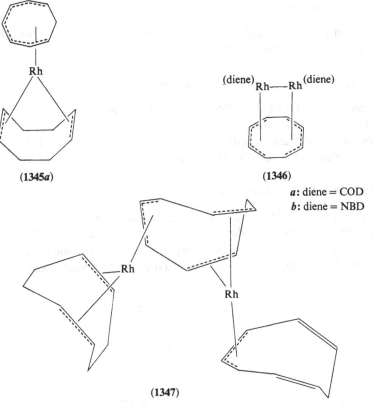

(1345*a*)   (1346)

a: diene = COD
b: diene = NBD

(1347)

For the use of 2Li⁺ COT²⁻ as a dechlorinating agent in the preparation of e.g. (COD)₂Pd from (COD) PdCl₂, see ref. 1235.

Scandium and lanthanide derivatives

90 The scandium complex (1338) (p. 399) reacts with sodium cyclopentadienide, yielding a product (1348) (up to 55 %)[1224] (*cf.* (552)).

$$(COT)Sc(CPD) \quad (1348)$$

The complex (COT)₂Ce (551) reacts with potassium to form the salt (1349) (84 %) (*cf.* (550), p. 184); with an excess of the alkali metal, further electron transfer results in the product (1350) (50 %).[1170]

(COT)₂Ce (551)

K / MeOCH₂CH₂OMe, 30° → [K(glyme)]⁺ [(COT)₂Ce]⁻ (1349)

K (excess) / MeOCH₂CH₂OMe, 60° → 2[K(glyme)]⁺ [(COT)₂Ce]²⁻ (1350)

Actinide derivatives

191 The action of microwave-discharged argon on $(COT)_2U$ vapour provides a (relatively) low-temperature generation of uranium vapour.[1236]

Titanium, zirconium and hafnium derivatives

193 Reaction of (COT)Ti(CPD) (**165**) with n-butyl-lithium leads to metallation, predominantly in the cyclopentadienyl ring.[1237]
For discussion of the i.r. and p.m.r. spectra of the complexes $(COT)M(allyl)_2$, $(COT)M(methylallyl)_2$ and $(COT)M(crotyl)_2$ (M = Zr, Hf), see ref. 1238.

Niobium and tantalum derivatives

194 Complexes $(COT)_2MR$ (**1340**) (M = Nb, Ta) and $(COT)TaMe_3$ (**1341**) (p. 400) form adducts with donor ligands such as $Me_2P[CH_2]_2PMe_2$ and $o\text{-}(Me_2As)_2.C_6H_4$; when R = Ph, reaction in refluxing toluene results in the transfer of the phenyl group from the metal to the coordinated COT, giving products such as (**1351**).[1230]

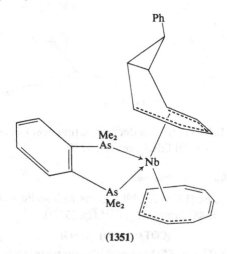

(**1351**)

Molybdenum derivatives

194 For u.v. and i.r. spectroscopic data on $(COT)Mo(PBu_3)_2(CO)_2$, see ref. 1239.

Iron, ruthenium and osmium derivatives

196 For a review of the iron, ruthenium and osmium carbonyl complexes of COT and its derivatives, see ref. 1240.

198 Hydrolysis of the iron tricarbonyl complex of methoxycyclo-octatetraene, (**298**; R = OMe), affords the cyclo-octatrienone complex (**300**) (90 %).[1198]

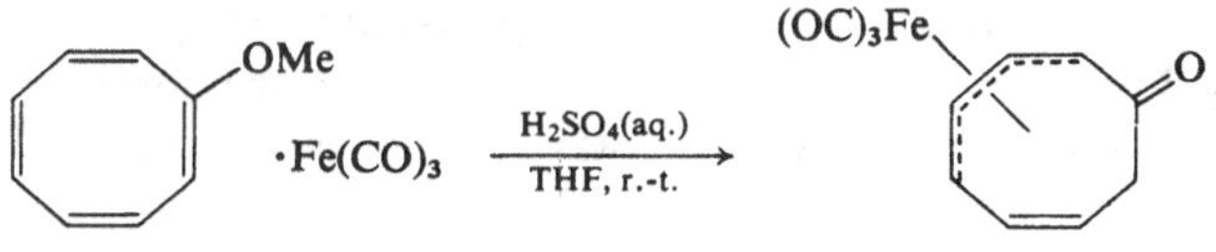

age

(298; R = OMe)   **(300)**

Full details of the thermal isomerisation of the iron tricarbonyl complexes of substituted COTs are given in ref. 1241.

199 For pulsed n.m.r. studies of solid $(COT)Fe_2(CO)_5$ **(612)**, see ref. 1181.

200 For reactions of protonated $(COT)Fe(CO)_3$, **(614)**, with nucleophiles, see ref. 1242.

201 A high-yield (90%) preparation of the salt **(621)** is described in ref. 1183.

Treatment of the salt **(621)** with sodium borohydride has been shown to result in a mixture of the complexes **(622)** (72%) and **(1352)** (18%).[1183, 1243]

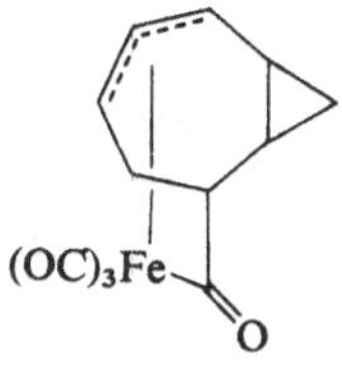

(1352)

(For further reactions of **(622)**, see ref. 1183.)

The preparation of the salt **(1353)** (*cf.* **(621)**) and its reaction with sodium borohydride are described in ref. 1184.

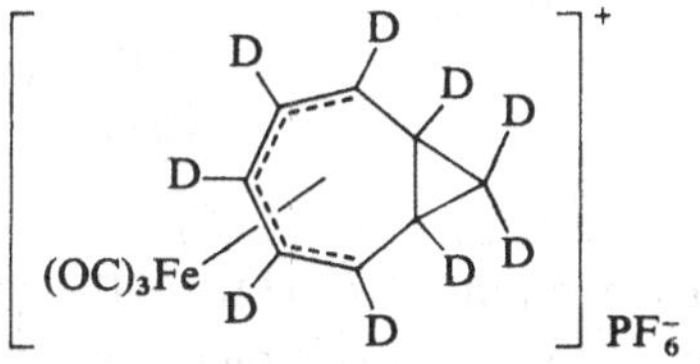

(1353)

The complex **(177)** reacts with dry hydrogen chloride in ether to give the product **(1354)** (57%).[1244]

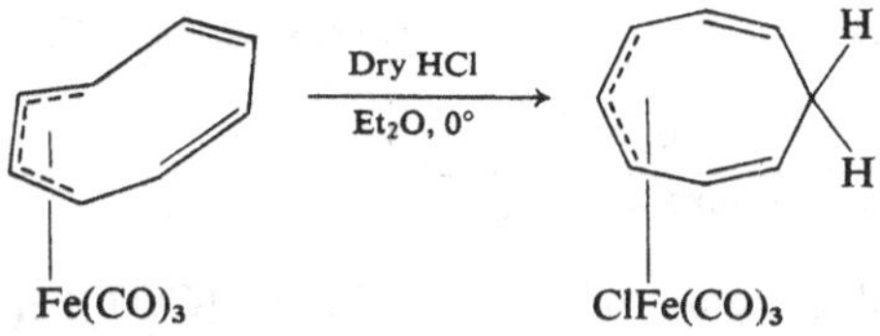

Fe(CO)₃   **ClFe(CO)₃**
(177)   **(1354)**

page
202 Full details of the formation of complexes (**624**) have been published.[1245]
   Treatment of (COT)Fe(CO)$_3$ (**177**) with boron trifluoride etherate in
chloroform–acetic acid gives an unstable product of probable structure
(**1355**).[1245]

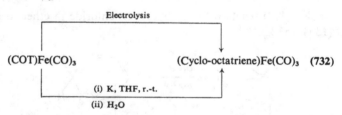

(**1355**)

204 The iron tricarbonyl complex (**177**) reacts with trityl tetrafluoroborate to
afford the salt (**1356**) (81 %); treatment with triethylamine or sodium methoxide then leads to the complex (**1289**) of tritylcyclo-octatetraene (74 or 59 %
respectively).[1196]

(For further reactions of (**1289**), see ref. 1196.)
   The electrochemical reduction of (COT)Fe(CO)$_3$ (**177**) affords the cyclo-
octatriene complex (**732**) (100 %); alkali metal reduction proceeds in much
lower yield (30 %).[1246]

For the displacement of COT from (COT)Fe(CO)$_3$ (**177**) by
Ph$_2$P[CH$_2$]$_3$PPh$_2$ and Me$_2$P.CF$_2$.CH$_2$.PMe$_2$, see ref. 1247.

Sodium bis(trimethylsilyl)amide effects the replacement of a CO ligand by CN^-, forming the product (1357) (70%).[1248]

$$(COT)Fe(CO)_3 \xrightarrow[C_6H_6,\ 120°]{NaN(SiMe_3)_2} Na^+ \ [(COT)Fe(CO)_2CN]^-$$

(177)   (1357)

Reaction of $(COT)Fe(CO)_3$ (177) with the Simmons–Smith reagent affords the product (1358) (25%).[1249]

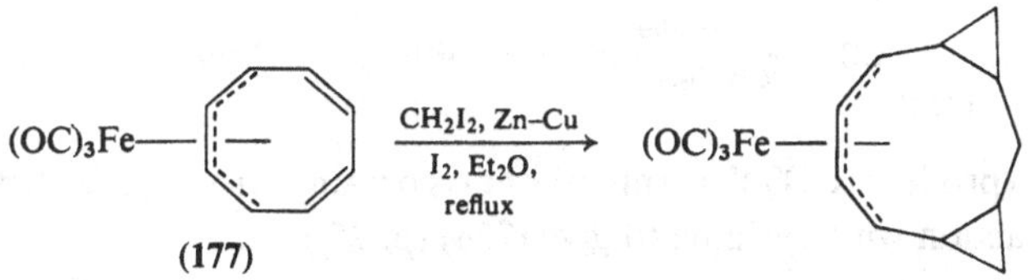

(1358)

(Stereochemistry
uncertain.)

207 Attempted application of the olefin metathesis reaction to $(COT)Fe(CO)_3$ (177) results in several products, one of which contains dimeric COT.[1250]

$$(COT)Fe(CO)_3 \xrightarrow[Toluene,\ r.-t.]{WCl_6,\ EtOH,\ EtAlCl_2} (C_{16}H_{16})Fe_2(CO)_6$$

(177)   (1359)

+ other products

The structure of (1359) is as shown.

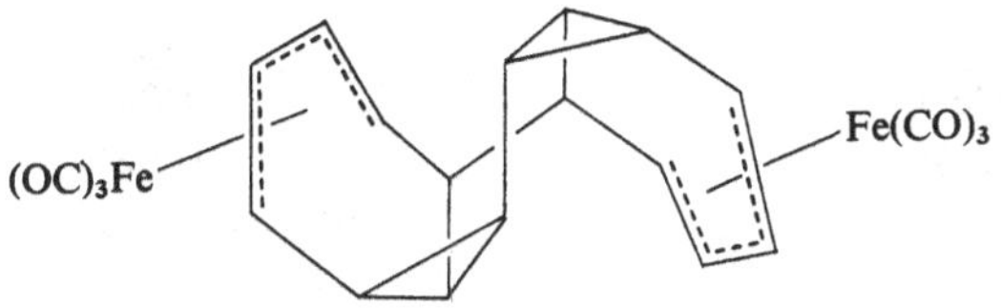

(1359)

08 For full details of the thermal isomerisation of ruthenium tricarbonyl complexes derived from substituted COTs, see ref. 1241.

Reaction of the complex (184) with dry hydrogen chloride in ether gives the product (1360)[1244] (*cf.* (1354), p. 405).

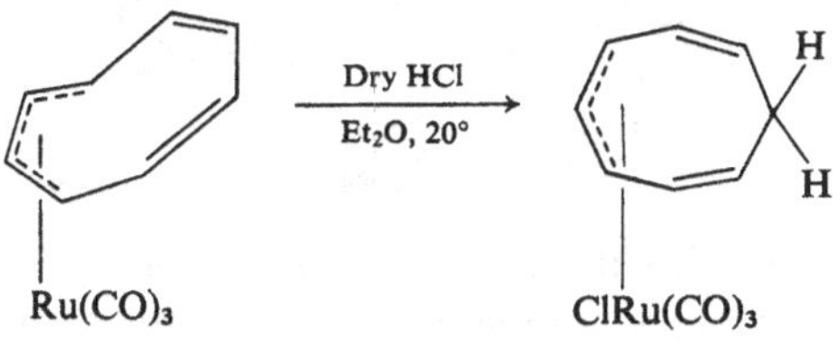

Cobalt, rhodium and iridium derivatives

211 Protonation of the rhodium complexes (**1343a**) and (**1344a**) (p. 402) with trifluoroacetic acid gives products similar to (**655b**) and (**656b**).[1233]

212 The iridium complexes (**1343b**) and (**1344b**) (p. 402) protonate in a fashion similar to that of (**654c**).[1233]

Reaction of the rhodium complex (**208**) with piperidine gives the product (**1361**)[1251] (*cf.* scheme 72).

$$[(COT)RhCl]_2 \xrightarrow[CH_2Cl_2,\ r.-t.]{Piperidine} (COT)RhCl(piperidine)$$
$$(208) \qquad\qquad\qquad\qquad (1361)$$

An analogous product (COT)IrCl(piperidine) is obtained from [(COT)IrCl]$_n$ (**215**) (*cf.* the reaction with pyridine to give (**216**) (p. 75)).

213 On successive treatment with COT, bipyridyl and ammonium hexafluorophosphate, the iridium complex (**215**) forms the salt (**1362**).[1251]

$$[(COT)IrCl]_n \xrightarrow[\substack{(ii)\ Bipyridyl \\ (iii)\ NH_4PF_6(aq.)}]{(i)\ COT,\ CH_2Cl_2} [(COT)Ir(bipyridyl)]^+\ PF_6^-$$
$$(215) \qquad\qquad\qquad\qquad\qquad (1361)$$

Cyclopropanation of (COT)Co(CPD) (**205**) may be effected by the Simmons–Smith reagent to give the product (**1363**) (19 %).[1249]

Co
(CPD)
(205)

CH₂I₂, Zn–Cu
I₂, Et₂O, reflux

Co
(CPD)
(1363)
(Stereochemistry uncertain.)

Cyclo-octatrienes

217 Epoxidation of cyclo-octa-1,3,5-triene with peracetic acid yields the isomers (**1364**) and (**1365**) (ratio 95:5; 24 %).[1252]

CH₃CO₃H
CH₂Cl₂,
Na₂CO₃,
15–20°

(1364)   (1365)

With hydrogen peroxide in acetonitrile–methanol, cyclo-octa-1,3,6-triene gives the products (**1366**) and (**1367**) (*ca.* 1:1); with perbenzoic acid, only 20 % of (**1366**) is obtained.[1253]

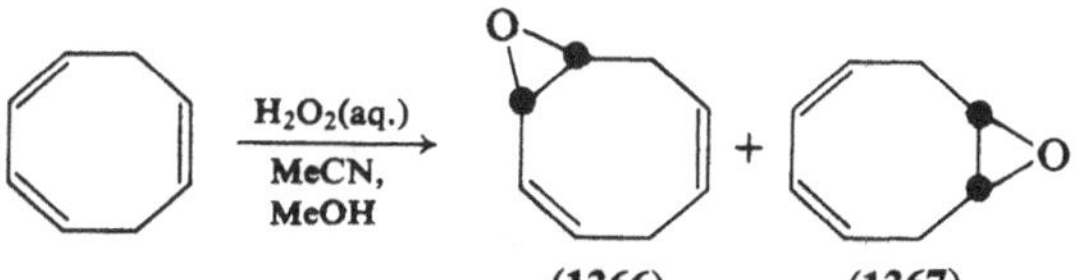

(1366)   (1367)

218 For the cyclopropanation of cyclo-octatrienes with di-iodomethane-diethyl-zinc, see ref. 1193.

219 Cyclo-octa-1,3,5-triene reacts with sulphur dioxide at 100° to give the adduct (1368); at slightly higher temperatures the triene is regenerated.[1254, 1255]

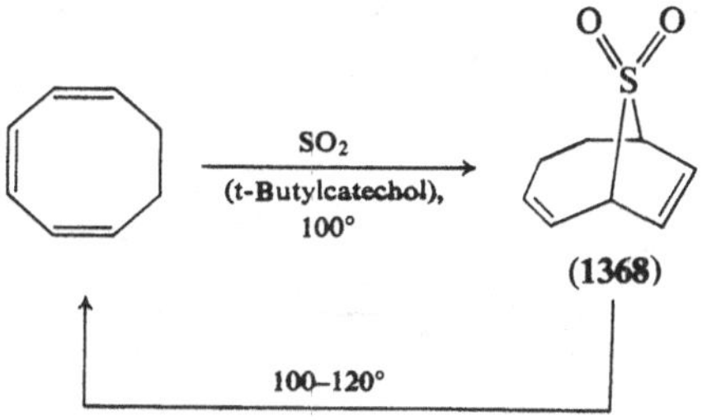

(1368)

100–120°

27 For ^{13}C spectral studies on (cyclo-octatriene)M(CO)$_3$ (725; M = Cr, Mo), see ref. 1176.

Cyclo-octa-1,3,6-triene reacts with (MeCN)$_3$Cr(CO)$_3$ to form the complex (1369; M = Cr).[1256]

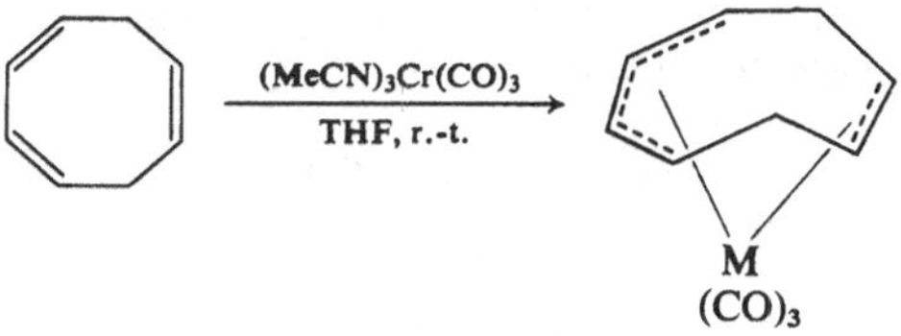

M
(CO)$_3$
(1369)

Similarly, reaction with (diglyme)Mo(CO)$_3$ gives the complex (1369; M = Mo) (45%); in this case the product rearranges quantitatively at 80° to afford the complex of cyclo-octa-1,3,5-triene, (725; M = Mo).[1256]

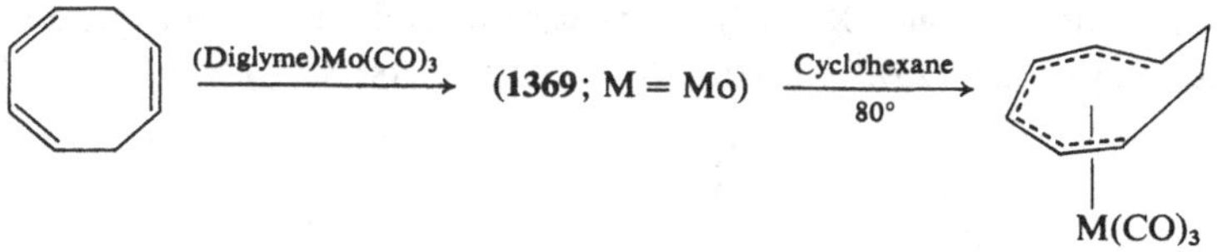

M(CO)$_3$
(725)

Reaction of cyclo-octa-1,3,5-triene with chromium(III) chloride and iso-propylmagnesium bromide, in the presence of cyclo-octa-1,3-diene, leads to the complex (1370) (up to 48%).[1173]

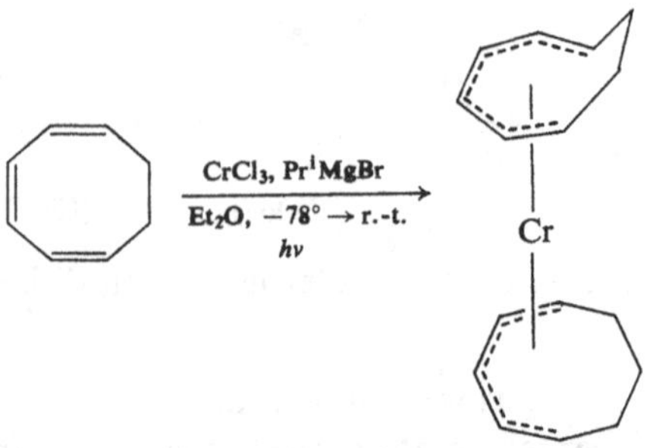

(1370)

228 Cyclo-octa-1,3,5-triene reacts with (CPD)Mn(CO)$_3$ under the influence of u.v. light to give the thermally unstable complex (1371).[1179]

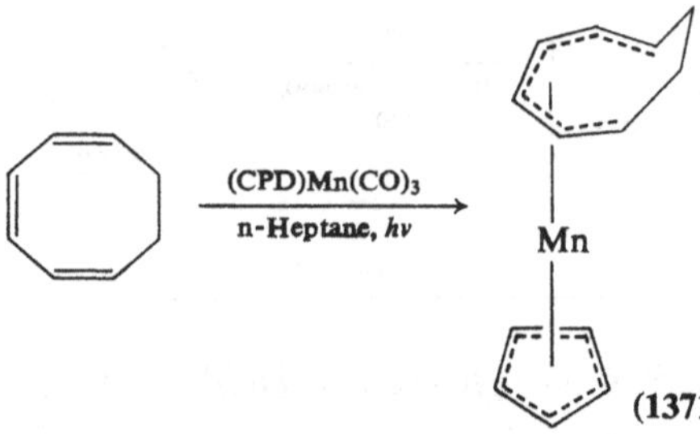

(1371)

230 For details of the reaction of cyclo-octa-1,3,5-triene with (COD)Ru(CO)$_3$ to yield the complex (bicyclo-octadiene)Ru(CO)$_3$ (735) (42%), see ref. 1186.

232 For the photochemical transformation of (cyclo-octatriene)Rh(CPD) (745) into the binuclear complex (1372), see ref. 1257.

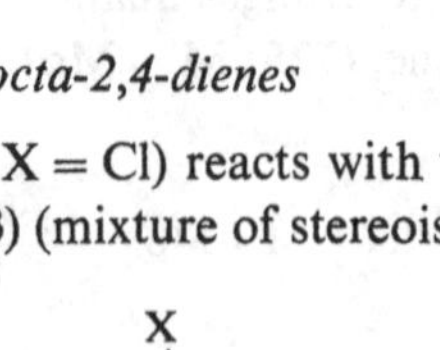

(1372)

Substituted bicyclo[4.2.0]octa-2,4-dienes

248 The dichloride (799/800; X = Cl) reacts with tetrachlorocyclopropene at 75° to yield the product (1373) (mixture of stereoisomers; 83%).[1258]

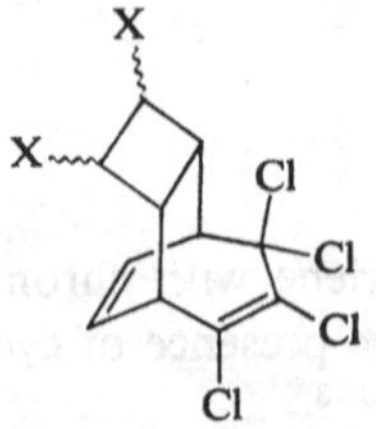

(1373)

Cyclo-octa-2,4,6-trienones

254 The trienone (821) may be prepared from methoxycyclo-octatetraene (see p. 385); for 8-substituted derivatives, see p. 419.

For the (low-temperature) preparation of the bicyclic dienone (822), see ref. 1259.

262 Reaction of cyclo-octatrienone (821) with (MeCN)Cr(CO)$_3$ leads to the complex (1374) (26%); with Cr(CO)$_6$, the products are (1375) (4%) and (1376) (11%).[1198]

Reaction of cyclo-octatrienone (821) with Fe(CO)$_5$, induced by u.v. radiation, gives the complex (300) (36%).[1259]

The structure of this product is apparently as shown.[1198, 1259]

With (benzylidene-acetone)Fe(CO)$_3$[1198, 1259] or (pent-3-en-2-one) Fe(CO)$_3$,[1259] the complex (1377) derived from the valence tautomer (822) is obtained (up to 60%); demetallation of this product (at low temperature) gives the bicyclic dienone (822) (29%).[1259]

(821)   (822)

Ce(NO₃)₄·2NH₄NO₃–Al₂O₃
Et₂O, pentane, −30°

e.g. (PhCH=CHCOCH₃) Fe(CO)₃ | C₆H₆, 60°

(OC)₃Fe—

(1377)   (Stereochemistry uncertain.)

Bicyclo[6.1.0]nona-2,4,6-trienes

272 The formation of the homosemibullvalene (1378) (30%) by photolysis of bicyclo[6.1.0]nonatriene (82*a*) has been reported.[1260]

(1378)

277 For details of the action of potassium in liquid ammonia on the *syn*- and *anti*-9-methylbicyclo[6.1.0]nonatrienes, (82*g*) and (82*q*), and on the 9,9-dimethyl-derivative (82*n*), see ref. 1261.

For the deprotonation of compounds (82; R² = H) to form anions of structure (1379), see the superscribed refs.

(82; R² = H)   (1379)

R = Ph[1193] R = 1-naphthyl, 2-naphthyl, 9-anthryl[1154]

278 For an n.m.r. spectral study of topomerisation in the anion (922), see ref. 1262.

286 Reaction of bicyclo[6.1.0]nonatriene with 3,6-diphenyl-*s*-tetrazine gives the product (1380) (*ca.* 40%), which is readily dehydrogenated to (1381) (67%)[1263] (*cf.* pp. 419, 420).

The chromium complex (1382) is formed by the action of (MeCN)₃Cr(CO)₃ on bicyclo[6.1.0]nonatriene; an analogue is produced from the *anti*-9-ethoxycarbonyl-derivative (82*h*) (yields 50–70%).[1256] Thermal rearrangement of the ligand in (1382) leads to the product (1383) (89%)[1256] (*cf.* the molybdenum complex (964), p. 288).

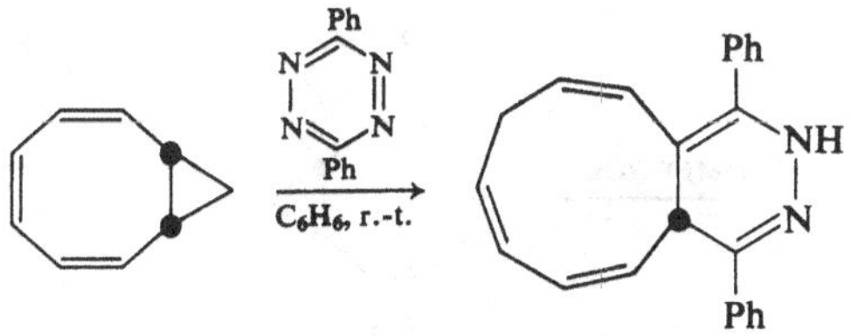

(1380)

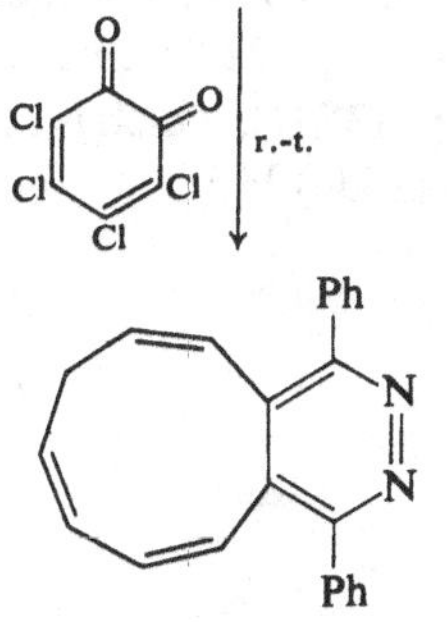

(1381)

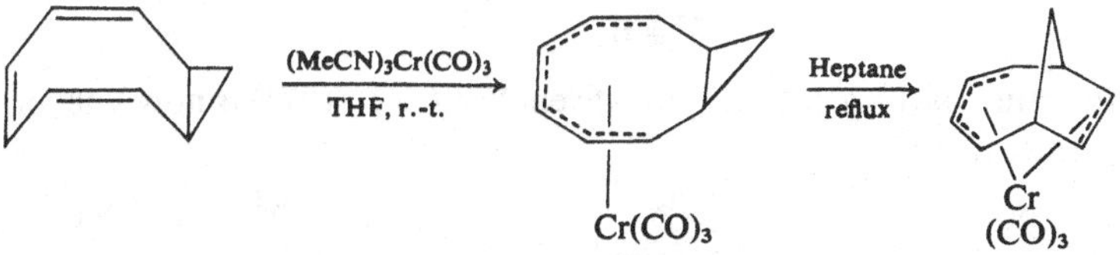

(1382)   (1383)

Molybdenum tricarbonyl complexes (**1384**) are formed from 9-substituted bicyclo[6.1.0]nonatrienes (**82***f–h*) and (diglyme)Mo(CO)$_3$[1264] (table 142).

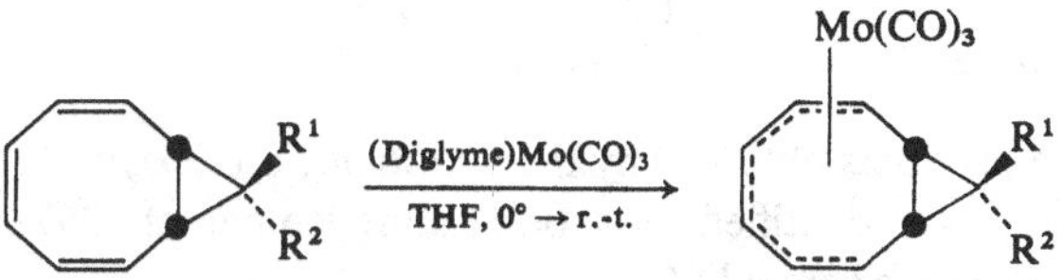

(**82**)   (**1384**)

Table 142

	R^1	R^2	Yield (%)
f	Cl	H	85–95
g	H	Me	45
h	CO$_2$Et	H	85–95

The tungsten complex (**1385**) results (25–30%) from bicyclo[6.1.0]nona-triene and (DMF)W(CO)$_3$.[1264]

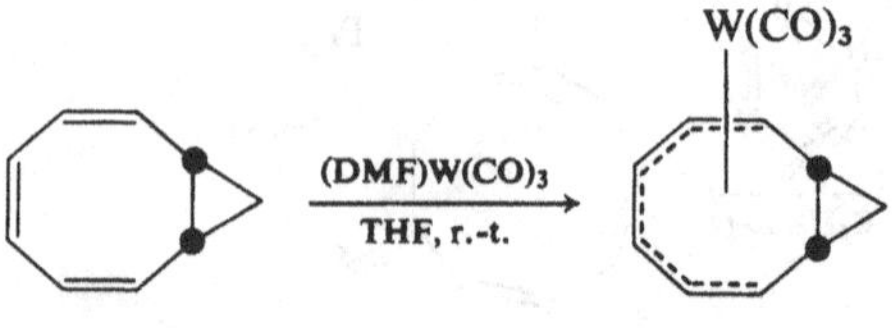

(1385)

The complex (**964**) readily absorbs carbon monoxide[1264] (*cf.* (cyclo-octatriene)Mo(CO)$_3$ (**725**; M = Mo), p. 227); this behaviour is also shown by (**1384*f***) and (**1384*h***), but not by (**1384*g***), which is consistent with the illustrated stereochemistry of (**1384**) and (**1386**).[1264]

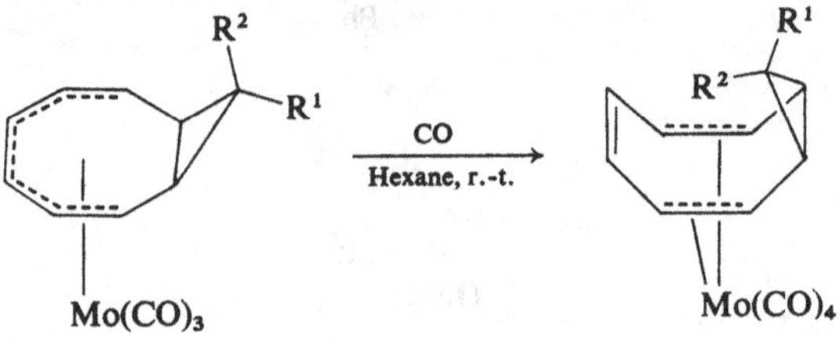

(1384) (1386)

R^1 = H, Cl, CO_2Et
R^2 = H

289 The structure of (bicyclo[6.1.0]nonatriene)$Fe_2(CO)_6$ (**971**) is now known to be as illustrated.[1265]

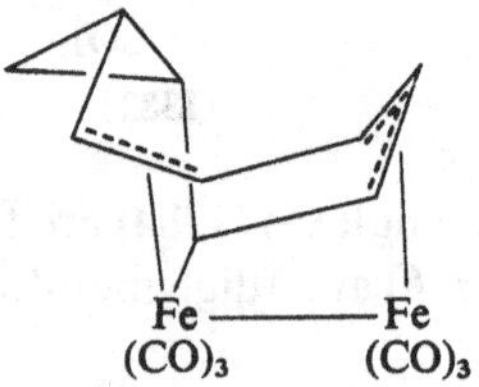

(971)

Another of the products obtained by the action of $Fe_2(CO)_9$ on bicyclo[6.1.0]-nonatriene has been identified as a bond-shift isomer of (**969**), having the structure (**1387**) shown below.[1266]

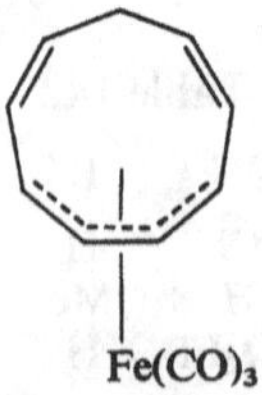

(1387)

290 At *ca.* 100°, the complex (**1387**) rearranges to (**970**)[1266] (*cf.* (**969**) and (**971**)).

291 Bicyclo[6.1.0]nonatriene reacts with (PhCN)$_2$PdCl$_2$ to give the complex (1388) (85%); treatment with methanol yields the dimeric product (1389), which with hydrogen chloride reverts to (1388).[1267]

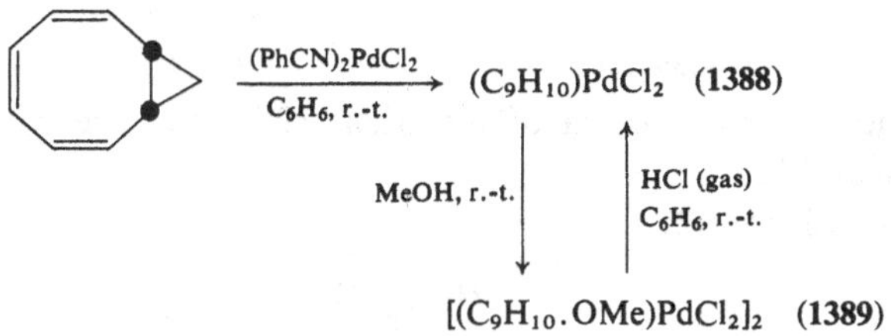

The probable structures of (1388) and (1389) are shown.

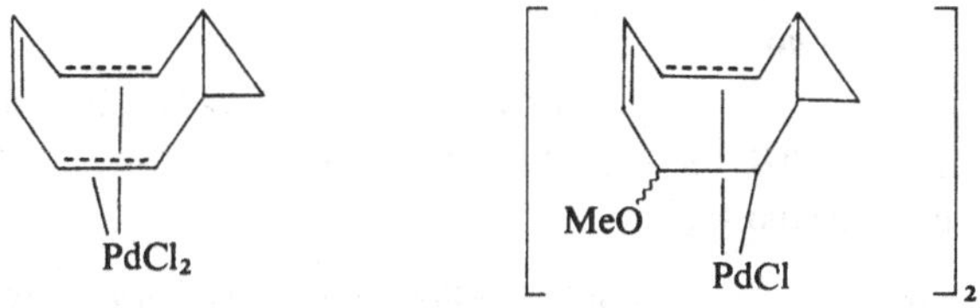

(1388)   (1389)

Bicyclo[4.2.1]nona-2,4,7-trienes

292 Addition to scheme 91:

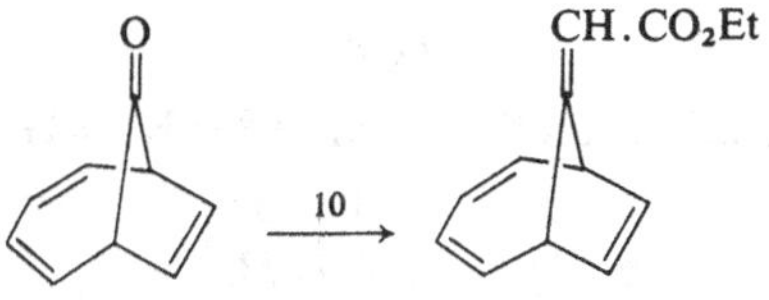

(540)

Reaction	Conditions	Yield (%)	Ref.
10	(EtO)$_2$PO.CH$_2$CO$_2$Et, NaH	—	1268

(For conversion of (540) into the (CH)$_{12}$ hydrocarbon (1390), see ref. 1268.)

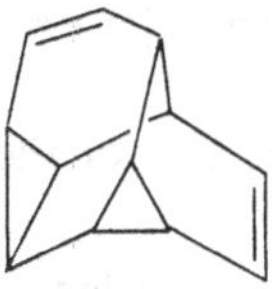

(1390)

For details of reaction 6, see ref. 1269. Conversion of the ketal (1391) into the dienone (1392) is also described in ref. 1269.

(1391)   (1392)

293 For the photo-electron spectra of the triene (966) and the trienone (540) see refs. 1270 and 1271 respectively.

Addition to scheme 92:

(540)

Reaction	Conditions	Yield (%)	Ref.
11	$Al(OPr^i)_3$, Me_2CO, xylene, reflux	77[b]	1272

[b] Product ratio 4:1.

For the generation of the complex (1393) from (540) and tin(IV) chloride in liquid sulphur dioxide, see ref. 1272.

(1393)

295 For theoretical calculations on the cation (985; $R^2 = H$), see ref. 1272.

296 For further mechanistic discussion of the solvolysis of the epimeric tosylates (990) and (991; H instead of D), see ref. 1273.

297 Attempted reduction of the ketone (540) *via* the hydrazone leads to the rearranged hydrocarbon (1378) (95%).[1274]

(540)   (1378)

For thermolysis of the sodium salt of the tosylhydrazone (997), see also ref. 1275.

299 Hydroboration of bicyclo[4.2.1]nonatriene, and subsequent treatment with hydrogen peroxide, affords the dienol (1394) (75–80%).[1276]
(For further transformations of (1394), see ref. 1276.)

For full details of the preparation of the products (1009) and (1010), see ref. 1277.

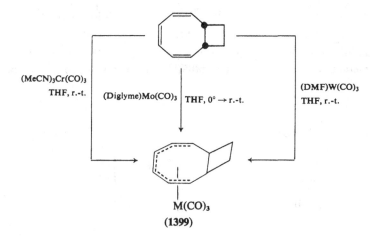

$$\text{(i) (Me}_2\text{CH.CHMe)}_2\text{BH} \atop \text{(ii) H}_2\text{O}_2$$

(1394)

Bicylo[6.2.0]deca-2,4,6-trienes

303   Treatment of bicyclo[6.2.0]decatriene (515*a*) with potassium in liquid ammonia gives the cyclodecatrienyl anion (1395); when the temperature is raised, rearrangement occurs *via* (1396) and (1397) to give (1398) as the final product.[1278]

(515*a*)   (1395)   (1396)

$\xrightarrow[\text{NH}_3\text{(liq.)}]{\text{K}}$ $\xrightarrow{-41°}$ $\Big\downarrow{-17.5°}$

(1398)   (1397)

$\xleftarrow{31°}$

(The products obtained by 'quenching' these anions are as expected.)

04   Bicyclo[6.2.0]decatriene reacts with (MeCN)$_3$Cr(CO)$_3$ to give the complex (1399; M = Cr),[1256] with (diglyme)Mo(CO)$_3$ to give (1399; M = Mo) (85–95%),[1264] and with (DMF)W(CO)$_3$ to give (1399; M = W) (25–30%).[1264]

(MeCN)$_3$Cr(CO)$_3$
THF, r.-t.

(Diglyme)Mo(CO)$_3$ THF, 0° → r.-t.

(DMF)W(CO)$_3$
THF, r.-t.

M(CO)$_3$

(1399)

Thermal rearrangement of the molybdenum complex (**1399**; M = Mo) leads to a derivative of bicyclo[4.2.2]decatriene, (**1400**) (55%) (*cf.* (**964**), p. 288); carbon monoxide is readily absorbed with the formation of the product (**1401**) (90–95%) (*cf.* (**725**; M = Mo), p. 227).[1264]

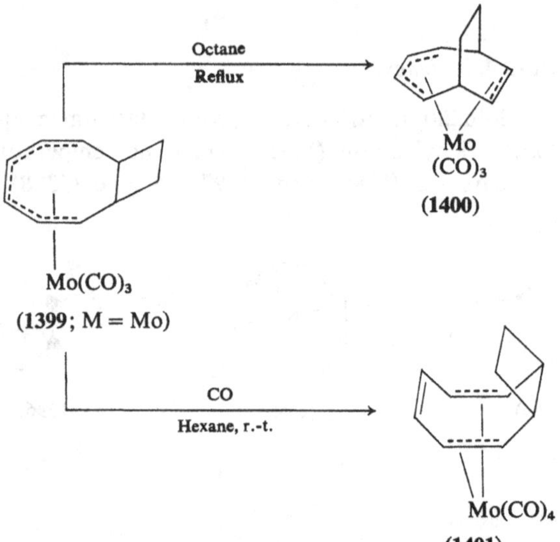

Octane
Reflux

Mo
(CO)₃

(**1400**)

Mo(CO)₃

(**1399**; M = Mo)

CO
Hexane, r.-t.

Mo(CO)₄

(**1401**)

For a study of the fluxional process in the complex (**1026**), using ¹³C n.m.r. spectroscopy, see ref. 1279.

305 Bicyclo[6.2.0]decatriene (**515a**) reacts with Ru₃(CO)₁₂ to afford the complexes (**1402**) (47%) and (**1403**) (20%), together with some minor products [1280] (*cf.* the analogous iron complexes (**1026**) and (**1027**), p. 304).

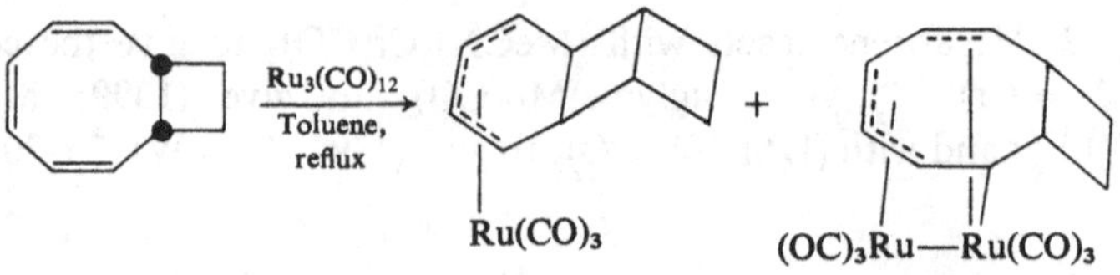

Ru₃(CO)₁₂
Toluene,
reflux

Ru(CO)₃

(OC)₃Ru—Ru(CO)₃

(**515a**)   (**1402**)   (**1403**)

Reaction with (PhCN)₂PdCl₂ yields the complex (**1404**) (83%), which with methanol gives the product (**1405**)[1267] (*cf.* the analogous reaction of bicyclo[6.1.0]nonatriene, p. 415).

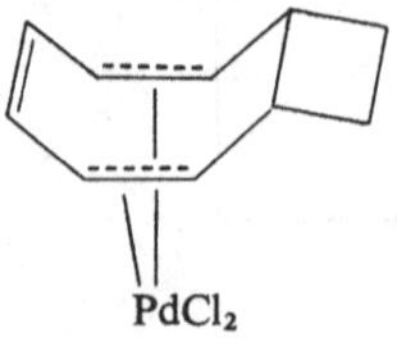

PdCl₂

(**1404**)

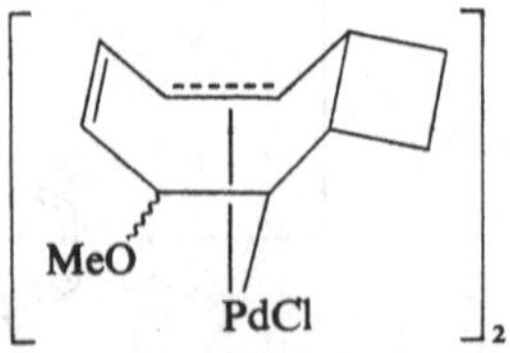

MeO

PdCl

(**1405**)

9-Azabicyclo[6.1.0]nona-2,4,6-trienes

310 Reaction of the *N*-acetyl-compound (92; R = Ac) with 3,6-diphenyl-*s*-tetrazine gives the product (1406) (75%), which is readily dehydrogenated to yield (1407) (85%)[1263] (*cf.* pp. 413, 420).

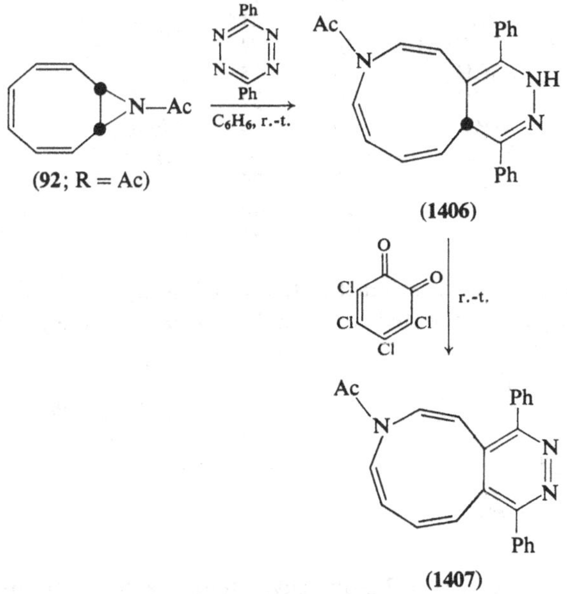

(92; R = Ac)

(1406)

(1407)

9-Oxabicyclo[6.1.0]nona-2,4,6-trienes

14 Deprotonation of the epoxide (27) with lithium di-isopropylamide generates the enolate anion (1291) (p. 385), which may be subjected to alkylation, acylation, etc. with the formation of the substituted trienones (1408) and/or the oxygenated COT (1290)[1197] (table 143). Reaction of the enolate anion (1291) with diphenyl disulphide or benzeneselenenyl chloride affords the acetophenone derivatives (1409) and (1410).[1197] Cyclo-octatrienone is a

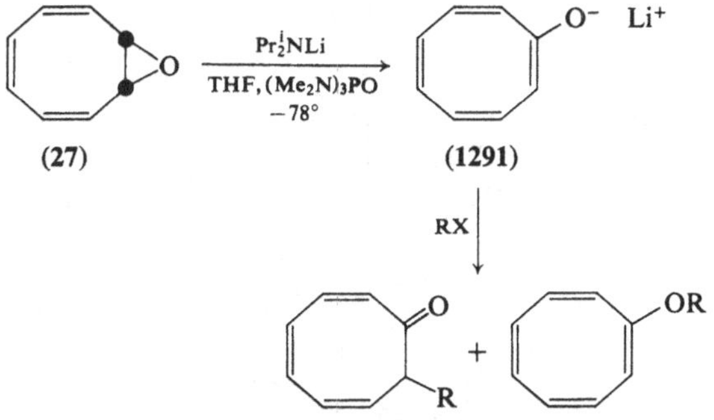

(27)

(1291)

(1408)

(1290)

Table 143

R	X	Yield (%) (1408)	(1290)
Me	I	78	—
Et	Br	16	11
Et	I	81	—
CH₂=CHCH₂	Br	74	—
PhCH₂	Br	15	—
PhCH₂	I	37	—
MeCO	Cl	—	Up to 26
EtO₂C	Cl	—	Up to 63
PhS	Cl	26	—
PhSe	Br	40	—
Br	Br	51	—

major product in some of these reactions; with CH_3CO_2D at 0°, the enolate anion (1291) gives the deuteriated compound (1408; R = D) (76%).[1197]

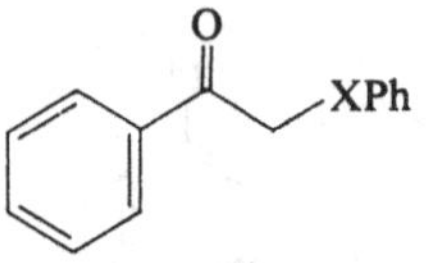

(1409)

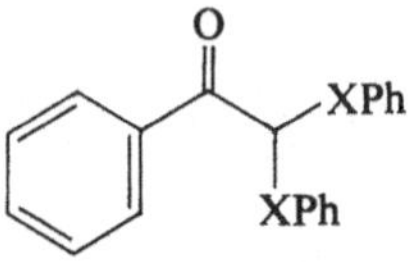

X = S, Se

(1410)

315 The epoxide (27) reacts with 3,6-diphenyl-*s*-tetrazine to yield the adduct (1411) (27% based on recovered tetrazine), which may be dehydrogenated to give the product (1412) (60%)[1281] (*cf.* pp. 413, 419).

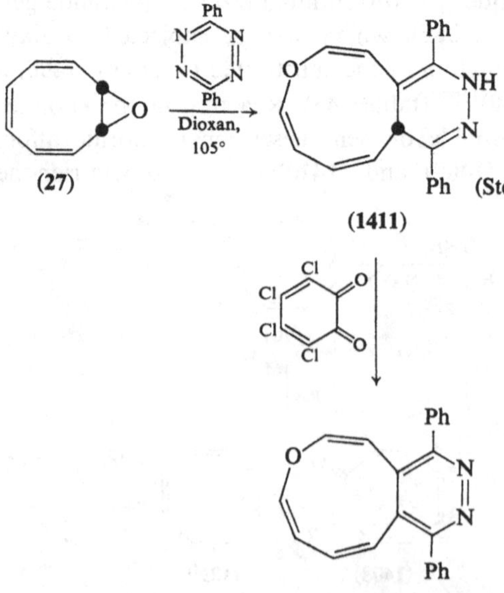

(27)

(1411) (Stereochemistry not certain.)

(1412)

age

9-Thiabicyclo[4.2.1]nona-2,4,7-trienes

324 For the crystal structure of the sulphone (108), see ref. 1282.

325 For further details of the deuteriation and methylation of the sulphone (108), see ref. 1157.

In liquid sulphur dioxide, the sulphone (108) reacts with antimony(v) fluoride (>2 equivalents) to form the *endo*-homotropylium-8-sulphinate complex (1413) (see also scheme 16, p. 40); stereochemical inversion occurs when the temperature is allowed to rise.[1157]

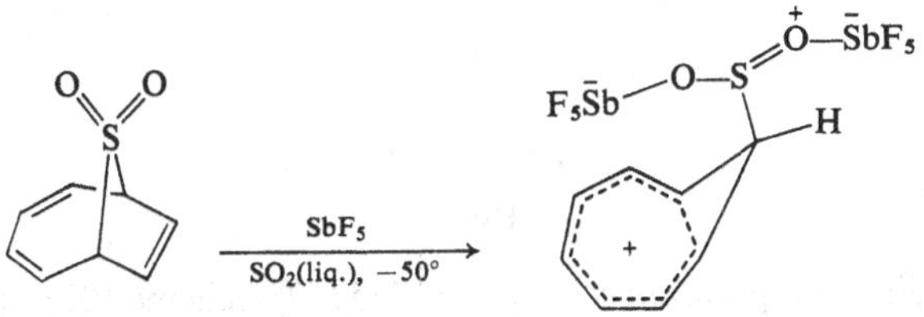

(108)   (1413)

Similar products are formed from the mono-and di-methyl derivatives, (1414) and (312c) (p. 112) respectively.[1157]

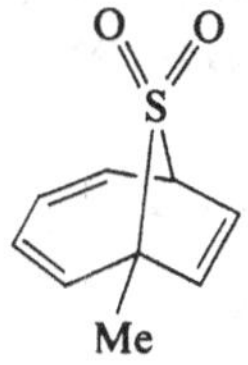

(1414)

9-Aza-10-oxobicyclo[4.2.2]deca-2,4,7-trienes

327 For the further reduction of the saturated lactam (1105) with lithium aluminium hydride, leading to the amine (1415), see ref. 1283.

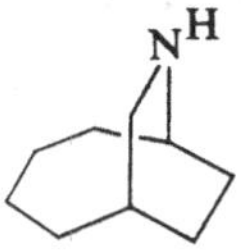

(1415)

Tricyclo[4.2.2.0²,⁵]deca-3,7-dienes

328 Addition to scheme 102:

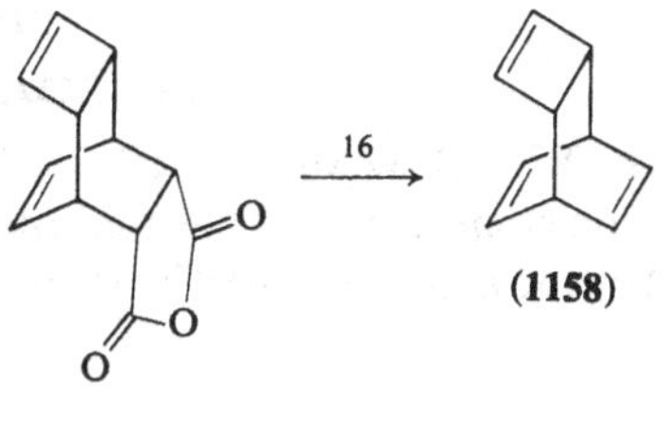

(1158)

Reaction	Reagents and conditions	Yield (%)	Ref.
16	$(Ph_3P)_2Ni(CO)_2$, diglyme, reflux	73	1160

Another preparation of 'Nenitzescu's hydrocarbon' (1158) (p. 421) (five steps, overall yield 44%) starts from the acryloyl chloride adducts (1270b) and (1271b) (p. 378).[1159]

For the conversion of the acrylonitrile adducts (1270a) and (1271a) into the ketone (1416) (50%), see ref. 1158.

(1416)

For a modified preparation of (1106) and (1107) (scheme 102), see ref. 1284.

331 The epoxides (1067) and (1113) have been prepared using m-chloroperbenzoic acid; a minor product of the reaction with the di-ester (1108) is apparently a dimer of (1113).[1285]

Treatment of the epoxide (1113) with dry hydrogen chloride in methanol gives the hydroxy-lactone-ester (1417).[1285]

(1417)

334 The dimethyl ester (1108) reacts with iodine azide to give the product (1418) (80–90%), together with some iodo-lactone-ester (1122; R = Me).[1284, 1286, 1287]

(1108) → (1418)

+ (1122; R = Me)

Other derivatives (1157; $R^1 = R^2 = Me$ or $R^1R^2 = CH_2OCH_2$), however, afford (1419) in dichloromethane, and in acetonitrile participation of the solvent results in (1420) as the major product[1284] (table 144).

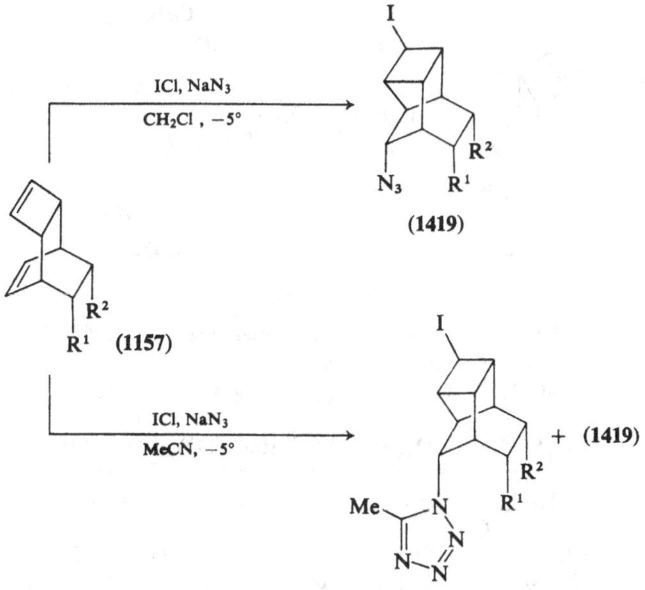

(1419)

R¹ (1157)

+ (1419)

(1420)

Table 144

R¹	R²	Solvent	Yield (%) (1419)	Yield (%) (1420)
Me	Me	CH₂Cl₂	88	—
		MeCN	(Not given)	46
H₂C﹏O﹏CH₂		CH₂Cl₂	94	—
		MeCN	18	56

35 Treatment of the anhydride (117) with mercury(II) acetate in methanol followed by aqueous sodium chloride, apparently yields the lactone-ester (1421) (40%).[1288]

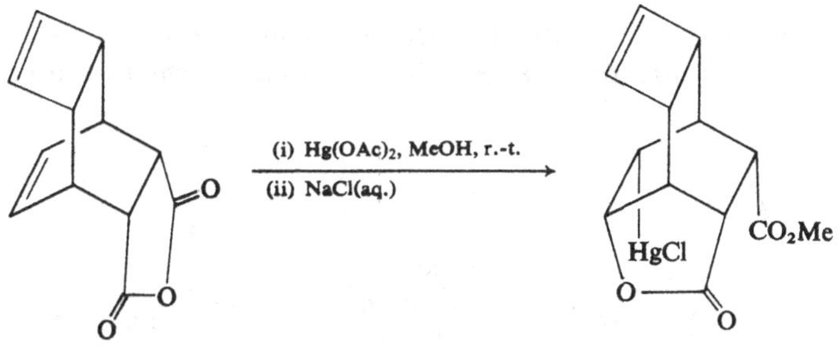

(i) Hg(OAc)₂, MeOH, r.-t.
(ii) NaCl(aq.)

(117)   (1421)

With other derivatives (1157), however, products of structure (1422) are formed (table 145). In the presence of sodium azide (in methanol), the isolated product (from the di-ester) has structure (1423).[1284, 1287]

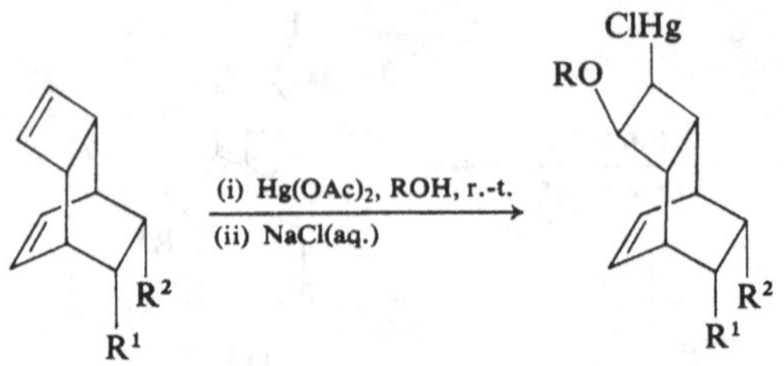

(i) Hg(OAc)₂, ROH, r.-t.
(ii) NaCl(aq.)

(1157) (1422)

Table 145

R¹		R²	R	Yield (%)	Ref.
Me		Me	H	85ᵃ	1284
H₂C	O	CH₂	H	81ᵃ	1284
CO₂Me		CO₂Me	H	100ᵃ, (12ᵇ)	1284, (1288)
Me		Me	Me	82	1284
H₂C	O	CH₂	Me	82	1284
CO₂Me		CO₂Me	Me	(96), 99	(1284), 1288
CO₂Me		CO₂Me	Ac	76	1288

ᵃ Reaction in aqueous THF.
ᵇ Reaction in aqueous acetone.

For the demercuration of the compounds (1422; R = H) with sodium boro-hydride, resulting in products (1424), see ref. 1284.

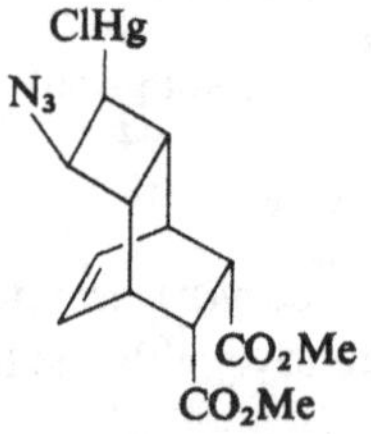

(1423) 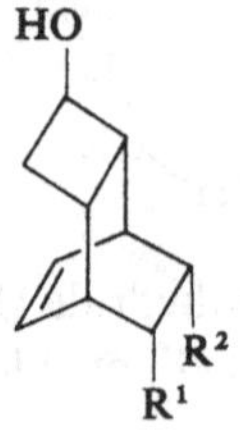 (1424)

The addition of nitrosyl chloride (from isoamyl nitrite and concentrated hydrochloric acid) to the di-ester (1108) leads to the chloro-ketone (1425) (90 %).[1284]

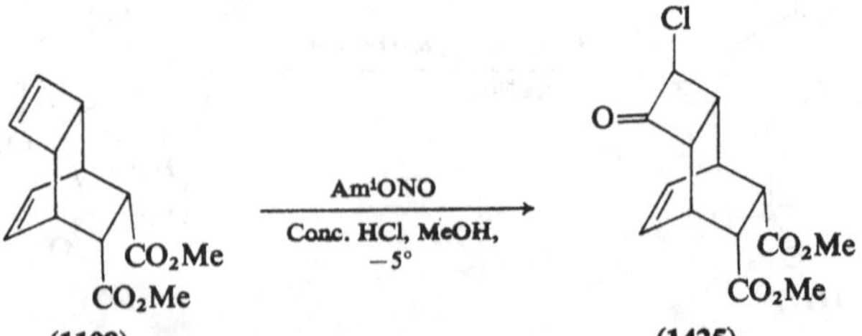

(1108) (1425)

Hydroboration of the di-ester (1108) may be used to prepare the compound (1424; R¹ = R² = CO₂Me) (96 %).[1284]

With benzenesulphenyl chloride, the di-ester (**1108**) forms a mixture of the products (**1426**) and (**1427**) in quantitative yield (ratio 3:1).[1289]

(**1108**)   (**1426**)   (**1427**)

A similar result is obtained with the cyclic ether (**1107**).[1289]

39   Addition of hemicyclone to the anhydride (**117**) affords the product (**1428**) (formulated as the *exo*-adduct) (98%), which on photolysis gives the diene-anhydride (**1429**) (49%).[1160]

(**117**)   (**1428**)   (**1429**)

Similar adducts are formed from the anhydride (**117**) and other cyclones, including tetracyclone, phencyclone and 2,5-diphenyl-3,4-di(2'-pyridyl)cyclopentadienone.[1290]

For reaction of the di-ester (**1108**) with the COT dimer (**10**), see p. 427.

Tricyclo[4.2.2.0²,⁵]deca-3,7,9-trienes

44   The bis(trifluoromethyl)-derivative (**1430**) affords the caged isomer (**1431**) (*ca.* 80% conversion) on u.v. irradiation; the cycloaddition is reversed by heat. [1291]

(**1430**)   (**1431**)

349 For the reaction of 'Nenitzescu's hydrocarbon' (**1158**) (p. 343) with hemi-cyclone, and photolytic conversion of the product into barrelene (24% overall yield from COT), see ref. 1160.

For reactions of the di-ester (**7**) with acecyclone etc., see p. 389.

The bis(trifluoromethyl)-compound (**1430**) reacts with tetrachloro-*o*-benzoquinone to give a 4:3 mixture of the *endo*- and *exo*-adducts (**1432**) and (**1433**); at 210° the cyclobutenes (**1310**) and (**1311**) are obtained. [1208]

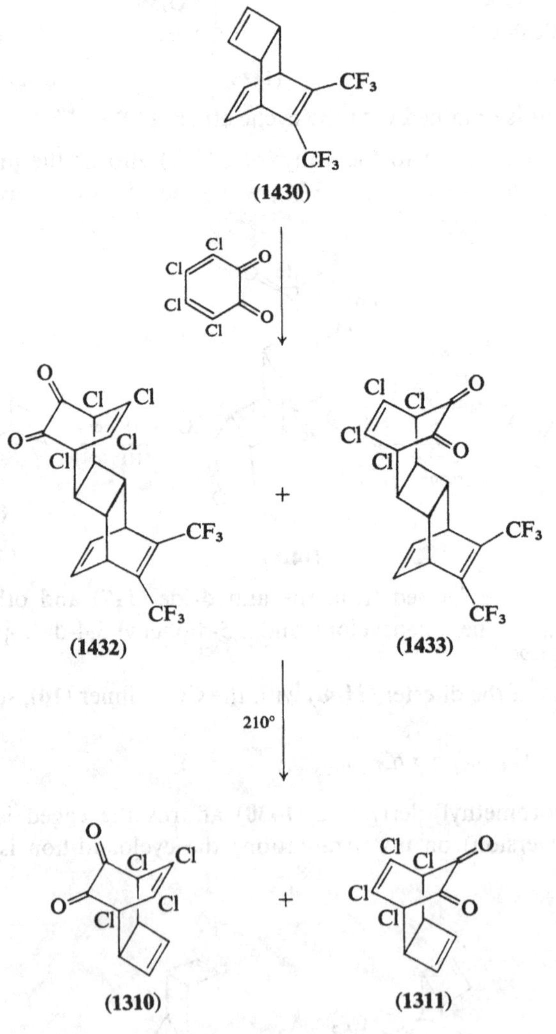

(**1430**)

(**1432**)   +   (**1433**)

210°

(**1310**)   +   (**1311**)

For reaction of the di-ester (**7**) with the COT dimer (**10**), see p. 427.

351 The bis(trifluoromethyl)-compound (**1430**) reacts with Mo(CO)$_6$ to give (in low yield) the complex (**1434**). [1161]

(1430)   (1434)

9,10-Diazatricyclo[4.2.2.0^{2,5}]deca-3,7-dienes

354 For transformations of the octamethyl-diester (1435) analogous with those of (809; R = Et) (scheme 105), see ref. 1213.

(1435)

355 For photo-caging of the octamethyl-derivative (1435), and the corresponding N-phenyltriazolinedione adduct, see ref. 1213.

Oligomers of cyclo-octatetraene

361 Treatment of dimer (9) with potassium t-butoxide leads to the isomer (1436) (77%).[1292]

(9)   (1436)

369 Additions to table 139 (*cf.* ref. 1293):

Dienophile	Reaction conditions	Yield (%)	Ref.
COT–dimethyl maleate adduct (1108)	100[a]	31	164
COT–dimethyl acetylenedicarboxylate adduct (7)	100[a]	10	164

[a] (See footnote to table 139.)

For details of the reaction of dimer (**10**) with vinylene carbonate, and subsequent transformations of the adduct (**1247**), see ref. 1293.

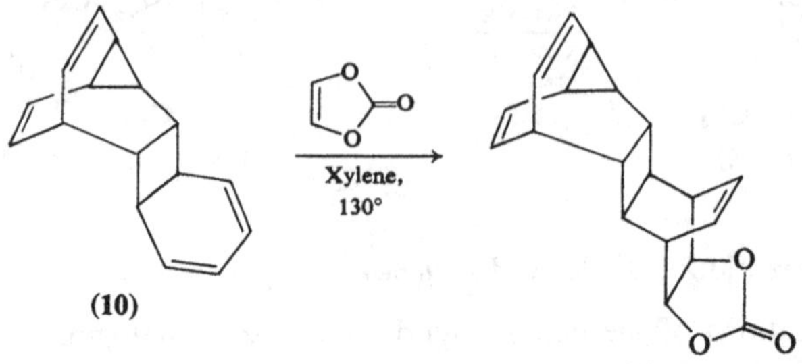

(**10**)

Xylene,
130°

(**1247**)

References

1 R. Willstätter and E. Waser, *Ber.*, 1911, **44**, 3423.
2 R. Willstätter and M. Heidelberger, *Ber.*, 1913, **46**, 517.
3 A. C. Cope and C. G. Overberger, *J. Amer. Chem. Soc.*, 1948, **70**, 1433.
4 W. Baker, *J. Chem. Soc.*, 1945, 258.
5 A. C. Cope and C. G. Overberger, *J. Amer. Chem. Soc.*, 1947, **69**, 976.
6 A. C. Cope and W. J. Bailey, *J. Amer. Chem. Soc.*, 1948, **70**, 2305.
7 K. Ziegler and H. Wilms, *Annalen*, 1950, **567**, 1.
8 G. Eglinton, R. A. Raphael and R. G. Willis, *Proc. Chem. Soc.*, 1962, 334.
9 G. Eglinton, W. McCrae, R. A. Raphael and J. A. Zabkiewicz, *J. Chem. Soc.* (C), 1969, 474.
10 C. G. Cardenas, A. N. Khafaji, C. L. Osborn and P. D. Gardner, *Chem. and Ind.*, 1965, 345.
11 W. Reppe, O. Schlichting, K. Klager and T. Toepel, *Annalen*, 1948, **560**, 1.
12 N. Hagihara, *J. Chem. Soc. Japan, Pure Chem. Sect.*, 1952, **73**, 237; *Chem. Abs.*, 1953, **47**, 9954i.
13 N. Hagihara, *J. Chem. Soc. Japan, Pure Chem. Sect.*, 1952, **73**, 323; *Chem. Abs.*, 1953, **47**, 10490i.
14 S. N. Ushakov and O. F. Solomon, *Izvest. Akad. Nauk S.S.S.R., Otdel. khim. Nauk*, 1954, 694; *Chem. Abs.*, 1955, **49**, 10868d.
15 G. N. Schrauzer and S. Eichler, *Chem. Ber.*, 1962, **95**, 550.
16 G. N. Schrauzer, *Chem. Ber.*, 1961, **94**, 1403.
17 W. Reppe, *Experientia*, 1949, **5**, 93.
18 G. N. Schrauzer, P. Glockner and S. Eichler, *Angew. Chem. Internat. Edn*, 1964, **3**, 185.
19 C. W. Bird, *Transition Metal Intermediates in Organic Synthesis*, Logos Press, London, 1967, ch. 1.
20 E.-A. Reinsch, *Theor. Chim. Acta*, 1968, **11**, 296.
21 E.-A. Reinsch, *Theor. Chim. Acta*, 1970, **17**, 309.
22 H. C. Longuet-Higgins and L. E. Orgel, *J. Chem. Soc.*, 1956, 1969.
23 W. Reppe, O. Schlichting and H. Meister, *Annalen*, 1948, **560**, 93.
24 L. E. Craig and C. E. Larrabee, *J. Amer. Chem. Soc.*, 1951, **73**, 1191.
25 A. C. Cope and S. W. Fenton, *J. Amer. Chem. Soc.*, 1951, **73**, 1195.
26 D. S. Withey, *J. Chem. Soc.*, 1952, 1930.
27 E. R. Lippincott, R. C. Lord and R. S. McDonald, *J. Amer. Chem. Soc.*, 1951, **73**, 3370.
28 Z. Kuri, *Bull. Chem. Soc. Japan*, 1953, **26**, 280.
29 Z. Kuri, *Bull. Chem. Soc. Japan*, 1954, **27**, 330.
30 Z. Kuri and S. Shida, *Bull. Chem. Soc. Japan*, 1952, **25**, 116.
31 Z. Kuri, *J. Chem. Soc. Japan, Pure Chem. Sect.*, 1955, **76**, 944; *Chem. Abs.*, 1956, **50**, 3093d.
32 E. Vogel, H. Kiefer and W. R. Roth, *Angew. Chem. Internat. Edn*, 1964, **3**, 442.
33 T. J. Katz and E. W. Turnblom, *J. Amer. Chem. Soc.*, 1970, **92**, 6701.
34 G. Schröder and W. Martin, *Angew. Chem. Internat. Edn*, 1966, **5**, 130.

430   *References*

35   M. Avram, I. G. Dinulescu, E. Marica, G. Mateescu, E. Sliam and C. D. Nenitzescu, *Chem. Ber.*, 1964, **97**, 382.
36   H. M. Frey, H.-D. Martin and M. Hekman, *Chem. Comm.*, 1975, 204.
37   W. Merk and R. Pettit, *J. Amer. Chem. Soc.*, 1967, **89**, 4788.
38   D. Bryce-Smith, A. Gilbert and J. Grzonka, *Chem. Comm.*, 1970, 498.
39   R. E. Benson and T. L. Cairns, *J. Amer. Chem. Soc.*, 1950, **72**, 5355.
40   R. Askani, *Chem. Ber.*, 1967, **102**, 3304.
41   R. M. Moriarty, C.-L. Yeh and N. Ishibi, *J. Amer. Chem. Soc.*, 1971, **93**, 3085.
42   L. A. Paquette, R. K. Russell and R. E. Wingard, *Tetrahedron Letters*, 1973, 1713.
43   L. A. Paquette, S. V. Ley, R. H. Meisinger, R. K. Russell and M. Oku, *J. Amer. Chem. Soc.*, 1974, **96**, 5806.
44   H. E. Zimmerman, R. W. Binkley, R. S. Givens, G. L. Grunewald and M. A. Sherwin, *J. Amer. Chem. Soc.*, 1969, **91**, 3316.
45   J. Meinwald and H. Tsuruta, *J. Amer. Chem. Soc.*, 1970, **92**, 2579.
46   G. E. Gream, L. R. Smith and J. Meinwald, *J. Org. Chem.*, 1974, **39**, 3461.
47   A. G. Anastassiou and B. Y.-H. Chao, *Chem. Comm.*, 1971, 979.
48   J. Gasteiger and R. Huisgen, *J. Amer. Chem. Soc.*, 1972, **94**, 6541.
49   L. A. Paquette, M. Oku, W. E. Heyd and R. H. Meisinger, *J. Amer. Chem. Soc.*, 1974, **96**, 5815.
50   T. Mukai, K. Kurabayashi and T. Hagiwara, *Sci. Reports Tohoku Univ.*, Ser. *1*, 1972, **55**, 224; *Chem. Abs.*, 1973, **79**, 37074j.
51   K. Kurabayashi and T. Mukai, *Tetrahedron Letters*, 1972, 1049.
52   T. A. Antkowiak, D. C. Sanders, G. B. Trimitsis, J. B. Press and H. Shechter, *J. Amer. Chem. Soc.*, 1972, **94**, 5366.
53   D. G. Farnum and J. P. Snyder, unpublished work; quoted by D. W. McNeil, M. E. Kent, E. Hedaya, P. F. D'Angelo and P. O. Schissel, *J. Amer. Chem. Soc.*, 1971, **93**, 3817.
54   L. A. Paquette, U. Jacobsson and M. Oku, *Chem. Comm.*, 1975, 115.
55   E. L. Allred and B. R. Beck, *J. Amer. Chem. Soc.*, 1973, **95**, 2393.
56   R. D. Miller and E. Hedaya, *J. Amer. Chem. Soc.*, 1969, **91**, 5401.
57   E. Hedaya, R. D. Miller, D. W. McNeil, P. F. D'Angelo and P. Schissel, *J. Amer. Chem. Soc.*, 1969, **91**, 1875.
58   F. A. Wodley, *J. Appl. Polymer Sci.*, 1971, **15**, 835.
59   J. Schormüller and H. J. Kochmann, *Z. Lebensm.-Untersuch.*, 1969, **141**, 1; *Chem. Abs.*, 1969, **71**, 122582c.
60   B. H. Eccleston, H. J. Coleman and N. G. Adams, *J. Amer. Chem. Soc.*, 1950, **72**, 3866.
61   G. Schomburg, *J. Chromatog.*, 1966, **23**, 18.
62   D. J. Brookman and D. T. Sawyer, *Analyt. Chem.*, 1968, **40**, 2013.
63   A. C. Cope and F. A. Hochstein, *J. Amer. Chem. Soc.*, 1950, **72**, 2515.
64   W. Schlenk, *Annalen*, 1951, **573**, 142.
65   R. D. Allendoerfer and P. H. Rieger, *J. Amer. Chem. Soc.*, 1965, **87**, 2336.
66   D. W. Scott, M. E. Gross, G. D. Oliver and H. M. Huffman, *J. Amer. Chem. Soc.*, 1949, **71**, 1634.
67   H. D. Springall, T. R. White and R. C. Cass, *Trans. Faraday Soc.*, 1954, **50**, 815.
68   E. J. Prosen, W. H. Johnson and F. D. Rossini, *J. Amer. Chem. Soc.*, 1950, **72**, 626.
69   R. B. Turner, W. R. Meador, W. von E. Doering, L. H. Knox, J. R. Mayer and D. W. Wiley, *J. Amer. Chem. Soc.*, 1957, **79**, 4127.

70 K. Watanabe, T. Nakayama and J. Mottl, *J. Quant. Spectroscopy Radiative Transfer*, 1962, **2**, 369.

71 C. Batich, P. Bischof and E. Heilbronner, *J. Electron Spectroscopy*, 1972/73, **1**, 333.

72 M. I. Al-Joboury and D. W. Turner, *J. Chem. Soc.*, 1964, 4434.

73 J. L. Franklin and S. R. Carroll, *J. Amer. Chem. Soc.*, 1969, **91**, 5940

74 W. E. Wentworth and W. Ristau, *J. Phys. Chem.*, 1969, **73**, 2126.

75 L. B. Anderson, J. F. Hansen, T. Kakihana and L. A. Paquette, *J. Amer. Chem. Soc.*, 1971, **93**, 161.

76 H. Lehmkuhl, E. Janssen, S. Kintopf, W. Leuchte and K. Mehler, *Chem.-Ing.-Tech.*, 1972, **44**, 170.

77 D. R. Thielen and L. B. Anderson, *J. Amer. Chem. Soc.*, 1972, **94**, 2521.

78 L. B. Anderson and L. A. Paquette, *J. Amer. Chem. Soc.*, 1972, **94**, 4915.

79 H. Lehmkuhl, S. Kintopf and E. Janssen, *J. Organometallic Chem.*, 1973, **56**, 41.

80 R. D. Rieke and R. A. Copenhafer, *J. Electroanalyt. Chem. Interfacial Electrochem.*, 1974, **56**, 409.

81 R. D. Allendoerfer, *J. Amer. Chem. Soc.*, 1975, **97**, 218.

82 A. J. Fry, C. S. Hutchins and L. L. Chung, *J. Amer. Chem. Soc.*, 1975, **97**, 591.

83 S. Shida and S. Fuji, *Bull. Chem. Soc. Japan*, 1951, **24**, 173.

84 H. J. Dauben, J. D. Wilson and J. L. Laity, *J. Amer. Chem. Soc.*, 1968, **90**, 811.

85 J. F. Labarre and O. Chalvet, *Tetrahedron Letters*, 1967, 5053.

86 J. L. Franklin, *Ind. and Eng. Chem.*, 1949, **41**, 1070.

87 W. F. Yates, *J. Phys. Chem.*, 1961, **65**, 185.

88 J. D. Cox, *Tetrahedron*, 1963, **19**, 1175.

89 B. D. Kybett, S. Carroll, P. Natalis, D. W. Bonnell, J. L. Margrave and J. L. Franklin, *J. Amer. Chem. Soc.*, 1966, **88**, 626.

90 E. R. Lippincott and R. C. Lord, *J. Amer. Chem. Soc.*, 1951, **73**, 3889.

91 S. W. Benson, F. R. Cruickshank, D. M. Golden, G. R. Haugen, H. E. O'Neal, A. S. Rodgers, R. Shaw and R. Walsh, *Chem. Rev.*, 1969, **69**, 279.

92 P. Schiess and A. Pullman, *J. Chim. phys.*, 1956, **53**, 101.

93 H. Grasshof, *Chem. Ber.*, 1951, **84**, 916.

94 S. Miyakawa, I. Tanaka and T. Uemura, *Bull. Chem. Soc. Japan*, 1951, **24**, 136.

95 N. L. Allinger, M. A. Miller, L. W. Chow, R. A. Ford and J. C. Graham, *J. Amer. Chem. Soc.*, 1965, **87**, 3430.

96 F. A. Van-Catledge and N. L. Allinger, *J. Amer. Chem. Soc.*, 1969, **91**, 2582.

97 D. W. Turner, unpublished work; quoted by H. C. Longuet-Higgins and K. L. McEwen, *J. Chem. Phys.*, 1957, **26**, 719.

98 B. Briat, D. A. Schooley, R. Records, E. Bunnenberg and C. Djerassi, *J. Amer. Chem. Soc.*, 1967, **89**, 7062.

99 H. P. Fritz and H. Keller, *Chem. Ber.*, 1962, **95**, 158.

100 F. A. Savin, *Optika i Spektroskopiya*, 1965, **19**, 743; *Chem. Abs.*, 1966, **64**, 10612f.

101 E. R. Lippincott, J. P. Sibilia and R. D. Fisher, *J. Opt. Soc. Amer.*, 1959, **49**, 83.

102 J. D. Roberts, *Angew. Chem. Internat. Edn*, 1963, **2**, 53.

103 I. J. Lawrenson and F. A. Rushworth, *Nature*, 1958, 391.

104 J. F. M. Oth, *Pure Appl. Chem.*, 1971, **25**, 573.

105 W. D. Larson and F. A. L. Anet, unpublished work; quoted by M. A. Cooper, D. D. Elleman, C. D. Pearce and S. L. Manatt, *J. Chem. Phys.*, 1970, **53**, 2343.

106 F. A. L. Anet, *J. Amer. Chem. Soc.*, 1962, **84**, 671.

107   H. Spiesecke and W. G. Schneider, *Tetrahedron Letters*, 1961, 468.
108   F. A. L. Anet and G. Schenck, *J. Amer. Chem. Soc.*, 1971, **93**, 556.
109   T. Ast, J. H. Beynon and R. G. Cooks, *Org. Mass Spectrometry*, 1972, **6**, 749.
110   K. Levsen and H. D. Beckey, *Org. Mass Spectrometry*, 1974, **9**, 570.
111   N. Bodor, M. J .S. Dewar and S. D. Worley, *J. Amer. Chem. Soc.*, 1970, **92**, 19.
112   J. H. D. Eland, *J. Mass Spectrometry Ion Phys.*, 1969, **2**, 471.
113   F. W. E. Knoop, J. Kistemaker and L. J. Oosterhoff, *Chem. Phys. Letters*, 1969, **3**, 73.
114   W. F. Frey, R. N. Compton, W. T. Naff and H. C. Schweinler, *J. Mass Spectrometry Ion Phys.*, 1973, **12**, 19.
115   F. J. Davis, R. N. Compton and D. R. Nelson, *J. Chem. Phys.*, 1973, **59**, 2324.
116   J. Bordner, R. G. Parker and R. H. Stanford, *Acta Cryst.*, 1972, **B28**, 1069.
117   I. L. Karle, *J. Chem. Phys.*, 1952, **20**, 65.
118   O. Bastiansen, L. Hedberg and K. Hedberg, *J. Chem. Phys.*, 1957, **27**, 1311.
119   M. Traetteberg, *Acta Chem. Scand.*, 1966, **20**, 1724.
120   O. Bastiansen, H. M. Seip and J. E. Boggs, *Perspectives in Structural Chem.*, 1971, **4**, 60.
121   K. L. McEwen and H. C. Longuet-Higgins, *J. Chem. Phys.*, 1956, **24**, 771.
122   H. C. Longuet-Higgins and K. L. McEwen, *J. Chem. Phys.*, 1957, **26**, 719.
123   F. A. L. Anet, A. J. R. Bourn and Y. S. Lin, *J. Amer. Chem. Soc.*, 1964, **86**, 3576.
124   M. J. S. Dewar, A. Harget and E. Haselbach, *J. Amer. Chem. Soc.*, 1969, **91**, 7521.
125   G. Wipff, U. Wahlgren, E. Kochanski and J. M. Lehn, *Chem. Phys. Letters*, 1971, **11**, 350.
126   C. J. Finder, D. Chung and N. L. Allinger, *Tetrahedron Letters*, 1972, 4677.
127   N. L. Allinger, J. T. Sprague and C. J. Finder, *Tetrahedron*, 1973, **29**, 2519.
128   G. Schröder, J. F. M. Oth and R. Merényi, *Angew. Chem. Internat. Edn*, 1965, **4**, 752.
129   Z. Luz and M. Meiboom, *J. Chem. Phys.*, 1973, **59**, 1077.
130   N. L. Allinger, *J. Org. Chem.*, 1962, **27**, 443.
131   C. A. Coulson and W. T. Dixon, *Tetrahedron*, 1962, **17**, 215.
132   W. T. Dixon, *Tetrahedron*, 1962, **18**, 875.
133   L. C. Snyder, *J. Phys. Chem.*, 1962, **66**, 2299.
134   R. Huisgen and F. Mietzsch, *Angew. Chem. Internat. Edn*, 1964, **3**, 83.
135   R. Huisgen, F. Mietzsch, G. Boche and H. Seidl, *Chem. Soc. Special Publ. no. 19*, 1965, p. 3.
136   R. Huisgen, G. Boche, A. Dahmen and W. Hechtl, *Tetrahedron Letters*, 1968, 5215.
137   A. Streitwieser, *Molecular Orbital Theory for Organic Chemists*, Wiley, 1961.
138   L. Salem, *The Molecular Orbital Theory of Conjugated Systems*, Benjamin, 1966.
139   H. P. Figeys, *Topics in Carbocyclic Chem.*, 1969, **1**, 269.
140   W. B. Person, G. C. Pimental and K. S. Pitzer, *J. Amer. Chem. Soc.*, 1952, **74**, 3437.
141   B. Bak and L. Hansen-Nygaard, *J. Chem. Phys.*, 1960, **33**, 418.
142   D. H. Lo and M. A. Whitehead, *J. Amer. Chem. Soc.*, 1969, **91**, 238.
143   M. Traetteberg, G. Hagen and S. J. Cyvin, *Z. Naturforsch.*, 1970, **25b**, 134.
144   P. von R. Schleyer, J. E. Williams and K. R. Blanchard, *J. Amer. Chem. Soc.*, 1970, **92**, 2377.
145   H. Iwamura, K. Morio and T. L. Kunii, *Chem. Comm.*, 1971, 1408.

References   433

146 R. B. Turner, B. J. Mallon, M. Tichy, W. von E. Doering, W. R. Roth and G. Schröder, *J. Amer. Chem. Soc.*, 1973, **95**, 8605.
147 Z. B. Maksić, K. Kovačević and M. Eckert-Maksić, *Tetrahedron Letters*, 1975, 101.
148 H. Iwamura, K. Morio, M. Oki and T. L. Kunii, *Tetrahedron Letters*, 1970, 4575.
149 Z. B. Maksić and M. Randić, *J. Amer. Chem. Soc.*, 1970, **92**, 424.
150 M. J. S. Dewar and G. J. Gleicher, *J. Amer. Chem. Soc.*, 1965, **87**, 685.
151 D. Peters, *J. Chem. Soc.*, 1958, 1023.
152 J. Kruszewski and T. M. Krygowski, *Tetrahedron Letters*, 1972, 3839.
153 F. A. Van-Catledge, *J. Amer. Chem. Soc.*, 1971, **93**, 4365.
154 N. C. Baird, *J. Amer. Chem. Soc.*, 1972, **94**, 4941.
155 W. O. Jones, *J. Chem. Soc.*, 1953, 2036.
156 W. O. Jones, *Chem. and Ind.*, 1955, 16.
157 J. Roemer-Mähler, D. Bienick and F. Korte, *Z. Naturforsch.*, 1975, **30b**, 290.
158 R. C. Lord and R. W. Walker, *J. Amer. Chem. Soc.*, 1954, **76**, 2518.
159 G. Schröder, *Chem. Ber.*, 1964, **97**, 3131.
160 G. Schröder, G. Kirsch and J. F. M. Oth, *Chem. Ber.*, 1974, **107**, 460.
161 S. C. Nyburg and J. Hilton, *Chem. and Ind.*, 1957, 1072.
162 S. C. Nyburg and J. Hilton, *Acta Cryst.*, 1959, **12**, 116.
163 L. A. Paquette, *Accounts Chem. Res.*, 1971, **4**, 280.
164 G. I. Fray and R. G. Saxton, unpublished work.
165 H. W. Moore, *J. Amer. Chem. Soc.*, 1964, **86**, 3398.
166 R. Merényi, J. F. M. Oth and G. Schröder, *Chem. Ber.*, 1964, **97**, 3150.
167 H. Nakanishi and O. Yamamoto, *Chem. Letters*, 1973, 1273.
168 K. Grohmann, J. B. Grutzner and J. D. Roberts, *Tetrahedron Letters*, 1969, 917.
169 G. Schröder and J. F. M. Oth, *Angew. Chem. Internat. Edn*, 1967, **6**, 414.
170 H. Iwamura and K. Morio, *Bull. Chem. Soc. Japan*, 1972, **45**, 3599.
171 H. E. Zimmerman, *Acounts Chem. Res.*, 1971, **4**, 272.
172 L. Hoesch, A. S. Dreiding and J. F. M. Oth, *Israel J. Chem.*, 1972, **10**, 439.
173 M. Jones and L. O. Schwab, *J. Amer. Chem. Soc.*, 1968, **90**, 6549.
174 I. Tanaka, *J. Chem. Soc. Japan, Pure Chem. Sect.*, 1954, **75**, 212; *Chem. Abs.*, 1954, **48**, 4984b.
175 C. D. Nenitzescu and F. Badea, *Comun. Acad. Rep. populare Romine*, 1959, **9**, 245; *Chem. Abs.*, 1960, **54**, 22417f.
176 I. Tanaka, S. Miyakawa and S. Shida, *Bull. Chem. Soc. Japan*, 1951, **24**, 119.
177 I. Tanaka and M. Okuda, *J. Chem. Phys.*, 1954, **22**, 1780.
178 H. Yamazaki and S. Shida, *J. Chem. Phys.*, 1956, **24**, 1278.
179 H. Yamazaki, *Bull. Chem. Soc. Japan*, 1958, **31**, 677.
180 G. J. Fonken, *Chem. and Ind.*, 1963, 1625.
181 E. Migirdicyan and S. Leach, *Bull. Soc. chim. Belg.*, 1962, **71**, 845.
182 E. H. White, E. W. Friend, R. L. Stern and H. Maskill, *J. Amer. Chem. Soc.*, 1969, **91**, 523.
183 H. E. Zimmerman and H. Iwamura, *J. Amer. Chem. Soc.*, 1970, **92**, 2015.
184 H. Iwamura, *Tetrahedron Letters*, 1973, 369.
185 G. S. Hammond and P. A. Leermakers, *J. Phys. Chem.*, 1962, **66**, 1148.
186 D. N. Dempster, T. Morrow and M. F. Quinn, *J. Photochem.*, 1973–74, **2**, 343.
187 M. Yamashita and H. Kashiwagi, *J. Phys. Chem.*, 1974, **78**, 2006.
188 A. Hirth, J. Faure and D. Lougnot, *Opt. Comm.*, 1973, **8**, 318.
189 S. Shida, H. Yamazaki and S. Arai, *J. Chem. Phys.*, 1958, **29**, 245.
190 L. P. Ellinger, *J. Appl. Chem.*, 1962, **12**, 387.

434   *References*

191   T. Shida and S. Iwata, *J. Amer. Chem. Soc.*, 1973, **95**, 3473.
192   A. R. McIntosh, D. R. Gee and J. K. S. Wan, *Spectrosc. Letters*, 1971, **4**, 217.
193   G. Foldiak, G. Cserep, V. Stenger and L. Wojnarovits, *Kem. Kozlem.*, 1969, **31**, 415; *Chem. Abs.*, 1970, **72**, 7892n.
194   G. Foldiak and L. Wojnarovits, *Acta Chim. (Budapest)*, 1970, **65**, 59; *Chem. Abs.*, 1970, **73**, 81413u.
195   L. Wojnarovits and P. Fejes, *Acta Chim. (Budapest)*, 1971, **69**, 177; *Chem. Abs.*, 1971, **75**, 124956j.
196   T. Matsuura, A. Horinaka and R. Nakashima, *Chem. Letters*, 1973, 887.
197   J. P. Wibaut and F. L. J. Sixma, *Rec. Trav. chim.*, 1954, **73**, 797.
198   N. A. Milas, J. T. Nolan and P. Ph. H. L. Otto, *J. Org. Chem.*, 1958, **23**, 624.
199   A. C. Cope, P. T. Moore and W. R. Moore, *J. Amer. Chem. Soc.*, 1958, **80**, 5505.
200   A. C. Cope and B. D. Tiffany, *J. Amer. Chem. Soc.*, 1951, **73**, 4158.
201   S. L. Friess and V. Boekelheide, *J. Amer. Chem. Soc.*, 1949, **71**, 4145.
202   T. Matsuda and M. Sugishita, *Bull. Chem. Soc. Japan*, 1967, **40**, 174.
203   V. D. Azatyan and G. T. Esayan, *Zhur. obschei Khim.*, 1956, **26**, 599; *Chem. Abs.*, 1956, **50**, 13858a.
204   C. R. Ganellin and R. Pettit, *J. Amer. Chem. Soc.*, 1957, **79**, 1767.
205   T. Kobayashi, J. Furukawa and N. Hagihara, *Yuki Gosei Kagaku Kyokai Shi*, 1962, **20**, 551; *Chem. Abs.*, 1963, **58**, 4436d.
206   A. C. Cope, N. A. Nelson and D. S. Smith, *J. Amer. Chem. Soc.*, 1954, **76**, 1100.
207   R. M. Carlson and R. K. Hill, *Org. Synth.*, 1970, **50**, 24.
208   M. Finkelstein, *Chem. Ber.*, 1957, **90**, 2097.
209   L. Eberson, K. Nyberg, M. Finkelstein, R. C. Petersen, S. D. Ross and J. J. Uebel, *J. Org. Chem.*, 1967, **32**, 16.
210   W. Kitching, K. A. Henzel and L. A. Paquette, *J. Amer. Chem. Soc.*, 1975, **97**, 4643.
211   R. M. Dessau, *J. Amer. Chem. Soc.*, 1970, **92**, 6356.
212   T. Kobayashi, J. Furukawa and N. Hagihara, *Yuki Gosei Kagaku Kyokai Shi*, 1962, **20**, 555; *Chem. Abs.*, 1963, **58**, 4436e.
213   J. L. von Rosenberg, J. E. Mahler and R. Pettit, *J. Amer. Chem. Soc.*, 1962, **84**, 2842.
214   S. Winstein, H. D. Kaesz, C. G. Kreiter and E. C. Friedrich, *J. Amer. Chem. Soc.*, 1965, **87**, 3267.
215   C. E. Keller and R. Pettit, *J. Amer. Chem. Soc.*, 1966, **88**, 604.
216   C. E. Keller and R. Pettit, *J. Amer. Chem. Soc.*, 1966, **88**, 606.
217   P. Warner, D. L. Harris, C. H. Bradley and S. Winstein, *Tetrahedron Letters*, 1970, 4013.
218   S. Winstein, C. G. Kreiter and J. I. Brauman, *J. Amer. Chem. Soc.*, 1966, **88**, 2047.
219   H. J. Dauben, J. Laity and S. Winstein, unpublished work; quoted by S. Winstein, *Quart. Rev.*, 1969, **23**, 141.
220   J. M. Bollinger and G. A. Olah, *J. Amer. Chem. Soc.*, 1969, **91**, 3380.
221   S. Winstein, *Chem. Soc. Special Publ. no. 21*, 1967, p. 5.
222   S. Winstein, *Quart. Rev.*, 1969, **23**, 141.
223   W. J. Hehre, *J. Amer. Chem. Soc.*, 1974, **96**, 5207.
224   R. C. Haddon, *Tetrahedron Letters*, 1975, 863.
225   R. Huisgen, G. Boche and H. Huber, *J. Amer. Chem. Soc.*, 1967, **89**, 3345.
226   J. A. Berson and J. A. Jenkins, *J. Amer. Chem. Soc.*, 1972, **94**, 8907.
227   W. J. Hehre, *J. Amer. Chem. Soc.*, 1972, **94**, 8908.

228 M. Avram, I. Dinulescu, M. Elian, M. Farcasiu, E. Marica, G. Mateescu and C. D. Nenitzescu, *Chem. Ber.*, 1964, **97**, 372.
229 R. Huisgen, G. Boche, W. Hechtl and H. Huber, *Angew. Chem. Internat. Edn*, 1966, **5**, 585.
230 R. Huisgen and J. Gasteiger, *Angew. Chem. Internat. Edn*, 1972, **11**, 1104.
231 R. Huisgen and G. Boche, *Tetrahedron Letters*, 1965, 1769.
232 G. Boche and R. Huisgen, *Tetrahedron Letters*, 1965, 1775.
233 M. Kröner, *Chem. Ber.*, 1967, **100**, 3162.
234 C. G. Overberger, M. A. Klotz and H. Mark, *J. Amer. Chem. Soc.*, 1953, **75**, 3186.
235 A. C. Cope, T. A. Liss and D. S. Smith, *J. Amer. Chem. Soc.*, 1957, **79**, 240.
236 T. Sasaki, K. Kanematsu and Y. Yukimoto, *J. Org. Chem.*, 1972, **37**, 890.
237 P. Y. Blanc, P. Diehl, H. Fritz and P. Schläpfer, *Experientia*, 1967, **23**, 896.
238 F. Lautenschlaeger, *J. Org. Chem.*, 1968, **33**, 2627.
239 H. Brintzinger and M. Langheck, *Chem. Ber.*, 1953, **86**, 557.
240 H. Brintzinger and H. Ellwanger, *Chem. Ber.*, 1954, **87**, 300.
241 W. H. Mueller and P. E. Butler, *Chem. Comm.*, 1966, 646.
242 F. Joy, M. F. Lappert and B. Prokai, *J. Organometallic Chem.*, 1966, **5**, 506.
243 T. J. Katz and H. L. Strauss, *J. Chem. Phys.*, 1960, **32**, 1873.
244 T. J. Katz, *J. Amer. Chem. Soc.*, 1960, **82**, 3784.
245 T. J. Katz, *J. Amer. Chem. Soc.*, 1960, **82**, 3785.
246 H. L. Strauss, T. J. Katz and G. K. Fraenkel, *J. Amer. Chem. Soc.*, 1963, **85**, 2360.
247 A. Carrington and P. F. Todd, *Mol. Phys.*, 1963–64, **7**, 533.
248 P. I. Kimmel and H. L. Strauss, *J. Phys. Chem.*, 1968, **72**, 2813.
249 T. J. Katz, W. H. Reinmuth and D. E. Smith, *J. Amer. Chem. Soc.*, 1962, **84**, 802.
250 A. Carrington, H. C. Longuet-Higgins, R. E. Moss and P. F. Todd, *Mol. Phys.*, 1965, **9**, 187.
251 F. J. Smentowski and G. R. Stevenson, *J. Amer. Chem. Soc.*, 1967, **89**, 5120.
252 F. J. Smentowski and G. R. Stevenson, *J. Phys. Chem.*, 1969, **73**, 340.
253 F. J. Smentowski and G. R. Stevenson, *J. Amer. Chem. Soc.*, 1969, **91**, 7401.
254 G. R. Stevenson and J. G. Concepción, *J. Phys. Chem.*, 1972, **76**, 2176.
255 H. van Willigen, *J. Amer. Chem. Soc.*, 1972, **94**, 7966.
256 C. A. Coulson, *Tetrahedron*, 1961, **12**, 193.
257 A. D. McLachlan and L. C. Snyder, *J. Chem. Phys.*, 1962, **36**, 1159.
258 R. E. Moss, *Mol. Phys.*, 1966, **10**, 501.
259 G. Wittig and D. Wittenberg, *Annalen*, 1957, **606**, 1.
260 A. C. Cope, A. C. Haven, F. L. Ramp and E. R. Trumbull, *J. Amer. Chem. Soc.*, 1952, **74**, 4867.
261 L. E. Craig, R. M. Elofson and I. J. Ressa, *J. Amer. Chem. Soc.*, 1953, **75**, 480.
262 R. C. Lord, *J. Chem. Phys.*, 1953, **21**, 378.
263 W. O. Jones, *J. Chem. Soc.*, 1954, 312.
264 W. O. Jones, *J. Chem. Soc.*, 1954, 1808.
265 W. Sanne and O. Schlichting, *Angew. Chem.*, 1963, **75**, 156.
266 R. E. Frank and D. L. McMasters, *Analyt. Chem.*, 1959, **31**, 2111.
267 M. Feldman and W. C. Flythe, *J. Amer. Chem. Soc.*, 1971, **93**, 1547.
268 B. J. Huebert and D. E. Smith, *J. Electroanalyt. Chem. Interfacial Electrochem.*, 1971, **31**, 333.
269 R. M. Elofson, *Analyt. Chem.*, 1949, **21**, 917.
270 J. H. Glover and H. W. Hodgson, *Analyst*, 1952, **77**, 473.
271 J. P. Petrovich, *Electrochim. Acta*, 1967, **12**, 1429.

272 A. J. Fry and A. Shuettenberg, *J. Org. Chem.*, 1974, **39**, 2452.
273 B. S. Jensen, A. Ronlan and V. D. Parker, *Acta Chem. Scand.*, 1975, **29B**, 394.
274 V. D. Azatyan, R. S. Gyuli-Kevkhyan, L. Kh. Freidlin and B. D. Polkovnikov, *Izvest. Akad. Nauk Armyan S.S.R.*, *Ser. khim. Nauk*, 1957, **10**, 55; *Chem. Abs.*, 1958, **52**, 1132f.
275 R. B. Turner and W. R. Meador, *J. Amer. Chem. Soc.*, 1957, **79**, 4133.
276 I. Jardine and F. J. McQuillin, *J. Chem. Soc.* (C), 1966, 458.
277 A. C. Cope and L. L. Estes, *J. Amer. Chem. Soc.*, 1950, **72**, 1128.
278 S. Akiyoshi, T. Matsuda and S. Tsunawaki, *J. Chem. Soc. Japan, Ind. Chem. Sect.*, 1954, **57**, 467; *Chem. Abs.*, 1955, **49**, 14656b.
279 A. Miyake and H. Kondo, *Angew. Chem. Internat. Edn.*, 1968, **7**, 631.
280 W. Strohmeier and N. Iglauer, *Z. phys. Chem. (Frankfurt)*, 1968, **61**, 29.
281 M. St.-Jaques and R. Prud'homme, *Tetrahedron Letters*, 1970, 4833.
282 T. Kauffmann, C. Kosel and W. Schoeneck, *Chem. Ber.*, 1963, **96**, 999.
283 G. Schröder, *Cyclooctatetraen*, Verlag Chemie, Weinheim, 1965.
284 G. Schröder, *Angew. Chem. Internat. Edn*, 1963, **2**, 45.
285 A. C. Cope and M. R. Kinter, *J. Amer. Chem. Soc.*, 1951, **73**, 3424.
286 A. C. Cope and H. O. Van Orden, *J. Amer. Chem. Soc.*, 1952, **74**, 175.
287 J. Gresser, A. Rajbenbach and M. Swarc, *J. Amer. Chem. Soc.*, 1961, **83**, 3005.
288 E. S. Ferdinandi, W. P. Garby and D. G. L. James, *Can. J. Chem.*, 1964, **42**, 2568.
289 J. L. Kice and T. S. Cantrell, *J. Amer. Chem. Soc.*, 1963, **85**, 2298.
290 H. Shechter, J. J. Gardikes, T. S. Cantrell and G. V. D. Tiers, *J. Amer. Chem. Soc.*, 1967, **89**, 3005.
291 T. S. Cantrell, *J. Org. Chem.*, 1967, **32**, 911.
292 R. M. Pike and P. M. McDonagh, *J. Chem. Soc.*, 1963, 4058.
293 R. W. Bradshaw, *Tetrahedron Letters*, 1966, 5711.
294 S. N. Ushakov and O. F. Solomon, *Zhur. priklad. Khim.*, 1954, **27**, 959; *Chem. Abs.*, 1955, **49**, 10868a.
295 K. Bittler, N. von Kutepow, D. Neubauer and H. Reis, *Angew. Chem. Internat. Edn*, 1968, **7**, 329.
296 R. Huisgen, *Angew. Chem. Internat. Edn*, 1968, **7**, 321.
297 E. Vogel, *Angew. Chem.*, 1961, **73**, 548.
298 E. Vogel, *Angew. Chem. Internat. Edn*, 1963, **2**, 1.
299 E. Vogel, W. Wiedemann, H. Kiefer and W. F. Harrison, *Tetrahedron Letters*, 1963, 673.
300 E. V. Dehmlow and J. Schönefeld, *Annalen*, 1971, **744**, 42.
301 T. Sasaki, K. Kanematsu and Y. Yukimoto, *J. Org. Chem.*, 1974, **39**, 455.
302 E. V. Dehmlow, H. Klabuhn and E.-C. Hass, *Annalen*, 1973, 1063.
303 E. V. Dehmlow and G. C. Ezimora, *Tetrahedron Letters*, 1970, 4047.
304 E. Ciganek, *J. Amer. Chem. Soc.*, 1966, **88**, 1979.
305 A. G. Anastassiou, R. P. Cellura and E. Ciganek, *Tetrahedron Letters*, 1970, 5267.
306 E. A. LaLancette and R. E. Benson, *J. Amer. Chem. Soc.*, 1963, **85**, 2853.
307 T. J. Katz and P. J. Garratt, *J. Amer. Chem. Soc.*, 1964, **86**, 4876.
308 T. Matsuda, K. Furuno and S. Akiyoshi, *J. Chem. Soc. Japan, Ind. Chem. Sect.*, 1955, **58**, 438; *Chem. Abs.*, 1956, **50**, 4804g.
309 S. Akiyoshi and T. Matsuda, *J. Amer. Chem. Soc.*, 1955, **77**, 2476.
310 D. D. Philips, *J. Amer. Chem. Soc.*, 1955, **77**, 5179.
311 K. F. Bangert and V. Boekelheide, *J. Amer. Chem. Soc.*, 1964, **86**, 905.
312 S. Masamune, C. G. Chin, K. Hojo and R. T. Seidner, *J. Amer. Chem. Soc.*, 1967, **89**, 4804.

313 M. B. Sohn, M. Jones and B. Fairless, *J. Amer. Chem. Soc.*, 1972, **94**, 4774.
314 C. J. Cheer, W. Rosen and J. J. Uebel, *Tetrahedron Letters*, 1974, 4045.
315 D. Schönleber, *Chem. Ber.*, 1969, **102**, 1789.
316 H. Dürr and H. Kober, *Annalen*, 1970, **740**, 74.
317 A. G. Anastassiou, *J. Amer. Chem. Soc.*, 1968, **90**, 1527.
318 A. G. Anastassiou and R. P. Cellura, *J. Org. Chem.*, 1972, **37**, 3126.
319 S. Masamune and N. T. Castellucci, *Angew. Chem. Internat. Edn*, 1964, **3**, 582.
320 G. C. Tustin, C. E. Monken and W. H. Okamura, *J. Amer. Chem. Soc.*, 1972, **94**, 5112.
321 T. J. Barton and M. Juvet, *Tetrahedron Letters*, 1975, 3893.
322 E. A. Chernyshev, N. G. Komalenkova, S. A. Bashkirova, A. V. Kisin and V. I. Pchelintsev, *Zhur. obshchei Khim.*, 1975, **45**, 2221; *Chem. Abs.*, 1976, **84**, 44247z.
323 G. O. Spessard and S. Masamune, unpublished work; quoted by S. Masamune and N. Darby, *Accounts Chem. Res.*, 1972, **5**, 272.
324 G. Schröder and Th. Martini, *Angew. Chem. Internat. Edn*, 1967, **6**, 806.
325 G. Schröder, S. R. Ramadas and P. Nikoloff, *Chem. Ber.*, 1972, **105**, 1072.
326 D. Bryce-Smith and A. Gilbert, *Proc. Chem. Soc.*, 1964, 87.
327 D. Bryce-Smith, A. Gilbert and M. G. Johnson, *J. Chem. Soc.* (*C*), 1967, 383.
328 A. G. Anastassiou, J. C. Wetzel and B. Y.-H. Chao, *J. Amer. Chem. Soc.*, 1975, **97**, 1124.
329 R. Huisgen and M. Christl, *Angew. Chem. Internat. Edn*, 1967, **6**, 456.
330 M. Christl and R. Huisgen, *Tetrahedron Letters*, 1968, 5209.
331 G. Bianchi, R. Gandolfi and P. Grünanger, *Tetrahedron*, 1972, **26**, 5113.
332 K. Bast, M. Christl, R. Huisgen, W. Mack and R. Sustmann, *Chem. Ber.*, 1973, **106**, 3258.
333 R. Huisgen and M. Christl, *Chem. Ber.*, 1973, **106**, 3291.
334 G. Bianchi, R. Gandolfi and P. Grünanger, *Chimica e Industria*, 1967, **49**, 757.
335 G. Bianchi, A. Gamba and R. Gandolfi, *Tetrahedron*, 1972, **28**, 1601.
336 G. Bianchi, R. Gandolfi and P. Grünanger, *Tetrahedron*, 1973, **29**, 2405.
337 P. Brown and R. C. Cookson, *Tetrahedron*, 1968, **24**, 2551.
338 A. S. Bailey and J. E. White, *J. Chem. Soc.* (B), 1966, 819.
339 E. W. Abel, M. A. Bennett and G. Wilkinson, *J. Chem. Soc.*, 1959, 3178.
340 M. Avram, G. Mateescu and C. D. Nenitzescu, *Annalen*, 1960, **636**, 174.
341 M. Avram, E. Sliam and C. D. Nenitzescu, *Annalen*, 1960, **636**, 184.
342 R. C. Cookson, J. Hudec and J. Marsden, *Chem. and Ind.*, 1961, 21.
343 E. Grovenstein, D. V. Rao and J. W. Taylor, *J. Amer. Chem. Soc.*, 1961, **83**, 1705.
344 G. Filippini, G. Induni and M. Simonetta, *Acta Cryst.*, 1973, **B29**, 2471.
345 G. Schröder, unpublished work; quoted by G. Schröder, *Cyclooctatetraen*, Verlag Chemie, Weinheim, 1965, p. 50.
346 L. A. Paquette and J. C. Philips, *Chem. Comm.*, 1969, 680.
347 P. Scheiner and W. R. Vaughan, *J. Org. Chem.*, 1961, **26**, 1923.
348 R. C. Cookson, E. Crundwell and J. Hudec, *Chem. and Ind.*, 1958, 1003.
349 D. M. Bratby and G. I. Fray, *J. C. S. Perkin I*, 1972, 195.
350 G. I. Fray and R. W. McCabe, unpublished work.
351 G. Mehta and P. S. Venkataramani, *Indian J. Chem.*, 1973, **11**, 822.
352 M. Avram, C. D. Nenitzescu and E. Marica, *Chem. Ber.*, 1957, **90**, 1857.
353 A. G. Anderson and D. R. Fagerburg, *Tetrahedron*, 1973, **29**, 2973.
354 C. D. Weis, *J. Org. Chem.*, 1963, **28**, 74.
355 G. O. Schenk, J. Kuhls and C. H. Krauch, *Z. Naturforsch.*, 1965, **20b**, 635.
356 G. O. Schenk, J. Kuhls and C. H. Krauch, *Annalen*, 1966, **693**, 20.
357 E. Vedejs, *Tetrahedron Letters*, 1968, 2633.

358   E. Vedejs and R. A. Shepherd, *Tetrahedron Letters*, 1970, 1863.
359   P. Warner, *Tetrahedron Letters*, 1971, 723.
360   L. A. Paquette, *Chem. Comm.*, 1971, 1076.
361   D. C. England and C. G. Krespan, *J. Org. Chem.*, 1970, **35**, 3300.
362   R. C. Cookson, S. S. H. Gilani and I. D. R. Stevens, *J. Chem. Soc.* (C), 1967, 1905.
363   A. B. Evnin, R. D. Miller and G. R. Evanega, *Tetrahedron Letters*, 1968, 5863.
364   R. Huisgen, W. E. Konz and U. Schnegg, *Angew. Chem. Internat. Edn*, 1972, **11**, 715.
365   P. Wegener, *Tetrahedron Letters*, 1967, 4985.
366   L. A. Paquette and T. J. Barton, *J. Amer. Chem. Soc.*, 1967, **89**, 5480.
367   L. A. Paquette, J. R. Malpass and T. J. Barton, *J. Amer. Chem. Soc.*, 1969, **91**, 4714.
368   E. M. Burgess and W. M. Williams, *J. Org. Chem.*, 1973, **38**, 1249.
369   E. J. Gardner, R. H. Squire, R. C. Elder and R. M. Wilson, *J. Amer. Chem. Soc.*, 1973, **95**, 1693.
370   R. M. Wilson, E. J. Gardner, R. C. Elder, R. H. Squire and L. R. Florian, *J. Amer. Chem. Soc.*, 1974, **96**, 2955.
371   L. A. Paquette and L. M. Leichter, unpublished work; quoted by L. A. Paquette, *Tetrahedron*, 1975, **31**, 2855.
372   G. I. Fray and D. P. S. Smith, *J. Chem. Soc.* (C), 1969, 2710.
373   V. Mark, *Chem. Comm.*, 1973, 910.
374   M. Avram, I. G. Dinulescu, E. Marica and C. D. Nenitzescu, *Chem. Ber.*, 1962, **95**, 2248.
375   H. P. Fritz and H. Keller, *Z. Naturforsch.*, 1961, **16b**, 231.
376   R. H. Cox, L. W. Harrison and W. K. Austin, *J. Phys. Chem.*, 1973, **77**, 200.
377   T. S. Cantrell and H. Shechter, *J. Amer. Chem. Soc.*, 1967, **89**, 5868.
378   T. A. Antkowiak and H. Shechter, *J. Amer. Chem. Soc.*, 1972, **94**, 5361.
379   D. A. Bak and K. Conrow, *J. Org. Chem.*, 1966, **31**, 3958.
380   T. J. Katz, C. P. Nicholson and C. A. Reilly, *J. Amer. Chem. Soc.*, 1966, **88**, 3832.
381   A. Streitwieser, U. Müller-Westerhoff, G. Sonnichsen, F. Mares, D. G. Morrell, K. O. Hodgson and C. A. Harmon, *J. Amer. Chem. Soc.*, 1973, **95**, 8644.
382   J. F. Monthony and W. H. Okamura, *Tetrahedron*, 1972, **28**, 4273.
383   D. N. Kursanov, Z. V. Todres, N. T. Toffe and Z. N. Parnes, *Doklady Akad. Nauk S.S.S.R.*, 1967, **174**, 362; *Chem. Abs.*, 1968, **68**, 2616h.
384   P. G. Farrell and S. F. Mason, *Z. Naturforsch.*, 1961, **16b**, 848.
385   J. H. Noordik, Th. E. M. Van den Hark, J. J. Mooij and A. A. K. Klaassen, *Acta Cryst.*, 1974, **B30**, 833.
386   J. H. Noordik, H. M. L. Degens and J. J. Mooij, *Acta Cryst.*, 1975, **B31**, 2144.
387   G. R. Stevenson and I. Ocasio, *J. Phys. Chem.*, 1975, **79**, 1387.
388   G. N. Schrauzer and S. Eichler, *Chem. Ber.*, 1962, **95**, 260.
389   M. B. Dines, *Inorg. Chem.*, 1972, **11**, 2949.
390   H. L. Haight, J. R. Doyle, N. C. Baenziger and G. F. Richards, *Inorg. Chem.*, 1963, **2**, 1301.
391   B. W. Cook, R. G. J. Miller and P. F. Todd, *J. Organometallic Chem.*, 1969, **19**, 421.
392   N. C. Baenziger, G. F. Richards and J. R. Doyle, *Inorg. Chem.*, 1964, **3**, 1529.
393   R. G. Salomon and J. K. Kochi, *Chem. Comm.*, 1972, 559.
394   R. G. Salomon and J. K. Kochi, *J. Amer. Chem. Soc.*, 1973, **95**, 1889.

395 R. G. Salomon and J. K. Kochi, *J. Organometallic Chem.*, 1972, **43**, C7.
396 H. W. Quinn and R. L. VanGilder, *Canad. J. Chem.*, 1970, **48**, 2435.
397 W. Partenheimer and E. H. Johnson, *Inorg. Chem.*, 1972, **11**, 2840.
398 F. S. Mathews and W. N. Lipscomb, *J. Amer. Chem. Soc.*, 1958, **80**, 4745.
399 F. S. Mathews and W. N. Lipscomb, *J. Phys. Chem.*, 1959, **63**, 845.
400 C. D. M. Beverwijk and J. P. C. M. van Dongen, *Tetrahedron Letters*, 1972, 4291.
401 P. Tauchner and R. Hüttel, *Chem. Ber.*, 1974, **107**, 3761.
402 H. Lehmkuhl, S. Kintopf and K. Mehler, *J. Organometallic Chem.*, 1972, **46**, Cl.
403 B. L. Kalsotra, R. K. Multani and B. D. Jain, *Indian J. Chem.*, 1972, **10**, 556.
404 R. G. Hayes and J. L. Thomas, *J. Amer. Chem. Soc.*, 1969, **91**, 6876.
405 D. F. Starks and A. Streitwieser, *J. Amer. Chem. Soc.*, 1973, **95**, 3423.
406 K. M. Sharma, S. K. Anand, R. K. Multani and B. D. Jain, *J. Organometallic Chem.*, 1970, **25**, 447.
407 H. Lehmkuhl and K. Mehler, *J. Organometallic Chem.*, 1970, **25**, C44.
408 H. Dietrich and M. Soltwisch, *Angew. Chem. Internat. Edn*, 1969, **8**, 765.
409 L. Hocks, J. Goffart, G. Duyckaerts and P. Teyssié, *Spectrochim. Acta*, 1974, **30A**, 907.
410 J. Schwartz and J. E. Sadler, *Chem. Comm.*, 1973, 172.
411 H. Lehmkuhl, *Synthesis*, 1973, 377.
412 D. J. Brauer and C. Krüger, *J. Organometallic Chem.*, 1972, **42**, 129.
413 D. J. Brauer and C. Krüger, *Inorg. Chem.*, 1975, **14**, 3053.
414 H. Dietrich and H. Dierks, *Angew. Chem. Internat. Edn*, 1966, **5**, 899.
415 H. Dierks and H. Dietrich, *Acta Cryst.*, 1968, **B24**, 58.
416 H. Breil and G. Wilke, *Angew. Chem Internat. Edn*, 1966, **5**, 898.
417 H.-J. Kablitz and G. Wilke, *J. Organometallic Chem.*, 1973, **51**, 241.
418 H. O. van Oven, *J. Organometallic Chem.*, 1973, **55**, 309.
419 P. A. Kroon and R. B. Helmholdt, *J. Organometallic Chem.*, 1970, **25**, 451.
420 J. L. Thomas and R. G. Hayes, *Inorg. Chem.*, 1972, **11**, 348.
421 S. Evans, J. C. Green, S. E. Jackson and B. Higginson, *J. C. S. Dalton*, 1974, 304.
422 J. Müller and W. Goll, *Chem. Ber.*, 1974, **107**, 2084.
423 L. J. Guggenberger and R. R. Schrock, *J. Amer. Chem. Soc.*, 1975, **97**, 6693.
424 R. B. King, *J. Organometallic Chem.*, 1967, **8**, 139.
425 R. B. King and A. Fronzaglia, *Chem. Comm.*, 1965, 547.
426 R. B. King and A. Fronzaglia, *Inorg. Chem.*, 1966, **5**, 1837.
427 C. G. Kreiter, A. Maasbol, F. A. L. Anet, H. D. Kaesz and S. Winstein, *J. Amer. Chem. Soc.*, 1966, **88**, 3444.
428 J. S. McKechnie and I. C. Paul, *J. Amer. Chem. Soc.*, 1966, **88**, 5927.
429 R. B. King, *J. Organometallic Chem.*, 1967, **8**, 129.
430 E. W. Randall, E. Rosenberg and L. Milone, *J. C. S. Dalton*, 1973, 1672.
431 F. A. Cotton, D. L. Hunter and P. Lahuerta, *J. Amer. Chem. Soc.*, 1974, **96**, 4723.
432 F. A. Cotton, D. L. Hunter and P. Lahuerta, *J. Amer. Chem. Soc.*, 1974, **96**, 7926.
433 R. B. King, *Appl. Spectroscopy*, 1969, **23**, 536.
434 J. Müller and H. Menig, *J. Organometallic Chem.*, 1975, **96**, 83.
435 M. A. Bennett and G. Wilkinson, *Chem. and Ind.*, 1959, 1516.
436 M. A. Bennett, L. Pratt and G. Wilkinson, *J. Chem. Soc.*, 1961, 2037.

440   *References*

437   K. M. Sharma, S. K. Anand, R. K. Multani and B. D. Jain, *J. Organometallic Chem.*, 1970, **22**, 685.
438   T. H. Coffield, K. G. Ihrman and W. Burns, *J. Amer. Chem. Soc.*, 1960, **82**, 1251.
439   R. B. King and M. N. Ackermann, *Inorg. Chem.*, 1974, **13**, 637.
440   M. R. Churchill, F. J. Rotella, R. B. King and M. N. Ackermann, *J. Organometallic Chem.*, 1975, **99**, C15.
441   A. Carbonaro, A. Greco and G. Dall'Asta, *J. Organometallic Chem.*, 1969, **20**, 177.
442   D. H. Gerlach, W. G. Peet and E. L. Muetterties, *J. Amer. Chem. Soc.*, 1972, **94**, 4545.
443   D. H. Gerlach and R. A. Schunn, *Inorg. Synth.*, 1974, **15**, 2.
444   G. Allegra, A. Colombo, A. Immirzi and I. W. Bassi, *J. Amer. Chem. Soc.*, 1968, **90**, 4455.
445   A. Carbonaro, A. L. Segre, A. Greco, C. Tosi and G. Dall'Asta, *J. Amer. Chem. Soc.*, 1968, **90**, 4453.
446   G. Allegra, A. Colombo and E. R. Mognaschi, *Gazzetta*, 1972, **102**, 1060.
447   A. Chierico and E. R. Mognaschi, *J.C.S. Faraday II*, 1973, **69**, 433.
448   P. Mag, L. Korecz, A. Carbonaro and K. Burger, *Radiochem. Radioanalyt. Letters*, 1972, **9**, 137.
449   T. A. Manuel and F. G. A. Stone, *Proc. Chem. Soc.*, 1959, 90.
450   T. A. Manuel and F. G. A. Stone, *J. Amer. Chem. Soc.*, 1960, **82**, 366.
451   R. B. King, *Organometallic Synth.*, 1965, **1**, 126.
452   A. Nakamura and N. Hagihara, *Bull. Chem. Soc. Japan*, 1959, **32**, 880.
453   M. D. Rausch and G. N. Schrauzer, *Chem. and Ind.*, 1959, 957.
454   C. E. Keller, G. F. Emerson and R. Pettit, *J. Amer. Chem. Soc.*, 1965, **87**, 1388.
455   A. J. Campbell, C. A. Fyfe and E. Maslowsky, *J. Amer. Chem. Soc.*, 1972, **94**, 2690.
456   Y. Shvo and E. Hazum, *Chem. Comm.*, 1975, 829.
457   B. Dickens and W. N. Lipscomb, *J. Amer. Chem. Soc.*, 1961, **83**, 4862.
458   B. Dickens and W. N. Lipscomb, *J. Chem. Phys.*, 1962, **37**, 2084.
459   F. A. Cotton, *Accounts Chem. Res.*, 1968, **1**, 257.
460   A. J. Campbell, C. A. Fyfe and E. Maslowsky, *Chem. Comm.*, 1971, 1032.
461   G. Rigatti, G. Boccalon, A. Ceccon and G. Giacometti, *Chem. Comm.*, 1972, 1165.
462   R. T. Bailey, E. R. Lippincott and D. Steele, *J. Amer. Chem. Soc.*, 1965, **87**, 5346.
463   F. A. Cotton, A. Davison and J. W. Faller, *J. Amer. Chem. Soc.*, 1966, **88**, 4507.
464   M. I. Bruce, *J. Mass Spectrometry Ion Phys.*, 1969, **2**, 349.
465   D. F. Hunt, J. W. Russell and R. L. Torian, *J. Organometallic Chem.*, 1972, **43**, 175.
466   G. K. Wertheim and R. H. Herber, *J. Amer. Chem. Soc.*, 1962, **74**, 2274.
467   R. H. Herber, R. B. King and M. N. Ackermann, *J. Amer. Chem. Soc.*, 1974, **96**, 5437.
468   B. Dickens and W. N. Lipscomb, *J. Amer. Chem. Soc.*, 1961, **83**, 489.
469   G. N. Schrauzer and S. Eichler, *Angew. Chem. Internat. Edn*, 1962, **1**, 454.
470   G. N. Schrauzer and P. W. Glockner, *J. Amer. Chem. Soc.*, 1968, **90**, 2800.
471   A. Robson and M. R. Truter, *Tetrahedron Letters*, 1964, 3079.
472   F. A. Cotton, A. Davison and A. Musco, *J. Amer. Chem. Soc.*, 1967, **89**, 6796.

473 F. A. Cotton and W. T. Edwards, *J. Amer Chem. Soc.*, 1968, **90**, 5412.
474 A. Davison, W. McFarlane, L. Pratt and G. Wilkinson, *J. Chem. Soc.*, 1962, 4821.
475 W. McFarlane and G. Wilkinson, *Inorg. Synth.*, 1966, **8**, 184.
476 G. N. Schrauzer, P. Glockner, and R. Merényi, *Angew. Chem. Internat. Edn*, 1964, **3**, 509.
477 J. Ashley-Smith, D. V. Howe, B. F. G. Johnson, J. Lewis and I. E. Ryder, *J. Organometallic Chem.*, 1974, **82**, 257.
478 U. Zahn and G. Harbottle, *J. Inorg. Nuclear Chem.*, 1966, **28**, 925.
479 M. I. Bruce, M. Cooke and M. Green, *J. Organometallic Chem.*, 1968, **13**, 227.
480 F. A. Cotton, A. Davison, T. J. Marks and A. Musco, *J. Amer. Chem. Soc.*, 1969, **91**, 6598.
481 J. A. K. Howard, S. A. R. Knox, V. Riera, F. G. A. Stone and P. Woodward, *Chem. Comm.*, 1974, 452.
482 S. A. R. Knox and F. G. A. Stone, *Accounts Chem. Res.*, 1974, **7**, 321.
483 R. Goddard, A. P. Humphries, S. A. R. Knox and P. Woodward, *Chem. Comm.*, 1975, 507.
484 R. Goddard, A. P. Humphries, S. A. R. Knox and P. Woodward, *Chem. Comm.*, 1975, 508.
485 F. A. Cotton and R. Eiss, *J. Amer. Chem. Soc.*, 1969, **91**, 6593.
486 M. I. Bruce, M. Cooke, M. Green and F. G. A. Stone, *Chem. Comm.*, 1967 523.
487 W. K. Bratton, F. A. Cotton, A. Davison, A. Musco and J. W. Faller, *Proc. Nat. Acad. Sci. U.S.A.*, 1967, **58**, 1324.
488 F. A. Cotton, B. G. DeBoer and T. J. Marks, *J. Amer. Chem. Soc.*, 1971, **93**, 5069.
489 R. Bau and B. C.-K. Chou, unpublished work; quoted by S. A. R. Knox and F. G. A. Stone, *Accounts Chem. Res.*, 1974, **7**, 321.
490 M. J. Bennett, F. A. Cotton and P. Legzdins, *J. Amer. Chem. Soc.*, 1968, **90**, 6335.
491 C. E. Cottrell, C. A. Fyfe and C. V. Senoff, *J. Organometallic Chem.*, 1972, **43**, 203.
492 A. J. P. Domingos, B. F. G. Johnson and J. Lewis, *J. Organometallic Chem.*, 1973, **49**, C33.
493 J. D. Edwards, R. Goddard, S. A. R. Knox, R. J. McKinney, F. G. A. Stone and P. Woodward, *Chem. Comm.*, 1975, 828.
494 A. Brookes, J. Howard, S. A. R. Knox, F. G. A. Stone and P. Woodward, *Chem. Comm.*, 1973, 587.
495 M. I. Bruce, M. Cooke and M. Green, *Angew. Chem. Internat. Edn*, 1968, **7**, 639.
496 M. I. Bruce, M. Cooke, M. Green and D. J. Westlake, *J. Chem. Soc. (A)*, 1969, 987.
497 M. Cooke, R. J. Goodfellow, M. Green, J. P. Maher and J. R. Yandle, *Chem. Comm.*, 1970, 565.
498 A. Greco, M. Green and F. G. A. Stone, *J. Chem. Soc. (A)*, 1971, 285.
499 T. Kitamura and T. Joh, *J. Organometallic Chem.*, 1974, **65**, 235.
500 A. Nakamura and N. Hagihara, *Bull. Chem. Soc. Japan*, 1960, **33**, 425.
501 A. Nakamura and N. Hagihara, *Nippon Kagaku Zasshi*, 1961, **82**, 1392; *Chem. Abs.*, 1963, **59**, 2854g.
502 H. P. Fritz and H. Keller, *Z. Naturforsch.*, 1961, **16b**, 348.
503 E. Paulus, W. Hoppe and R. Huber, *Naturwiss.*, 1967, **54**, 67.

504 B. H. Robinson and J. Spencer, *J. Organometallic Chem.*, 1971, **33**, 97.
505 M. D. Brice, R. J. Dellaca, B. R. Penfold and J. L. Spencer, *Chem. Comm.*, 1971, 72.
506 J. Chatt and L. M. Venanzi, *J. Chem. Soc.*, 1957, 4735.
507 M. A. Bennett and J. D. Saxby, *Inorg. Chem.*, 1968, **7**, 321.
508 H. C. Volger, M. M. P. Gaasbeek, H. Hogeveen and K. Vrieze, *Inorg. Chim. Acta*, 1969, **3**, 145.
509 W. Partenheimer and E. F. Hoy, *J. Amer. Chem. Soc.*, 1973, **95**, 2840.
510 R. Grigg and J. L. Jackson, *Tetrahedron*, 1973, **29**, 3903.
511 R. R. Schrock and J. A. Osborn, *Inorg. Chem.*, 1970, **9**, 2339.
512 K. S. Brenner, E. O. Fischer, H. P. Fritz and C. G. Kreiter, *Chem. Ber.*, 1963, **96**, 2632.
513 C. Cocevar, G. Mestroni and A. Camus, *J. Organometallic Chem.*, 1972, **35**, 389.
514 A. L. Onderdelinden and A. van der Ent, *Inorg. Chim. Acta*, 1972, **6**, 420.
515 H. Yamazaki, *Bull. Chem. Soc. Japan*, 1971, **44**, 582.
516 H. Lehmkuhl and W. Leuchte, *J. Organometallic Chem.*, 1970, **23**, C30.
517 H. Lehmkuhl, W. Leuchte and W. Eisenbach, *Annalen*, 1973, 692.
518 B. Bogdanović, M. Kröner and G. Wilke, *Annalen*, 1966, **699**, 1.
519 G. Wilke, *Angew. Chem.*, 1960, **72**, 581.
520 G. N. Schrauzer and H. Thyret, *Z. Naturforsch.*, 1961, **16b**, 353.
521 G. N. Schrauzer and H. Thyret, *Theor. Chim. Acta*, 1963, **1**, 172.
522 J. Müller and W. Goll, *Chem. Ber.*, 1973, **106**, 1129.
523 J. L. Davidson, M. Green, F. G. A. Stone and A. J. Welch, *J. Amer. Chem. Soc.*, 1975, **97**, 7490.
524 S. D. Robinson and B. L. Shaw, *J. Chem. Soc.*, 1964, 5002.
525 H. Frye, E. Kuljian and J. Viebrock, *Z. Naturforsch.*, 1965, **20b**, 269.
526 W. Partenheimer, *Inorg. Chem.*, 1972, **11**, 743.
527 C. V. Goebel, *Diss. Abs.*, 1967, **28B**, 625.
528 B. F. G. Johnson, J. Lewis and D. A. White, *J. Amer. Chem. Soc.*, 1969, **91**, 5186.
529 K. A. Jensen, *Acta Chem. Scand.*, 1953, **7**, 868.
530 H. P. Fritz and D. Sellman, *Spectrochim. Acta*, 1967, **23A**, 1991.
531 J. P. Yesinowski and T. L. Brown, *J. Mol. Structure*, 1971, **9**, 474.
532 A. C. Cope and H. C. Campbell, *J. Amer. Chem. Soc.*, 1951, **73**, 3536.
533 A. C. Cope and H. C. Campbell, *J. Amer. Chem. Soc.*, 1952, **74**, 179.
534 A. C. Cope and D. F. Rugen, *J. Amer. Chem. Soc.*, 1953, **75**, 3215.
535 A. C. Cope and R. M. Pike, *J. Amer. Chem. Soc.*, 1953, **75**, 3220.
536 D. Bryce-Smith and J. E. Lodge, *J. Chem. Soc.*, 1963, 695.
537 G. R. Stevenson, J. G. Concepción and L. Echegoyen, *J. Amer. Chem. Soc.*, 1974, **96**, 5452.
538 A. C. Cope and M. R. Kinter, *J. Amer. Chem. Soc.*, 1950, **72**, 630.
539 G. R. Stevenson and L. Echegoyen, *J. Phys. Chem.*, 1975, **79**, 929.
540 J. Gasteiger, G. E. Gream, R. Huisgen, W. E. Konz and U. Schnegg, *Chem. Ber.*, 1971, **104**, 2412.
541 C. A. Harmon and A. Streitwieser, *J. Org. Chem.*, 1973, **38**, 549.
542 L. A. Paquette, R. S. Beckley and W. B. Farnham, *J. Amer. Chem. Soc.*, 1975, **97**, 1089.
543 L. A. Paquette and R. S. Beckley, *J. Amer. Chem. Soc.*, 1975, **97**, 1084.
544 A. C. Cope and D. J. Marshall, *J. Amer. Chem. Soc.*, 1953, **75**, 3208.
545 L. A. Paquette and K. A. Henzel, *J. Amer. Chem. Soc.*, 1973, **95**, 2726.
546 L. A. Paquette and K. A. Henzel, *J. Amer. Chem. Soc.*, 1975, **97**, 4649.

547 G. E. Gream and M. Mular, *Austral. J. Chem.*, 1975, **28**, 2227.
548 R. P. Houghton and E. S. Waight, *J. Chem. Soc.* (C), 1969, 978.
549 A. C. Cope, M. Burg and S. W. Fenton, *J. Amer. Chem. Soc.*, 1952, **74**, 173.
550 D. E. Gwynn, G. M. Whitesides and J. D. Roberts, *J. Amer. Chem. Soc.*, 1965, **87**, 2862.
551 M. Cooke, C. R. Russ and F. G. A. Stone, *J. C. S. Dalton*, 1975, 256.
552 B. F. G. Johnson, J. Lewis and G. L. P. Randall, *J. Chem. Soc.* (A), 1971, 422.
553 A. C. Cope and M. Burg, *J. Amer. Chem. Soc.*, 1952, **74**, 168.
554 J. F. M. Oth, R. Merényi, Th. Martini and G. Schröder, *Tetrahedron Letters*, 1966, 3087.
555 A. Krebs, *Angew. Chem. Internat. Edn*, 1965, **4**, 953.
556 C. L. Osborn, T. C. Shields, B. A. Shoulders, J. F. Krause, H. V. Cortez and P. D. Gardner, *J. Amer. Chem. Soc.*, 1965, **87**, 3158.
557 R. Bloch, F. Leyendecker and N. Toshima, *Tetrahedron Letters*, 1973, 1025.
558 J. H. Bowie, G. E. Gream and M. Mular, *Austral. J. Chem.*, 1972, **25**, 1107.
559 A. C. Cope, R. M. Pike and D. F. Rugen, *J. Amer. Chem. Soc.*, 1954, **76**, 4945.
560 L. A. Paquette and K. A. Henzel, *J. Amer. Chem. Soc.*, 1973, **95**, 2724.
561 K. A. Henzel, unpublished work; quoted by L. A. Paquette and R. S.Beckley, *J. Amer. Chem. Soc.*, 1975, **97**, 1084.
562 G. Schröder, G. Kirsch, J. F. M. Oth, R. Huisgen, W. E. Konz and U. Schnegg, *Chem. Ber.*, 1971, **104**, 2405.
563 D. P. Shoemaker, H. Kindler, W. G. Sly and R. C. Srivastava, *J. Amer. Chem. Soc.*, 1965, **87**, 482.
564 G. W. Buchanan, *Tetrahedron Letters*, 1972, 665.
565 W. E. Konz, W. Hechtl and R. Huisgen, *J. Amer. Chem. Soc.*, 1970, **92**, 4104.
566 R. Huisgen and W. E. Konz, *J. Amer. Chem. Soc.*, 1970, **92**, 4102.
567 R. Huisgen, W. E. Konz and G. E. Gream, *J. Amer. Chem. Soc.*, 1970, **92**, 4105.
568 M. S. Brookhart and M. A. M. Atwater, *Tetrahedron Letters*, 1972, 4399.
569 L. A. Paquette, D. R. James and G. H. Birnberg, *J. Amer. Chem. Soc.*, 1974, **96**, 7454.
570 A. Carrington and P. F. Todd, *Mol. Phys.*, 1964, **8**, 299.
571 G. R. Stevenson and J. G. Concepción, *J. Phys. Chem.*, 1974, **78**, 90.
572 A. Carrington, R. E. Moss and P. F. Todd, *Mol. Phys.*, 1966, **12**, 95.
573 R. D. Rieke and R. A. Copenhafer, *Tetrahedron Letters*, 1971, 4097.
574 G. R. Stevenson and J. G. Concepción, *J. Amer. Chem. Soc.*, 1973, **95**, 5692.
575 T. H. Brown and M. Karplus, *J. Chem. Phys.*, 1967, **46**, 870.
576 G. R. Stevenson, M. Colón, J. G. Concepción and A. M. Block, *J. Amer. Chem. Soc.*, 1974, **96**, 2283.
577 L. A. Paquette, J. F. Hansen, T. Kakihana and L. B. Anderson, *Tetrahedron Letters*, 1970, 533.
578 L. A. Paquette, M. J. Broadhurst, C. Lee and J. Clardy, *J. Amer. Chem. Soc.*, 1973, **95**, 4647.
579 J. F. M. Oth, unpublished work; quoted by P. J. Garratt and M. V. Sargent, in *Nonbenzenoid Aromatics* (ed. J. P. Snyder), vol. 2, Academic Press, 1971, p. 225.
580 G. Schröder, G. Kirsch and J. F. M. Oth, *Tetrahedron Letters*, 1969, 4575.
581 I. Pogány, Gh. Mihai and C. D. Nenitzescu, *Acad. Rep. populare Romine, Studii Cercetari Chim.*, 1959, **7**, 235; *Chem. Abs.*, 1960, **54**, 7584i.
582 J. Gasteiger and R. Huisgen, *Angew. Chem. Internat. Edn*, 1972, **11**, 716.

444   References

583   L. A. Paquette, W. Kitching, W. E. Heyd and R. H. Meisinger, J. Amer. Chem. Soc., 1974, 96, 7371.
584   M. Brookhart, E. R. Davis and D. L. Harris, J. Amer. Chem. Soc., 1972, 94, 7853.
585   C. E. Keller, B. A. Shoulders and R. Pettit, J. Amer. Chem. Soc., 1966, 88, 4760.
586   F. A. L. Anet, J. Amer. Chem. Soc., 1967, 89, 2491.
587   M. Green, S. Heathcock and D. C. Wood, J. C. S. Dalton, 1973, 1564.
588   L. A. Paquette, S. V. Ley, S. Maiorana, D. F. Schneider, M. J. Broadhurst and R. A. Boggs, J. Amer. Chem. Soc., 1975, 97, 4658.
589   A. Nakamura and N. Hagihara, Nippon Kagaku Zasshi, 1961, 82, 1387; Chem. Abs., 1963, 59, 2855a.
590   B. F. G. Johnson, J. Lewis, P. McArdle and G. L. P. Randall, J. C. S. Dalton, 1972, 2076.
591   Y. Becker, A. Eisenstadt and Y. Shvo, Chem. Comm., 1972, 1156.
592   F. A. L. Anet, H. D. Kaesz, A. Maasbol and S. Winstein, J. Amer. Chem. Soc., 1967, 89, 2489.
593   L. A. Bock, unpublished work; quoted by L. A. Paquette, S. V. Ley, S. Maiorana, D. F. Schneider, M. J. Broadhurst and R. A. Boggs, J. Amer. Chem. Soc., 1975, 97, 4658.
594   J. E. Alsop and R. Davis, J. C. S. Dalton, 1973, 1686.
595   V. Riera, S. A. R. Knox and F. G. A. Stone, unpublished work; quoted by S. A. R. Knox and F. G. A. Stone, Acounts Chem. Res., 1974, 7, 321.
596   H. Frye and G. B. McCauley, Inorg. Nuclear Chem. Letters, 1968, 4, 347.
597   A. C. Cope and J. E. Meili, J. Amer. Chem. Soc., 1967, 89, 1883.
598   A. C. Cope and W. R. Moore, J. Amer. Chem. Soc., 1955, 77, 4939.
599   F. A. L. Anet and B. Gregorovich, Tetrahedron Letters, 1966, 5961.
600   R. S. H. Liu and C. G. Krespan, J. Org. Chem., 1969, 34, 1271.
601   E. Grovenstein and D. V. Rao, Tetrahedron Letters, 1961, 148.
602   E. Grovenstein, T. C. Campbell and T. Shibata, J. Org. Chem., 1969, 34, 2418.
603   L. A. Paquette and M. Oku, J. Amer. Chem. Soc., 1974, 96, 1219.
604   L. A. Paquette, R. H. Meisinger and R. E. Wingard, J. Amer. Chem. Soc., 1973, 95, 2230.
605   L. A. Paquette, R. E. Wingard and R. H. Meisinger, J. Amer. Chem. Soc., 1971, 93, 1047.
606   L. A. Paquette, R. H. Meisinger and R. E. Wingard, J. Amer. Chem. Soc., 1972, 94, 9224.
607   R. Breslow, W. Vitale and K. Wendel, Tetrahedron Letters, 1965, 365.
608   F. A. L. Anet and L. A. Bock, J. Amer. Chem. Soc., 1968, 90, 7130.
609   E. Le Goff and R. B. LaCount, Tetrahedron Letters, 1965, 2787.
610   J. A. Elix, M. V. Sargent and F. Sondheimer, J. Amer. Chem. Soc., 1970, 92, 973.
611   R. Breslow, W. Horspool, H. Sugiyama and W. Vitale, J. Amer. Chem. Soc., 1966, 88, 3677.
612   J. A. Elix, M. V. Sargent and F. Sondheimer, J. Amer. Chem. Soc., 1970, 92, 962.
613   D. Bryce-Smith, A. Gilbert and J. Grzonka, Angew. Chem. Internat. Edn, 1971, 10, 746.
614   D. A. Wright, K. Seff and D. P. Shoemaker, J. Cryst. Mol. Structure, 1972, 2, 41.
615   N. N. Podgornova, E. S. Lipina and V. V. Perekalin, Zhur. org. Khim., 1975, 11, 213; Chem. Abs., 1975, 82, 139464r.

References 445

616 G. R. Stevenson, M. Colón, I. Ocasio, J. G. Concepción and A. M. Block, *J. Phys. Chem.*, 1975, **79**, 1685.
617 A. C. Cope and D. S. Smith, *J. Amer. Chem. Soc.*, 1952, **74**, 5136.
618 K. Saito and T. Mukai, *Tetrahedron Letters*, 1973, 4885.
619 R. Grubbs, R. Breslow, R. Herber and S. J. Lippard, *J. Amer. Chem. Soc.*, 1967, **89**, 6864.
620 J. A. Elix, M. V. Sargent and F. Sondheimer, *J. Amer. Chem. Soc.*, 1967, **89**, 5080.
621 J. A. Elix and M. V. Sargent, *J. Amer. Chem. Soc.*, 1969, **91**, 4734.
622 J. A. Elix, M. V. Sargent and F. Sondheimer, *Chem. Comm.*, 1966, 509.
623 A. Krebs, unpublished work; quoted by R. W. Hoffmann, *Dehydrobenzene and Cycloalkynes*, Academic Press, New York, 1967, p. 349.
624 A. Krebs and D. Byrd, *Annalen*, 1967, **707**, 66.
625 R. D. Miller and V. Y. Abraitys, *Tetrahedron Letters*, 1971, 891.
626 W. Lippke, W. Ferree and H. Morrison, *J. Amer. Chem. Soc.*, 1974, **96**, 2134.
627 J. A. Elix, M. V. Sargent and F. Sondheimer, *J. Amer. Chem. Soc.*, 1967, **89**, 180.
628 J. A. Elix, M. V. Sargent and F. Sondheimer, *J. Amer. Chem. Soc.*, 1970, **92**, 969.
629 L. A. Paquette, R. E. Wingard and J. M. Photis, *J. Amer. Chem. Soc.*, 1974, **96**, 5801.
630 L. A. Paquette and R. E. Wingard, *J. Amer. Chem. Soc.*, 1972, **94**, 4398.
631 L. A. Paquette and J. M. Photis, *Tetrahedron Letters*, 1975, 1145.
632 L. A. Paquette, J. C. Philips and R. E. Wingard, *J. Amer. Chem. Soc.*, 1971, **93**, 4516.
633 L. A. Paquette, J. M. Photis, J. Fayos and J. Clardy, *J. Amer. Chem. Soc.*, 1974, **96**, 1217.
634 L. A. Paquette, J. M. Photis, K. B. Gifkins and J. Clardy, *J. Amer. Chem. Soc.*, 1975, **97**, 3536.
635 R. K. Russell, R. E. Wingard and L. A. Paquette, *J. Amer. Chem. Soc.*, 1974, **96**, 7483.
636 J. G. Atkinson, D. E. Ayer, G. Büchi and E. W. Robb, *J. Amer. Chem. Soc.*, 1963, **85**, 2257.
637 H. Prinzbach and J. Rivier, *Angew. Chem. Internat. Edn*, 1967, **6**, 1069.
638 V. K. Shitikov, T. N. Kolosova, V. A. Sergeev, V. V. Korshak and P. O. Okulevich, *Zhur. org. Khim.*, 1974, **10**, 1007; *Chem. Abs.*, 1974, **81**, 49311p.
639 P. Chini, N. Palladino and A. Santambrogio, *J. Chem. Soc.* (C), 1967, 836.
640 J. R. Leto and M. F. Leto, *J. Amer. Chem. Soc.*, 1961, **83**, 2944.
641 T. Sakakibara, S. Nishimura, K. Kimura, S. Fujioka and Y. Odaira, *Tetrahedron Letters*, 1971, 4719.
642 J. F. H. Braams, H. J. T. Bos and J. F. Arens, *Rec. Trav. chim.*, 1968, **87**, 193.
643 G. A. Chukhadzhyan, E. L. Sarkisyan and T. S. Elbakyan, *Zhur. org. Khim.*, 1972, **8**, 1119; *Chem. Abs.*, 1972, **77**, 125828x.
644 G. A. Chukhadzhyan, E. L. Sarkisyan and T. S. Elbakyan, *Zhur. obshchei Khim.*, 1973, **43**, 2302; *Chem. Abs.*, 1974, **80**, 70486p.
645 G. A. Chukhadzhyan, E. L. Sarkisyan and T. S. Elbakyan, *Zhur. org. Khim.*, 1974, **10**, 1408; *Chem. Abs.*, 1975, **82**, 98478c.
646 B. V. Rozynov, M. M. Teplyakov, V. P. Chebotarev and V. V. Korshak, *Izvest. Akad. Nauk S.S.S.R., Ser. khim.*, 1974, 1602; *Chem. Abs.*, 1974, **81**, 120154x.
647 P. de Mayo and R. W. Yip, *Proc. Chem. Soc.*, 1964, 84.
648 F. A. Cotton, J. W. Faller and A. Musco, *J. Amer. Chem. Soc.*, 1968, **90**, 1438.

649 A. Padwa and R. Hartman, *J. Amer. Chem. Soc.*, 1964, **86**, 4212.
650 I. W. McKay and R. N. Warrener, *Tetrahedron Letters*, 1970, 4779.
651 I. W. McKay and R. N. Warrener, *Tetrahedron Letters*, 1970, 4783.
652 R. Criegee, W. Eberius and H.-A. Brune, *Chem. Ber.*, 1968, **101**, 94.
653 G. Maier and U. Mende, *Angew. Chem. Internat. Edn*, 1969, **8**, 132.
654 E. H. White and H. C. Dunathan, *J. Amer. Chem. Soc.*, 1964, **86**, 453.
655 L. A. Paquette, J. M. Photis and G. D. Ewing, *J. Amer. Chem. Soc.*, 1975, **97**, 3538.
656 G. Avitabile, P. Ganis and V. Petraccone, *J. Phys. Chem.*, 1969, **73**, 2378.
657 P. Ganis, A. Musco and P. A. Temussi, *J. Phys. Chem.*, 1969, **73**, 3201.
658 L. A. Paquette, T. Kakihana and J. F. Hansen, *Tetrahedron Letters*, 1970, 529.
659 L. A. Paquette, J. F. Hansen and T. Kakihana, *J. Amer. Chem. Soc.*, 1971, **93**, 168.
660 E. H. White and R. L. Stern, *Tetrahedron Letters*, 1964, 193.
661 A. Padwa and R. Hartman, *J. Amer. Chem. Soc.*, 1966, **88**, 1518.
662 J. G. Concepción and G. Vincow, *J. Phys. Chem.*, 1975, **79**, 2042.
663 A. Streitwieser and R. Walker, *J. Organometallic Chem.*, 1975, **97**, C41.
664 E. Rigamonti and G. Bajo, *Chimica e Industria*, 1973, **55**, 702.
665 S. Z. Goldberg, K. N. Raymond, C. A. Harmon and D. H. Templeton, *J. Amer. Chem. Soc.*, 1974, **96**, 1348.
666 F. A. Cotton, J. W. Faller and A. Musco, *J. Amer. Chem. Soc.*, 1966, **88**, 4506.
667 M. J. Bennett, F. A. Cotton and J. Takats, *J. Amer. Chem. Soc.*, 1968, **90**, 903.
668 F. A. Cotton and A. Musco, *J. Amer. Chem. Soc.*, 1968, **90**, 1444.
669 F. A. Cotton and M. D. LaPrade, *J. Amer. Chem. Soc.*, 1968, **90**, 2026.
670 F. A. Cotton and J. Takats, *J. Amer. Chem. Soc.*, 1968, **90**, 2031.
671 J. M. Photis and L. A. Paquette, unpublished work; quoted by L. A. Paquette, *Tetrahedron*, 1975, **31**, 2855.
672 R. D. Rieke and R. A. Copenhafer, *Tetrahedron Letters*, 1971, 879.
673 M. Tsutsui, *Chem. and Ind.*, 1962, 780.
674 H.-P. Throndsen and H. Zeiss, *J. Organometallic Chem.*, 1963–4, **1**, 301.
675 G. Büchi, C. W. Perry and E. W. Robb, *J. Org. Chem.*, 1962, **27**, 4106.
676 R. Criegee, G. Schröder, G. Maier and H.-G. Fischer, *Chem. Ber.*, 1960, **93**, 1553.
677 R. Criegee and G. Schröder, *Annalen*, 1959, **623**, 1.
678 H. H. Freedman and D. R. Peterson, *J. Amer. Chem. Soc.*, 1962, **84**, 2837.
679 E. H. Braye, W. Hübel and I. Caplier, *J. Amer. Chem. Soc.*, 1961, **83**, 4406.
680 H. H. Freedman, *J. Amer. Chem. Soc.*, 1961, **83**, 2195.
681 D. F. Pollock and P. M. Maitlis, *J. Organometallic Chem.*, 1971, **26**, 407.
682 P. M. Maitlis and F. G. A. Stone, *Proc. Chem. Soc.*, 1962, 330.
683 R. C. Cookson and D. W. Jones, *Proc. Chem. Soc.*, 1963, 115.
684 A. E. van der Hout-Lodder and H. M. Buck, *Rec. Trav. chim.*, 1972, **91**, 667.
685 P. M. Maitlis, D. Pollock, M. L. Games and W. J. Pryde, *Canad. J. Chem.*, 1965, **43**, 470.
686 L. F. Pelosi, *Diss. Abs.*, 1973, **34B**, 1428.
687 M. Neuenschwander and A. Niederhauser, *Helv. Chim. Acta*, 1970, **53**, 519.
688 J. Ficini, A.-M. Touzin and A. Krief, *Bull. Soc. chim. France*, 1972, 2388.
689 R. Criegee and R. Huber, *Chem. Ber.* 1970, **103**, 1862.
690 T. J. Meyers and K. V. Scherer, unpublished work; quoted by R. West, *Accounts Chem. Res.*, 1970, **3**, 130.

691 A. Roedig, *Angew. Chem. Internat. Edn*, 1969, **8**, 150.
692 A. Roedig, R. Helm, R. West and R. M. Smith, *Tetrahedron Letters*, 1969, 2137.
693 A. Roedig and V. Kimmel, *Annalen*, 1972, **755**, 122.
694 A. Roedig, B. Heinrich and V. Kimmel, *Annalen*, 1975, 1195.
695 A. Roedig, G. Bonse, R. Helm and R. Kohlhaupt, *Chem. Ber.*, 1971, **104**, 3378.
696 A. Roedig, V. Kimmel and W. Lippert, *Tetrahedron Letters*, 1971, 1219.
697 G. Maier and U. Mende, *Angew. Chem. Internat. Edn*, 1968, **7**, 537.
698 G. Maier and M. Schneider, unpublished work; quoted by G. Maier, *Angew. Chem. Internat. Edn*, 1974, **13**, 425.
699 H. P. Throndsen, P. J. Wheatley and H. Zeiss, *Proc. Chem. Soc.*, 1964, 357.
700 G. S. Pawley, W. N. Lipscomb and H. H. Freedman, *J. Amer. Chem. Soc.*, 1964, **86**, 4725.
701 P. J. Wheatley, *J. Chem. Soc.*, 1965, 3136.
702 H. H. Freedman and R. S. Gohlke, *Proc. Chem. Soc.*, 1963, 249
703 J. Haase and P. Widmann, *Z. Naturforsch.*, 1974, **29a**, 533.
704 R. Criegee and G. Louis, *Chem. Ber.*, 1957, **90**, 417.
705 R. Criegee, W.-D. Wirth, W. Engel and H.-A. Brune, *Chem. Ber.*, 1963, **96**, 2230.
706 R. Criegee and R. Askani, *Angew. Chem. Internat. Edn*, 1968, **7**, 537.
707 A. Roedig, G. Bonse and R. Helm, *Chem. Ber.*, 1973, **106**, 2156.
708 G. Maier, *Chem. Ber.*, 1963, **96**, 2238.
709 A. Roedig, G. Bonse, R. Ganns and V. Kimmel, *Annalen*, 1973, 2025.
710 A. S. Lankey and M. A. Ogliaruso, *J. Org. Chem.*, 1971, **36**, 3339.
711 T. J. Katz and M. Rosenberger, *J. Amer. Chem. Soc.*, 1962, **84**, 865.
712 T. J. Katz, M. Rosenberger and R. K. O'Hara, *J. Amer. Chem. Soc.*, 1964, **86**, 249.
713 T. J. Katz and J. J. Mrowca, *J. Amer. Chem. Soc.*, 1967, **89**, 1105.
714 H. E. Zimmerman and G. L. Grunewald, *J. Amer. Chem. Soc.*, 1966, **88**, 183.
715 L. A. Paquette, D. R. James and G. H. Birnberg, *Chem. Comm.*, 1974, 722.
716 D. R. James, G. H. Birnberg and L. A. Paquette, *J. Amer. Chem. Soc.*, 1974, **96**, 7465.
717 L. A. Paquette, G. H. Birnberg, J. Clardy and B. Parkinson, *Chem. Comm.*, 1973, 129.
718 R. M. Moriarty, C.-L. Yeh, E.-L. Yeh and K. C. Ramey, *J. Amer. Chem. Soc.*, 1972, **94**, 9229.
719 R. Aumann, *J. Organometallic Chem.*, 1974, **66**, C6.
720 D. Ehntholt, A. Rosan and M. Rosenblum, *J. Organometallic Chem.*, 1973, **56**, 315.
721 R. M. Moriarty, C.-L. Yeh and K. C. Ramey, *J. Amer. Chem. Soc.*, 1971, **93**, 6709.
722 R. Aumann, *Angew. Chem. Internat. Edn*, 1972, **11**, 522.
723 R. M. Moriarty, C.-L. Yeh, K. N. Chen, E.-L. Yeh, K. C. Ramey and C. W. Jefford, *J. Amer. Chem. Soc.*, 1973, **95**, 4756.
724 Z. V. Todres, V. Ya. Bespalov, *Zhur. org. Khim.*, 1972, **8**, 19; *Chem. Abs.*, 1972, **76**, 112451y.
725 T. S. Cantrell and H. Shechter, *J. Amer. Chem. Soc.*, 1967, **89**, 5877.
726 D. N. Kursanov, Z. V. Todres, Yu. I. Lyakhovetskii and Z. G. Dremina, *Izvest. Akad. Nauk S.S.S.R.*, *Ser. khim.*, 1967, 2197; *Chem. Abs.*, 1968, **68**, 48829h.

727 E. E. Gol'teuzen, Z. V. Todres, A. Ya. Kaminskii, S. S. Gitis and D. N. Kursanov, *Izvest. Akad. Nauk S.S.S.R.*, *Ser. khim.*, 1972, 1083; *Chem. Abs.*, 1972, **77**, 87205g.

728 Z. V. Todres and D. N. Kursanov, *Doklady Akad. Nauk S.S.S.R.*, 1972, **205**, 1117; *Chem. Abs.*, 1972, **77**, 151559e.

729 Z. V. Todres, M. P. Starodubtseva and D. N. Kursanov, *Izvest. Akad. Nauk S.S.S.R.*, *Ser. khim.*, 1974, 230; *Chem. Abs.*, 1974, **80**, 107655w.

730 D. N. Kursanov and Z. V. Todres, *Doklady Akad. Nauk S.S.S.R.*, 1967, **172**, 1086; *Chem Abs.*, 1967, **67**, 6148j.

731 Z. V. Todres, Yu. I. Lyakhovetskii and D. N. Kursanov, *Izvest. Akad. Nauk S.S.S.R.*, *Ser. khim.*, 1969, 1455; *Chem. Abs.*, 1969, **71**, 112131r.

732 Z. V. Todres and S. P. Avagyan, *Zhur. Vsesoyuz. Khim. obshch. im. D. I. Mendeleeva*, 1973, **18**, 478; *Chem. Abs.*, 1973, **79**, 145479d.

733 Z. V. Todres and S. P. Avagyan, *Internat. J. Sulfur Chem.*, 1973, **8**, 373.

734 R. W. Murray and M. L. Kaplan, *J. Org. Chem.*, 1966, **31**, 962.

735 K. Conrow and P. C. Radlick, *J. Org. Chem.*, 1961, **26**, 2260.

736 W. R. Roth, *Annalen*, 1964, **671**, 25.

737 J. I. Brauman, J. Schwartz and E. E. van Tamelen, *J. Amer. Chem. Soc.*, 1968, **90**, 5328.

738 V. D. Azatyan, *Doklady Akad. Nauk S.S.S.R.*, 1954, **98**, 403; *Chem. Abs.*, 1955, **49**, 12318i.

739 V. D. Azatyan and R. S. Gyuli-Kevkhyan, *Doklady Akad. Nauk Armyan. S.S.R.*, 1955, **20**, 81; *Chem. Abs.*, 1956, **50**, 4051a.

740 J. M. Bellama and J. B. Davison, *J. Organometallic Chem.*, 1975, **86**, 69.

741 V. D. Azatyan, *Izvest. Akad. Nauk Armyan. S.S.R.*, *Khim. Nauki*, 1964, **17**, 706; *Chem. Abs.*, 1965, **63**, 4323f.

742 J. M. Bellama and J. B. Davison, *Synthetic Reactions Inorg. Metal-Org. Chem.*, 1975, **5**, 87.

743 T. J. Katz and P. J. Garratt, *J. Amer. Chem. Soc.*, 1964, **86**, 5194.

744 S. W. Staley and T. J. Henry, *J. Amer. Chem. Soc.*, 1969, **91**, 1239.

745 T. J. Katz and P. J. Garratt, *J. Amer. Chem. Soc.*, 1963, **85**, 2852.

746 A. G. Anastassiou and R. C. Griffith, *J. Amer. Chem. Soc.*, 1973, **95**, 2379.

747 J. M. Brown and M. M. Ogilvy, *J. Amer. Chem. Soc.*, 1974, **96**, 292.

748 A. G. Anastassiou and R. C. Griffith, *Tetrahedron Letters*, 1973, 3067.

749 C. P. Lewis and M. Brookhart, *J. Amer. Chem. Soc.*, 1975, **97**, 651.

750 E. W. Turnblom and T. J. Katz, *J. Amer. Chem. Soc.*, 1973, **95**, 4292.

751 S. W. Staley and T. J. Henry, *J. Amer. Chem. Soc.*, 1970, **92**, 7612.

752 F. A. Cotton and G. Deganello, *J. Amer. Chem. Soc.*, 1972, **94**, 2142.

753 F. A. Cotton and G. Deganello, *J. Amer. Chem. Soc.*, 1973, **95**, 396.

754 V. D. Azatyan and R. S. Gyuli-Kevkhyan, *Izvest. Akad. Nauk Armyan. S.S.R.*, *Khim. Nauki*, 1961, **14**, 451; *Chem. Abs.*, 1963, **58**, 3327a.

755 T. S. Cantrell, *Tetrahedron Letters*, 1968, 5635.

756 T. S. Cantrell, *J. Amer. Chem. Soc.*, 1970, **92**, 5480.

757 A. Diaz and J. Fulcher, *J. Amer. Chem. Soc.*, 1974, **96**, 7954.

758 M. Sakai, R. F. Childs and S. Winstein, *J. Org. Chem.*, 1972, **37**, 2517.

759 R. W. Hoffmann, H. Kurz, M. T. Reetz and R. Schüttler, *Chem. Ber.*, 1975, **108**, 109.

760 Z. V. Todres, F. M. Stoyanovich, Yu. L. Gol'farb and D. N. Kursanov, *Khim. geterotsikl. Soedinenii*, 1973, 632; *Chem. Abs.*, 1973, **79**, 66117f.

761 F. Mares, K. Hodgson and A. Streitwieser, *J. Organometallic Chem.*, 1971, **28**, C24.

762 K. O. Hodgson, F. Mares, D. F. Starks and A. Streitwieser, *J. Amer. Chem. Soc.*, 1972, **95**, 8650.

763 D. G. Karraker and M. P. Palm, *Nuclear Science Abs.*, 1974, **29**, 9681.
764 K. O. Hodgson and K. N. Raymond, *Inorg. Chem.*, 1972, **11**, 171.
765 F. Mares, K. Hodgson and A. Streitwieser, *J. Organometallic Chem.*, 1970, **24**, C68.
766 K. O. Hodgson and K. N. Raymond, *Inorg. Chem.*, 1972, **11**, 3030.
767 B. L. Kalsotra, R. K. Multani and B. D. Jain, *Chem. and Ind.*, 1972, 339.
768 K. D. Warren, *Inorg. Chem.*, 1975, **14**, 3095.
769 J. D. Jamerson, A. P. Masino and J. Takats, *J. Organometallic Chem.*, 1974, **65**, C33.
770 A. Streitwieser and N. Yoshida, *J. Amer. Chem. Soc.*, 1969, **91**, 7528.
771 J. Goffart, J. Fuger, B. Gilbert, B. Kanellakopulos and G. Duyckaerts, *Inorg. Nuclear Chem. Letters*, 1972, **8**, 403.
772 J. Goffart, J. Fuger, D. Brown and G. Duyckaerts, *Inorg. Nuclear Chem. Letters*, 1974, **10**, 413.
773 D. F. Starks, T. C. Parsons, A. Streitwieser and N. Edelstein, *Inorg. Chem.*, 1974, **13**, 1307.
774 A. Streitwieser and U. Müller-Westerhoff, *J. Amer. Chem. Soc.*, 1968, **90**, 7364.
775 D. G. Karraker, J. A. Stone, E. R. Jones and N. Edelstein, *J. Amer. Chem. Soc.*, 1970, **92**, 4841.
776 S. E. Anderson, *J. Organometallic Chem.*, 1974, **71**, 263.
777 A. Streitwieser and C. A. Harmon, *Inorg. Chem.*, 1973, **12**, 1102.
778 D. G. Karraker, *Inorg. Chem.*, 1973, **12**, 1105.
779 A. Streitwieser, D. Dempf, G. N. La Mar, D. G. Karraker and N. Edelstein, *J. Amer. Chem. Soc.*, 1971, **93**, 7343.
780 N. Edelstein, G. N. La Mar, F. Mares and A. Streitwieser, *Chem. Phys. Letters*, 1971, **8**, 399.
781 A. Zalkin and K. N. Raymond, *J. Amer. Chem. Soc.*, 1969, **91**, 5667.
782 A. Avdeef, K. N. Raymond, K. O. Hodgson and A. Zalkin, *Inorg. Chem.*, 1972, **11**, 1083.
783 K. O. Hodgson, D. Dempf and K. N. Raymond, *Chem. Comm.*, 1971, 1592.
784 K. O. Hodgson and K. N. Raymond, *Inorg. Chem.*, 1973, **12**, 458.
785 R. G. Hayes and N. Edelstein, *J. Amer. Chem. Soc.*, 1972, **94**, 8688.
786 A. Streitwiesser, *Topics Nonbenzenoid Aromatic Chem.*, 1973, **1**, 221.
787 D. G. Karraker and J. A. Stone, *J. Amer. Chem. Soc.*, 1974, **96**, 6885.
788 N. Kumar and R. K. Multani, *J. Organometallic Chem.*, 1973, **63**, 47.
789 H. O. van Oven and H. J. de Liefde Meijer, *J. Organometallic Chem.*, 1969, **19**, 373.
790 J. Goffart and G. Duyckaerts, *J. Organometallic Chem.*, 1975, **94**, 29.
791 H. K. Hofstee, H. O. van Oven and H. J. de Liefde Meijer, *J. Organometallic Chem.*, 1972, **42**, 405.
792 H. R. van der Wal, F. Overzet, H. O. van Oven, J. L. de Boer, H. J. de Liefde Meijer and F. Jellinek, *J. Organometallic Chem.*, 1975, **92**, 329.
793 H.-J. Kablitz, R. Kallweit and G. Wilke, *J. Organometallic Chem.*, 1972, **44**, C49.
794 R. R. Schrock and J. Lewis, *J. Amer. Chem. Soc.*, 1973, **95**, 4102.
795 M. Green, J. A. Howard, J. L. Spencer and F. G. A. Stone, *Chem. Comm.*, 1975, 3.
796 T. J. Barton and M. Juvet, *Tetrahedron Letters*, 1975, 2561.
797 J. Dunogues, R. Calas, J. Dedier and F. Pisciotti, *J. Organometallic Chem.*, 1970, **25**, 51.
798 N. Kumar and R. K. Sharma, *Chem. and Ind.*, 1974, 261.
799 H. K. Hofstee, C. J. Groenenboom, H. O. van Oven and H. J. de Liefde Meijer, *J. Organometallic Chem.*, 1975, **85**, 193.

450   References

800   M. E. E. Veldman and H. O. van Oven, *J. Organometallic Chem.*, 1975, **84**, 247.
801   R. Hubin and J. Goffart, *C.R. Hebd. Séances Acad. Sci.*, *Ser. C.* 1974, **279**, 903; *Chem. Abs.*, 1975, **82**, 105071v.
802   C. G. Salentine and M. F. Hawthorne, *Chem. Comm.*, 1975, 848.
803   J. Knol, A. Westerhof, H. O. van Oven and H. J. de Liefde Meijer, *J. Organometallic Chem.*, 1975, **96**, 257.
804   H. D. Kaesz, S. Winstein and C. G. Kreiter, *J. Amer. Chem. Soc.*, 1966, **88**, 1319.
805   A. Maasbol, unpublished work; quoted by S. Winstein, *Chem. Soc. Special Publ. no. 21*, 1967, p. 5.
806   A. Carbonaro, A. Greco and G. Dall'Asta, *J. Org. Chem.*, 1968, **33**, 3948.
807   A. Carbonaro, A. Greco and G. Dall'Asta, *Tetrahedron Letters*, 1968, 5129.
808   A. Carbonaro, A. Greco and G. Dall'Asta, *Tetrahedron Letters*, 1967, 2037.
809   A. Carbonaro and A. Greco, *J. Organometallic Chem.*, 1970, **25**, 477.
810   I. W. Bassi and R. Scordamaglia, *J. Organometallic Chem.*, 1972, **37**, 353.
811   M. Cooke, J. A. K. Howard, C. R. Russ, F. G. A. Stone and P. Woodward, *J. Organometallic Chem.*, 1974, **78**, C43.
812   J. Schwartz, *Chem. Comm.*, 1972, 814.
813   E. B. Fleischer, A. L. Stone, R. B. K. Dewar, J. D. Wright, C. E. Keller and R. Pettit, *J. Amer. Chem. Soc.*, 1966, **88**, 3158.
814   R. E. Davis, T. A. Dodds, T.-H. Hseu, J. C. Wagnon, T. Devon, J. Tancrede, J. S. McKennis and R. Pettit, *J. Amer. Chem. Soc.*, 1974, **96**, 7562.
815   A. Davison, W. McFarlane, L. Pratt and G. Wilkinson, *Chem. and Ind.*, 1961, 553.
816   G. N. Schrauzer, *J. Amer. Chem. Soc.*, 1961, **83**, 2966.
817   M. Brookhart and E. R. Davis, *J. Amer. Chem. Soc.*, 1970, **92**, 7622.
818   M. Brookhart and E. R. Davis, *Tetrahedron Letters*, 1971, 4349.
819   A. Davison, W. McFarlane and G. Wilkinson, *Chem. and Ind.*, 1962, 820.
820   J. D. Holmes and R. Pettit, *J. Amer. Chem. Soc.*, 1963, **85**, 2531.
821   V. Heil, B. F. G. Johnson, J. Lewis and D. J. Thompson, *Chem. Comm.*, 1974, 270.
822   B. F. G. Johnson, J. Lewis, A. W. Parkins and G. L. P. Randall, *Chem. Comm.*, 1969, 595.
823   R. E. Dessy and L. Wieczorek, *J. Amer. Chem. Soc.*, 1969, **91**, 4963.
824   A. Nakamura and N. Hagihara, *Mem. Inst. Sci. Ind. Res.*, *Osaka Univ.*, 1960, **17**, 187; *Chem. Abs.*, 1960, **55**, 6457f.
825   F. Faraone, F. Zingales, P. Uguagliati and U. Belluco, *Inorg. Chem.*, 1968, **7**, 2362.
826   L. A. Paquette, S. V. Ley, M. J. Broadhurst, D. Truesdell, J. Fayos and J. Clardy, *Tetrahedron Letters*, 1973, 2943.
827   D. J. Ehntholt and R. C. Kerber, *J. Organometallic Chem.*, 1972, **38**, 139.
828   L. A. Paquette, *Trans. New York Acad. Sci.*, 1974, **36**, 357.
829   L. A. Paquette, *Tetrahedron*, 1975, **31**, 2855.
830   M. Green and D. C. Wood, *Chem. Comm.*, 1967, 1062.
831   M. Green and D. C. Wood, *J. Chem. Soc. (A)*, 1969, 1172.
832   U. Kruerke, *Angew. Chem. Internat. Edn*, 1967, **6**, 79.
833   D. M. Roe and A. G. Massey, *J. Organometallic Chem.*, 1970, **23**, 547.
834   Y. Shvo and E. Hazum, *Chem. Comm.*, 1974, 336.
835   E. W. Abel and S. Moorhouse, *Inorg. Nuclear Chem. Letters*, 1970, **6**, 621.
836   M. Cooke, P. T. Draggett, M. Green, B. F. G. Johnson, J. Lewis and D. J. Yarrow, *Chem. Comm.*, 1971, 621.

837   F. Faraone, F. Cusmano and R. Pietropaolo, *J. Organometallic Chem.*, 1971, **26**, 147.
838   R. R. Schrock, B. F. G. Johnson and J. Lewis, *J. C. S. Dalton*, 1974, 951.
839   S. Otsuka and A. Nakamura, *Inorg. Chem.*, 1966, **5**, 2059.
840   J. Evans, B. F. G. Johnson, J. Lewis and D. J. Yarrow, *J. C. S. Dalton*, 1974, 2375.
841   L. Busetto, A. Palazzi and R. Ros, *Inorg. Chem.*, 1970, **9**, 2792.
842   H. I. Heitner and S. J. Lippard, *Inorg. Chem.*, 1972, **11**, 1447.
843   R. J. W. Le Fèvre, D. V. Radford and J. D. Saxby, *Inorg. Chem.*, 1969, **8**, 1532.
844   A. Pidcock and G. C. Roberts, *J. Chem. Soc.* (A), 1970, 2922.
845   A. Schott, H. Schott, G. Wilke, J. Brandt, H. Hoberg and E. G. Hoffmann, *Annalen*, 1973, 508.
846   J. R. Doyle, J. H. Hutchinson, N. C. Baenziger and L. W. Tresselt, *J. Amer. Chem. Soc.*, 1961, **83**, 2768.
847   C. R. Kistner, J. H. Hutchinson, J. R. Doyle and J. C. Storlie, *Inorg. Chem.*, 1963, **2**, 1255.
848   L. A. Paquette, M. J. Broadhurst, P. Warner, G. A. Olah and G. Liang, *J. Amer. Chem. Soc.*, 1973, **95**, 3386.
849   G. A. Olah, J. S. Staral and G. Liang, *J. Amer. Chem. Soc.*, 1974, **96**, 6233.
850   R. Huisgen and J. Gasteiger, *Tetrahedron Letters*, 1972, 3661.
851   H. Hogeveen and C. J. Gassbeck, *Rec. Trav. chim.*, 1970, **89**, 1079.
852   P. A. Christensen, Y. Y. Huang, A. Meesters and T. S. Sorensen, *Can. J. Chem.*, 1974, **52**, 3424.
853   W. R. Roth and B. Peltzer, *Annalen*, 1965, **685**, 56.
854   H. Kloosterziel and A. P. ter Borg, *Rec. Trav. chim.*, 1965, **84**, 1305.
855   D. S. Glass, J. Zirner and S. Winstein, *Proc. Chem. Soc.*, 1963, 276.
856   R. Huisgen F. Mietzsch, G. Boche and H. Seidl, *Angew. Chem. Internat. Edn*, 1965, 4, 368.
857   F. A. L. Anet and I. Yavari, *Tetrahedron Letters*, 1975, 4221.
858   W. Sanne, *Festschrift Carl Wurster zum 60 Geburtstag*, 1960, 79; *Chem. Abs.*, 1962, **56**, 9995b.
859   W. von E. Doering and W. R. Roth, *Tetrahedron*, 1963, **19**, 715.
860   J. Zirner and S. Winstein, *Proc. Chem. Soc.*, 1964, 235.
861   O. L. Chapman, G. W. Borden, R. W. King and B. Winkler, *J. Amer. Chem. Soc.*, 1964, **86**, 2660.
862   T. D. Goldfarb and L. Lindqvist, *J. Amer. Chem. Soc.*, 1967, **89**, 4588.
863   P. Datta, T. D. Goldfarb and R. S. Boikess, *J. Amer. Chem. Soc.*, 1969, **91**, 5429.
864   P. Ahlberg, D. L. Harris, M. Roberts, P. Warner, P. Seidl, M. Sakai, D. Cook, A. Diaz, J. P. Dirlam, H. Hamberger and S. Winstein, *J. Amer. Chem. Soc.*, 1972, **94**, 7063.
865   A. C. Cope, C. L. Stevens and F. A. Hochstein, *J. Amer. Chem. Soc.*, 1950, **72**, 2510.
866   M. Oda, T. Sato and Y. Kitahara, *Synthesis*, 1974, 721.
867   H. Kloosterziel and E. Zwanenburg, *Rec. Trav. chim.*, 1969, **88**, 1373.
868   P. Radlick and S. Winstein, *J. Amer. Chem. Soc.*, 1963, **85**, 344.
869   W. R. Roth, *Annalen*, 1964, **671**, 10.
870   M. S. Baird and C. B. Reese, *Chem. Comm.*, 1970, 1519.
871   R. S. Boikess and S. Winstein, *J. Amer. Chem. Soc.*, 1963, **85**, 343.
872   D. I. Schuster and F.-T. Lee, *Tetrahedron Letters*, 1965, 4119.
873   A. C. Cope, H. R. Nace and L. L. Estes, *J. Amer. Chem. Soc.*, 1950, **72**, 1123.
874   E. L. Allred and K. J. Voorhees, *J. Amer. Chem. Soc.*, 1973, **95**, 620.

875   P. Crews and J. Beard, *J. Org. Chem.*, 1973, **38**, 522.
876   G. Kresze and H. Bathelt, *Tetrahedron*, 1973, **29**, 2219.
877   J. R. Malpass, *Chem. Comm.*, 1972, 1246.
878   J. R. Malpass and N. J. Tweddle, *Chem. Comm.*, 1972, 1247.
879   I. A. Akhtar and G. I. Fray, *J. Chem. Soc.* (C), 1971, 2800.
880   I. A. Akhtar and G. I. Fray, *J. Chem. Soc.* (C), 1971, 2802.
881   J. C. Chadwick and G. I. Fray, unpublished work.
882   D. M. Bratby, G. I. Fray and D. J. Neville, unpublished work.
883   J. Müller and W. Goll, *J. Organometallic Chem.*, 1974, **71**, 257.
884   E. O. Fischer and C. Palm, *Z. Naturforsch.*, 1959, **14b**, 347.
885   E. O. Fischer, C. Palm and H. P. Fritz, *Chem. Ber.*, 1959, **92**, 2645.
886   R. Aumann and S. Winstein, *Tetrahedron Letters*, 1970, 903.
887   V. S. Armstrong and C. K. Prout, *J. Chem. Soc.*, 1962, 3770.
888   A. Nakamura and N. Hagihara, *Nippon Kagaku Zasshi*, 1961, **82**, 1389; *Chem. Abs.*, 1963, **59**, 2855h.
889   M. Brookhart, N. M. Lippman and E. J. Reardon, *J. Organometallic Chem.*, 1973, **54**, 247.
890   T. A. Manuel and F. G. A. Stone, *J. Amer. Chem. Soc.*, 1960, **82**, 6240.
891   W. McFarlane, L. Pratt and G. Wilkinson, *J. Chem. Soc.*, 1963, 2162.
892   W. Slegeir, R. Case, J. S. McKennis and R. Pettit, *J. Amer. Chem. Soc.*, 1974, **96**, 287.
893   R. B. King, *Inorg. Chem.*, 1963, **2**, 807.
894   G. F. Emerson, J. E. Mahler, R. Pettit and R. Collins, *J. Amer. Chem. Soc.*, 1964, **86**, 3590.
895   F. A. Cotton and W. T. Edwards, *J. Amer. Chem. Soc.*, 1969, **91**, 843.
896   F. A. Cotton, D. L. Hunter and P. Lahuerta, *J. Amer. Chem. Soc.*, 1975, **97**, 1046.
897   A. C. Szary, S. A. R. Knox and F. G. A. Stone, *J. C. S. Dalton*, 1974, 662.
898   J. Müller and E. O. Fischer, *J. Organometallic Chem.*, 1966, **5**, 275.
899   J. Müller, C. G. Kreiter, B. Mertschenck and S. Schmitt, *Chem. Ber.*, 1975, **108**, 273.
900   G. Huttner and V. Bejenke, *Chem. Ber.*, 1974, **107**, 156.
901   E. O. Fischer and J. Müller, *Chem. Ber.*, 1963, **96**, 3217.
902   E. O. Fischer and C. Palm, *Z. Naturforsch.*, 1959, **14b**, 598.
903   A. Nakamura and N. Hagihara, *Bull. Chem. Soc. Japan*, 1961, **34**, 452.
904   R. B. King, P. M. Treichel and F. G. A. Stone, *J. Amer. Chem. Soc.*, 1961, **83**, 3593.
905   J. Evans, B. F. G. Johnson and J. Lewis, *J. C. S. Dalton*, 1972, 2668.
906   M. Kröner, *Chem. Ber.*, 1967, **100**, 3172.
907   H. Straub, J. M. Rao and E. Müller, *Annalen*, 1973, 1352.
908   G. Boche, W. Hechtl, H. Huber and R. Huisgen, *J. Amer. Chem. Soc.*, 1967, **89**, 3344.
909   J. Gasteiger and R. Huisgen, *Tetrahedron Letters*, 1972, 3665.
910   J. A. Elix, M. V. Sargent and F. Sondheimer, *J. Amer. Chem. Soc.*, 1967, **89**, 5081.
911   J. B. Davison and J. M. Bellama, *Inorg. Chim. Acta*, 1975, **14**, 263.
912   J. A. K. Howard, S. A. R. Knox, F. G. A. Stone, A. C. Szary and P. Woodward, *Chem. Comm.*, 1974, 788.
913   M. Oda, Y. Kayama and Y. Kitahara, *Tetrahedron Letters*, 1974, 2019.
914   M. Oda, Y. Kayama, H. Miyazaki and Y. Kitahara, *Angew. Chem. Internat. Edn*, 1975, **14**, 418.
915   V. Georgian, L. Georgian and A. V. Robertson, *Tetrahedron*, 1963, **19**, 1219.

916 E. Vogel, *Annalen*, 1958, **615**, 14.
917 M. Avram, E. Marica and C. D. Nenitzescu, *Chem. Ber.*, 1959, **92**, 1088.
918 A. T. Blomquist and A. G. Cooke, *Chem. and Ind.*, 1960, 873.
919 T. Sasaki, K. Kanematsu and Y. Yukimoto, *J. C. S. Perkin I*, 1973, 375.
920 E. L. Allred, B. R. Beck and K. J. Voorhees, *J. Org. Chem.*, 1975, **39**, 1426.
921 N. L. Allinger, M. A. Miller and L. A. Tushaus, *J. Org. Chem.*, 1963, **28**, 2555.
922 M. Finkelstein, R. C. Petersen and S. D. Ross, *Tetrahedron*, 1967, **23**, 3875.
923 C. R. Ganellin and R. Pettit, *J. Chem. Soc.*, 1958, 576.
924 D. H. R. Barton, *Helv. Chim. Acta*, 1959, **42**, 2604.
925 R. Anet, *Tetrahedron Letters*, 1961, 720.
926 H. Hoever, *Tetrahedron Letters*, 1962, 255.
927 E. Müller, H. Straub and J. M. Rao, *Tetrahedron Letters*, 1970, 773.
928 H. Straub, J. M. Rao and E. Müller, *Annalen*, 1973, 1339.
929 R. Pettit and J. Henery, *Org. Synth.*, 1970, **50**, 36.
930 J. P. Snyder, V. T. Bandurco, F. Darack and H. Olsen, *J. Amer. Chem. Soc.*, 1974, **96**, 5158.
931 D. G. Farnum and J. P. Snyder, *Tetrahedron Letters*, 1965, 3861.'
932 G. I. Fray and D. A. Johnson, unpublished work.
933 L. A. Paquette and J. C. Stowell, *J. Amer. Chem. Soc.*, 1971, **93**, 5735.
934 L. A. Paquette, M. J. Kukla and J. C. Stowell, *J. Amer. Chem. Soc.*, 1972, **94**, 4920.
935 C. Ganter, S. M. Pokras and J. D. Roberts, *J. Amer. Chem. Soc.*, 1966, **88**, 4235.
936 G. Büchi and E. M. Burgess, *J. Amer. Chem. Soc.*, 1962, **84**, 3104.
937 L. L. Barber, O. L. Chapman and J. D. Lassila, *J. Amer. Chem. Soc.*, 1969, **91**, 531.
938 M. Brookhart, M. Ogliaruso and S. Winstein, *J. Amer. Chem. Soc.*, 1967, **89**, 1965.
939 M. Ogawa, M. Takagi and T. Matsuda, *Chem. Letters*, 1972, 527.
940 J. R. Dobbelaere and H. M. Buck, *Rec. Trav. chim.*, 1974, **93**, 159.
941 A. C. Cope, S. F. Schaeren and E. R. Trumbull, *J. Amer. Chem. Soc.*, 1954, **76**, 1096.
942 M. Ogawa and T. Matsuda, *Chem. Letters*, 1975, 47.
943 K. Tomisawa and T. Mukai, *J. Amer. Chem. Soc.*, 1973, **95**, 5405.
944 A.-H. Khuthier and J. C. Robertson, *J. Org. Chem.*, 1970, **35**, 3760.
945 M. Ogawa, M. Takagi and T. Matsuda, *Tetrahedron*, 1973, **29**, 3813.
946 S. Masamune, H. Zenda, M. Wiesel, N. Nakatsuka and G. Bigam, *J. Amer. Chem. Soc.*, 1968, **90**, 2727.
947 T. L. Burkoth, *J. Org. Chem.*, 1966, **31**, 4259.
948 M. Jones, S. D. Reich and L. T. Scott, *J. Amer. Chem. Soc.*, 1970, **92**, 3118.
949 P. Radlick, W. Fenical and G. Alford, *Tetrahedron Letters*, 1970, 2707.
950 W. Grimme, *Chem. Ber.*, 1967, **100**, 113.
951 E. Vogel, *Angew. Chem.*, 1962, **74**, 829.
952 P. Radlick and W. Fenical, *J. Amer. Chem. Soc.*, 1969, **91**, 1560.
953 A. G. Anastassiou and R. C. Griffith, *Chem. Comm.*, 1971, 1301.
954 A. G. Anastassiou and R. C. Griffith, *Chem. Comm.*, 1972, 399.
955 G. J. Fonken and W. Moran, *Chem. and Ind.*, 1963, 1841.
956 E. A. LaLancette and R. E. Benson, *J. Amer. Chem. Soc.*, 1965, **87**, 1941.
957 J. C. Barborak, T.-M. Su, P. von R. Schleyer, G. Boche and G. Schneider, *J. Amer. Chem. Soc.*, 1971, **93**, 279.
958 S. W. Staley and T. J. Henry, *J. Amer. Chem. Soc.*, 1969, **91**, 7787.

454   *References*

959   G. Boche and G. Schneider, *Tetrahedron Letters*, 1974, 2449.
960   J. E. Baldwin and A. H. Andrist, *J. Amer. Chem. Soc.*, 1971, **93**, 4055.
961   L. A. Paquette and M. J. Epstein, *J. Amer. Chem. Soc.*, 1971, **93**, 5936.
962   L. A. Paquette and M. J. Epstein, *J. Amer. Chem. Soc.*, 1973, **95**, 6717.
963   S. W. Staley, *Intra-Sci. Chem. Reports*, 1971, **5**, 149.
964   J. E. Baldwin, A. H. Andrist and R. K. Pinschmidt, *J. Amer. Chem. Soc.*, 1972, **94**, 5845.
965   R. Grigg, R. Hayes and A. Sweeney, *Chem. Comm.*, 1971, 1248.
966   F.-G. Klärner, *Angew. Chem. Internat. Edn*, 1972, **11**, 832.
967   F.-G. Klärner, *Tetrahedron Letters*, 1971, 3611.
968   M. Hendrick, unpublished work; quoted by M. B. Sohn, M. Jones and B. Fairless, *J. Amer. Chem. Soc.*, 1972, **94**, 4774.
969   M. Jones and L. T. Scott, *J. Amer. Chem. Soc.*, 1967, **89**, 150.
970   S. Masamune, P. M. Baker and K. Hojo, *Chem. Comm.*, 1969, 1203.
971   E. Vogel, W. Grimme and E. Dinné, *Tetrahedron Letters*, 1965, 391.
972   S. Winstein, G. Moshuk, R. Rieke and M. Ogliaruso, *J. Amer. Chem. Soc.*, 1973, **95**, 2624.
973   A. G. Anastassiou and E. Yakali, *J. Amer. Chem. Soc.*, 1971, **93**, 3803.
974   P. Warner and S. Winstein, *J. Amer. Chem. Soc.*, 1971, **93**, 1284.
975   T. Sasaki, K. Kanematsu and Y. Yukimoto, *Chem. Letters*, 1972, 1005.
976   T. Sasaki, K. Kanematsu and Y. Yukimoto, *J. C. S. Perkin I*, 1973, 375.
977   W. H. Okamura, T. I. Ito and P. M. Kellett, *Chem. Comm.*, 1971, 1317.
978   M. Ogliaruso and S. Winstein, *J. Amer. Chem. Soc.*, 1967, **89**, 5290.
979   M. A. Ogliaruso, *J. Amer. Chem. Soc.*, 1970, **92**, 7490.
980   R. Rieke, M. Ogliaruso, R. McClung and S. Winstein, *J. Amer. Chem. Soc.*, 1966, **88**, 4729.
981   F. J. Smentowski, R. M. Owens and B. D. Faubion, *J. Amer. Chem. Soc.*, 1968, **90**, 1537.
982   T. J. Katz and C. Talcott, *J. Amer. Chem. Soc.*, 1966, **88**, 4732.
983   R. M. Owens, unpublished work; quoted by F. J. Smentowski, R. M. Owens and B. D. Faubion, *J. Amer. Chem. Soc.*, 1968, **90**, 1537.
984   M. Ogliaruso, R. Rieke and S. Winstein, *J. Amer. Chem. Soc.*, 1966, **88**, 4731.
985   S. V. Ley and L. A. Paquette, *J. Amer. Chem. Soc.*, 1974, **96**, 6670.
986   L. B. Anderson, M. J. Broadhurst and L. A. Paquette, *J. Amer. Chem. Soc.*, 1973, **95**, 2198.
987   D. A. Brewer, J. C. Schug and M. A. Ogliaruso, *Tetrahedron*, 1975, **31**, 69.
988   T. I. Ito, F. C. Baldwin and W. H. Okamura, *Chem. Comm.*, 1971, 1440.
989   S. W. Staley and G. M. Cramer, *J. Amer. Chem. Soc.*, 1973, **95**, 5051.
990   P. Radlick and G. Alford, *J. Amer. Chem. Soc.*, 1969, **91**, 6529.
991   A. G. Anastassiou, V. Orfanos and J. H. Gebrian, *Tetrahedron Letters*, 1969, 4491.
992   G. Boche, H. Böhme and D. Martens, *Angew. Chem. Internat. Edn*, 1969, **8**, 594.
993   G. Boche and D. Martens, *Angew. Chem. Internat. Edn*, 1972, **11**, 724.
994   G. Boche, D. Martens and W. Danzer, *Angew. Chem. Internat. Edn*, 1969, **8**, 984.
995   G. Moshuk, G. Petrowski and S. Winstein, *J. Amer. Chem. Soc.*, 1968, **90**, 2179.
996   W. H. Okamura and T. W. Osborn, *J. Amer. Chem. Soc.*, 1970, **92**, 1061.
997   C. S. Baxter and P. J. Garratt, *J. Amer. Chem. Soc.*, 1970, **92**, 1062.
998   C. S. Baxter and P. J. Garratt, *Tetrahedron*, 1971, **27**, 3285.

999   J. Clardy, L. K. Read, M. J. Broadhurst and L. A. Paquette, *J. Amer. Chem. Soc.* 1972, **94**, 2904.

1000   L. A. Paquette, M. J. Broadhurst, L. K. Read and J. Clardy, *J. Amer. Chem. Soc.*, 1973, **95**, 4639.

1001   J. E. Baldwin and R. K. Pinschmidt, *Chem. Comm.*, 1971, 820.

1002   G. Boche, H. Weber and J. Benz, *Angew. Chem. Internat. Edn*, 1974, **13**, 207.

1003   L. A. Paquette and M. J. Broadhurst, *J. Amer. Chem. Soc.*, 1972, **94**, 632.

1004   L. A. Paquette, M. J. Broadhurst, C. Lee and J. Clardy, *J. Amer. Chem. Soc.*, 1972, **94**, 630.

1005   J. E. Baldwin and D. B. Bryan, *J. Amer. Chem.* 1974, **96**, 319.

1006   A. G. Anastassiou and R. C. Griffith, *J. Amer. Chem. Soc.*, 1971, **93**, 3083.

1007   T. Sasaki, K. Kanematsu and Y. Yukimoto, *Heterocycles*, 1974, **2**, 1.

1008   A. G. Anastassiou, S. S. Libsch and R. C. Griffith, *Tetrahedron Letters*, 1973, 3103.

1009   M. Brookhart, D. L. Harris and R. C. Dammann, *Chem. Comm.*, 1973, 187.

1010   E. J. Reardon and M. Brookhart, *J. Amer. Chem. Soc.*, 1973, **95**, 4311.

1011   G. Deganello, H. Maltz and J. Kozarich, *J. Organometallic Chem.*, 1973, **60**, 323.

1012   G. Deganello, *J. Organometallic Chem.*, 1973, **59**, 329.

1013   G. Scholes, C. R. Graham and M. Brookhart, *J. Amer. Chem. Soc.*, 1974, **96**, 5665.

1014   G. Deganello and L. Toniolo, *J. Organometallic Chem.*, 1974, **74**, 255.

1015   A. Eisenstadt, unpublished work; quoted by E. J. Reardon and M. Brookhart, *J. Amer. Chem. Soc.*, 1973, **95**, 4311.

1016   W. Grimme, *J. Amer. Chem. Soc.*, 1973, **95**, 2381.

1017   D. C. Sanders and H. Shechter, *J. Amer. Chem. Soc.*, 1973, **95**, 6858.

1018   M. T. Reetz, R. W. Hoffmann, W. Schäfer and A. Schweig, *Angew. Chem. Internat. Edn*, 1973, **12**, 81.

1019   L. A. Paquette and M. J. Broadhurst, *J. Org. Chem.*, 1973, **38**, 1893.

1020   E. Vedejs, M. F. Salomon and P. D. Weeks, *J. Amer. Chem. Soc.*, 1973, **95**, 6770.

1021   D. I. Schuster and C. W. Kim, *J. Amer. Chem. Soc.*, 1974, **96**, 7437.

1022   A. S. Kende and T. L. Bogard, *Tetrahedron Letters*, 1967, 3383.

1023   A. F. Diaz, J. Fulcher, M. Sakai and S. Winstein, *J. Amer. Chem. Soc.*, 1974, **96**, 1264.

1024   W. Kirmse and G. Voigt, *J. Amer. Chem. Soc.*, 1974, **96**, 7598.

1025   J. A. Berson, R. R. Boettcher and J. J. Vollmer, *J. Amer. Chem. Soc.*, 1971, **93**, 1540.

1026   R. E. Leone and P. von R. Schleyer, *Angew. Chem. Internat. Edn*, 1970, **9**, 860.

1027   L. A. Paquette, M. Oku, W. B. Farnham, G. A. Olah and G. Liang, *J. Org. Chem.*, 1975, **40**, 700.

1028   A. S. Kende, unpublished work; quoted by R. E. Leone and P. von R. Schleyer, *Angew. Chem. Internat. Edn*, 1970, **9**, 860.

1029   L. G. Cannell, *Tetrahedron Letters*, 1966, 5967.

1030   H. Tsuruta, K. Kurabayashi and T. Mukai, *Bull. Chem. Soc. Japan*, 1972, **45**, 2822.

1031   H. Tsuruta, T. Kumagai and T. Mukai, *Chem. Letters*, 1972, 981.

1032   M. Sakai, A. Diaz and S. Winstein, *J. Amer. Chem. Soc.*, 1970, **92**, 4452.

1033   J. B. Press and H. Shechter, *Tetrahedron Letters*, 1972, 2677.

1034   L. A. Paquette and M. J. Broadhurst, *J. Org. Chem.*, 1973, **38**, 1886.

1035   G. Schröder and H. Röttele, *Angew. Chem. Internat. Edn*, 1968, **7**, 635.

1036 H. Röttele, P. Nikoloff, J. F. M. Oth and G. Schröder, *Chem. Ber.*, 1969, 102, 3367.
1037 S. W. Staley and T. J. Henry, *J. Amer. Chem. Soc.*, 1971, 93, 1292.
1038 S. W. Staley, G. M. Cramer and A. W. Orvedal, *J. Amer. Chem. Soc.*, 1974, 96, 7433.
1039 R. Wheland and P. D. Bartlett, *J. Amer. Chem. Soc.*, 1973, 95, 4003.
1040 S. W. Staley, G. N. Cramer and W. G. Kingsley, *J. Amer. Chem. Soc.*, 1973, 95, 5052.
1041 F. A. Cotton and G. Deganello, *J. Organometallic Chem.*, 1972, 38, 147.
1042 F. A. Cotton, B. A. Frenz, G. Deganello and A. Shaver, *J. Organometallic Chem.*, 1973, 50, 227.
1043 F. A. Cotton, B. A. Frenz, J. M. Troup and G. Deganello, *J. Organometallic Chem.*, 1973, 59, 317.
1044 F. A. Cotton, B. A. Frenz and J. M. Troup, *J. Organometallic Chem.*, 1973, 61, 337.
1045 F. A. Cotton and J. M. Troup, *J. Amer. Chem. Soc.*, 1973, 95, 3798.
1046 F. A. Cotton, V. W. Day, B. A. Frenz, K. I. Hardcastle and J. M. Troup, *J. Amer. Chem. Soc.*, 1973, 95, 4522.
1047 A. G. Anastassiou, S. W. Eachus, R. L. Elliott and E. Yakali, *Chem. Comm.*, 1972, 531.
1048 A. G. Anastassiou, R. L. Elliott, H. W. Wright and J. Clardy, *J. Org. Chem.*, 1973, 38, 1959.
1049 A. G. Anastassiou, R. L. Elliott and A. Lichtenfeld, *Tetrahedron Letters*, 1972, 4569.
1050 A. G. Anastassiou and R. L. Elliott, *Chem. Comm.*, 1973, 601.
1051 S. Masamune, K. Hojo and S. Takada, *Chem. Comm.*, 1969, 1204.
1052 A. G. Anastassiou and J. H. Gebrian, *J. Amer. Chem. Soc.*, 1969, 91, 4011.
1053 A. G. Anastassiou and J. H. Gebrian, *Tetrahedron Letters*, 1970, 825.
1054 A. G. Anastassiou, S. W. Eachus, R. P. Cellura and J. H. Gebrian, *Chem. Comm.*, 1970, 1133.
1055 W. Grimme and K. Seel, *Angew. Chem. Internat. Edn*, 1973, 12, 507.
1056 J. M. Holovka, P. D. Gardner, C. B. Strow, M. L. Hill and T. V. Van Auken, *J. Amer. Chem. Soc.*, 1968, 90, 5041.
1057 A. G. Anastassiou and R. P. Cellura, *Chem. Comm.*, 1969, 1521.
1058 J. M. Holovka, R. R. Grabbe, P. D. Gardner, C. B. Strow, M. L. Hill and T. V. Van Auken, *Chem. Comm.*, 1969, 1522.
1059 S. Masamune, S. Takada and R. T. Seidner, *J. Amer. Chem. Soc.*, 1969, 91, 7769.
1060 A. G. Anastassiou and R. P. Cellura, *Chem. Comm.*, 1969, 903.
1061 A. G. Anastassiou and E. Reichmanis, *J. Org. Chem.*, 1973, 38, 2421.
1062 C. R. Ganellin and R. Pettit, *Chem. Ber.*, 1957, 90, 2951.
1063 C. R. Ganellin and R. Pettit, *J. Chem. Soc.*, 1958, 576.
1064 M. Ogawa, M. Sugishita, M. Takagi and T. Matsuda, *Tetrahedron*, 1975, 31, 299.
1065 J. Hambrecht, H. Straub and E. Müller, *Chem. Ber.*, 1974, 107, 2985.
1066 G. Strukul, P. Viglino, R. Ros and M. Graziani, *J. Organometallic Chem.*, 1974, 74, 307.
1067 R. Aumann and H. Averbeck, *J. Organometallic Chem.*, 1975, 85, C4.
1068 H. Maltz and G. Deganello, *J. Organometallic Chem.*, 1971, 27, 383.
1069 W. L. Mock and P. A. H. Isaac, *J. Amer. Chem. Soc.*, 1972, 94, 2749.
1070 A. G. Anastassiou and H. Yamamoto, *Chem. Comm.*, 1973, 840.
1071 A. G. Anastassiou and R. P. Cellura, *Chem. Comm.*, 1967, 762.

1072   A. G. Anastassiou and R. M. Lazarus, *Chem. Comm.*, 1970, 373.

1073   A. G. Anastassiou, A. E. Winston and E. Reichmanis, *Chem. Comm.*, 1973, 779.

1074   T. J. Katz, J. C. Carnahan, G. M. Clarke and N. Acton, *J. Amer. Chem. Soc.*, 1970, **92**, 734.

1075   L. A. Paquette, R. H. Meisinger and R. Gleiter, *J. Amer. Chem. Soc.*, 1973, **95**, 5414.

1076   U. Jacobson, unpublished work; quoted by L. A. Paquette, *Tetrahedron*, 1975, **31**, 2855.

1077   L. A. Paquette, J. R. Malpass, G. R. Krow and T. J. Barton, *J. Amer. Chem. Soc.*, 1969, **91**, 5296.

1078   G. R. Krow and J. Reilly, *Tetrahedron Letters*, 1973, 3075.

1079   G. R. Krow and J. Reilly, *J. Org. Chem.*, 1975, **40**, 136.

1080   C. D. Nenitzescu, M. Avram, I. I. Pogany, G. D. Mateescu and M. Farcasiu, *Acad. Rep. populare Romine, Studii Cercetari Chim.*, 1963, **11**, 7; *Chem. Abs.*, 1964, **60**, 6764h.

1081   H. H. Westberg and H. J. Dauben, *Tetrahedron Letters*, 1968, 5123.

1082   A. G. Akehurst and G. I. Fray, unpublished work.

1083   M. Avram, I. G. Dinulescu, F. Chiraleu and C. D. Nenitzescu, *Rev. Roumaine Chim.*, 1973, **18**, 863; *Chem. Abs.*, 1973, **79**, 65885t.

1084   S. Masamune, H. Cuts and M. G. Hogben, *Tetrahedron Letters*, 1966, 1017.

1085   W. G. Dauben and D. L. Whalen, *Tetrahedron Letters*, 1966, 3743.

1086   W. G. Dauben, M. G. Buzzolini, C. H. Schallhorn, D. L. Whalen and K. J. Palmer, *Tetrahedron Letters*, 1970, 787.

1087   R. C. Cookson, E. Crundwell, R. R. Hill and J. Hudec, *J. Chem. Soc.*, 1964, 3062.

1088   T. Sasaki, K. Kanematsu and A. Kondo, *J. Org. Chem.*, 1974, **39**, 2246.

1089   G. I. Fray, G. R. Geen, D. I. Davies, L. T. Parfitt and M. J. Parrott, *J. C. S. Perkin I*, 1974, 729.

1090   I. A. Akhtar, G. I. Fray and B. R. Kettlewell, unpublished work.

1091   I. A. Akhtar and G. I. Fray, unpublished work.

1092   D. N. Butler and R. A. Snow, *Canad. J. Chem.*, 1974, **52**, 447.

1093   C. M. Anderson, J. B. Bremner, H. H. Westberg and R. N. Warrener, *Tetrahedron Letters*, 1969, 1585.

1094   E. E. Nunn, W. S. Wilson and R. N. Warrener, *Tetrahedron Letters*, 1972, 175.

1095   J. A. Elix, W. S. Wilson and R. N. Warrener, *Tetrahedron Letters*, 1970, 1837.

1096   W. S. Wilson and R. N. Warrener, *Tetrahedron Letters*, 1970, 4787.

1097   S. Masamune and N. Darby, *Accounts Chem. Res.*, 1972, **5**, 272.

1098   W. von E. Doering and J. W. Rosenthal, *J. Amer. Chem. Soc.*, 1966, **88**, 2078.

1099   E. Vedejs, *Tetrahedron Letters*, 1970, 4963.

1100   L. A. Paquette and J. C. Stowell, *Tetrahedron Letters*, 1970, 2259.

1101   S. F. Nelsen and J. P. Gillespie, *Tetrahedron Letters*, 1969, 5059.

1102   C. M. Anderson, J. B. Bremner, I. W. McCay and R. N. Warrener, *Tetrahedron Letters*, 1968, 1255.

1103   W. G. Dauben and L. N. Reitman, *J. Org. Chem.*, 1975, **40**, 835.

1104   W. G. Dauben and L. N. Reitman, *J. Org. Chem.*, 1975, **40**, 841.

1105   C. M. Anderson, I. W. McCay and R. N. Warrener, *Tetrahedron Letters*, 1970, 2735.

1106   R. N. Warrener, J. A. Elix and W. S. Wilson, *Austral. J. Chem.*, 1973, **26**, 389.
1107   I. Fleming and E. Wildsmith, *Chem. Comm.*, 1970, 223.
1108   R. Pettit and J. Henery, *Org. Synth.*, 1970, **50**, 21.
1109   R. Pettit, *Pure Appl. Chem.*, 1969, **17**, 253.
1110   E. Vogel and K. Hasse, *Annalen*, 1958, **615**, 22.
1111   R. Criegee, W. Hörauf and W. D. Schellenberg, *Chem. Ber.*, 1953, **86**, 126.
1112   H. Olsen and J. P. Snyder, *J. Amer. Chem. Soc.*, 1974, **96**, 7839.
1113   G. Maier, *Chem. Ber.*, 1969, **102**, 3310.
1114   L. A. Paquette, *J. Amer. Chem. Soc.*, 1970, **92**, 5765.
1115   R. Askani, *Tetrahedron Letters*, 1970, 3349.
1116   G. Wilke, *Plenary Lecture, Internat. Conf. Organometallic Chem.*, Kyoto, 1977.
1117   H. Tanida, S. Teratake, Y. Hata and M. Watanabe, *Tetrahedron Letters*, 1969, 5341.
1118   G. Schröder and J. F. M. Oth, *Tetrahedron Letters*, 1966, 4083.
1119   G. Schröder, W. Martin and J. F. M. Oth, *Angew. Chem. Internat. Edn*, 1967, **6**, 870.
1120   J. F. M. Oth, D. M. Smith, U. Prange and G. Schröder, *Angew. Chem. Internat. Edn*, 1973, **12**, 327.
1121   M. R. Willcott, J. F. M. Oth, J. Thio, G. Plinke and G. Schröder, *Tetrahedron Letters*, 1971, 1579.
1122   G. Schröder, G. Plinke and J. F. M. Oth, *Angew. Chem. Internat. Edn*, 1972, **11**, 424.
1123   J. G. Concepción and G. Vincow, *J. Phys. Chem.*, 1975, **79**, 2037.
1124   G. Schröder, G. Plinke, D. M. Smith and J. F. M. Oth, *Angew. Chem. Internat. Edn*, 1973, **12**, 325.
1125   G. Schröder, G. Heil, H. Röttele and J. F. M. Oth, *Angew. Chem. Internat. Edn*, 1972, **11**, 426.
1126   G. Schröder, R. Neuberg and J. F. M. Oth, *Angew. Chem. Internat. Edn*, 1972, **11**, 51.
1127   G. I. Fray and R. G. Saxton, *Tetrahedron Letters*, 1973, 3579.
1128   R. R. Deshpande, A. Gilbert, G. I. Fray and R. G. Saxton, *Tetrahedron Letters*, 1971, 2163.
1129   J. F. M. Oth, H. Röttele and G. Schröder, *Tetrahedron Letters*, 1970, 61.
1130   A. H.-J. Wang, I. C. Paul and G. N. Schrauzer, *Chem. Comm.*, 1972, 736.
1131   G. Schröder, *Chem. Ber.*, 1964, **97**, 3140.
1132   G. Schröder, *Angew. Chem. Internat. Edn*, 1965, **4**, 695.
1133   J. Daub and V. Trautz, *Tetrahedron Letters*, 1970, 3265.
1134   J. Daub and U. Erhardt, *Tetrahedron*, 1972, **28**, 181.
1135   U. Erhardt and J. Daub, *Chem. Comm.*, 1974, 83.
1136   J. N. Labows, J. Meinwald, H. Röttele and G. Schröder, *J. Amer. Chem. Soc.*, 1967, **89**, 612.
1137   V. Z. Williams, P. von R. Schleyer, G. J. Gleicher and L. B. Rodewald, *J. Amer. Chem. Soc.*, 1966, **88**, 3862.
1138   P. von R. Schleyer, E. Osawa and M. G. B. Drew, *J. Amer. Chem. Soc.*, 1968, **90**, 5034.
1139   R. Breslow, R. W. Johnson and A. Krebs, *Tetrahedron Letters*, 1975, 3443.
1140   R. C. Bingham, M. J. S. Dewer and D. H. Lo, *J. Amer. Chem. Soc.*, 1975, **97**, 1294.
1141   Z. Meić and M. Randić, *Croat. Chem. Acta*, 1968, **40**, 43.
1142   J. Aihara, *Bull. Chem. Soc. Japan*, 1975, **48**, 517.
1143   K. Kovačević, M. Eckert-Maksić and Z. B. Maksić, *Croat. Chem. Acta*, 1974, **46**, 249.

1144 W. C. Herndon, *J. Amer. Chem. Soc.*, 1976, **98**, 887.
1145 Yu. B. Vysotskii, *Zhur. strukt. Khim.*, 1974, **15**, 566; *Chem. Abs.*, 1974, **81**, 62988j.
1146 R. C. Haddon, *Tetrahedron Letters*, 1975, 863.
1147 G. R. Stevenson, R. Concepción and I. Ocasio, *J. Phys. Chem.*, 1976, **80**, 861.
1148 J. J. Mooij and E. de Boer, *Mol. Phys.*, 1976, **32**, 113.
1149 V. Dvořák and J. Michl, *J. Amer. Chem. Soc.*, 1976, **98**, 1080.
1150 T. Watanabe, T. Shida and S. Iwata, *Chem. Phys.*, 1976, **13**, 65.
1151 D. A. Kleier, D. A. Dixon and W. N. Lipscomb, *Theor. Chim. Acta*, 1975, **40**, 33.
1152 C. Eaborn, B. C. Pant, E. R. A. Peeling and S. C. Taylor, *J. Chem. Soc.* (C), 1969, 2823.
1153 J. E. Bloor, A. C. R. Brown and D. G. L. James, *J. Phys. Chem.*, 1966, **70**, 2191.
1154 S. W. Staley and G. E. Linkowski, *J. Amer. Chem. Soc.*, 1976, **98**, 5010.
1155 A. G. Anastassiou and E. Reichmanis, *J. Amer. Chem. Soc.*, 1976, **98**, 8267.
1156 A. G. Anastassiou, *Pure Appl. Chem.*, 1974, **44**, 691 (footnote 25).
1157 L. A. Paquette, U. Jacobsson and S. V. Ley, *J. Amer. Chem. Soc.*, 1976, **98**, 152.
1158 P. K. Freeman, D. M. Balls and D. J. Brown, *J. Org. Chem.*, 1968, **33**, 2211.
1159 T. W. Wickersham, J. P. Li and E. J. Warawa, *Synthesis*, 1975, 399.
1160 W. G. Dauben, G. T. Rivers, R. J. Twieg and W. T. Zimmerman, *J. Org. Chem.*, 1976, **41**, 887.
1161 A. R. L. Bursics, E. Bursics-Szekeres, M. Murray and F. G. A. Stone, *J. Fluorine Chem.*, 1976, **7**, 619.
1162 V. P. Arya and S. J. Shenoy, *Indian J. Chem.*, 1976, **14B**, 883.
1163 T. S. Cantrell, *J. Org. Chem.*, 1974, **39**, 853.
1164 G. R. Stevenson, I. Ocasio and A. Bonilla, *J. Amer. Chem., Soc.*, 1976, **98**, 5469.
1165 G. R. Stevenson and I. Ocasio, *Tetrahedron Letters*, 1976, 427.
1166 G. R. Stevenson and I. Ocasio, *J. Amer. Chem. Soc.*, 1976, **98**, 890.
1167 D. A. Edwards and R. Richards, *J. Organometallic Chem.*, 1975, **86**, 407.
1168 S. R. Ely, T. E. Hopkins and C. W. DeKock, *J. Amer. Chem. Soc.*, 1976, **98**, 1624.
1169 S. R. Ely, *Diss. Abs.*, 1976, **36B**, 5058.
1170 A. Greco, S. Cesca and G. Bertolini, *J. Organometallic Chem.*, 1976, **113**, 321.
1171 L. Hocks, R. Hubin and J. Goffart, *J. Organometallic Chem.*, 1976, **104**, 199.
1172 D. J. Brauer and C. Krüger, *Inorg. Chem.*, 1975, **14**, 3053.
1173 J. Müller, W. Holzinger and F. H. Köhler, *Chem. Ber.*, 1976, **109**, 1222.
1174 P. L. Timms and T. W. Turney, *J. C. S. Dalton*, 1976, 2021.
1175 D. J. Brauer and C. Krüger, *Inorg. Chem.*, 1976, **15**, 2511.
1176 C. G. Kreiter, M. Lang and H. Strack, *Chem. Ber.*, 1975, **108**, 1502.
1177 M. L. H. Green and J. Knight, *J. C. S. Dalton*, 1976, 213.
1178 R. Goddard, S. A. R. Knox, F. G. A. Stone, M. J. Winter and P. Woodward, *Chem. Comm.*, 1976, 559.
1179 P. L. Pauson and J. A. Segal, *J. C. S. Dalton*, 1975, 2387.
1180 F. A. Cotton and D. L. Hunter, *J. Amer. Chem. Soc.*, 1976, **98**, 1413.
1181 A. J. Campbell, C. E. Cottrell, C. A. Fyfe and K. R. Jeffrey, *Inorg. Chem.*, 1976, **15**, 1321.
1182 J. C. Green, P. Powell and J. Van Tilborg, *J. C. S. Dalton*, 1976, 1974.
1183 R. Aumann and J. Knecht, *Chem. Ber.*, 1976, **109**, 174.
1184 R. Aumann, *Chem. Ber.*, 1976, **109**, 168.

460   *References*

1185   R. Bau, B.C.-K. Chou, S. A. R. Knox, V. Riera and F. G. A. Stone, *J. Organometallic Chem.*, 1974, **82**, C43.
1186   A. J. P. Domingos, B. F. G. Johnson and J. Lewis, *J. C. S. Dalton*, 1975, 2288.
1187   P. J. Harris, J. A. K. Howard, S. A. R. Knox, R. P. Phillips, F. G. A. Stone and P. Woodward, *J. C. S. Dalton*, 1976, 377.
1188   P. V. Rinze, *J. Organometallic Chem.*, 1975, **90**, 343.
1189   D. J. Brauer and C. Krüger, *J. Organometallic Chem.*, 1976, **122**, 265.
1190   F. A. L. Anet and P. M. Hendrichs, *Tetrahedron Letters*, 1970, 829.
1191   G. L. Grunewald, J. M. Grindel, P. N. Patil and K. N. Salman, *J. Medicin. Chem.*, 1976, **19**, 10.
1192   G. L. Grunewald and J. M. Grindel, *Org. Photochem. Synth.*, 1976, **2**, 20.
1193   S. W. Staley, G. E. Linkowski and A. S. Heyn, *Tetrahedron*, 1975, **31**, 1131.
1194   L. A. Paquette, C. D. Wright, S. G. Traynor, D. L. Taggart and G. D. Ewing, *Tetrahedron*, 1976, **32**, 1885.
1195   J. D. Edwards, J. A. K. Howard, S. A. R. Knox, V. Riera, F. G. A. Stone and P. Woodward, *J. C. S. Dalton*, 1976, 75.
1196   B. F. G. Johnson, J. Lewis and J. W. Quail, *J. C. S. Dalton*, 1975, 1252.
1197   P. A. Chaloner and A. B. Holmes, *J. C. S. Perkin I*, 1976, 1838.
1198   B. F. G. Johnson, J. Lewis and D. Wege, *J. C. S. Dalton*, 1976, 1874.
1199   L. A. Paquette and R. S. Beckley, *Org. Photochem. Synth.*, 1976, **2**, 45.
1200   A. J. H. Klunder and B. Zwanenburg, *Tetrahedron*, 1975, **31**, 1419.
1201   K. Saito and T. Mukai, *Bull. Chem. Soc. Japan*, 1975, **48**, 2334.
1202   J. H. Hammons, C. T. Kresge and L. A. Paquette, *J. Amer. Chem. Soc.*, 1976, **98**, 8172.
1203   L. A. Paquette and J. M. Photis, *J. Amer. Chem. Soc.*, 1976, **98**, 4936.
1204   F. Wagner and H. Meier, *Tetrahedron*, 1974, **30**, 773.
1205   G. R. Geen and G. I. Fray, unpublished work.
1206   N. L. Bauld, F. R. Farr and C. E. Hudson, *J. Amer. Chem. Soc.*, 1974, **96**, 5634.
1207   L. A. Paquette, J. M. Gardlik and J. M. Photis, *J. Amer. Chem. Soc.*, 1976, **98**, 7096.
1208   R. N. Warrener, E. E. Nunn and M. N. Paddon-Row, *Tetrahedron Letters*, 1976, 2355.
1209   R. J. Atkins and G. I. Fray, unpublished work.
1210   E. H. White, R. L. Stern, T. J. Lobl, S. H. Smallcombe, H. Maskill and E. W. Friend, *J. Amer. Chem. Soc.*, 1976, **98**, 3247.
1211   G. A. Olah, J. S. Staral and L. A. Paquette, *J. Amer. Chem. Soc.*, 1976, **98**, 1267.
1212   L. F. Pelosi and W. T. Miller, *J. Amer. Chem. Soc.*, 1976, **98**, 4311.
1213   R. Askani and M. Wieduwilt, *Chem. Ber.*, 1976, **109**, 1887.
1214   G. Maier, G. Fritschi and B. Hoppe, *Tetrahedron Letters*, 1971, 1463.
1215   M. J. Gerace, D. M. Lemal and H. Ertl, *J. Amer. Chem. Soc.*, 1975, **97**, 5584.
1216   R. Riemschneider, *Z. Naturforsch.*, 1961, **16b**, 759.
1217   E. Wenkert, E. W. Hagaman, L. A. Paquette, R. E. Wingard and R. K. Russell, *Chem. Comm.*, 1973, 135.
1218   A. K. Cheng, F. A. L. Anet, J. Mioduski and J. Meinwald, *J. Amer. Chem. Soc.*, 1974, **96**, 2887.
1219   Y. C. Toong, W. T. Borden and A. Gold, *Tetrahedron Letters*, 1975, 1549.
1220   Z. V. Todres and S. P. Avagyan, *Zhur. Vsesoyuz. Khim. obshch. im. D. I. Mendeleeva*, 1975, **20**, 717; *Chem. Abs.*, 1976, **84**, 73314z.
1221   Z. V. Todres, S. P. Avagyan and D. N. Kursanov, *J. Organometallic Chem.*, 1975, **97**, 139.

1222 G. Kaupp and K. Rösch, *Angew. Chem. Internat. Edn*, 1976, **15**, 163.
1223 Z. V. Todres, S. P. Avagyan and D. N. Kursanov, *Zhur. org. Khim.*, 1975, **11**, 2457; *Chem. Abs.*, 1976, **84**, 73348p.
1224 A. Westerhof and H. J. de Liefde Meijer, *J. Organometallic Chem.*, 1976, **116**, 319.
1225 D. W. Clack and K. D. Warren, *J. Organometallic Chem.*, 1976, **122**, C28.
1226 H.-D. Amberger, R. D. Fischer and B. Kanellakopulos, *Theor. Chim. Acta*, 1975, **37**, 105.
1227 N. Edelstein, A. Streitwieser, D. G. Morrell and R. Walker, *Inorg. Chem.*, 1976, **15**, 1397.
1228 L. K. Templeton, D. H. Templeton and R. Walker, *Inorg. Chem.*, 1976, **15**, 3000.
1229 H. R. van der Wal, F. Overzet, H. O. van Oven, J. L. Boer, H. J. de Liefde Meijer and F. Jellinek, *J. Organometallic Chem.*, 1975, **92**, 329.
1230 R. R. Schrock, L. J. Guggenberger and A. D. English, *J. Amer. Chem. Soc.*, 1976, **98**, 903.
1231 F. A. Cotton and J. R. Kolb, *J. Organometallic Chem.*, 1976, **107**, 113.
1232 M. A. Bennett, T. W. Matheson, G. B. Robertson, A. K. Smith and P. A. Tucker, *J. Organometallic Chem.*, 1976, **121**, C18.
1233 A. K. Smith and P. M. Maitlis, *J. C. S. Dalton*, 1976, 1773.
1234 J. Müller, H.-O. Stühler, G. Huttner and K. Scherzer, *Chem. Ber.*, 1976, **109**, 1211.
1235 M. Green, J. A. K. Howard, J. L. Spencer and F. G. A. Stone, *Chem. Comm.*, 1975, 449.
1236 H. U. Lee and R. N. Zare, *J. Chem. Phys.*, 1976, **64**, 431.
1237 M. Vliek, C. J. Groenenboom, H. J. de Liefde Meijer and F. Jellinek, *J. Organometallic Chem.*, 1975, **97**, 67.
1238 E. G. Hoffmann, R. Kallweit, G. Schroth, K. Seevogel, W. Stempfle and G. Wilke, *J. Organometallic Chem.*, 1975, **97**, 183.
1239 F. Hohmann, H. T. Dieck, K. D. Franz and K. A. O. Starzewski, *J. Organometallic Chem.*, 1973, **55**, 321.
1240 G. Deganello, P. Uguagliati, L. Calligaro, P. L. Sandrini and F. Zingales, *Inorg. Chim. Acta*, 1975, **13**, 247.
1241 M. Cooke, J. A. K. Howard, C. R. Russ, F. G. A. Stone and P. Woodward, *J. C. S. Dalton*, 1976, 70.
1242 R. Aumann, *J. Organometallic Chem.*, 1974, **78**, C31.
1243 R. Aumann, *Angew. Chem. Internat. Edn*, 1973, **12**, 574.
1244 A. D. Charles, P. Diversi, B. F. G. Johnson and J. Lewis, *J. Organometallic Chem.*, 1976, **116**, C25.
1245 B. F. G. Johnson, J. Lewis, D. J. Thompson and B. Heil, *J. C. S. Dalton*, 1975, 567.
1246 N. El Murr, M. Riveccié and E. Laviron, *Tetrahedron Letters*, 1976, 3339.
1247 G. R. Langford, M. Akhtar, P. D. Ellis, A. G. MacDiarmid and J. D. Odom, *Inorg. Chem.*, 1975, **14**, 2937.
1248 H. Behrens and M. Moll, *Z. anorg. Chem.*, 1975, **416**, 193.
1249 D. L. Reger and A. Gabrielli, *J. Amer. Chem. Soc.*, 1975, **97**, 4421.
1250 H. A. Bockmeulen, R. G. Holloway, A. W. Parkins and B. R. Penfold, *Chem. Comm.*, 1976, 298.
1251 G. Zassinovich, G. Mestroni and A. Camus, *J. Organometallic Chem.*, 1975, **91**, 379.
1252 N. Heap, G. R. Green and G. H. Whitham, *J. Chem. Soc. (C)*, 1969, 160.
1253 P. Radlick and S. Winstein, *J. Amer. Chem. Soc.*, 1964, **86**, 1866.
1254 W. L. Mock, *J. Amer. Chem. Soc.*, 1970, **92**, 3807.

462   *References*

1255   W. L. Mock and J. H. McCausland, *J. Org. Chem.*, 1976, **41**, 242.
1256   A. Salzer, *J. Organometallic Chem.*, 1976, **117**, 245.
1257   J. Evans, B. F. G. Johnson, J. Lewis and R. Watt, *J. C. S. Dalton*, 1974, 2368.
1258   J. T. Groves and C. A. Bernhardt, *J. Org. Chem.*, 1975, **40**, 2806.
1259   M. S. Brookhart, G. W. Koszalka, G. O. Nelson, G. Scholes and R. A. Watson, *J. Amer. Chem. Soc.*, 1976, **98**, 8155.
1260   G. Petrowski, unpublished work; quoted by M. Sakai, D. L. Harris and S. Winstein, *Chem. Comm.*, 1972, 861.
1261   L. A. Paquette, S. V. Ley, S. G. Traynor, J. T. Martin and J. M. Geckle, *J. Amer. Chem. Soc.*, 1976, **98**, 8162.
1262   G. Boche, A. Bieberbach and H. Weber, *Angew. Chem. Internat. Edn*, 1976, **14**, 562.
1263   A. G. Anastassiou and E. Reichmanis, *Chem. Comm.*, 1976, 313.
1264   A. Salzer, *J. Organometallic Chem.*, 1976, **107**, 79.
1265   J. Takats, *J. Organometallic Chem.*, 1975, **90**, 211.
1266   M. Airoldi, G. Deganello and J. Kozarich, *Inorg. Chim. Acta*. 1976, **20**, L5.
1267   P. L. Sandrini, R. A. Michelin, G. Deganello and U. Belluco, *Inorg. Chim. Acta*, 1976, **17**, 65.
1268   E. Vedejs and R. A. Shepherd, *J. Org. Chem.*, 1976, **41**, 742.
1269   D. I. Schuster and C. W. Kim, *J. Org. Chem.*, 1975, **40**, 505.
1270   P. Bischof, R. Gleiter and E. Heilbronner, *Helv. Chim. Acta*, 1970, **53**, 1425.
1271   W. Schäfer, H. Schmidt, A. Schweig, R. W. Hoffmann and H. Kurz, *Tetrahedron Letters*, 1974, 1953.
1272   A. Diaz, J. Fulcher, R. Cetina, M. Rubio and R. Reynoso, *J. Org. Chem.*, 1975, **40**, 2459.
1273   A. Diaz and J. Fulcher, *J. Amer. Chem. Soc.*, 1976, **98**, 798.
1274   M. Sakai, D. L. Harris and S. Winstein, *Chem. Comm.*, 1972, 861.
1275   T. V. R. Babu and H. Shechter, *J. Amer. Chem. Soc.*, 1976, **98**, 8261.
1276   S. Masamune, C. U. Kim, K. E. Wilson, G. O. Spessard, P. E. Georghiou and G. S. Bates, *J. Amer. Chem. Soc.*, 1975, **97**, 3512.
1277   J. B. Press and H. Shechter, *J. Org. Chem.*, 1975, **40**, 2446.
1278   S. W. Staley and A. S. Heyn, *J. Amer. Chem. Soc.*, 1975, **97**, 3852.
1279   F. A. Cotton and D. L. Hunter, *J. Amer. Chem. Soc.*, 1975, **97**, 5739.
1280   G. Deganello, P. L. Sandrini, R. A. Michelin and L. Toniolo, *J. Organometallic Chem.*, 1975, **90**, C31.
1281   A. G. Anastassiou and S. J. Girgenti, *Angew. Chem. Internat. Edn*, 1975, **14**, 814.
1282   B. Prelesnik and W. Nowacki, *Z. Krist.*, 1976, **143**, 252.
1283   D. R. Battiste and J. G. Traynham, *J. Org. Chem.*, 1975, **40**, 1239.
1284   G. Mehta and P. N. Pandey, *J. Org. Chem.*, 1975, **40**, 3631.
1285   T. Sasaki, K. Kanematsu and A. Kondo, *J. Org. Chem.*, 1975, **40**, 1642.
1286   G. Mehta, P. K. Dutta and P. N. Pandey, *Tetrahedron Letters*, 1975, 445.
1287   T. Sasaki, K. Kanematsu and A. Kondo, *Tetrahedron*, 1975, **31**, 2215.
1288   T. Sasaki, K. Kanematsu, A. Kondo and Y. Nishitani, *J. Org. Chem.*, 1974, **39**, 3569.
1289   G. Mehta and P. N. Pandey, *Tetrahedron Letters*, 1975, 3567.
1290   I. A. Akhtar, R. J. Atkins, G. I. Fray and G. R. Geen, unpublished work.
1291   R. S. Liu, *Tetrahedron Letters*, 1969, 1409.
1292   E. Osawa, H. Henke and G. Schröder, *Tetrahedron Letters*, 1976, 847.
1293   J. Daub, U. Erhardt and V. Trautz, *Chem. Ber.*, 1976, **109**, 2197.

Author Index

Numbers indicate references, not pages

Abel, E. W. 339, 835
Abraitys, V. Y. 625
Ackermann, M. N. 439, 440, 467
Acton, N. 1074
Adams, N. G. 60
Ahlberg, P. 864
Aihara, J. 1142
Airoldi, M. 1266
Akehurst, A. G. 1082
Akhtar, I. A. 879, 880, 1090, 1091, 1290
Akhtar, M. 1247
Akiyoshi, S. 278, 308, 309
Alford, G. 949, 990
Al-Joboury, M. I. 72
Allegra, G. 444, 446
Allendoerfer, R. D. 65, 81
Allinger, N. L. 95, 96, 126, 127, 130, 921
Allred, E. L. 55, 874, 920
Alsop, J. E. 594
Amberger, H.-D. 1226
Anand, S. K. 406, 437
Anastassiou, A. G. 47, 305, 317, 318, 328, 746, 748, 953, 954, 973, 991, 1006, 1008, 1047, 1048, 1049, 1050, 1052, 1053, 1054, 1057, 1060, 1061, 1070, 1071, 1072, 1073, 1155, 1156, 1263, 1281
Anderson, A. G. 353
Anderson, C. M. 1093, 1102, 1105
Anderson, L. B. 75, 77, 78, 577, 986
Anderson, S. E. 776
Andrist, A. H. 960, 964
Anet, F. A. L. 105, 106, 108, 123, 427, 586, 592, 599, 608, 857, 1190, 1218
Anet, R. 925
Antkowiak, T. A. 52, 378
Arai, S. 189
Arens, J. F. 642
Armstrong, V. S. 887
Arya, V. P. 1162
Ashley-Smith, J. 477
Askani, R. 40, 706, 1115, 1213
Ast, T. 109
Atkins, R. J. 1209, 1290
Atkinson, J. G. 636
Atwater, M. A. M. 568

Aumann, R. 719, 722, 886, 1067, 1183, 1184, 1242, 1243
Austin, W. K. 376
Avagyan, S. P. 732, 733, 1220, 1221, 1223
Avdeef, A. 782
Averbeck, H. 1067
Avitabile, G. 656
Avram, M. 35, 228, 340, 341, 352, 374, 917, 1080, 1083
Ayer, D. E. 636
Azatyan, V. D. 203, 274, 738, 739, 741, 754

Babu, T. V. R. 1275
Badea, F. 175
Baenziger, N. C. 390, 392, 846
Bailey, A. S. 338
Bailey, R. T. 462
Bailey, W. J. 6
Baird, M. S. 870
Baird, N. C. 154
Bajo, G. 664
Bak, B. 141
Bak, D. A. 379
Baker, P. M. 970
Baker, W. 4
Baldwin, F. C. 988
Baldwin, J. E. 960, 964, 1001, 1005
Balls, D. M. 1158
Bandurco, V. T. 930
Bangert, K. F. 311
Barber, L. L. 937
Barborak, J. C. 957
Bartlett, P. D. 1039
Barton, D. H. R. 924
Barton, T. J. 321, 366, 367, 796, 1077
Bashkirova, S. A. 322
Bassi, I. W. 444, 810
Bast, K. 332
Bastiansen, O. 118, 120
Bates, G. S. 1276
Bathelt, H. 876
Batich, C. 71
Battiste, D. R. 1283
Bau, R. 489, 1185
Bauld, N. L. 1206
Baxter, C. S. 997, 998

Beard, J. 875
Beck, B. R. 55, 920
Becker, Y. 591
Beckey, H. D. 110
Beckley, R. S. 542, 543, 561, 1199
Behrens, H. 1248
Bejenke, V. 900
Bellama, J. M. 740, 742, 911
Belluco, U. 825, 1267
Bennett, M. A. 339, 435, 436, 507, 1232
Bennett, M. J. 490, 667
Benson, R. E. 39, 306, 956
Benson, S. W. 91
Benz, J. 1002
Bernhardt, C. A. 1258
Berson, J. A. 226, 1025
Bertolini, G. 1170
Bespalov, V. Ya. 724
Beverwijk, C. D. M. 400
Beynon, J. H. 109
Bianchi, G. 331, 334, 335, 336
Bieberbach, A. 1262
Bienick, D. 157
Bigam, G. 946
Bingham, R. C. 1140
Binkley, R. W. 44
Bird, C. W. 19
Birnberg, G. H. 569, 715, 716, 717
Bischof, P. 71, 1270
Bittler, K. 295
Blanc, P. Y. 237
Blanchard, K. R. 144
Bloch, R. 557
Block, A. M. 576, 616
Blomquist, A. T. 918
Bloor, J. E. 1153
Boccalon, G. 461
Boche, G. 135, 136, 225, 229, 231, 232, 856,
    908, 957, 959, 992, 993, 994, 1002, 1262
Bock, L. A. 593, 608
Bockmeulen, H. A. 1250
Bodor, N. 111
Boekelheide, V. 201, 311
Boer, J. L. 1229
Boettcher, R. R. 1025
Bogard, T. L. 1022
Bogdanović, B. 518
Boggs, J. E. 120
Boggs, R. A. 588, 593
Böhme, H. 992
Boikess, R. S. 863, 871
Bollinger, J. M. 220
Bonilla, A. 1164
Bonnell, D. W. 89
Bonse, G. 695, 707, 709
Borden, G. W. 861
Borden, W. T. 1219
Bordner, J. 116
Bos, H. J. T. 642

Bourn, A. J. R. 123
Bowie, J. H. 558
Braams, J. F. H. 642
Bradley, C. H. 217
Bradshaw, R. W. 293
Brandt, J. 845
Bratby, D. M. 349, 882
Bratton, W. K. 487
Brauer, D. J. 412, 413, 1172, 1175, 1189
Brauman, J. I. 218, 737
Bray, E. H. 679
Breil, H. 416
Bremner, J. B. 1093, 1102
Brenner, K. S. 512
Breslow, R. 607, 611, 619, 1139
Brewer, D. A. 987
Briat, B. 98
Brice, M. D. 505
Brintzinger, H. 239, 240
Broadhurst, M. J. 578, 588, 593, 826, 848,
    986, 999, 1000, 1003, 1004, 1019, 1034
Brookes, A. 494
Brookhart, M. S. 568, 584, 749, 817, 818,
    889, 938, 1009, 1010, 1013, 1015, 1259
Brookman, D. J. 62
Brown, A. C. R. 1153
Brown, D. 772
Brown, D. J. 1158
Brown, J. M. 747
Brown, P. 337
Brown, T. H. 575
Brown, T. L. 531
Bruce, M. I. 464, 479, 486, 495, 496
Brune, H.-A. 652, 705
Bryan, D. B. 1005
Bryce-Smith, D. 38, 326, 327, 536, 613
Buchanan, G. W. 564
Büchi, G. 636, 675, 936
Buck, H. M. 684, 940
Bunnenberg, E. 98
Burg, M. 549, 553
Burger, K. 448
Burgess, E. M. 368, 936
Burkoth, T. L. 947
Burns, W. 438
Bursics, A. R. L. 1161
Bursics-Szekeres, E. 1161
Busetto, L. 841
Butler, D. N. 1092
Butler, P. E. 241
Buzzolini, M. G. 1086
Byrd, D. 624

Cairns, T. L. 39
Calas, R. 797
Calligaro, L. 1240
Campbell, A. J. 455, 460, 1181
Campbell, H. C. 532, 533
Campbell, T. C. 602

Camus, A. 513, 1251
Cannell, L. G. 1029
Cantrell, T. S. 289, 290, 291, 377, 725, 755, 756, 1163
Caplier, I. 679
Carbonaro, A. 441, 445, 448, 806, 807, 808, 809
Cardenas, C. G. 10
Carlson, R. M. 207
Carnahan, J. C. 1074
Carrington, A. 247, 250, 570, 572
Carroll, S. R. 73, 89
Case, R. 892
Cass, R. C. 67
Castellucci, N. T. 319
Ceccon, A. 461
Cellura, R. P. 305, 318, 1054, 1057, 1060, 1071
Cesca, S. 1170
Cetina, R. 1272
Chadwick, J. C. 881
Chaloner, P. A. 1197
Chalvet, O. 85
Chao, B. Y.-H. 47, 328
Chapman, O. L. 861, 937
Charles, A. D. 1244
Chatt, J. 506
Chebotarev, V. P. 646
Cheer, C. J. 314
Chen, K. N. 723
Cheng, A. K. 1218
Chernyshev, E. A. 322
Chierico, A. 447
Childs, R. F. 758
Chin, C. G. 312
Chini, P. 639
Chiraleu, F. 1083
Chou, B. C.-K. 489, 1185
Chow, L. W. 95
Christensen, P. A. 852
Christl, M. 329, 330, 332, 333
Chukhadzhyan, G. A. 643, 644, 645
Chung, D. 126
Chung, L. L. 82
Churchill, M. R. 440
Ciganek, E. 304, 305
Clack, D. W. 1225
Clardy, J. 578, 633, 634, 717, 826, 999, 1000, 1004, 1048
Clarke, G. M. 1074
Cocevar, C. 513
Coffield, T. H. 438
Coleman, H. J. 60
Collins, R. 894
Colombo, A. 444, 446
Colón, M. 576, 616
Compton, R. N. 114, 115
Concepción, J. G. 254, 537, 571, 574, 576, 616, 662, 1123

Concepción, R. 1147
Conrow, K. 379, 735
Cook, B. W. 391
Cook, D. 864
Cooke, A. G. 918
Cooke, M. 479, 486, 495, 496, 497, 551, 811, 836, 1241
Cooks, R. G. 109
Cookson, R. C. 337, 342, 348, 362, 683, 1087
Cooper, M. A. 105
Cope, A. C. 3, 5, 6, 25, 63, 199, 200, 206, 235, 260, 277, 285, 286, 532, 533, 534, 535, 538, 544, 549, 553, 559, 597, 598, 617, 865, 873, 941
Copenhafer, R. A. 80, 573, 672
Cortez, H. V. 556
Cotton, F. A. 431, 432, 459, 463, 472, 473, 480, 485, 487, 488, 490, 648, 666, 667, 668, 669, 670, 752, 753, 895, 896, 1041, 1042, 1043, 1044, 1045, 1046, 1180, 1231, 1279
Cottrell, C. E. 491, 1181
Coulson, C. A. 131, 256
Cox, J. D. 88
Cox, R. H. 376
Craig, L. E. 24, 261
Cramer, G. M. 989, 1038, 1040
Crews, P. 875
Criegee, R. 652, 676, 677, 689, 704, 705, 706, 1111
Cruickshank, F. R. 91
Crundwell, E. 348, 1087
Cserep, G. 193
Cusmano, F. 837
Cuts, H. 1084
Cyvin, S. J. 143

Dahmen, A. 136
Dall'Asta, G. 441, 445, 806, 807, 808
Dammann, R. C. 1009
D'Angelo, P. F. 53, 57
Danzer, W. 994
Darack, F. 930
Darby, N. 323, 1097
Datta, P. 863
Daub, J. 1133, 1134, 1135, 1293
Dauben, H. J. 84, 219, 1081
Dauben, W. G. 1085, 1086, 1103, 1104, 1160
Davidson, J. L. 523
Davies, D. I. 1089
Davis, E. R. 584, 817, 818
Davis, F. J. 115
Davis, R. 594
Davis, R. E. 814
Davison, A. 463, 472, 474, 480, 487, 815, 819
Davison, J. B. 740, 742, 911
Day, V. W. 1046
DeBoer, B. G. 488
de Boer, E. 1148
de Boer, J. L. 792

Dedier, J. 797
Deganello, G. 752, 753, 1011, 1012, 1014, 1041, 1042, 1043, 1068, 1240, 1266, 1267, 1280
Degens, H. M. L. 386
Dehmlow, E. V. 300, 302, 303
DeKock, C. W. 1168
de Liefde Meijer, H. J. 789, 791, 792, 799, 803, 1224, 1229, 1237
Dellaca, R. J. 505
de Mayo, P. 647
Dempf, D. 779, 783
Dempster, D. N. 186
Deshpande, R. R. 1128
Dessau, R. M. 211
Dessy, R. E. 823
Devon, T. 814
Dewar, M. J. S. 111, 124, 150, 1140
Dewar, R. B. K. 813
Diaz, A. 757, 864, 1023, 1032, 1272, 1273
Dickens, B. 457, 458, 468
Dieck, H. T. 1239
Diehl, P. 237
Dierks, H. 414, 415
Dietrich, H. 408, 414, 415
Dines, M. B. 389
Dinné, E. 971
Dinulescu, I. G. 35, 228, 374, 1083
Dirlam, J. P. 864
Diversi, P. 1244
Dixon, D. A. 1151
Dixon, W. T. 131, 132
Djerassi, C. 98
Dobbelaere, J. R. 940
Dodds, T. A. 814
Doering, W. von E. 69, 146, 859, 1098
Domingos, A. J. P. 492, 1186
Doyle, J. R. 390, 392, 846, 847
Draggett, P. T. 836
Dreiding, A. S. 172
Dremina, Z. G. 726
Drew, M. G. B. 1138
Dunathan, H. C. 654
Dunogues, J. 797
Dürr, H. 316
Dutta, P. K. 1286
Duyckaerts, G. 409, 771, 772, 790
Dvořák, V. 1149

Eaborn, C. 1152
Eachus, S. W. 1047, 1054
Eberius, W. 652
Eberson, L. 209
Eccleston, B. H. 60
Echegoyen, L. 537, 539
Eckert-Maksić, M. 147, 1143
Edelstein, N. 773, 775, 779, 780, 785, 1227
Edwards, D. A. 1167
Edwards, J. D. 493, 1195

Edwards, W. T. 473, 895
Eglinton, G. 8, 9
Ehntholt, D. 720, 827
Eichler, S. 15, 18, 388, 469
Eisenbach, W. 517
Eisenstadt, A. 591, 1015
Eiss, R. 485
Eland, J. H. D. 112
Elbakyan, T. S. 643, 644, 645
Elder, R. C. 369, 370
Elian, M. 228
Elix, J. A. 610, 612, 613, 620, 621, 622, 627, 628, 910, 1095, 1106
Elleman, D. D. 105
Ellinger, L. P. 190
Elliott, R. L. 1047, 1048, 1049, 1050
Ellis, P. D. 1247
Ellwanger, H. 240
El Murr, N. 1246
Elofson, R. M. 261, 269
Ely, S. R. 1168, 1169
Emerson, G. F. 454, 894
Engel, W. 705
England, D. C. 361
English, A. D. 1230
Epstein, M. J. 961, 962
Erhardt, U. 1134, 1135, 1293
Ertl, H. 1215
Esayan, G. T. 203
Estes, L. L. 277, 873
Evanega, G. R. 363
Evans, J. 840, 905, 1257
Evans, S. 421
Evnin, A. B. 363
Ewing, G. D. 655, 1194
Ezimora, G. C. 303

Fagerburg, D. R. 353
Fairless, B. 313, 968
Faller, J. W. 463, 487, 648, 666
Faraone, F. 825, 837
Farcasiu, M. 228, 1080
Farnham, W. B. 542, 1027
Farnum, D. G. 53, 931
Farr, F. R. 1206
Farrell, P. G. 384
Faubion, B. D. 981, 983
Faure, J. 188
Fayos, J. 633, 826
Fejes, P. 195
Feldman, M. 267
Fenical, W. 949, 952
Fenton, S. W. 25, 549
Ferdinandi, E. S. 288
Ferree, W. 626
Ficini, J. 688
Figeys, H. P. 139
Filippini, G. 344
Finder, C. J. 126, 127

Finkelstein, M. 208, 209, 922
Fischer, E. O. 512, 884, 885, 898, 901, 902
Fischer, H.-G. 676
Fisher, R. D. 101, 1226
Fleischer, E. B. 813
Fleming, I. 1107
Florian, L. R. 370
Flythe, W. C. 267
Foldiak, G. 193, 194
Fonken, G. J. 180, 955
Ford, R. A. 95
Fraenkel, G. K. 246
Frank, R. E. 266
Franklin, J. L. 73, 86, 89
Franz, K. D. 1239
Fray, G. I. 164, 349, 350, 372, 879, 880, 881, 882, 932, 1082, 1089, 1090, 1091, 1127, 1128, 1205, 1209, 1290
Freedman, H. H. 678, 680, 700, 702
Freeman, P. K. 1158
Freidlin, L. Kh. 274
Frenz, B. A. 1042, 1043, 1044, 1046
Frey, H. M. 36
Frey, W. F. 114
Friedrich, E. C. 214
Friend, E. W. 182, 1210
Friess, S. L. 201
Fritschi, G. 1214
Fritz, H. 237
Fritz, H. P. 99, 375, 502, 512, 530, 885
Fronzaglia, A. 425, 426
Fry, A. J. 82, 272
Frye, H. 525, 596
Fuger, J. 772
Fuji, S. 83
Fujioka, S. 641
Fulcher, J. 757, 1023, 1272, 1273
Furukawa, J. 205, 212
Furuno, K. 308
Fyfe, C. A. 455, 460, 491, 1181

Gaasbeek, M. M. P. 508
Gabrielli, A. 1249
Gamba, A. 335
Games, M. L. 685
Gandolfi, R. 331, 334, 335, 336
Ganellin, C. R. 204, 923, 1062, 1063
Ganis, P. 656, 657
Ganns, R. 709
Ganter, C. 935
Garby, W. P. 288
Gardikes, J. J. 290
Gardlik, J. M. 1207
Gardner, F. J. 369, 370
Gardner, P. D. 10, 556, 1056, 1058
Garratt, P. J. 307, 579, 743, 745, 997, 998
Gassbeck, C. J. 851
Gasteiger, J. 48, 540, 582, 850, 909
Gebrian, J. H. 991, 1052, 1053, 1054

Geckle, J. M. 1261
Gee, D. R. 192
Geen, G. R. 1089, 1205, 1290
Georghiou, P. E. 1276
Georgian, L. 915
Georgian, V. 915
Gerace, M. J. 1215
Gerlach, D. H. 442, 443
Giacometti, G. 461
Gifkins, K. B. 634
Gilani, S. S. H. 362
Gilbert, A. 38, 326, 327, 613, 1128
Gilbert, B. 771
Gillespie, J. P. 1101
Girgenti, S. J. 1281
Gitis, S. S. 727
Givens, R. S. 44
Glass, D. S. 855
Gleicher, G. J. 150, 1137
Gleiter, R. 1075, 1270
Glockner, P. W. 18, 470, 476
Glover, J. H. 270
Goddard, R. 483, 484, 493, 1178
Goebel, C. V. 527
Goffart, J. 409, 771, 790, 801, 1171
Gohlke, R. S. 702
Gold, A. 1219
Goldberg, S. Z. 665
Golden, D. M. 91
Goldfarb, T. D. 862, 863
Gol'farb, Yu. L. 760
Goll, W. 422, 522, 883
Gol'teuzen, E. E. 727
Goodfellow, R. J. 497
Grabbe, R. R. 1058
Graham, C. R. 1013
Graham, J. C. 95
Grasshof, H. 93
Graziani, M. 1066
Gream, G. E. 46, 540, 547, 558, 567
Greco, A. 441, 445, 498, 806, 807, 808, 809, 1170
Green, G. R. 1252
Green, J. C. 421, 1182
Green, M. 479, 486, 495, 496, 497, 498, 523, 587, 795, 830, 831, 836, 1235
Green, M. L. H. 1177
Gregorovich, B. 599
Gresser, J. 287
Griffith, R. C. 746, 748, 953, 954, 1006, 1008
Grigg, R. 510, 965
Grimme, W. 950, 971, 1016, 1055
Grindel, J. M. 1191, 1192
Groenenboom, C. J. 799, 1237
Grohmann, K. 168
Gross, M. E. 66
Grovenstein, E. 343, 601, 602
Groves, J. T. 1258
Grubbs, R. 619

Grünanger, P. 331, 334, 336
Grunewald, G. L. 44, 714, 1191, 1192
Grutzner, J. B. 168
Grzonka, J. 38, 613
Guggenberger, L. J. 423, 1230
Gwynn, D. E. 550
Gyuli-Kevkhyan, R. S. 274, 739, 754

Haase, J. 703
Haddon, R. C. 224, 1146
Hagaman, E. W. 1217
Hagen, G. 143
Hagihara, N. 12, 13, 205, 212, 452, 500, 501,
    589, 824, 888, 903
Hagiwara, T. 50
Haight, H. L. 390
Hamberger, H. 864
Hambrecht, J. 1065
Hammond, G. S. 185
Hammons, J. H. 1202
Hansen, J. F. 75, 577, 658, 659
Hansen-Nygaard, L. 141
Harbottle, G. 478
Hardcastle, K. I. 1046
Harget, A. 124
Harmon, C. A. 381, 541, 665, 777
Harris, D. L. 217, 584, 864, 1009, 1260, 1274
Harris, P. J. 1187
Harrison, L. W. 376
Harrison, W. F. 299
Hartman, R. 649, 661
Haselbach, E. 124
Hass, E.-C. 302
Hasse, K. 1110
Hata, Y. 1117
Haugen, G. R. 91
Haven, A. C. 260
Hawthorne, M. F. 802
Hayes, R. 965
Hayes, R. G. 404, 420, 785
Hazum, E. 456, 834
Heap, N. 1252
Heathcock, S. 587
Hechtl, W. 136, 229, 565, 908
Hedaya, E. 53, 56, 57
Hedberg, K. 118
Hedberg, L. 118
Hehre, W. J. 223, 227
Heidelberger, M. 2
Heil, B. 1245
Heil, G. 1125
Heil, V. 821
Heilbronner, E. 71, 1270
Heitner, H. I. 842
Hekman, M. 36
Helm, R. 692, 695, 707
Helmholdt, R. B. 419
Hendrichs, P. M. 1190
Hendrick, M. 968

Henery, J. 929, 1108
Henke, H. 1292
Henry, T. J. 744, 751, 958, 1037
Henzel, K. A. 210, 545, 546, 560, 561
Herber, R. 619
Herber, R. H. 466, 467
Herndon, W. C. 1144
Heyd, W. E. 49, 583
Heyn, A. S. 1193, 1278
Higginson, B. 421
Hill, M. L. 1056, 1058
Hill, R. K. 207
Hill, R. R. 1087
Hilton, J. 161, 162
Hirth, A. 188
Hoberg, H. 845
Hochstein, F. A. 63, 865
Hocks, L. 409, 1171
Hodgson, H. W. 270
Hodgson, K. O. 381, 761, 762, 764, 765, 766,
    782, 783, 784
Hoesch, L. 172
Hoever, H. 926
Hoffman, E. G. 845, 1238
Hoffman, R. W. 623, 759, 1018, 1271
Hofstee, H. K. 791, 799
Hogben, M. G. 1084
Hogeveen, H. 508, 851
Hohmann, F. 1239
Hojo, K. 312, 970, 1051
Holloway, R. G. 1250
Holmes, A. B. 1197
Holmes, J. D. 820
Holovka, J. M. 1056, 1058
Holzinger, W. 1173
Hopkins, T. E. 1168
Hoppe, B. 1214
Hörauf, W. 1111
Horinaka, A. 196
Horspool, W. 611
Houghton, R. P. 548
Howard, J. A. K. 481, 795, 811, 912, 1187,
    1195, 1235, 1241
Howe, D. V. 477
Hoy, E. F. 509
Hseu, T.-H. 814
Huang, Y. Y. 852
Hübel, W. 679
Huber, H. 225, 229, 908
Huber, R. 503, 689
Hubin, R. 801, 1171
Hudec, J. 342, 348, 1087
Hudson, C. E. 1206
Huebert, B. J. 268
Huffman, H. M. 66
Huisgen, R. 48, 134, 135, 136, 225, 229, 230,
    231, 232, 296, 329, 330, 332, 333, 364, 540,
    562, 565, 566, 567, 582, 850, 856, 908, 909
Humphries, A. P. 483, 484, 494

Hunt, D. F. 465
Hunter, D. L. 431, 432, 896, 1180, 1279
Hutchins, C. S. 82
Hutchinson, J. H. 846, 847
Hüttel, R. 401
Huttner, G. 900, 1234

Iglauer, N. 280
Ihrman, K. G. 438
Immirzi, A. 444
Induni, G. 344
Isaac, P. A. H. 1069
Ishibi, N. 41
Ito, T. I. 977, 988
Iwamura, H. 145, 148, 170, 183, 184
Iwata, S. 191, 1150

Jackson, J. L. 510
Jackson, S. E. 421
Jacobsson, U. 54, 1076, 1157
Jain, B. D. 403, 406, 437, 767
Jamerson, J. D. 769
James, D. G. L. 288, 1153
James, D. R. 569, 715, 716
Janssen, E. 76, 79
Jardine, I. 276
Jefford, C. W. 723
Jeffrey, K. R. 1181
Jellinek, F. 792, 1229, 1237
Jenkins, J. A. 226
Jensen, B. S. 273
Jensen, K. A. 529
Joh, T. 499
Johnson, B. F. G. 477, 492, 528, 552, 590,
    821, 822, 836, 838, 840, 905, 1186, 1196,
    1198, 1244, 1245, 1257
Johnson, D. A. 932
Johnson, E. H. 397
Johnson, M. G. 327
Johnson, R. W. 1139
Johnson, W. H. 68
Jones, D. W. 683
Jones, E. R. 775
Jones, M. 173, 313, 948, 968, 969
Jones, W. O. 155, 156, 263, 264
Joy, F. 242
Juvet, M. 321, 796

Kablitz, H.-J. 417, 793
Kaesz, H. D. 214, 427, 592, 804
Kakihana, T. 75, 577, 658, 659
Kallweit, R. 793, 1238
Kalsotra, B. L. 403, 767
Kaminskii, A. Y. 727
Kanellakopulos, B. 771, 1226
Kanematsu, K. 236, 301, 919, 975, 976,
    1007, 1088, 1285, 1287, 1288
Kaplan, M. L. 734
Karle, I. L. 117

Karplus, M. 575
Karraker, D. G. 763, 775, 778, 779, 787
Kashiwagi, H. 187
Katz, T. J. 33, 243, 244, 245, 246, 249, 307,
    380, 711, 712, 713, 743, 745, 750, 982,
    1074
Kauffmann, T. 282
Kaupp, G. 1222
Kayama, Y. 913, 914
Keller, C. E. 215, 216, 454, 585, 813
Keller, H. 99, 375, 502
Kellett, P. M. 977
Kende, A. S. 1022, 1028
Kent, M. E. 53
Kerber, R. C. 827
Kettlewell, B. R. 1090
Khafaji, A. N. 10
Khuthier, A.-H. 944
Kice, J. L. 289
Kiefer, H. 32, 299
Kim, C. U. 1276
Kim, C. W. 1021, 1269
Kimmel, P. I. 248
Kimmel, V. 693, 694, 696, 709
Kimura, K. 641
Kindler, H. 563
King, R. B. 424, 425, 426, 429, 433, 439, 440,
    451, 467, 893, 904
King, R. W. 861
Kingsley, W. G. 1040
Kinter, M. R. 285, 538
Kintopf, S. 76, 79, 402
Kirmse, W. 1024
Kirsch, G. 160, 562, 580
Kisin, A. V. 322
Kistemaker, J. 113
Kistner, C. R. 847
Kitahara, Y. 866, 913, 914
Kitamura, T. 499
Kitching, W. 210, 583
Klaassen, A. A. K. 385
Klabuhn, H. 302
Klager, K. 11
Klärner, F.-G. 966, 967
Kleier, D. A. 1151
Kloosterziel, H. 854, 867
Klotz, M. A. 234
Klunder, A. J. H. 1200
Knecht, J. 1183
Knight, J. 1177
Knol, J. 803
Knoop, F. W. E. 113
Knox, L. H. 69
Knox, S. A. R. 481, 482, 483, 484, 489, 493,
    494, 595, 897, 912, 1178, 1185, 1187, 1195
Kobayashi, T. 205, 212
Kober, H. 316
Kochanski, E. 125
Kochi, J. K. 393, 394, 395

Kochman, H. J. 59
Köhler, F. H. 1173
Kohlhaupt, R. 695
Kolb, J. R. 1231
Kolosova, T. N. 638
Komalenkova, N. G. 322
Kondo, A. 1088, 1285, 1287, 1288
Kondo, H. 279
Konz, W. E. 364, 540, 562, 565, 566, 567
Korecz, L. 448
Korshak, V. V. 638, 646
Korte, F. 157
Kosel, C. 282
Koszalka, G. W. 1259
Kovačević, K. 147, 1143
Kozarich, J. 1011, 1266
Krauch, C. H. 355, 356
Krause, J. F. 556
Krebs, A. 555, 623, 624, 1139
Kreiter, C. G. 214, 218, 427, 512, 804, 899,
    1176
Kresge, C. T. 1202
Krespan, C. G. 361, 600
Kresz, G. 876
Krief, A. 688
Kröner, M. 233, 518, 906
Kroon, P. A. 419
Krow, G. R. 1077, 1078, 1079
Krüerke, U. 832
Krüger, C. 412, 413, 1172, 1175, 1189
Kruszewski, J. 152
Krygowski, T. M. 152
Kuhls, J. 355, 356
Kukla, M. J. 934
Kuljian, E. 525
Kumagai, T. 1031
Kumar, N. 788, 798
Kunii, T. L. 145, 148
Kurabayashi, K. 50, 51, 1030
Kuri, Z. 28, 29, 30, 31
Kursanov, D. N. 383, 726, 727, 728, 729,
    730, 731, 760, 1221, 1223
Kurz, H. 759, 1271
Kybett, B. D. 89

Labarre, J. F. 85
Labows, J. N. 1136
LaCount, R. B. 609
Lahuerta, P. 431, 432, 896
Laity, J. L. 84, 219
LaLancette, E. A. 306, 956
La Mar, G. N. 779, 780
Lang, M. 1176
Langford, G. R. 1247
Langheck, M. 239
Lankey, A. S. 710
Lappert, M. F. 242
LaPrade, M. D. 669
Larrabee, C. E. 24

Larson, W. D. 105
Lassila, J. D. 937
Lautenschlaeger, F. 238
Laviron, E. 1246
Lawrenson, I. J. 103
Lazarus, R. M. 1072
Leach, S. 181
Lee, C. 578, 1004
Lee, F.-T. 872
Lee, H. U. 1236
Leermakers, P. A. 185
Le Fèvre, R. J. W. 843
Le Goff, E. 609
Legzdins, P. 490
Lehmkuhl, H. 76, 79, 402, 407, 411, 516, 517
Lehn, J. M. 125
Leichter, L. M. 371
Lemal, D. M. 1215
Leone, R. E. 1026, 1028
Leto, J. R. 640
Leto, M. F. 640
Leuchte, W. 76, 516, 517
Levsen, K. 110
Lewis, C. P. 749
Lewis, J. 477, 492, 528, 552, 590, 794, 821,
    822, 836, 838, 840, 905, 1186, 1196, 1198,
    1244, 1245, 1257
Ley, S. V. 43, 588, 593, 826, 985, 1157, 1261
Leyendecker, F. 557
Li, J. P. 1159
Liang, G. 848, 849, 1027
Liösch, S. S. 1008
Lichtenfeld, A. 1049
Lin, Y. S. 123
Lindqvist, L. 862
Linkowski, G. E. 1154, 1193
Lipina, E. S. 615
Lippard, S. J. 619, 842
Lippert, W. 696
Lippincott, E. R. 27, 90, 101, 462
Lippke, W. 626
Lippman, N. M. 889
Lipscomb, W. N. 398, 399, 457, 458, 468,
    700, 1151
Liss, T. A. 235
Liu, R. S. H. 600, 1291
Lo, D. H. 142, 1140
Lobl, T. J. 1210
Lodge, J. E. 536
Longuet-Higgins, H. C. 22, 97, 121, 122, 250
Lord, R. C. 27, 90, 158, 262
Lougnot, D. 188
Louis, G. 704
Luz, Z. 129
Lyakhovetskii, Yu. I. 726, 731

Maasbol, A. 427, 592, 805
McArdle, P. 590
McCabe, R. W. 350

McCauley, G. B. 596
McCausland, J. H. 1255
McCay, I. W. 650, 651, 1102, 1105
McClung, R. 980
McCrae, W. 9
MacDiarmid, A. G. 1247
McDonagh, P. M. 292
McDonald, R. S. 27
McEwen, K. L. 97, 121, 122
McFarlane, W. 474, 475, 815, 819, 891
McIntosh, A. R. 192
Mack, W. 332
McKechnie, J. S. 428
McKennis, J. S. 814, 892
McKinney, R. J. 493
McLachlan, A. D. 257
McMasters, D. L. 266
McNeil, D. W. 53, 57
McQuillin, F. J. 276
Mag, P. 448
Maher, J. P. 497
Mahler, J. E. 213, 894
Maier, G. 653, 676, 697, 698, 708, 1113, 1214
Maiorana, S. 588, 593
Maitlis, P. M. 681, 682, 685, 1233
Maksić, Z. B. 147, 149, 1143
Mallon, B. J. 146
Malpass, J. R. 367, 877, 878, 1077
Maltz, H. 1011, 1068
Manatt, S. L. 105
Manuel, T. A. 449, 450, 890
Mares, F. 381, 761, 762, 765, 780
Margrave, J. L. 89
Marica, E. 35, 228, 352, 374, 917
Mark, H. 234
Mark, V. 373
Marks, T. J. 480, 488
Marsden, J. 342
Marshall, D. J. 544
Martens, D. 992, 993, 994
Martin, H.-D. 36
Martin, J. T. 1261
Martin, W. 34, 1119
Martini, Th. 324, 554
Masamune, S. 312, 319, 323, 946, 970, 1051, 1059, 1084, 1097, 1276
Masino, A. P. 769
Maskill, H. 182, 1210
Maslowsky, E. 455, 460
Mason, S. F. 384
Massey, A. G. 833
Mateescu, G. D. 35, 228, 340, 1080
Matheson, T. W. 1232
Mathews, F. S. 398, 399
Matsuda, T. 202, 278, 308, 309, 939, 942, 945, 1064
Matsuura, T. 196
Mayer, J. R. 69

Meador, W. R. 69, 275
Meesters, A. 852
Mehler, K. 76, 402, 407
Mehta, G. 351, 1284, 1286, 1289
Meiboom, M. 129
Meić, Z. 1141
Meier, H. 1204
Meili, J. E. 597
Meinwald, J. 45, 46, 1136, 1218
Meisinger, R. H. 43, 49, 583, 604, 605, 606, 1075
Meister, H. 23
Mende, U. 653, 697
Menig, H. 434
Merényi, R. 128, 166, 476, 554
Merk, W. 37
Mertschenck, B. 899
Mestroni, G. 513, 1251
Meyers, T. J. 690
Michelin, R. A. 1267, 1280
Michl, J. 1149
Mietzsch, F. 134, 135, 856
Migirdicyan, E. 181
Mihai, Gh. 581
Milas, N. A. 198
Miller, M. A. 95, 921
Miller, R. D. 56, 57, 363, 625
Miller, R. G. J. 391
Miller, W. T. 1212
Milone, L. 430
Mioduski, J. 1218
Miyakawa, S. 94, 176
Miyake, A. 279
Miyazaki, H. 914
Mock, W. L. 1069, 1254, 1255
Mognashi, E. R. 446, 447
Moll, M. 1248
Monken, C. E. 320
Monthony, J. F. 382
Mooij, J. J. 385, 386, 1148
Moore, H. W. 165
Moore, P. T. 199
Moore, W. R. 199, 598
Moorhouse, S. 835
Moran, W. 955
Moriarty, R. M. 41, 718, 721, 723
Morio, K. 145, 148, 170
Morrell, D. G. 381, 1227
Morrison, H. 626
Morrow, T. 186
Moshuk, G. 972, 995
Moss, R. E. 250, 258, 572
Mottl, J. 70
Mrowca, J. J. 713
Mueller, W. H. 241
Muetterties, E. L. 442
Mukai, T. 50, 51, 618, 943, 1030, 1031, 1201
Mular, M. 547, 558
Müller, E. 907, 927, 928, 1065

Müller, J. 422, 434, 522, 883, 898, 899, 901, 1173, 1234
Müller-Westerhoff, U. 381, 774
Multani, R. K. 403, 406, 437, 767, 788
Murray, M. 1161
Murray, R. W. 734
Musco, A. 472, 480, 487, 648, 657, 666, 668

Nace, H. R. 873
Naff, W. T. 114
Nakamura, A. 452, 500, 501, 589, 824, 839, 888, 903
Nakanishi, H. 167
Nakashima, R. 196
Nakatsuka, N. 946
Nakayama, T. 70
Natalis, P. 89
Nelsen, S. F. 1101
Nelson, D. R. 115
Nelson, G. O. 1259
Nelson, N. A. 206
Nenitzescu, C. D. 35, 175, 228, 340, 341, 352, 374, 581, 917, 1080, 1083
Neubauer, D. 295
Neuberg, R. 1126
Neuenschwander, M. 687
Neville, D. J. 882
Nicholson, C. P. 380
Niederhauser, A. 687
Nikoloff, P. 325, 1036
Nishimura, S. 641
Nishitani, Y. 1288
Nolan, J. T. 198
Noordik, J. H. 385, 386
Nowacki, W. 1282
Nunn, E. E. 1094, 1208
Nyberg, K. 209
Nyburg, S. C. 161, 162

Ocasio, I. 387, 616, 1147, 1164, 1165, 1166
Oda, M. 866, 913, 914
Odaira, Y. 641
Odom, J. D. 1247
Ogawa, M. 939, 942, 945, 1064
Ogilvy, M. M. 747
Ogliaruso, M. A. 710, 938, 972, 978, 979, 980, 984, 987
O'Hara, R. K. 712
Okamura, W. H. 320, 382, 977, 988, 996
Oki, M. 148
Oku, M. 43, 49, 54, 603, 1027
Okuda, M. 177
Okulevich, P. O. 638
Olah, G. A. 220, 848, 849, 1027, 1211
Oliver, G. D. 66
Olsen, H. 930, 1112
Onderdelinden, A. L. 514
O'Neal, H. E. 91
Oosterhoff, L. J. 113

Orfanos, V. 991
Orgel, L. E. 22
Orvedal, A. W. 1038
Osawa, E. 1138, 1292
Osborn, C. L. 10, 556
Osborn, J. A. 511
Osborn, T. W. 996
Oth, J. F. M. 104, 128, 160, 166, 169, 172, 554, 562, 579, 580, 1036, 1118, 1119, 1120, 1121, 1122, 1124, 1125, 1126, 1129
Otsuka, S. 839
Otto, P. Ph. H. L. 198
Overberger, C. G. 3, 5, 234
Overzet, F. 792, 1229
Owens, R. M. 981, 983

Paddon-Row, M. N. 1208
Padwa, A. 649, 661
Palazzi, A. 841
Palladino, N. 639
Palm, C. 884, 885, 902
Palm, M. P. 763
Pandey, P. N. 1284, 1286, 1289
Pant, B. C. 1152
Paquette, L. A. 42, 43, 49, 54, 75, 78, 163, 210, 346, 360, 366, 367, 371, 542, 543, 545, 546, 560, 561, 569, 577, 578, 583, 588, 593, 603, 604, 605, 606, 629, 630, 631, 632, 633, 634, 635, 655, 658, 659, 671, 715, 716, 717, 826, 828, 829, 848, 933, 934, 961, 962, 985, 986, 999, 1000, 1003, 1004, 1019, 1027, 1034, 1075, 1076, 1077, 1100, 1114, 1157, 1194, 1199, 1202, 1203, 1207, 1211, 1217, 1261
Parfitt, L. T. 1089
Parker, R. G. 116
Parker, V. D. 273
Parkins, A. W. 822, 1250
Parkinson, B. 717
Parnes, Z. N. 383
Parrott, M. J. 1089
Parsons, T. C. 773
Partenheimer, W. 397, 509, 526
Patil, P. N. 1191
Paul, I. C. 428, 1130
Paulus, E. 503
Pauson, P. L. 1179
Pawley, G. S. 700
Pchelintsev, V. I. 322
Pearce, C. D. 105
Peeling, E. R. A. 1152
Peet, W. G. 442
Pelosi, L. F. 686, 1212
Peltzer, B. 853
Penfold, B. R. 505, 1250
Perekalin, V. V. 615
Perry, C. W. 675
Person, W. B. 140
Peters, D. 151

Petersen, D. R. 678
Petersen, R. C. 209, 922
Petraccone, V. 656
Petrovich, J. P. 271
Petrowski, G. 995, 1260
Pettit, R. 37, 204, 213, 215, 216, 454, 585, 813, 814, 820, 892, 894, 923, 929, 1062, 1063, 1108, 1109
Philips, J. C. 346, 632
Phillips, D. D. 310
Phillips, R. P. 1187
Photis, J. M. 629, 631, 633, 634, 655, 671, 1203, 1207
Pidcock, A. 844
Pietropaolo, R. 837
Pike, R. M. 292, 535, 559
Pimental, G. C. 140
Pinschmidt, R. K. 964, 1001
Pisciotti, F. 797
Pitzer, K. S. 140
Plinke, G. 1121, 1122, 1124
Podgornova, N. N. 615
Pogány, I. I. 581, 1080
Pokras, S. M. 935
Polkovnikov, B. D. 274
Pollock, D. F. 681, 685
Powell, P. 1182
Prange, U. 1120
Pratt, L. 436, 474, 815, 891
Prelesnik, B. 1282
Press, J. B. 52, 1033, 1277
Prinzbach, H. 637
Prokai, B. 242
Prosen, E. J. 68
Prout, C. K. 887
Prud'homme, R. 281
Pryde, W. J. 685
Pullman, A. 92

Quail, J. W. 1196
Quinn, H. W. 396
Quinn, M. F. 186

Radford, D. V. 843
Radlick, P. C. 735, 868, 949, 952, 990, 1253
Rajbenbach, A. 287
Ramadas, S. R. 325
Ramey, K. C. 718, 721, 723
Ramp, F. L. 260
Randall, E. W. 430
Randall, G. L. P. 552, 590, 822
Randić, M. 149, 1141
Rao, D. V. 343, 601
Rao, J. M. 907, 927, 928
Raphael, R. A. 8, 9
Rausch, M. D. 453
Raymond, K. N. 665, 764, 766, 781, 782, 783, 784
Read, L. K. 999, 1000

Reardon, E. J. 889, 1010, 1015
Records, R. 98
Reese, C. B. 870
Reetz, M. T. 759, 1018
Reger, D. L. 1249
Reich, S. D. 948
Reichmanis, E. 1061, 1073, 1155, 1263
Reilly, C. A. 380
Reilly, J. 1078, 1079
Reinmuth, W. H. 249
Reinsch, E.-A. 20, 21
Reis, H. 295
Reitman, L. N. 1103, 1104
Reppe, W. 11, 17, 23
Ressa, I. J. 261
Reynoso, R. 1272
Richards, G. F. 390, 392
Richards, R. 1167
Rieger, P. H. 65
Rieke, R. D. 80, 573, 672, 972, 980, 984
Riemschneider, R. 1216
Riera, V. 481, 595, 1185, 1195
Rigamonti, E. 664
Rigatti, G. 461
Rinze, P. V. 1188
Ristau, W. 74
Riveccié, M. 1246
Rivers, G. T. 1160
Rivier, J. 637
Robb, E. W. 636, 675
Roberts, G. C. 844
Roberts, J. D. 102, 168, 550, 935
Roberts, M. 864
Robertson, A. V. 915
Robertson, G. B. 1232
Robertson, J. C. 944
Robinson, B. H. 504
Robinson, S. D. 524
Robson, A. 471
Rodewald, L. B. 1137
Rodgers, A. S. 91
Roe, D. M. 833
Roedig, A. 691, 692, 693, 694, 695, 696, 707, 709
Roemer-Mähler, J. 157
Ronlan, A. 273
Ros, R. 841, 1066
Rosan, A. 720
Rösch, K. 1222
Rosen, W. 314
Rosenberg, E. 430
Rosenberger, M. 711, 712
Rosenblum, M. 720
Rosenthal, J. W. 1098
Ross, S. D. 209, 922
Rossini, F. D. 68
Rotella, F. J. 440
Roth, W. R. 32, 146, 736, 853, 859, 869
Röttele, H. 1035, 1036, 1125, 1129, 1136

Rozynov, B. V. 646
Rubio, M. 1272
Rugen, D. F. 534, 559
Rushworth, F. A. 103
Russ, C. R. 551, 811, 1241
Russell, J. W. 465
Russell, R. K. 42, 43, 635, 1217
Ryder, I. E. 477

Sadler, J. E. 410
St-Jaques, M. 281
Saito, K. 618, 1201
Sakai, M. 758, 864, 1023, 1032, 1260, 1274
Sakakibara, T. 641
Salem, L. 138
Salentine, C. G. 802
Salman, K. N. 1191
Salomon, M. F. 1020
Salomon, R. G. 393, 394, 395
Salzer, A. 1256, 1264
Sanders, D. C. 52, 1017
Sandrini, P. L. 1240, 1267, 1280
Sanne, W. 265, 858
Santambrogio, A. 639
Sargent, M. V. 579, 610, 612, 613, 620, 621, 622, 627, 628, 910
Sarkisyan, E. L. 643, 644, 645
Sasaki, T. 236, 301, 919, 975, 976, 1007, 1088, 1285, 1287, 1288
Sato, T. 866
Savin, F. A. 100
Sawyer, D. T. 62
Saxby, J. D. 507, 843
Saxton, R. G. 164, 1127, 1128
Schaeren, S. F. 941
Schäfer, W. 1018, 1271
Schallhorn, C. H. 1086
Scheiner, P. 347
Schellenberg, W. D. 1111
Schenck, G. E. 108
Schenck, G. O. 355, 356
Scherer, K. V. 690
Scherzer, K. 1234
Schiess, P. 92
Schissel, P. O. 53, 57
Schläpfer, P. 237
Schlenk, W. 64
Schleyer, P. von R. 144, 957, 1026, 1028, 1137, 1138
Schlichting, O. 11, 23, 265
Schmidt, H. 1271
Schmitt, S. 899
Schnegg, U. 364, 540, 562
Schneider, D. F. 588, 593
Schneider, G. 957, 959
Schneider, M. 698
Schneider, W. G. 107
Schoeneck, W. 282
Scholes, G. 1013, 1259

Schomburg, G. 61
Schönefeld, J. 300
Schönleber, D. 315
Schooley, D. A. 98
Schormüller, J. 59
Schott, A. 845
Schott, H. 845
Schrauzer, G. N. 15, 16, 18, 388, 453, 469, 470, 476, 520, 521, 816, 1130
Schrock, R. R. 423, 511, 794, 838, 1230
Schröder, G. 34, 128, 146, 159, 160, 166, 169, 283, 284, 324, 325, 345, 554, 562, 580, 676, 677, 1035, 1036, 1118, 1119, 1120, 1121, 1122, 1124, 1125, 1126, 1129, 1131, 1132, 1136, 1292
Schroth, G. 1238
Schug, J. C. 987
Schunn, R. A. 443
Schuster, D. I. 872, 1021, 1269
Schüttler, R. 759
Schwab, L. O. 173
Schwartz, J. 410, 737, 812
Schweig, A. 1018, 1271
Schweinler, H. C. 114
Scordamaglia, R. 810
Scott, D. W. 66
Scott, L. T. 948, 969
Seel, K. 1055
Seevogel, K. 1238
Seff, K. 614
Segal, J. A. 1179
Segre, A. L. 445
Seidl, H. 135, 856
Seidl, P. 864
Seidner, R. T. 312, 1059
Seip, H. M. 120
Sellman, D. 530
Senoff, C. V. 491
Sergeev, V. A. 638
Sharma, K. M. 406, 437
Sharma, R. K. 798
Shaver, A. 1042
Shaw, B. L. 524
Shaw, R. 91
Shechter, H. 52, 290, 377, 378, 725, 1017, 1033, 1275, 1277
Shenoy, S. J. 1162
Shepherd, R. A. 358, 1268
Sherwin, M. A. 44
Shibata, T. 602
Shida, S. 30, 83, 176, 178, 189
Shida, T. 191, 1150
Shields, T. C. 556
Shitikov, V. K. 638
Shoemaker, D. P. 563, 614
Shoulders, B. A. 556, 585
Shuettenberg, A. 272
Shvo, Y. 456, 591, 834
Sibilia, J. P. 101

Simonetta, M. 344
Sixma, F. L. J. 197
Slegeir, W. 892
Sliam, E. 35, 341
Sly, W. G. 563
Smallcombe, S. H. 1210
Smentowski, F. J. 251, 252, 253, 981, 983
Smith, A. K. 1232, 1233
Smith, D. E. 249, 268
Smith, D. M. 1120, 1124
Smith, D. P. S. 372
Smith, D. S. 206, 235, 617
Smith, L. R. 46
Smith, R. M. 692
Snow, R. A. 1092
Snyder, J. P. 53, 930, 931, 1112
Snyder, L. C. 133, 257
Sohn, M. B. 313, 968
Solomon, O. F. 14, 294
Soltwisch, M. 408
Sondheimer, F. 610, 612, 613, 620, 622, 627, 628, 910
Sonnichsen, G. 381
Sorensen, T. S. 852
Spencer, J. L. 504, 505, 795, 1235
Spessard, G. O. 323, 1276
Spiesecke, H. 107
Sprague, J. T. 127
Springall, H. D. 67
Squire, R. H. 369, 370
Srivavasta, R. C. 563
Staley, S. W. 744, 751, 958, 963, 989, 1037, 1038, 1040, 1154, 1193, 1278
Stanford, R. H. 116
Staral, J. S. 849, 1211
Starks, D. F. 405, 762, 773
Starodubtseva, M. P. 729
Starzewski, K. A. O. 1239
Steele, D. 462
Stempfle, W. 1238
Stenger, V. 193
Stern, R. L. 182, 660, 1210
Stevens, C. L. 865
Stevens, I. D. R. 362
Stevenson, G. R. 251, 252, 253, 254, 387, 537, 539, 571, 574, 576, 616, 1147, 1164, 1165, 1166
Stone, A. L. 813
Stone, F. G. A. 449, 450, 481, 482, 486, 489, 493, 494, 498, 523, 551, 595, 682, 795, 811, 890, 897, 904, 912, 1161, 1178, 1185, 1187, 1195, 1235, 1241
Stone, J. A. 775, 787
Storlie, J. C. 847
Stowell, J. C. 933, 934, 1100
Stoyanovich, F. M. 760
Strack, H. 1176
Straub, H. 907, 927, 928, 1065
Strauss, H. L. 243, 246, 248

Streitwieser, A. 137, 381, 405, 541, 663, 761, 762, 765, 770, 773, 774, 777, 779, 780, 786, 1227
Strohmeier, W. 280
Strow, C. B. 1056, 1058
Strukul, G. 1066
Stühler, H.-O. 1234
Su, T.-M. 957
Sugishita, M. 202, 1064
Sugiyama, H. 611
Sustmann, R. 332
Swarc, M. 287
Sweeney, A. 965
Szary, A. C. 897, 912

Taggart, D. L. 1194
Takada, S. 1051, 1059
Takagi, M. 939, 945, 1064
Takats, J. 667, 670, 769, 1265
Talcott, C. 982
Tanaka, I. 94, 174, 176, 177
Tancrede, J. 814
Tanida, H. 1117
Tauchner, P. 401
Taylor, J. W. 343
Taylor, S. C. 1152
Templeton, D. H. 665, 1228
Templeton, L. K. 1228
Temussi, P. A. 657
Teplyakov, M. M. 646
Teratake, S. 1117
ter Borg, A. P. 854
Teyssié, P. 409
Thielen, D. R. 77
Thio, J. 1121
Thomas, J. L. 404, 420
Thompson, D. J. 821, 1245
Throndsen, H. P. 674, 699
Thyret, H. 520, 521
Tichy, M. 146
Tiers, G. V. D. 290
Tiffany, B. D. 200
Timms, P. L. 1174
Todd, P. F. 247, 250, 391, 570, 572
Todres, Z. V. 383, 724, 726, 727, 728, 729, 730, 731, 732, 733, 760, 1220, 1221, 1223
Toepel, T. 11
Toffe, N. T. 383
Tomisawa, K. 943
Toniolo, L. 1014, 1280
Toong, Y. C. 1219
Torian, R. L. 465
Toshima, N. 557
Tosi, C. 445
Touzin, A.-M. 688
Traetteberg, M. 119, 143
Trautz, V. 1133, 1293
Traynham, J. G. 1283
Traynor, S. G. 1194, 1261

Treichel, P. M. 904
Tresselt, L. W. 846
Trimitsis, G. B. 52
Troup, J. M. 1043, 1044, 1045, 1046
Truesdell, D. 826
Trumbull, E. R. 260, 941
Truter, M. R. 471
Tsunawaki, S. 278
Tsuruta, H. 45, 1030, 1031
Tsutsui, M. 673
Tucker, P. A. 1232
Turnblom, E. W. 33, 750
Turner, D. W. 72, 97
Turner, R. B. 69, 146, 275
Turney, T. W. 1174
Tushaus, L. A. 921
Tustin, G. C. 320
Tweddle, N. J. 878
Twieg, R. J. 1160

Uebel, J. J. 209, 314
Uemura, T. 94
Uguagliati, P. 825, 1240
Ushakov, S. N. 14, 294

Van Auken, T. V. 1056, 1058
Van-Catledge, F. A. 96, 153
Van den Hark, Th. E. M. 385
van der Ent, A. 514
van der Hout-Lodder, A. E. 684
van der Wal, H. R. 792, 1229
van Dongen, J. P. C. M. 400
VanGilder, R. L. 396
Van Orden, H. O. 286
van Oven, H. O. 418, 789, 791, 792, 799, 800, 803, 1229
van Tamelen, E. E. 737
Van Tilborg, J. 1182
van Willigen, H. 255
Vaughan, W. R. 347
Vedejs, E. 357, 358, 1020, 1099, 1268
Veldman, M. E. E. 800
Venanzi, L. M. 506
Venkataramani, P. S. 351
Viebrock, J. 525
Viglino, P. 1066
Vincow, G. 662, 1123
Vitale, W. 607, 611
Vliek, M. 1237
Vogel, E. 32, 297, 298, 299, 916, 951, 971, 1110
Voigt, G. 1024
Volger, H. C. 508
Vollmer, J. J. 1025
von Kutepow, N. 295
von Rosenberg, J. L. 213
Voorhees, K. J. 874, 920
Vrieze, K. 508
Vysotskii, Yu. B. 1145

Wagner, F. 1204
Wagnon, J. C. 814
Wahlgren, U. 125
Waight, E. S. 548
Walker, R. 663, 1227, 1228
Walker, R. W. 158
Walsh, R. 91
Wan, J. K. S. 192
Wang, A. H.-J. 1130
Warawa, E. J. 1159
Warner, P. 217, 359, 848, 864, 974
Warren, K. D. 768, 1225
Warrener, R. N. 650, 651, 1093, 1094, 1095, 1096, 1102, 1105, 1106, 1208
Waser, E. 1
Watanabe, K. 70
Watanabe, M. 1117
Watanabe, T. 1150
Watson, R. A. 1259
Watt, R. 1257
Weber, H. 1002, 1262
Weeks, P. D. 1020
Wege, D. 1198
Wegener, P. 365
Weis, C. D. 354
Welch, A. J. 523
Wendel, K. 607
Wenkert, E. 1217
Wentworth, W. E. 74
Wertheim, G. K. 466
West, R. 690, 692
Westberg, H. H. 1081, 1093
Westerhof, A. 803, 1224
Westlake, D. J. 496
Wetzel, J. C. 328
Whalen, D. L. 1085, 1086
Wheatley, P. J. 699, 701
Wheland, R. 1039
White, D. A. 528
White, E. H. 182, 654, 660, 1210
White, J. E. 338
White, T. R. 67
Whitehead, M. A. 142
Whitesides, G. M. 550
Whitham, G. H. 1252
Wibaut, J. P. 197
Wickersham, T. W. 1159
Widmann, P. 703
Wieczorek, L. 823
Wiedemann, W. 299
Wieduwilt, M. 1213
Wiesel, M. 946
Wildsmith, E. 1107
Wiley, D. W. 69
Wilke, G. 416, 417, 518, 519, 793, 845, 1238
Wilkinson, G. 339, 435, 436, 474, 475, 815, 819, 891
Willcott, M. R. 1121
Williams, J. E. 144

Williams, V. Z. 1137
Williams, W. M. 368
Willis, R. G. 8
Willstätter, R. 1, 2
Wilms, H. 7
Wilson, J. D. 84
Wilson, K. E. 1276
Wilson, R. M. 369, 370
Wilson, W. S. 1094, 1095, 1096, 1106
Wingard, R. E. 42, 604, 605, 606, 629, 630,
632, 635, 1217
Winkler, B. 861
Winstein, S. 214, 217, 218, 219, 221, 222,
427, 592, 758, 804, 805, 855, 860, 864, 868,
871, 886, 938, 972, 974, 978, 980, 984, 995,
1023, 1032, 1253, 1260, 1274
Winston, A. E. 1073
Winter, M. J. 1178
Wipff, G. 125
Wirth, W.-D. 705
Withey, D. S. 26
Wittenberg, D. 259
Wittig, G. 259
Wodley, F. A. 58
Wojnarovits, L. 193, 194, 195
Wood, D. C. 587, 830, 831
Woodward, P. 481, 483, 484, 493, 494, 811,
912, 1178, 1187, 1195, 1241
Worley, S. D. 111
Wright, C. D. 1194
Wright, D. A. 614
Wright, H. W. 1048
Wright, J. D. 813

Yakali, E. 973, 1047
Yakhovetskii, Y. I. 726
Yamamoto, H. 1070
Yamamoto, O. 167
Yamashita, M. 187
Yamazaki, H. 178, 179, 189, 515
Yandle, J. R. 497
Yarrow, D. J. 836, 840
Yates, W. F. 87
Yavari, I. 857
Yeh, C.-L. 41, 718, 721, 723
Yeh, E.-L. 718, 723
Yesinowski, J. P. 531
Yip, R. W. 647
Yoshida, N. 770
Yukimoto, Y. 236, 301, 919, 975, 976, 1007

Zabkiewicz, J. A. 9
Zahn, U. 478
Zalkin, A. 781, 782
Zare, R. N. 1236
Zassinovich, G. 1251
Zeiss, H. 674, 699
Zenda, H. 946
Ziegler, K. 7
Zimmerman, H. E. 44, 171, 183, 714
Zimmerman, W. T. 1160
Zingales, F. 825, 1240
Zirner, J. 855, 860
Zwanenburg, B. 1200
Zwanenburg, E. 867

Subject Index

[12]Annulene, 365
[16]Annulene, 359, 360
  anion radical, 360
  dianion, 360
  dication, 360
  substituted, 359
[18]Annulene, 1-chloro-2-fluoro, 364
[17]Annulenyl anion, 362
Aza[17]annulenes, 363
9-Azabicyclo[4.2.1]nona-2,4,7-triene, 318
  9-amino, 309
  9-benzyl, 319
  9-benzyl, 9-oxide, 319
  9-carboxylic acid amide, 318
  9-cyano, 37, 318, 319, 320, 378
  9-ethoxycarbonyl, 318
  9-formyl, 318
  9-hydroxy, 182, 318, 319
  9-methoxycarbonyl, 318
  9-(4-nitrobenzoyl), 318
  9-nitroso, 318
9-Azabicyclo[6.1.0]nona-2,4,6-triene, 308
  9-acetyl, 308, 309
  9-carboxylic acid dimethylamide, 308, 309
  9-chloro, 308
  9-cyano, 37, 309, 319
  9-ethoxycarbonyl, 37, 308, 309, 310; Rh
    complex, 310
  9-methoxycarbonyl, 309
  9-N-phthalimido, 37, 309
  9-sulphinic acid, phenyl ester, 308, 309
9-Aza-10-oxobicyclo[4.2.2]decane, 327, 421
9-Aza-10-oxobicyclo[4.2.2]deca-2,4,7-triene,
  207, 326, 327
  9-benzyl, 326
  9-chlorosulphonyl, 326
  9-chlorosulphonyl-5-methoxy, 105
  9-chlorosulphonyl-7-methoxy, 105
  9-chlorosulphonyl-5-(4-methoxyphenyl),
    105
  9-chlorosulphonyl-7-(4-methoxyphenyl),
    105
  9-chlorosulphonyl-5-methyl, 105
  9-chlorosulphonyl-7-methyl, 105
  9-chlorosulphonyl-5-phenyl, 105
  9-chlorosulphonyl-7-phenyl, 105
  Fe complex, 207

Azonine, 309
  N-ethoxycarbonyl, 309

Benzocyclo-octatetraene, 131, 164
  10,11-cyclobuteno, 131
  10,11-cyclobuteno-9,12-dihydro, 124, 131
  9,12-dihydro, 124, 131
  9,12-dihydro-10,11-dimethoxycarbonyl,
    131
  10,11-dimethoxycarbonyl, 131
  9,10,11,12-tetraphenyl, 164
Bicyclo[6.2.0]deca-1,3,5,7,9-pentaene
  9,10-bis(t-butoxy), 301
  9-t-butoxy, 301
  9-t-butoxy-10-chloro, 301
  9-chloro-10-methyl, 301
  9-methyl, 301
cis-Bicyclo[6.2.0]deca-2,4,6-triene
  9-chloromethyl, 175, 303
  9-chloro-9,10,10-trifluoro, 39, 301
  Cr complex, 417
  9,9-dichloro-10,10-difluoro, 39, 301, 303
  Fe complex (of valence tautomer), 304,
    418
  formation, 175
  Mo complexes, 417, 418
  Pd complex, 418
  photolysis, 302
  reaction: with base, 303; with K, 417;
    with transition metal derivatives, 304,
    305, 417, 418
  reduction, 302, 303
  Ru complex (of valence tautomer), 418
  9,9,10,10-tetracyano-2-phenoxy, 104
  thermolysis, 302
  W complex, 417
trans-Bicyclo[6.2.0]deca-2,4,6-triene, 302
Bicyclo[6.4.0]dodeca-1,3,5,7-tetraene, 4,5-
  dimethyl, 388-9
Bicyclo[6.4.0]dodeca-2,4,6-triene, 176, 307-
  308
  Fe complex, 308
  Mo complex, 307
Bicyclo[4.2.1]nona-2,4,7-triene
  anti-9-benzyl-9-hydroxy, 176
  anti-9-(4-bromophenyl-9-hydroxy, 4-
    bromobenzoate ester, 176

Bicyclo[4.2.1]nona-2,4,7-triene—cont.
  Cr complex, 42–13
  cycloaddition reaction, 299
  anti-9-deuterio-9-hydroxy, 4-toluene-
    sulphonate ester, 294, 296
  anti-9-deuterio-9-methoxy, 299
  9,9-dicyano, 36, 269
  9,9-dimethoxy, 292
  syn-9-dimethylamino, 265
  syn-9-fluoro, 265
  formation, 287, 293
  hydroboration, 416
  anti-9-hydroxy, 293; 4-toluenesulphonate
    ester, 293, 296, 416
  syn-9-hydroxy, 293, 298; acetate ester,
    293; sulphite ester, 293; 4-toluene-
    sulphonate ester, 293, 294, 295, 416
  syn-9-hydroxy-9-(4-methoxyphenyl), 178;
    4-nitrobenzoate ester, 296
  syn-9-hydroxy-9-methyl, 176, 177, 178,
    292, 295; acetate ester, 176
  syn-9-hydroxy-9-phenyl, 177, 178, 292,
    296; benzoate ester, 176, 292, 294; 4-
    nitrobenzoate ester, 296
  syn-9-methoxy, 293, 299
  9-methylene, 292, 295
  Mo complex, 286
  9-phenylmethylene, 292
  photo-electron spectrum, 416
  photolysis, 297
  2,3,4,5-tetradeuterio-syn-9-hydroxy-9-
    phenyl, 296
  thermal rearrangement, 294
cis-Bicyclo[6.1.0]nona-2,4,6-triene
  9-acetoxy-9-methyl, 176, 270, 282
  anti-9-(9-anthryl), 378, 412
  syn-9-bromo, 174, 264
  anti-9-(4-bromophenyl), 398
  anti-9-t-butyl, 264
  syn-9-t-butyl, 174, 264
  anti-9-carboxylic acid, 263, 264; amide,
    263; chloride, 263; dimethylamide, 263;
    hydrazide, 263
  syn-9-carboxylic acid, 262; chloride, 262
  anti-9-chloro: cycloaddition reactions,
    281, 284, 285; formation, 35, 36, 174,
    272; Mo complex, 413, 414; reaction,
    with transition metal, derivatives 412–
    14; reduction, 278; thermolysis, 264
  syn-9-chloro, 35, 36, 269, 272
  anti-9-chloro-9-deuterio, 264
  9-chloro-9-methyl, 174
  anti-9-(4-chlorophenyl), 398
  Cr complex, 413
  anti-9-cyano, 174, 264
  syn-9-cyano, 174, 264
  9-cyano-1,9-dimethyl, 268
  9-cyano-2,9-dimethyl, 268
  9-cyano-3,9-dimethyl, 268

  9-cyano-4,9-dimethyl, 268
  anti-9-cyano-9-methyl, 269
  syn-9-cyano-9-methyl, 269
  cycloaddition reactions, 280, 282–5, 286,
    412
  anti-9-deuterio, 264
  syn-9-deuterio, 264
  9,9-dibromo, 35, 269
  9,9-dichloro, 35, 174, 269, 279, 280
  9,9-dicyano, 35, 269
  3,6-dideuterio, 265
  9,9-dideuterio, 378
  9,9-dimethyl: formation, 174; reaction,
    with Rh derivatives, 268, 290, 291;
    reduction, 277; Rh complexes, 290,
    291; thermolysis, 264, 268
  anti-9-dimethylamino, 182, 264
  syn-9-dimethylamino, 182
  anti-9-dimethylaminomethyl, 263; N-
    oxide, 263, 268
  2,7-diphenyl, 265
  9,9-diphenyl, 378
  anti-9-ethoxycarbonyl: Cr complex, 412;
    cycloaddition reactions, 282; forma-
    tion, 35, 36, 264; functional group
    transformations, 263; Mo complexes,
    413, 414; photolysis, 273; reaction, with
    transition metal derivatives, 412, 413;
    thermal rearrangement, 264
  syn-9-ethoxycarbonyl, 35, 36, 262, 264
  anti-9-ethyl-9-methyl, 264
  syn-9-ethyl-9-methyl, 264
  Fe complexes, 288–9, 414
  anti-9-fluoro, 174, 264
  syn-9-fluoro, 264
  formation, 35, 164
  anti-9-formyl, 263; tosylhydrazone, 263,
    270; tosylhydrazone, sodium salt, 270,
    273
  syn-9-formyl, 262; tosylhydrazone, 262,
    271; tosylhydrazone, sodium salt, 271
  1,2,7,8,9,9-hexadeuterio, 265
  anti-9-hydroxymethyl, 263, 264
  syn-9-hydroxymethyl, 262
  4-methoxy, 175, 285
  9-methoxy, reaction with Fe derivatives,
    290
  anti-9-methoxy, 174, 264, 278
  syn-9-methoxy, 174, 264
  anti-9-methoxycarbonyl: formation, 35,
    36, 263, 264; hydrolysis, 263; proton-
    abstraction, 278; reduction, 263; ther-
    molysis, 264
  syn-9-methoxycarbonyl, 35, 36, 262, 264
  anti-9-(4-methoxyphenyl), 398
  1-methyl, 175, 277, 284
  2-methyl, 175, 277, 284
  3-methyl, 175, 277, 284
  4-methyl, 175, 277, 284

cis-Bicyclo[6.1.0]nona-2,4,6-triene—*cont.*
  anti-9-methyl: cycloaddition reactions, 283, 284; protonation, 274; reaction, with Rh derivative, 291; reduction, 277, 412; Rh complex, 291; thermolysis, 264
  syn-9-methyl: formation, 35, 36, 174; Mo complex, 413; reaction, with transition metal derivatives, 291, 413; reduction, 277; Rh complex, 291; thermolysis, 264
  9-methylene, 268
  Mo complexes, 286, 287–8, 414
  anti-9-(1-naphthyl), 378, 412
  anti-9-(2-naphthyl), 378, 412
  syn-9-(2-naphthyl), 378
  Pd complex, 415
  anti-9-phenyl, 398, 412
  photolysis, 272, 412
  protonation, 274
  reaction: with base, 84, 277; with Cr derivative, 412; with Fe derivatives, 288–90; with iodine azide, 274; with Mo derivative, 286–8; with Pd derivative, 415; with Rh derivatives, 268, 290–291; with W derivative, 413
  reduction, 275, 277
  Rh complexes, 290, 291
  1,3,5,7-tetramethyl, 175
  thermolysis, 264, 268
  anti-9-(*p*-tolyl), 398
  anti-9-(4-trifluoromethylphenyl), 398
  W complex, 413
trans-Bicyclo[6.1.0]nona-2,4,6-triene, 272, 279
  anion radical, 279
Bicyclo[4.2.1]nona-2,4,7-trienone
  cycloaddition reaction, 300
  7,8-dimethyl, 112
  formation, 181
  functional group transformations, 292, 293, 415, 416
  oxime, 85, 292
  photo-electron spectrum, 416
  photolysis, 5, 298
  reaction: with acids, 295; with diazomethane, 299, 416
  Sn complex, 416
  tosylhydrazone, 292, 297; sodium salt, 297, 299, 416
Bicyclo[3.3.0]octa-2,6-diene
  1,2,3,4,4,5,6,7,8,8-decachloro, 163, 166
  4,8-dibromo, 170
  4,5-dibromo-1,2,3,4,6,7,8,8-octachloro, 166
  1,2,3,5,6,7,8-heptamethyl-4-methylene, 160
  1,2,3,5,6,7-hexamethyl-4,8-dimethylene, 161, 170
Bicyclo[4.2.0]octa-2,4-diene
  7-acetoxy, 233

  cis-7-acetoxy-8-methyl, 21
  trans-7-acetoxy-8-methyl, 21
  acetoxymethyl-7,8-dibromo, 251
  3,8-bis(1-cyano-1-methylethyl), 238
  7,8-bis(1-hydroxyethyl), 180
  7,8-bis(1-hydroxy-1-methylethyl), 180, 242
  7,8-bis(1-hydroxy-1-phenylmethyl), 180
  3,8-bis(trimethylsilyl), Fe complex, 239
  7-bromo, 25, 233
  bromo-7,8-dibromo, 251
  7-chloro-8-methylthio, 27
  Co complex, 231
  cyano-7,8-dibromo, 251
  cycloaddition reactions, 219–21, 222–4, 225
  cis-7,8-diacetoxy, 21
  trans-7,8-diacetoxy: cycloaddition reactions, 248–50; formation, 20, 21, 240; photolysis, 246; reaction, with acids, 246; reduction, 242, 244, 246; valence tautomerism, 234
  cis-7,8-diazido, 26
  trans-7,8-dibromo: cycloaddition reactions, 240, 248–50; debromination, 245; dehydrobromination, 83, 245; formation, 24; oxidative cleavage, 240–241; reaction, with iodine azide, 242; reduction, 240, 243–4; ring-cleavage reactions, 247; valence tautomerism, 234
  7,8-dichloro, 3, 240, 243, 245–6, 410
  cis-7,8-dichloro: cycloaddition reactions, 248–50; formation, 23, 34; oxidative cleavage, 240–1; valence tautomerism, 234
  trans-7,8-dichloro: cycloaddition reactions, 248–50; formation, 23, 34; valence tautomerism, 234
  7,7-diethoxy, 257
  dimerisation, 225
  7,7-dimethoxy, 257
  3,8-dimethyl, 237
  cis-7,8-dimethyl, 234
  trans-7,8-dimethyl, 234
  7-(3,3-dimethylbutynyl), 234
  7-ethoxycarbonyl, 35
  Fe complexes, 229, 230
  fluoro-7,8-dibromo, 251
  formation, 29, 214
  7-hydroxy, 233
  7-hydroxy-7-methyl, 260
  methoxycarbonyl-7,8-dibromo, 251
  methoxymethyl-7,8-dibromo, 251
  methyl-7,8-dibromo, 251
  phenyl-7,8-dibromo, 251
  7-phenylethynyl, 234
  photolysis, 216
  Ru complexes, 229, 230

Bicyclo[4.2.0]octa-2,4-diene—*cont.*
   thermolysis, 215
   valence tautomerism, 29, 214
Bicyclo[5.1.0]octa-2,4-diene
   8-acetyl-6-methoxy, 203; Fe complex, 203
   Fe complex, 201
   6-hydroxy, Fe complex, 201
Bicyclo[4.2.0]octa-2,4-dien-7-one, 254, 255,
   261, 411–12
   Fe complex, 411–12
Bicyclo[5.1.0]octa-3,5-dien-2-one (2,3-
   homotropone), 201
   Fe complex, 201
Bicyclo[5.1.0]octadienyl cation
   8-acetyl, Fe complex, 203
   Co complex, 211
   Fe complex, 200, 201
   Ir complex, 211, 212
   methyl, Fe complex, 200
   perdeuterio, Fe complex, 405
   8-phenyl, Nb complex, 404
   Rh complexes, 211, 408
   Ru complex, 208
Bicyclo[3.3.0]octa-1,3,6-triene
   2,3,4,5,6,7-hexachloro-8,8-dimethoxy,167
   perchloro, 161, 165, 166, 167
Bicyclo[3.3.0]octa-1,3,7-triene, perchloro,
   161, 165, 166, 167
Bicyclo[3.3.0]octa-1,4,6-triene, 18, 168, 396
Bicyclo[4.2.0]octa-2,4,7-triene
   7-acetoxy, 101–3
   7-acetoxymethyl-8-methyl,121–2
   7-benzoyl, 101–3
   7,8-bis(acetoxymethyl), 121–2
   7-bromo, 101–3
   7-chloro, 101–3
   1-cyano-7,8-dibutyl, 133
   1-cyano-7,8-diethyl, 133
   7-cyclopropyl, 101–3
   1,4-dimethyl, 119
   1,5-dimethyl, 119
   2,7-dimethyl, 119
   3,7-dimethyl, 119
   7,8-dimethyl, 121–2
   1,6-dimethyl-7,8-diphenyl, 137, 143, 146
   2,5-dimethyl-3,4-diphenyl, 137
   2,7-dimethyl-1,8-diphenyl, 143, 147
   7-ethyl, 101–3
   7-fluoro, 101–3
   formation, 3, 18, 245, 354
   7-(1-ketobut-2-enyl), 101–3
   7-methoxy, 101–3
   7-methoxycarbonyl, 101–3
   7-methyl, 101–3
   octamethyl, 394, 395, 396
   1,2,4,7,8-pentamethyl, 152
   1,2,5,7,8-pentamethyl, 152
   7-phenyl, 101–3
   photolysis, 18

   1,2,7,8-tetrachloro, 394
   1,6,7,8-tetrachloro, 394
   2,3,4,5-tetrachloro, 391, 396
   3,4,7,8-tetramethoxycarbonyl, 136
   1,2,7,8-tetramethyl, 143, 148; Fe complex,
      148, 394
   1,3,5,7-tetramethyl, Fe complex, 149, 150
   1,6,7,8-tetramethyl, Fe complex, 148, 394
   1,3,4,6-tetraphenyl, 136
   2,5,7,8-tetraphenyl, 392, 393
   3,4,7-trimethoxycarbonyl, 133
   7-trimethylsilyl, Fe complex, 198
   1-trimethylsilyl-7-triphenylmethyl, Fe
      complex, 198
   7-triphenylmethyl, Fe complex, 198
   valence tautomerism, 2, 10
Bicyclo[3.3.0]octa-1,3,6-trien-8-one
   2,3,4,5,6,7-hexachloro, 167
   2,3,4,5,7-pentachloro-6-dimethylamino,
      167
   2,3,4,5,7-pentachloro-6-phenylamino, 167
Bicyclo[5.4.2]tetradeca-7,9,11,12-tetraene,
      128, 131, 132
   Mo complex, 132
Bicyclo[6.3.0]undeca-2,4,6-triene, 176, 306–
      307
   Mo complex, 307
   10-(2-tetrahydropyranyloxy), 176
Bis(cyclo-octa-1,3,5,7-tetraenyl)metal
   species, 53, 59

Cycloheptadecaoctaenes, 362
Z,Z,E,Z-Cyclonona-1,3,5,7-tetraenes (as
   intermediates), 267, 283
Z,Z,Z,E-Cyclonona-1,3,5,7-tetraene, 272
Z,Z,Z,E-Cyclonona-1,3,5,7-tetraenes (as in-
   termediates), 266
Z,Z,Z,Z-Cyclonona-1,3,5,7-tetraene, 272,
      277
   9-chloro, 272–3
   Fe complexes, 288, 289, 290, 414
Z,Z,Z,Z-Cyclonona-1,3,5,7-tetraenes (as intermediates), 267
Cyclononatetraenyl anion, 277–8
   chloro, 279
Z,Z,Z-Cyclonona-1,3,5-triene, 7,9-diazido,
      274
Z,Z,Z-Cyclonona-1,3,6-triene
   8-methyl, 277
   9-methyl, 277
Cyclononatrienyl anion, 275–6
   8-methyl, 276
Z,Z,Z,E-Cyclo-octa-1,3,5,7-tetraene, 18
   1,3,6,8-tetraphenyl, 145
Z,Z,Z,Z-Cyclo-octa-1,3,5,7-tetraene
   acetoxy, 92–3, 97–8, 101–3, 385, 419
   4-acetoxybutyl, 86
   2-acetoxyethyl, 87
   1-(2-acetoxyethyl)-6-methyl, 114

Z,Z,Z,Z-Cyclo-octatetraene—*cont.*
1-acetoxymethyl-2-methyl (1-acetoxy-methyl-8-methyl), 113, 121–2
3-acetoxypropyl, 86
acetyl, 79, 90; Fe complex, 202
acylation, 26
Ag complexes, 4, 49, 51–2, 189–90
allyldimethylsilyl, 81, 106–7; Fe complex, 106–7
4-aminobutyl, 86
3-aminopropyl, 87
anion radical, 28, 29, 377
Au complexes, 52
benzoyl: Ag complex, 106; cycloaddition reaction, 101–3; formation, 81, 92; hydrogenation, 101; reaction with phenylmagnesium bromide, 97
benzyl, 384, 398; dianion, 398
4-biphenyl, 80; anion radical, 100
1,2-bis(acetoxymethyl), 109, 121–2
1,8-bis(acetoxymethyl), 109, 114
1,8-bis(bromomethyl) (1,2-bis(bromo-methyl)), 114, 117
1,8-bis(chloromethyl) (1,2-bis(chloro-methyl)), 114, 117
1,8-bis(1-hydroxyethyl) (1,2-bis(1-hy-droxyethyl)), 112
1,8-bis(hydroxymethyl) (1,2-bis(hydroxy-methyl)), 114, 121, 123
bromo: bromination, 99; Fe complex, 106, 204; formation, 83; reaction with K, 100; substituted COTs from, 81–2, 83–4, 92, 387; thermal rearrangement, 97
bromodideuteriomethyl, 91
3-bromo-3,3-dideuteriopropyl, 88
2-bromoethyl, 87
bromomethyl, 90
1-bromo-4-methyl, 110
1-bromo-5-methyl, 110
1-bromomethyl-2,3,4,5,6,7-hexadeuterio, 91
t-butoxy, 83, 94; anion radical, 100; K salt, 385
butyl, 79, 80, 101; anion radical, 100; dianion, 185; Np complex, 185; Pu complex, 185; U complex, 185
t-butyl, 79–80, 384
4-t-butylphenyl, 80; anion radical, 100
carbonylation, 34
cation radical, 21
Ce complexes, 183–5, 379, 400, 403
chloro: cycloaddition reactions, 101–3, 104–5; formation, 83; hydrogenation, 101; rearrangement, 97–8
chlorodideuteriomethyl, 91
1-chlorodideuteriomethyl-2,3,4,5,6,7-hexadeuterio, 91
Z-2-chloro-1-fluorovinyl, 301

chloromethyl, 90, 95
1-chloromethyl-2,3,4,5,6,7- hexadeuterio, 91
Co complexes: cycloaddition reaction, 408; formation, 71, 72, 189, 211; protonation, 211; structures, 71, 72–3, 384
Cr complexes: formation, 57–8, 59, 188, 380; protonation, 194; structures, 58, 59, 380
Cu complexes, 49–51, 379
cyano, 81, 82, 85, 92, 104–5
3-cyano-1,2-diethyl (2-cyano-1,8-diethyl), 133, 134
2-cyanoethyl, 87
1-cyano-2,3,4,5,6,7-hexadeuterio, 80, 102–3
cyanohydroxymethyl, 385
cyanomethyl, 90, 96; Fe complex, 202
1-cyanomethyl-4-methyl, 387
3-cyanopropyl, 79, 86
cycloaddition reactions: with *p*-benzo-quinone, 39, 42, 46–7; with carbenes, 35–7, 378; with dienophiles, 42–6, 378; with 1,3-dipolar reagents, 40–2; with electron-deficient dienes, 47–8; with monosubstituted COTs, 101; with nitrenes, 37–8; with polyhaloethylenes, 39; with silylenes, 38; with sulphur, 38, 378; with sulphur dioxide, 40, 378; with sulphur monoxide, 39–40
cyclobuteno, 126, 127, 303; anion radical, 390; dianion, 303
cyclo-octatetraenyl (bicyclo-octatetra-enyl): anion radical, 100, 386; dianion, 386; Fe complexes, 386; formation, 81, 385; hydrogenation, 101; K salt, 385; Os complex, 387; reaction, with transition metal derivatives, 108, 386, 387; Ru complexes, 108, 386–7
cyclopropyl, 81, 101–3; dianion, 185; U complex, 185, 191
deuterio, 81; anion radical, 100
1,8-diacetyl (1,2-diacetyl), 112, 124
dianion: acylation, mechanism, 178, 179; as dechlorinating reagent, 403; as reducing agent, 172, 377, 398; formation, 28, 30–1, 48–9, 52, 53, 378; photo-ionisation, 379; protonation, 28, 173; structure, 28, 49, 377, 378
dianion, reaction: with actinide deriva-tives, 185, 187; with alkyl halides, 173; with carbon dioxide, 178, 179, 398–9; with carbonyl chloride, 181; with carbonyl compounds, 178, 180; with carboxylic acid anhydrides, 177–8; with carboxylic acid chlorides, 176–7; with carboxylic esters, 178, 181; with chlorodi- and chlorotrimethylsilane, 174; with Co derivatives, 189; with

Z,Z,Z,Z-Cyclo-octatetraene—*cont.*
dianion, reaction—*cont.*
  COT, 28; with dichlorophenylphosphine, 175; with dihalo-compounds, 174, 175–6, 398; with 1,1-dimethoxytrimethylamine, 182; with dimethylcarbamoyl chloride, 181; with group IVA metal derivatives, 187, 400; with group VA metal derivatives, 187–8, 400; with group VIA metal derivatives, 188; with Ir derivative, 401; with isoamyl nitrite, 182; with lanthanide derivatives, 183–5; with Ni derivatives, 189; with Os derivative, 189; with Pt derivative, 189; with Rh derivatives, 401–2; with Ru derivatives, 188–9, 400; with Sc derivatives, 399; with Th derivative, 190; with thiophene derivatives, 182–3, 399; with Tl derivative, 187; with U derivative, 190; with Y derivative, 184, 185
1,4-dibromo (1,6-dibromo), 110, 111, 118, 387, 388
1,5-dibromo, 111
1,2-dibromo-3,4,5,6,7,8-hexachloro, 158
1,6-dibromo-2,3,4,5,7,8-hexachloro, 158
1,2-dibutyl-3-cyano, 133
dication, 393
1,2-dicyano, 387
1,5-dicyano-2,6-bis(diethylamino)-3,4,7,8-tetraphenyl, 157
1,2-didehydro, 83–4, 124, 164, 245; anion radical, 100
1,4-dideuterio, 17
1,4-dideuterio-2,3-dimethyl, 112, 118, 121
4,4-dideuterio-4-hydroxybutyl, 87; 4-nitrobenzenesulphonate ester, 87
1,1-dideuterio-2-hydroxyethyl, 88
2,2-dideuterio-2-hydroxyethyl, 88
dideuteriohydroxymethyl, 91
3,3-dideuterio-3-hydroxypropyl, 88; 4-nitrobenzenesulphonate ester, 88
diethylamino, 84
1,2-diformyl (1,8-diformyl), 114, 115, 122, 123
dimers, *see*: tetracyclo[8.4.2.0^{2,9}.0^{3,8}]-hexadeca-4,6,11,13,15-pentaene (dimer (193)); tricyclo[8.6.0.0^{2,9}]hexadeca-3,5,7,11,13,15-hexaene (dimer (9)); heptacyclo[7.5.2.0^{2,7}.0^{3,13}.0^{6,12}.0^{8,10}.0^{11,14}]hexadeca-4,15-diene (dimer (11)); heptacyclo[10.2.2.0^{2,11}.0^{3,10}.0^{4,14}.0^{5,9}.0^{8,13}]hexadeca-6,15-diene (dimer (12)); pentacyclo[9.3.2.0^{2,9}.0^{3,8}.0^{10,12}]hexadeca-4,6,13,15-tetraene (dimer (10))
1,2-dimethoxycarbonyl, 109–10, 115, 120, 122, 387
1,8-dimethoxycarbonyl: formation, 109–110, 387; functional group transformations, 114; structure, 115
1,2-dimethyl (1,8-dimethyl): Ag complex, 122; cycloaddition reactions, 121–2; dianion, 120; formation, 109, 112, 114; hydrogenation, 121; p.m.r. spectrum, 388; thermolysis, 118
1,3-dimethyl, 111, 118; dianion, 120
1,4-dimethyl (1,6-dimethyl), 110, 112, 118; anion radical, 388; dianion, 120
1,5-dimethyl, 111, 118, 122; anion radical, 388; dianion, 120
dimethylamino, 84, 96
2-dimethylaminoethyl, 79, 87
dimethylaminomethyl, 90
4-dimethylaminophenyl, 80, 101; anion radical, 100
3-dimethylaminopropyl, 86
1,4-dimethyl-2,3-diphenyl, 143, 147
3,8-dimethyl-1,2-diphenyl, 137, 143, 146
1,4-dinitro (1,6-dinitro), 118, 239
1,2-diphenyl (1,8-diphenyl), 109, 110, 121
1,3-diphenyl, 110, 121
1,4-diphenyl (1,6-diphenyl), 109, 110, 121
1,5-diphenyl, 109, 110, 121
Er complex, 185, 379
ethoxy, 83, 94, 257, 385, 419
ethoxycarbonyloxy, 385, 419
ethoxymethyl, 90
ethyl: Ag complex, 106; anion radical, 100; cycloaddition reactions, 101–3; dianion, 185; Fe complex, 106; formation, 80; hydrogenation, 101; Np complex, 185; Pu complex, 185; U complex, 185, 191
Fe complexes: AlCl$_3$-catalysed reactions, 201, 202, 203; catalytic activity, 196; cycloaddition reactions, 204–7, 407; electrophilic substitution, 82, 202; formation, 60–2, 63, 64, 382; hydration, 406; ligand displacement reactions, 196, 197, 204, 406–7; photoreactions, 198, 199; protonation, 199–201, 405; reduction, 406; review, 404; structures, 61, 62, 63, 71, 199, 382, 405; tritylation, 204, 385, 406
fluoro: cycloaddition reactions, 101–3, 104–5; formation, 81, 92; structure, 94; thermolysis, 97
formation, 1–4
formyl, 81, 82, 90, 385; Fe complex, 81, 82, 202
1-formyl-2-methyl (1-formyl-8-methyl), 113
formyl phenyl, Fe complex, 203
Gd complex, 184
halogenation, 23–5, 83
1,2,4,5,6,7-hexachloro-3,8-dimethyl, 158
1,2,3,4,5,6-hexadeuterio-7-deuteriomethyl, 107; Fe complex, 106, 108

Z,Z,Z,Z-Cyclo-octatetraene—*cont.*
  1,2,3,4,5,6-hexadeuterio-7-dideuterio-
    hydroxymethyl, 91
  1,2,3,4,5,6-hexadeuterio-7-ethoxycar-
    bonyl, 80, 94
  1,2,3,4,5,6-hexadeuterio-7-(1-hydroxy-1-
    methylethyl), 91, 94
  1,2,3,4,5,6-hexadeuterio-7-methoxy-
    carbonyl, 80, 91, 97, 102–3
  1,2,3,4,5,6-hexadeuterio-7-methyl, 97,
    102–3
  1,2,3,4,5,6-hexadeuterio-7-phenyl, 80, 97,
    102–3
  Hf complexes, 53, 187, 193, 404
  Ho complex, 185
  hydroformylation, 34
  hydrohalogenation, 25
  1-hydroxybut-2-enyl, 92
  3-hydroxybut-1-enyl, 92
  1-hydroxybutyl, 92
  4-hydroxybutyl, 86; 4-nitrobenzenesul-
    phonate, 86, 95–6
  1-hydroxyethyl, 79, 90; Fe complex, 202
  2-hydroxyethyl: 4-bromobenzenesul-
    phonate, 95–6; formation, 79, 81, 82,
    90; functional group transformations,
    87, 88; hydrogenation, 101; 4-nitro-
    benzenesulphonate, 87, 89; 4-toluene-
    sulphonate, 87
  1-(2-hydroxyethyl)-8-methyl, 113
  1-(2-hydroxyethyl)-4-methyl (1-(2-
    hydroxyethyl)-6-methyl), 111, 123; Fe
    complex, 123; 4-nitrobenzoate, Fe
    complex, 123
  1-(2-hydroxyethyl)-5-methyl, 111
  hydroxymethyl, 79, 90, 101; Fe complex,
    202
  1-hydroxy-1-methylethyl, 79, 90
  1-hydroxymethyl-8-methyl, 113; α-meth-
    oxy-α-phenylacetate ester, 113, 115
  1-hydroxy-1-phenylmethyl, 81, 92, 96,
    101
  1-hydroxyprop-2-enyl, 81, 92
  3-hydroxyprop-1-enyl, 92
  3-hydroxypropyl, 79, 82, 86; 4-nitro-
    benzenesulphonate, 86, 95–6; 4-toluene-
    sulphonate, 86
  iodo, 104–5
  Ir complexes: formation, 74, 75, 401, 408;
    ligand displacement reaction, 213; pro-
    tonation, 211, 212
  isopropoxy, 94; dianion, 100
  isopropyl, 85, 94
  1-ketobut-2-enyl, 81, 92, 101–3
  1-ketobutyl, 92
  La complexes, 184, 379
  methoxy: cycloaddition reactions, 101–3,
    104, 105–6; dianion, 100; Fe complexes,
    106, 204, 386, 404; formation, 83; hydro-

  lysis, 385; protonation, 99; structure,
    94
  methoxycarbonyl: cycloaddition reac-
    tions, 101–3, 104–5; Fe complex, 106,
    204; formation, 79–80, 90, 384
  1-methoxycarbonyl-2-methyl (1-meth-
    oxycarbonyl-8-methyl), 109–10, 113,
    115
  1-methoxycarbonyl-2-phenyl (1-methoxy-
    carbonyl-8-phenyl), 109–10, 115
  1-methoxyethyl, Fe complex, 202
  methoxymethyl, 83, 104–5; Fe complex,
    83, 202
  4-methoxyphenyl, 81, 105–6
  2-methoxyvinyl, 90
  1-(2-methoxyvinyl)-8-methyl, 113
  methyl: Ag complex, 106; anion radical,
    100, 275; cycloaddition reactions, 101–
    103, 104–5, 105–6; dianion, 100; Fe
    complex, 106–8, 200, 204; formation,
    79, 81, 84–5; hydrogenation, 101;
    protonation, 99; reaction with transi-
    tion metal derivatives, 106, 108
  Mn complexes, 60, 381
  Mo complexes: formation, 57, 188, 195,
    380–1; i.r. and u.v. spectra, 404; pro-
    tonation, 195; structures, 58, 59, 380,
    381
  N-morpholinomethyl, Fe complex, 202
  Nb complexes, 188, 194, 400, 404
  Nd complexes, 183, 184, 190, 379
  Ni complexes: formation, 75, 76, 189,
    213; ligand displacement reaction, 213;
    structures, 76, 77, 384
  Np complexes, 185, 187
  octabromo, 157
  octachloro: chlorination, 163; formation,
    157, 158, 394; rearrangement, 161, 163
  octadeuterio, 2; Fe complex, 382
  octafluoro, 395
  octakis(4-chlorophenyl), 155
  octakis(4-methoxyphenyl), 155
  octakis(*p*-tolyl), 155
  octakis(trifluoromethyl), 155, 161, 394
  octamethyl: dianion, 186; epoxidation,
    161; formation, 153, 158, 159, 394;
    protonation, 161; reaction, with dieno-
    philes, 395; structure, 159, 160; thermal
    rearrangement, 160; U complex, 186
  octaphenyl, 153, 154, 155, 159
  oligomerisation, 11–17
  oligodeuterio, 31
  Os complexes: formation, 70, 189,
    382–3; ligand displacement reactions,
    211; protonation, 211; review, 404;
    structures, 70, 71
  oxidation, 19–21
  Pa complex, 185
  Pd complexes, 77

Z,Z,Z,Z-Cyclo-octatetraene—*cont.*
pentadeuteriophenyl, 80; anion radical, 100
1,2,3,4,6-pentamethyl, 151, 152, 394
1,2,3,5,8-pentamethyl, 151, 152, 394
phenoxy, 83, 104
phenyl: Ag complex, 106; anion radical, 100; arylation, 110; cycloaddition reactions, 101–6; dianion, 100, 185; Fe complex, 106, 203, 204; formation, 79, 80, 81; hydrogenation, 101; Pd complexes, 108; protonation, 99; Pt complexes, 108; reaction, with transition metal derivatives, 106, 108; Ru complex, 208; U complex, 185, 400; valence tautomerism, 94
phenyl triphenylmethyl, Fe complex, 204
photolysis, 18
physical properties, 6–11, 377
polymerisation, 34
Pr complexes, 183, 184
prop-2-enyl, 82
propyl, 79, 101, 106; Ag complex, 106; anion radical, 100
protonation, 21–2
Pt complexes, 77, 213
Pu complexes, 53, 185, 187
purification, 4
radiolysis, 18
reaction: with actinides, 53; with alkali metal alkoxides, 30–1; with alkali metals, 28–9, 48–9; with alkane-sulphenyl chlorides, 27; with boron trichloride, 27; with Co derivatives, 71–3; with Fe derivatives, 60–4; with free radicals, 32–4, 378; with group IB metal derivatives, 49–52, 379; with group IVA metal derivatives, 53–7; with group VA metal derivatives, 57; with group VIA metal derivatives, 57–9, 380–1; with group VIIA metal derivatives, 59–60; with iodine azide, 26; with Ir derivatives, 74–5, 408; with lanthanides and their derivatives, 52–3, 184, 379; with metal alkyls and aryls, 32, 80, 384; with Mg, 52, 190; with Ni derivatives, 75–6; with Os derivatives, 70, 382–3; with Pd derivatives, 77; with Pt derivative, 77; with Rh derivatives, 73–4; with Ru and its derivatives, 64–5, 67–9, 70, 382; with sulphur dichloride, 27; *see also* cycloaddition reactions *subentry*
reduction, 28–30, 377
Rh complexes: formation, 73, 74, 212, 401, 402, 408; ligand displacement reaction, 213; protonation, 211–12, 408; structures, 73, 74, 212, 402, 403

Ru complexes: cycloaddition reactions, 210; formation, 64–5, 67–9, 188–9, 208, 211, 382, 400; ligand displacement reaction, 210; protonation, 208, 407; review, 404; structures, 65–7, 69, 70, 208, 382, 400, 401
Sc complexes, 399, 403
Sm complexes, 183, 184
structure, 8–11
Ta complexes, 57, 194, 400, 404
Tb complex, 184
1,2,3,4-tetrachloro, 390, 392, 393
1,2,3,8-tetrachloro, 390, 392, 393
1,3,5,7-tetraethoxy, 134, 143
1,2,4,6-tetraethoxycarbonyl (1,3,5,8-tetra-ethoxycarbonyl), 134, 140, 144, 146
1,3,5,7-tetraethoxycarbonyl, 141, 146
1,3,5,7-tetraisopropenyl, 141
1,3,5,7-tetrakis(dimethylamino)-2,4,6,8-tetramethoxycarbonyl, 156
1,3,5,7-tetrakis(hydroxymethyl), 134
1,2,4,6-tetrakis(1-hydroxy-1-methylethyl), 134, 141
1,2,4,7-tetrakis(1-hydroxy-1-methylethyl), 134, 141
1,3,5,7-tetrakis(1-hydroxy-1-methylethyl), 141
1,2,4,7-tetrakis(*p*-methoxyphenyl)-3,5,6,8-tetrakis(*p*-tolyl), 155
tetramer, 14, 375–6
1,2,4,6-tetramethoxycarbonyl (1,3,5,8-tetramethoxycarbonyl), 134, 140, 146
1,2,5,6-tetramethoxycarbonyl (1,4,5,8-tetramethoxycarbonyl), 134, 136, 142
1,2,3,4-tetramethyl: dianion, 145; epoxidation, 393; formation, 137, 392; structure, 142, 143, 145
1,2,3,4-tetramethyl, reaction: with Fe derivative, 148; with 4-phenyltriazo-line-3,5-dione, 147, 393
1,2,3,8-tetramethyl: formation, 137, 392; structure, 142, 143, 145
1,2,3,8-tetramethyl, reaction: with Fe derivative, 148; with 4-phenyl-1,2,4-triazoline-3,5-dione, 147, 393
1,2,4,5-tetramethyl, 137, 147
1,2,4,7-tetramethyl (1,3,6,8-tetramethyl), 137, 147
1,2,5,6-tetramethyl, 137, 147, 392
1,3,5,7-tetramethyl: anion radical, 145, 393; Cr complex, 148; dianion, 145, 147, 175, 393; dication, 393; Fe complexes, 148–51; formation, 135, 137; hydrogenation, 146; Mo complex, 148; Np complex, 186; protonation, 393; structure, 142; thermal isomerisation, 144; U complex, 186, 400; W complex, 148
1,4,5,8-tetramethyl, 392

Z,Z,Z,Z-Cyclo-octatetraene—*cont.*
1,2,4,7-tetraphenyl, 390, 392, 393; *see also* 1,3,6,8-tetraphenyl *subentry*
1,3,5,7-tetraphenyl, 134, 137, 146, 390; dianion, 146; U complex, 186
1,3,6,8-tetraphenyl, 136, 137, 145, 146, 392
1,3,5,7-tetrapropoxy, 134
Th complex, 53, 185, 186, 379
thermolysis, 17–18
Ti complexes: formation, 53–7, 187, 192–193, 400; ligand metallation reaction, 404; spectra, 54, 57, 192, 380; structures, 54–7, 193
Tl complexes, 187, 191
p-tolyl, 80; anion radical, 100
1,2,5-trimethoxycarbonyl (1,4,5-trimethoxycarbonyl), 133
1,2,3-trimethyl, 390; dianion, 390
trimethylgermyl, 81, 106, 385; Fe complex, 106, 204
1-trimethylgermyl-3-triphenylmethyl, Fe complex, 204
2,4,6-trimethylphenyl, 80; anion radical, 100
trimethylsilyl, 81, 106, 385; Fe complex, 106, 198, 204; Ru complex, 208
1-trimethylsilyl-3-triphenylmethyl, Fe complex, 198, 204
trimethylstannyl, 81, 106, 385; Fe complex, 106
trityl, 106, 385; Fe complex, 106, 198, 204, 385, 406
U complex, 53, 185, 186, 379, 400, 404
V complex, 187, 400
valence tautomerism, 9–10
vinyl, 2, 81, 101, 103; dianion, 185; U complex, 185
W complexes, 57, 188, 195
Y complexes, 184, 185, 190
Zr complexes: formation, 53, 54, 56, 194; spectra, 404; structures, 55, 380, 400
Cyclo-octa-1,3,5,7-tetraenecarboxylic acid
formation, 81, 82, 90
hydrogenation, 101
2-hydroxymethyl, lactone, 123, 132; Fe complex, 122, 132
structure, 93
2,4,6,8-tetraphenyl, 152
Cyclo-octa-1,3,5,7-tetraene-1,2-dicarboxylic acid (cyclo-octa-1,3,5,7-tetraene-1,8-dicarboxylic acid), 114, 115, 116, 121
anhydride, 116
calcium salt, 116
Cyclo-octa-1,3,5,7-tetraene-1,3,5,7-tetracarboxylic acid, 141
Cyclo-octa-1,3,5,7-tetraene-1,3,5,8-tetracarboxylic acid, 140

anhydride, 140
Cyclo-octa-1,3,5,7-tetraenylacetic acid, 88
amide, 90
4-methyl, methyl ester, 387
4-Cyclo-octa-1,3,5,7-tetraenylbutanoic acid, 86
methyl ester, 86, 87
Cyclo-octa-1,3,5,7-tetraenyldideuterioacetic acid, methyl ester, 88
Cyclo-octa-1,3,5,7-tetraenyl-lithium, 81, 82, 385
Cyclo-octa-1,3,5,7-tetraenylmagnesium bromide, 82
3-Cyclo-octa-1,3,5,7-tetraenylpropanoic acid, 86, 87
amide, 87
ethyl ester, 81, 82, 86
methyl ester, 88
3-Cyclo-octa-1,3,5,7-tetraenylprop-2-en-1-al, 92
Cyclo-octa-1,3,5-triene
7-acetoxy, 233, 234
Ag complexes, 214, 226
7-azido, 235
trans-7,8-bis(4-bromophenylethynyl), 247
3,8-bis(1-cyano-1-methylethyl), 238
trans-7,8-bis(3,3-dimethylbutynyl), 247
7,8-bis(dimethylsilyl), 174
7,8-bis(diphenylhydroxymethyl), 180
7,8-bis(1-hydroxyethyl), 180
7,8-bis(1-hydroxy-1-methylethyl), 180
7,8-bis(1-hydroxy-1-phenylmethyl), 180
trans-7,8-bis(3-methylbut-3-enynyl), 247
7,8-bis(methylene), 118, 124, 125, 126
trans-7,8-bis(*p*-tolylethynyl), 247
trans-7,8-bis(2,4,6-trimethylphenylethynyl), 247
7-bromo, 25, 233, 234, 235
7-bromomethylene-8-methylene, 117, 126
butyl, 32
7-chloromethylene-8-methylene, 117, 125, 126
Co complexes, 231
Cr complexes, 194, 226, 227, 409
7-cyano, Co complex, 211
7-cyanoimino, 37, 42, 254
7-cyanomethylene, 96
cycloaddition reactions: with carbenes, 218, 219, 409; with chlorosulphonyl isocyanate, 222; with cyclopentadienones, 224; with dienophiles, 219–21; with electron-deficient dienes, 223; with sulphur dioxide, 409; with tropone, 225
7-(cyclo-octa-2,4,6-trienyloxy), 233, 234
trans-7,8-diacetoxy, 234
cis-7,8-diazido, 26
cis-7,8-dibromo, 24, 83, 236, 237
trans-7,8-dibromo, 24, 234, 247

Cyclo-octa-1,3,5-triene—*cont.*
  7,8-dibromo-1,4-dinitro, 118
  cis-7,8-dichloro, 23, 83, 234, 236
  trans-7,8-dichloro, 23, 234, 236
  7,8-dideuterio, 30, 173, 233
  7,7-diethoxy, 257
  trans-7,8-diethynyl, 247
  7,8-dihydroxy, 19
  dimerisation, 225
  7,7-dimethoxy, 257, 258
  3,8-dimethyl, 237
  cis-7,8-dimethyl, 173, 234
  trans-7,8-dimethyl, 173, 234
  4-dimethylaminophenyl, 32
  7-(3,3-dimethylbutynyl), 234
  trans-7,8-diphenylethynyl, 247
  trans-7,8-diprop-1-ynyl, 247
  epoxidation, 408
  7-ethoxy-8-methylene, 95
  ethyl, 32
  Fe complexes, 228, 229, 230, 406
  formation, 28, 29, 30
  halogenation, 1, 217
  hydrogenation, 217
  hydrohalogenation, 217
  7-hydroxy, 233, 234, 235, 257; Fe complex, 406
  7-(1-hydroxyethyl), 238
  7-hydroxy-7-methyl, 260
  7-hydroxy-8-methylene, 95
  7-hydroxy-8-phenylmethylene, 96
  Ir complexes, 233
  7-methoxy, Pd complex, 77
  7-methyl, Cr complex, 194
  7-methylene, 84, 277
  Mo complexes, 226, 227, 409, 414
  perdeuterio, 352
  phenyl, 32
  7-phenylethynyl, 234
  photolysis, 216
  protonation, 217
  purification, 214
  reaction: with base, 31, 217, 225; with Ag
    derivatives, 226; with Co derivatives,
    231, 232; with Fe derivatives, 228, 229,
    230; with group VIA metal deriva-
    tives, 226–7, 409; with Ir derivative,
    233; with Mn derivatives, 228, 410;
    with Rh derivatives, 232; with Ru
    derivatives, 229, 230, 231; with V
    derivative, 226; *see also* cycloaddition
    reactions *subentry*
  Rh complexes, 232, 410
  Ru complexes, 189, 229, 230, 231
  thermal isomerisation, 214
  thermolysis, 215
  3,7,8-tribromo, 111
  7-trichlorosilyl, 34
  7-triethylsilyl, 34

  1,3,5-trimethyl-7-methylene, Fe complex,
    149, 150
  V complex, 226
  valence tautomerism, 29, 214, 219
  W complex, 226, 227
Cyclo-octa-1,3,6-triene
  Ag complex, 214, 226
  5,8-bis(1-cyano-1-methylethyl), 33, 238
  5,8-bis(difluoroamino), 33
  5,8-bis(dimethylsilyl), 174
  5,8-bis(diphenylhydroxymethyl), 180, 238
  5,8-bis(1-hydroxyethyl), 180, 238
  5,8-bis(1-hydroxy-1-methylethyl), 180,
    238
  5,8-bis(1-hydroxy-1-phenylmethyl), 180
  5,8-bis(trimethylsilyl), 174, 239
  butyl, 32
  Cr complex, 409
  5-cyanomethylene, 96
  cycloaddition reactions, 218, 219–20, 221,
    224
  5,8-dideuterio, 173
  5-diethylsilyl, 34
  trans-5,8-dimethyl, 173, 237
  4-dimethylaminophenyl, 32
  5,8-dinitro, 33, 239
  epoxidation, 408
  ethyl, 32
  formation, 28, 29, 243
  hydrogenation, 217
  5-methylene, 277
  Mn complex, 228
  Mo complexes, 227, 409
  phenyl, 32
  photolysis, 216
  purification, 214
  reaction: with base, 29, 215, 225; with car-
    benes, 218, 219, 409; with dienophiles,
    219, 221; with electron-deficient dienes,
    224; with nitrosobenzenes, 221; with
    peroxy-compounds, 408; with transi-
    tion metal derivatives, 226, 228, 231,
    409
  thermal isomerisation, 29, 214
  5-triethylsilyl, 34
  2,5,8-tris(trimethylsilyl), 239
  W complex, 227
Cyclo-octa-1,3,5-triene-*trans*-7,8-di-
  carboxylic acid, 179
Cyclo-octa-1,3,6-triene-5,8-dicarboxylic
  acid, 179
Cyclo-octa-2,5,7-trien-1,4-dione, 240
Cyclo-octa-2,4,6-trienone
  acetals, 257
  8-benzyl, 419
  8-bromo, 419
  Cr complex, 411
  8-deuterio, 428
  enolate anion, 385, 419, 420

Cyclo-octa-2,4,6-trienone—*cont.*
 enolisation, 254
 8-ethyl, 419
 Fe complexes, 107, 262, 404, 411
 formation, 20, 96, 254, 385, 419
 8-methyl, 419
 oximes, 258; benzenesulphonate esters,
  258
 8-phenylseleno, 419
 8-phenylthio, 419
 photolysis, 255–6
 8-(prop-2-enyl), 419
 protonation, 256
 reaction: with amines, 259; with bases,
  257; with cyanoacetic esters, 259; with
  Grignard reagents, 260; with hydroxylamine, 258; with lithium diethylcuprate, 261; with malononitrile, 259;
  with metal alkyls, 260; with sodium
  azide, 259; with transition metal
  derivatives, 262, 411
 reduction, 257
 thermolysis, 254–5
 2,4,6-triethoxy, 143
 valence tautomerism, 254, 261, 411
Cyclo-octatrienyl anion, 173, 217

9,10-Diazatricyclo[4.2.2.0^{2,5}]deca-3,7-diene,
  3, 354
 9,10-dibenzoyl, 354, 357
 9,10-diethoxycarbonyl, 249, 354, 355,
  357
 9,10-dimethoxycarbonyl-1,2,3,4,5,6,7,8-
  octamethyl, 395
9,10-Diazatricyclo[4.2.2.0^{2,5}]deca-3,7-
  diene-9,10-dicarboxylic acid, *N*-
  methylimide, 249, 354, 355, 357
 1,4-dimethyl-2,3-diphenyl, 147
 2-fluoro, 105
 1-phenyl, 105
 3-phenyl, 105, 356
9,10-Diazatricyclo[4.2.2.0^{2,5}]deca-3,7-
  diene-9,10-dicarboxylic acid, *N*-
  phenylimide, 46, 249, 355, 357
 1-acetoxymethyl, 251, 356
 2-acetoxymethyl, 251, 356
 3-acetoxymethyl, 105, 356
 7-acetoxymethyl, 251, 356
 2-bromo, 105
 3-bromo, 105
 7-bromo, 251
 2-chloro, 105
 3-chloro, 105
 1-cyano, 105, 251, 356
 2-cyano, 105, 251, 356
 3-cyano, 105, 356
 7-cyano, 105, 251
 2-fluoro, 105
 3-iodo, 105

 1-methoxycarbonyl, 105, 251, 356
 3-methoxycarbonyl, 105
 7-methoxycarbonyl, 105
 1-methoxymethyl, 105, 251, 356
 2-methoxymethyl, 105, 251, 356
 3-methoxymethyl, 105, 356
 2-methyl, 251, 356
 3-methyl, 105, 356
 1,2,3,4,5,6,7,8-octamethyl, 395
 2-phenyl, 251, 356
 7-phenyl, 251, 356
 1,2,3,4-tetramethyl, 147
 2,3,4,5-tetramethyl, 147
[1,2:4,5]Dicyclo-octatetraenocyclohexa-1,4-
  diene, 125, 131

Heptacyclo[7.5.2.0^{2,7}.0^{3,13}.0^{6,12}.0^{8,10}.0^{11,14}]-
  hexadeca-4,15-diene (dimer (11))
 Ag complexes, 373
 formation, 11–14, 17
 photo-isomerisation, 372
 reduction, 372
 structure, 11, 13
 thermal isomerisation, 13–14, 372
Heptacyclo[10.2.2.0^{2,11}.0^{3,10}.0^{4,14}.0^{5,9}.0^{8,13}]-
  hexadeca-6,15-diene (dimer (12))
 Ag complex, 13, 375
 cycloaddition reactions, 373–5
 epoxidation, 373
 formation, 11–14, 17
 hydrogenation, 373
 structure, 11, 13–14
1,2-Hexamethylenecyclo-octa-1,3,5,7-
  tetraene, 124
 methyl, 124
Homotropylium ion, 21, 22, 214, 377
 endo-8-bromo, 189
 exo-8-bromo, 189, 236
 endo-8-chloro, 23, 24, 236
 exo-8-chloro, 236
 Cr complexes, 194, 227
 endo-8-deuterio, 22
 exo-8-deuterio, 22
 1-hydroxy, 256
 exo-8-hydroxy, 312
 1-methoxy, 99, 258
 1-methyl, 99
 Mo complex, 195, 227
 1-phenyl, 99
 endo-8-sulphinate–SbF$_5$ complex, 421;
  dimethyl, 421; methyl, 421
 1,3,5,7-tetramethyl, 393
 W complex, 195, 227

Oxa[17]annulenes, 360
9-Oxabicyclo[6.1.0]nona-2,4,6-triene
 Ag complex, 316
 Cu complex, 316
 epoxidation, 312

9-Oxabicyclo[6.1.0]nona-2,4,6-triene—*cont.*
formation, 19
Pd complex, 317
photolysis, 311
protonation, 312
Pt complex, 317
reaction, with bases, 314, 385, 419; with
3,6-diphenyl-*s*-tetrazine, 420; with
ethyl-lithium, 314; with Grignard
reagents, 314–15; with maleic anhydride, 315; with transition metal
derivatives, 315–17
reduction, 313
Rh complexes, 317
thermolysis, 311
valence tautomerism, 315
Z,Z,Z,E-Oxonin, 311
Z,Z,Z,Z-Oxonin, 311

Pentacyclo[9.3.2.0^{2,9}.0^{3,8}.0^{10,12}]hexadeca-
4,6,13,15-tetraene (dimer (**10**))
Ag complex, 371
dimerisation, 14
Fe complexes, 63, 367, 371–2
formation, 11–16, 375
hydrogenation, 368
n.m.r. spectra, 13, 377
photolysis, 368
reaction: with base, 368; with dienophiles,
368–9, 370, 428; with transition metal
derivatives, 371–2
structure, 13
thermal isomerisation, 12, 13, 368, 375
1,2-Pentamethylenecyclo-octa-1,3,5,7-
tetraene, 127, 128, 129
3,4-dideuterio, 129
9-Phosphabicyclo[4.2.1]nona-2,4,7-triene
anti-9-phenyl, 322, 323; 9-oxide, 322; Pd
complex, 323
syn-9-phenyl, 317, 322, 323; 9-oxide, 322;
Pd complex, 323
9-Phosphabicyclo[6.1.0]nona-2,4,6-triene,
9-phenyl, 175, 317
[4.3.2]Propella-2,4,10-triene, 128
[4.4.2]Propella-2,4,11-triene, 128, 129
11-deuterio-12-methyl, 128
2,5-dideuterio, 129
11,12-dideuterio, 128, 129
3,4-dimethyl, 129, 388
11,12-dimethyl, 128, 129
2-methyl, 128, 129
3-methyl, 128, 129
11-methyl, 129
[5.4.2]Propella-8,10,12-triene, 128, 129
12,13-dideuterio, 129

Semibullvalene, *see* Tricyclo[3.3.0.0^{2,8}]octa-
3,6-diene

9-Silabicyclo[4.2.1]nona-2,4,7-triene, 9,9-
dimethyl, 190, 320–1

Tetracyclo[8.4.2.0^{2,9}.0^{3,8}]hexadeca-
4,6,11,13,15-pentaene (dimer (**193**)), 67–8
Ru complex, 64, 66–8
1,2-Tetramethylenecyclo-octa-1,3,5,7-
tetraene, 127, 128, 129, 130, 132
3-deuterio-8-methyl, 128
4-deuterio-3-methyl, 129
3,4-dideuterio, 129
3,6-dideuterio, 129
3,8-dideuterio, 128, 131
4,7-dideuterio, 131
5,6-dideuterio, 131
3,4-dimethyl, 129
3,8-dimethyl, 128
4,5-dimethyl, 129
5,6-dimethyl, 388
3-methyl, 129
4-methyl, 128, 129
5-methyl, 128, 129
1,3-Tetramethylenecyclo-octa-1,3,5,7-
tetraene, 131
1,2-Tetramethylenecyclo-octa-1,3,5,7-
tetraene-2',3'-dicarboxylic acid
anhydride, 126; 1'-bromo, 126; 1'-chloro,
126
9-Thiabicyclo[4.2.1]nona-2,4,7-triene, 318,
324, 325
9-Thiabicyclo[4.2.1]nona-2,4,7-triene *anti*-
9-oxide, 324
9-Thiabicyclo[4.2.1]nona-2,4,7-triene *syn*-9-
oxide, 5, 39, 324, 325
9-Thiabicyclo[4.2.1]nona-2,4,7-triene 9,9-
dioxide, 5, 40, 324, 325, 421
anion, 325
dianion, 325
1,6-dideuterio, 324, 325, 421
1,6-dideuterio-7,8-dimethyl, 112
1,6-dimethyl, 112, 325, 421
7,8-dimethyl, 112
1-methyl, 421
9-Thiabicyclo[6.1.0]nona-2,4,6-triene, 38,
318, 325, 378
Tricyclo[4.2.2.0^{2,5}]deca-3,7-diene
3-acetoxy-9,9,10,10-tetracyano, 101–3
endo, *cis*-9,10-bis(hydroxymethyl), 328,
342; dibenzenesulphonate ester, 328;
Pd complex, 342
endo, *cis*-9,10-bis(3-methylbutoxy-
carbonyl), 328
3-bromo-9,9,10,10-tetracyano, 101–3
endo-9-chloroformyl, 378
exo-9-chloroformyl, 378
3-chloro-9,9,10,10-tetracyano, 101–3
9-cyano, 42–3
endo-9-cyano, 378, 422
exo-9-cyano, 378, 422

Tricyclo[4.2.2.0^{2,5}]deca-3,7-diene—*cont.*
  endo, cis-9,10-dibutoxycarbonyl, 328
  trans-9,10-dichloroformyl, 42–3, 328
  trans-9,10-dichloroformyl-3-phenyl, 101–103
  endo, cis-9,10-diethoxycarbonyl, 328
  endo, cis-9,10-di-isobutoxycarbonyl, 328
  endo, cis-9,10-dimethoxycarbonyl, *see separate entry below*
  trans-9,10-dimethoxycarbonyl, 328, 339
  endo, cis-9,10-dimethyl, 328, 342, 422, 423; Mo complex, 342; Pd complex, 342
  9,9,10,10-tetracyano, 42–3, 336
  9,9,10,10-tetracyano-3,4-dimethyl, 121–2
  9,9,10,10-tetracyano-3,7-dimethyl, 122
  9,9,10,10-tetracyano-3-ethyl, 101–3
  9,9,10,10-tetracyano-3-fluoro, 101–3
  9,9,10,10-tetracyano-3-methoxy, 101–3
  9,9,10,10-tetracyano-3-methoxycarbonyl, 101–3
  9,9,10,10-tetracyano-3-methyl, 101–3
  9,9,10,10-tetracyano-3-phenyl, 101–3
Tricyclo[4.2.2.0^{2,5}]deca-3,7-diene, *endo, cis*-9,10-dimethoxycarbonyl
  Ag complex, 341
  cycloaddition reactions, 336–7, 339–41, 427
  epoxidation, 331, 422
  formation, 328
  hydroboration, 424
  Pd complex, 342
  photolysis, 330
  reaction: with bases, 328; with benzenesulphenyl chloride, 425; with bromine, 333; with t-butyl hypochlorite, 331; with chlorotrimethylsilane, 328; with dienes, 339–41; with 1,3-dipolar reagents, 336–7; with iodine azide, 422; with lithium aluminium hydride, 328; with mercury(ɪɪ) acetate, 335, 424; with nitrosyl chloride, 424; with peracids, 331, 422; with sulphuric acid–methanol, 331; with transition metal derivatives, 341, 342
Tricyclo[4.2.2.0^{2,5}]deca-3,7-diene-9-carboxylic acid, 42–3
Tricyclo[4.2.2.0^{2,5}]deca-3,7-diene-*endo, cis*-9,10-dicarboxylic acid, 328
Tricyclo[4.2.2.0^{2,5}]deca,3-7-diene-*endo, cis*-9,10-dicarboxylic acid anhydride
  3-acetoxy, 101–3
  3-acetoxymethyl-4-methyl, 121–2
  3-benzoyl, 101–3
  3,4-bis(acetoxymethyl), 121–2
  cycloaddition reactions, 337–9, 425
  3-cyclopropyl, 101–3
  3,4-dimethyl, 121–2
  2,5-dimethyl-3,4-diphenyl, 146
  epoxidation, 331
  esterification, 328
  3-ethyl, 101–3
  formation, 42–3
  hydrolysis, 328
  3-(1-ketobut-2-enyl), 101–3
  3-methoxycarbonyl, 101–3
  3-methyl, 101–3
  3-phenyl, 101–3
  photolysis, 330
  reaction: with ammonia, 328; with benzonitrile oxide, 336–7; with bromine, 333; with t-butyl hypochlorite, 331; with chlorine, 331; with dienes, 337–9, 341, 425; with free radicals, 336; with hydrazine, 328; with iodine azide, 334; with iodine monochloride, 334; with mercury(ɪɪ) acetate, 335, 423; with perbenzoic acid, 331; with rhodium(ɪɪɪ) chloride, 342
  reduction, 335
  Rh complex, 342
  tricyclo[4.2.2.0^{2,5}]deca-3,7,9-triene from, 421
Tricyclo[4.2.2.0^{2,5}]deca-3,7-diene-*endo, cis*-9,10-dicarboxylic acid hydrazide, 328
Tricyclo[4.2.2.0^{2,5}]deca-3,7-diene-*endo, cis*-9,10-dicarboxylic acid imide, 328
  9,10-dicyano, 42–3
  9,10-dicyano-3-phenyl, 101–3
  N-phenyl, 42–3
Tricyclo[4.2.2.0^{2,5}]deca-3,7-diene-*trans*-9,10-dicarboxylic acid, 328, 332
Tricyclo[4.2.2.0^{2,5}]deca-7,9-diene
  1,2,3,3,4,4,5,6,7,8-decadeuterio-9,10-dimethoxycarbonyl, 352
  trans-3,4-diacetoxy-9,10-dimethoxycarbonyl, 353
  trans-3,4-dibromo-9,10-dimethoxycarbonyl, 248–50, 353
  cis-3,4-dichloro-9,10-dimethoxycarbonyl, 248–50, 352
  trans-3,4-dichloro-9,10-dimethoxycarbonyl, 248–50, 353
  cis-3,4-dideuterio-9,10-dimethoxycarbonyl, 352
  9,10-dimethoxycarbonyl, 220, 344, 352
  3,3-dimethoxy-9,10-dimethoxycarbonyl, 352
Tricyclo[4.2.2.0^{2,5}]deca-3,7-diene-9,10-dione, 1,6,7,8-tetrachloro, 391, 426
Tricyclo[4.2.2.0^{2,5}]deca,3-9-diene-7,8-dione, 1,6,9,10-tetrachloro, 391, 426
Tricyclo[4.2.2.0^{2,5}]deca-3,7-dien-9-one, 422
Tricyclo[4.2.2.0^{2,5}]deca-7,9-dien-3-one, 9,10-dimethoxycarbonyl, 261, 352
Tricyclo[4.2.2.0^{2,5}]deca-3,7,9-triene
  9,10-bis(trifluoromethyl), 425, 426; Mo complex, 426
  3-chloro-9,10-dicyano, 104

Tricyclo[4.2.2.0^{2,5}]deca-3,7,9-triene—*cont.*
  9,10-dicyano, 42–3
  9,10-dicyano-2-fluoro, 104
  9,10-diethoxycarbonyl, 42–3, 343
  9,10-dimethoxycarbonyl, *see separate
    entry below*
  formation, 328, 421
  reaction, with hemicyclone, 426; with
    3,6-bis(2'-pyridyl)-*s*-tetrazine, 351
  thermolysis, 343
Tricyclo[4.2.2.0^{2,5}]deca-3,7,9-triene, 9,10-
    dimethoxycarbonyl
  Ag complex, 351
  cycloaddition reactions, 344, 346–51, 388
  formation, 42–3
  Pd complex, 351
  photolysis, 344
  reaction: with dienes, 348–51, 388; with
    1,3-dipolar reagents, 344, 346–8; with
    transition metal derivatives, 351
  reduction, 344
  Rh complex, 351
  thermolysis, 343–4
Tricyclo[4.2.2.0^{2,5}]deca-3,7,9-triene-9,10-
    dicarboxylic acid, 343
  anhydride, 343
Tricyclo[8.6.0.0^{2,9}]hexadeca-3,5,7,11,13,15-
    hexaene (dimer (9))
  Ag complex, 367
  base-catalysed isomerisation, 427
  chloro, 102, 359
  cycloaddition reactions, 362–6
  dianion, 30–1, 361
  1,10-didehydro, 31
  epoxidation, 360
  Fe complexes (of valence tautomers), 63,
    64, 367, 368, 371–2
  fluoro, 102, 359

  formation, 11–13, 14, 30–1
  hydrogenation, 361
  methoxycarbonyl, 102, 359
  Os complex (of valence tautomer), 383
  phenyl, 102, 359
  photolysis, 359
  reaction: with alkali metals, 360–1; with
    bases, 361, 427; with dienophiles, 363–
    6; with ethoxycarbonylcarbene, 362;
    with ethoxycarbonylnitrene, 362–3;
    with peracetic acid, 360; with transition
    metal derivatives, 367–8
  Ru complex (of valence tautomer), 382
  structure, 11
  thermolysis, 13, 359
  valence tautomerism, 364
Tricyclo[3.3.0.0^{2,8}]octa-3,6-diene (semi-
    bullvalene)
  Ag complex, 170
  ^{13}C n.m.r. spectrum, 397
  1,3-dimethyl, 111
  1,5-dimethyl, 111
  Fe complexes, 171–2
  fluxional nature, 169, 397
  formation, 18
  hydrogenation, 170
  octamethyl, 160, 169, 170, 397–8
  reaction: with bromine, 170; with transi-
    tion metal derivatives, 4, 170–2
  1,3,5,7-tetramethyl, 137, 144
  thermolysis, 4
  W complex, 170
1,2-Trimethylenecyclo-octa-1,3,5,7-
    tetraene, 127, 128
  1'-chloro-3'-keto-1'-methyl, 124
  1'-hydroxy-3'-keto-1'-methyl, 124
  3'-keto-1'-methylene, 124
  3-methyl, 126